W9-CKM-185

# Quantum Chemistry and Molecular Spectroscopy

Clifford E. Dykstra

*Department of Chemistry*
*Indiana University-Purdue University*
*at Indianapolis*

Prentice Hall, Englewood Cliffs, New Jersey 07632

**Library of Congress Cataloging-in-Publication Data**

Dykstra, Clifford E.
Quantum chemistry and molecular spectroscopy / Clifford E. Dykstra.
p. cm.
Includes bibliographical refereces and index.
ISBN 0-13-747312-5
1. Quantum chemistry. 2. Molecular spectroscopy. I. Title.
QD462.D95 1991
541.2′8—dc20

Acquisition Editor: Dan Joraanstad
Editorial/production supervision: Judi Wisotsky
Editorial Assistant: Lynne Breitfeller
Copy Editor: Carol Dean
Cover design: Karen Stephens
Prepress Buyer: Paula Massenaro
Manufacturing Buyer: Lori Bulwin

A Simon & Schuster Company
Englewood Cliffs, New Jersey 07632



Printed in the United States of America
10 9 8 7 6 5 4 3 2 1

ISBN 0-13-747312-5

Prentice-Hall International (UK) Limited, *London*
Prentice-Hall of Australia Pty. Limited, *Sydney*
Prentice-Hall Canada Inc., *Toronto*
Prentice-Hall Hispanoamericana, S.A., *Mexico*
Prentice-Hall of India Private Limited, *New Delhi*
Prentice-Hall of Japan, Inc., *Tokyo*
Simon & Schuster Asia Pte. Ltd., *Singapore*
Editora Prentice-Hall do Brasil, Ltda., *Rio de Janeiro*

# Preface

This text evolved from the experiences of teaching several different physical chemistry courses at the University of Illinois at Urbana-Champaign for thirteen years, and from recent experience with similar courses at Indiana University – Purdue University at Indianapolis (IUPUI). It is primarily intended as a text for a junior or senior year course in physical chemistry, and in particular, a course that introduces quantum mechanics and spectroscopy. It is also intended for use in certain more advanced courses.

The aim is to provide a smooth and rigorous introduction to the subject. Certain aspects do not follow traditional approaches. For instance, the very early experiments and puzzlements of quantum phenomena are largely absent as they are probably better material for a history of science course. In other ways, I have attempted to streamline the subjects of quantum mechanics and spectroscopy, though without skirting the fundamentals. This has led to an arrangement of topics that is not entirely standard. One decision was to rely on the two-body oscillator as the first, primary example and to use it again and again instead of a smattering of introductory model problems. This makes it possible to move through a variety of topics expeditiously.

The usual goal of an introductory physical chemistry course is, in part, to explain the basis for a wide range of molecular spectroscopy experiments, to provide a theoretical basis for chemical bonding and electronic structure, and to introduce quantum mechanics. This is just as much a goal when there are six weeks allowed for this material as when there are sixteen weeks, though the depth of coverage differs. Thus, I have tried to make this text workable at several different levels, ranging from that of a short introduction for undergraduates of five or six weeks to that of a course for beginning graduate students. For instance, the theoretical basis of the first five chapters should be enough to jump ahead and select sections from most of the remaining chapters and thereby craft a short course on this material.

Several of the chapter sections are distinguished by the use of matrix quantum mechanical methods and linear algebra techniques, and they are designated with the symbol ✯. The others are almost always the more elementary sections. Problems that pertain to material in these sections are listed with a ✯ as well. Covering the text except for the ✯ sections should constitute a hearty fifteen-week introduction, whereas including some or all of the ✯ sections would make for an advanced or specialized class.

The experimental basis of quantum chemistry and molecular structure is a crucial aspect of the subject. The aim of the text has been to develop the quantum mechanics to the point of predicting and interpeting spectra. It is intended that students will be learn the connection between quantum mechanical analysis and the measurements they may make in a laboratory, but topics related to instrumentation are not considered.

The exercises at the end of each chapter bring out essential points of the material. Sometimes they amount to completing or extending a problem that had been discussed, and generally, they pose a concise challenge to a student's understanding. Notions that emerge from reading the text will be strengthened by completing the exercises. At the end of the text, solutions are given for a number of the problems.

An important goal of this text is to meet students' needs for learning how to exploit computers in working out chemical physics problems. Where possible, derivations and formulas have been written in a form that is most suggestive of a computational implementation. There is an added value from doing this, because the "knowns" should then seem more distinguishable from "unknowns" in various formulas. Actual computer solution of problems is not considered in the text, because from my experience, using a computer is not the major hurdle; knowing what you want a computer to do – what are the numbers that go in and what are the numbers that come out – can be. This text, and in particular the ✯ sections employing matrix methods, are meant to provide the type of understanding, the level of sophistication, and the motivation to translate quantum mechanical and spectroscopic problems into computational procedures.

I should like to thank Dr. Joseph Augspurger for preparing most of the solutions to the chapter exercises and for preparing several of the figures. I should also like to thank colleagues who provided critical comments during the course of drafting the text. They include Professors R. DeKock, K. D. Jordan, D. J. Malik, S. P. McGlynn, K. J. Miller, E. Oldfield, K. A. Schugart, and M. Thiemans. I also thank J. D. Augspurger, H. Le and K. D. Park for comments on certain sections of the text.

*C. E. D.*

Indianapolis

1991

# Contents

★ Designates sections intended for extended courses.

# Chapter 1

# Introduction

*One subject of this text, quantum mechanics, provides the physical basis for the other subject, molecular spectroscopy. Both fit into the general category of chemical physics or physical chemistry, and the development of both has occurred through a strong interplay of experimental investigation and theoretical analysis.*

## 1.1 CHEMICAL PHYSICS

Chemical physics is a broad area of molecular science, and its boundaries are not at all sharp. In many ways, it is at the core of chemical science because it is concerned with the most detailed view of molecules that is possible. This means it concerns the structure of molecules, starting from a description of electrons in molecules and the nature of chemical bonds. It also concerns dynamics or the changes in a molecular system in time, and it concerns properties of assemblies of atoms and molecules. Beyond that, the subject of chemical physics includes properties and phenomena of gases, liquids, and solids. To study physical chemistry or chemical physics is to pursue the most fundamental understanding of chemistry and molecular science that can be achieved.

Because it is developed from basic physical laws, chemical physics deals with most issues quantitatively and mathematically. Even the qualitative notions that emerge usually rely on mathematical arguments. Often the theories used in chemical physics are presented most concisely as mathematical expressions.

Chemical physics leads to understanding at such a basic level that it enters into molecular problems of most every type. It is a gateway of sorts to biophysics, biochemistry, materials science, and even medical technology. But that becomes clear only as one begins to study chemical physics.

The most detailed view of atoms and molecules is one that reveals their constituents: electrons, neutrons, and protons. These constituents are particles, very small entities that have mass. They are so small and so light that they are beyond the limits of our own senses and experience. Not surprisingly, the mechanics of systems of such particles are also outside everyday experience. Even so, there is a correspondence between our macroscopic world and the subatomic world, and we will analyze systems of particles in both worlds. The picture in the macroscopic world is generally referred to as a *classical* picture. It has been established by human perception and observation. The picture of small, light particles is termed a *quantum* picture because quantization of energy – the partitioning of energy into discrete chunks – and of other things is the distinction between this world and the macroscopic world. Because the discrete chunks are so tiny, they appear to constitute something continuous in the macroscopic world, and so there is at least this connection with our usual experience.

The modern atomic theory of matter is almost two centuries old. It was in the early 1800's that Dalton's work brought the first evidence that matter is not continuously divisible and that there is some fundamental type of particle, the atom. The line of thought that began with the atomic theory of matter took its next major step in the early 1900's when experiments pointed to the existence of subatomic particles. In a few more decades, it became clear that there are even smaller particles. Even today, superconducting supercollider experiments have been planned by physicists in the search for still more exotic particles. The general notion that stands out is that as matter is viewed with an ever more powerful microscope, a point is reached where we see all matter as composed of discrete building blocks (particles) rather than continuous materials.

Einstein's special theory of relativity in 1905 connected the property of mass with energy. Perhaps in a vague way this connection makes it less surprising that scientists in the early 1900's were finding that many strange observations could be explained if energy came in discrete chunks. In other words, it is not only matter but also energy that comes in discrete building blocks in the tiny world of atoms. The problems that led to the hypothesis of quantization of energy involved the absorption spectrum of hydrogen, the photoelectric effect, and the temperature dependence of the heat capacities of solids. One after another, unexplained phenomena were explained by a quantum hypothesis, and this hypothesis eventually grew into what we now refer to as *quantum mechanics*. After one becomes familiar with quantum mechanics, it is fascinating to look back at the early developments; they mark the start of a major scientific revolution.

## 1.2 THEORY AND EXPERIMENT

The pursuit of understanding in most branches of science is a process of observation and analysis. In chemical physics, laboratory experiments are the means for observation, which is to say that experiment is the means for probing and measuring molecular systems. The analysis of the data, though, may be carried out for different reasons. For one, the analysis might use the data with some generally accepted theory so as to deduce some useful quantity that is not directly measurable. We will find, for instance, that bond lengths are not measured the way an object's length is measured in our macroscopic world; instead they are often determined on the basis of measurements of energy changes in molecules. In this way, the established physical understanding provides the means for utilizing experimental information in examining molecular systems. So, quantum mechanical theories and spectroscopic experiments must go hand in hand.

Another purpose for carrying out an experiment and analyzing the data is to test one notion, concept, model, hypothesis, or theory, or possibly to select from among several competing theories. If the data do not conform to what is anticipated by one theory, then its validity has been challenged. In such a circumstance, one may devise a new notion, concept, or theory to better fit the data, or possibly reject one concept in favor of another. Whatever new idea emerges is then rightly tested in still further experiments. We will find that many problems are analyzed with approximations or idealizations that make the mathematics less complicated or that offer a more discernible physical picture. Then, experimental testing offers a validation or rejection of the approximation.

There is a serious interplay between observation and understanding, or between experiment and theory. Of course, this is true throughout science, but in chemical physics the interplay very much affects the way developments are viewed. The goal is always a physical understanding of how molecular systems exist and behave, and ultimately that is embodied in theories that at some point have been well tested by experiments. In many respects, a textbook presentation of our understanding of chemical physics is a presentation of theories, and yet, experiments are very much a part of the story. We can not properly explain our best physical picture of molecular behavior without knowing the means of observation (experiment). So, the direction for this text is to present our basic quantum mechanical picture of molecules (theories) integrated with the experiments (molecular spectroscopy) that reveal the picture. This is done without the details of instrumental technology; that is an important topic on its own.

## Bibliography

1. A. J. Ihde, *The Development of Modern Chemistry* (Harper & Row, New York, 1964).

2. T. S. Kuhn, *The Structure of Scientific Revolutions* (University of Chicago Press, Chicago, 1970). This is an insightful essay on what

constitutes a scientific revolution and how the thinking of scientists changes as a revolution unfolds.

3. T. S. Kuhn, *Black-Body Theory and the Quantum Discontinuity, 1894-1912* (Oxford University Press, New York, 1978). This provides a fascinating and critical account of the development of the origins of quantum mechanics. In particular, it challenges the account of many texts about the origin of the quantum hypothesis.
4. W. Kauzmann, *Quantum Chemistry* (Academic Press, New York, 1957). The introductory chapter of this text presents a very welcome philosophy and perspective for learning about modern chemical physics.

# Chapter 2

# Classical Mechanics

*Mechanics is the analysis of the motions of objects. In this chapter, we consider the mechanics of particles in our macroscopic world. This classical description will be connected to the very different description of microscopic particles that will be presented in later chapters. Classical mechanics allows for continuous variability in the momenta and energies of particles, whereas at the quantum or microsocopic level, this does not always happen.*

## 2.1 EQUATIONS OF MOTION

Differential equations that prescribe the motion of a particle or system of particles are *equations of motion*. In the mechanics of Isaac Newton, the equations of motion are one of Newton's laws: The total force acting on an object equals the mass of the object times the time rate of change of the object's velocity (i.e., the acceleration). Letting $\vec{F}$ be the vector of force, m the mass of the object, and $\vec{a}$ the acceleration vector, then the relation

$$\vec{F} = m\,\vec{a} \tag{2-1}$$

is an equation of motion. If one finds or knows $\vec{F}$, this equation can be used to find out everything about the object's motion in space and time. Solution of this differential equation will yield the position vector as a function of time, i.e., $\vec{r}(t)$. This is a *classical mechanics* picture; nothing appears quantized.

Newton's formulation is not the only way in which classical equations of motion can be found. Lagrange and Hamilton developed other means, and it is the formulation of Hamilton that is the most useful framework for developing the mechanics of quantum systems. It is important to realize that Newtonian, Lagrangian, and Hamiltonian mechanics offer equivalent descriptions of classical systems. The $\vec{r}(t)$ found with one formulation will be the same as that found with another, even if the route to that function appears different.

At this point, we shall restrict attention to the mechanics of systems of point-mass particles. This means that particles are assumed to have no volume. Their mass is in an infinitesimally small region of space, a point. And until later, we will restrict attention to *conservative* systems, which are those for which the potential energy has no explicit time dependence. Of course, these restrictions are not an aspect of a particular mechanical formulation; they are used as a convenience that is in keeping with the types of systems of most immediate interest.

The classical kinetic energy, T, of any particle is the square of the momentum divided by twice the particle's mass. Momentum is a vector: In a Cartesian space, there is an x-component, a y-component, and a z-component. Thus,

$$T = \frac{|\vec{p}|^2}{2m} = \frac{p_x^2 + p_y^2 + p_z^2}{2m} = \frac{\mathbf{p} \bullet \mathbf{p}}{2m} \tag{2-2}$$

where two forms of notation for a vector quantity have been used, namely $\vec{p}$ and $\mathbf{p}$. If there is a system of N different particles, a subscript serves to distinguish the masses and momenta of the particles, e.g., $m_i$ and $\mathbf{p}_i$ where $i = 1, \ldots, N$. The kinetic energy for the system is just the sum of the kinetic energies of each of the particles:

$$T = \sum_{i=1}^{N} \frac{\mathbf{p}_i \cdot \mathbf{p}_i}{2m_i} \tag{2-3}$$

Written in this way, the kinetic energy appears as an explicit function of the momentum components of each of the particles; that is, $T = T(\mathbf{p}_1, \mathbf{p}_2, ..., \mathbf{p}_N)$.

The potential energy, V, of a conservative system of particles must depend only on the position coordinates of the particles. For example, the potential energy for two electrically charged particles, one with charge $q_1$ and the other with charge $q_2$, is $q_1q_2/r$, where r is the separation distance between the particles. If a Cartesian coordinate system is used to specify particle positions, then the first particle's position may be given by $(x_1, y_1, z_1)$ and the second one's position by $(x_2, y_2, z_2)$. Thus, the potential energy in this case is a function of these six coordinates:

$$V = V(x_1, y_1, z_1, x_2, y_2, z_2) = \frac{q_1 q_2}{\sqrt{(x_2 - x_1)^2 + (y_2 - y_1)^2 + (z_2 - z_1)^2}}$$

It is not required that position coordinates be Cartesian coordinates, though this is often the most convenient. In some important examples, spherical polar coordinates will be used. It is a good idea, then, to carry out further mechanical developments without specifying the type of coordinate system, merely requiring that it be sufficient to specify the positions of all the particles in the system. This is usually termed a *generalized coordinate system*. It is typical notation to use q for a generalized coordinate, or $q_i$ for one from a set of generalized coordinates. The number of degrees of freedom is independent of the choice of coordinate system, and so for a single particle the three possible directions could be denoted by the coordinates $q_1$, $q_2$, and $q_3$. Each generalized position coordinate defines a direction in which there may be a component of momentum. (For every q there is a p.)

The *Hamiltonian*, H, is defined to be simply the sum of T and V, and so it is a function of position and momentum coordinates. For the systems being considered, the Hamiltonian is the total energy of the system.

$$H = T + V \tag{2-4}$$

For non-conservative systems, the potential will have an explicit dependence on time, and so the Hamiltonian will have to be a function of time as well.

Hamilton determined that for a generalized coordinate system, the equations of motion can be found from H:

$$\frac{\partial H}{\partial q_i} = -\dot{p}_i \tag{2-5}$$

$$\frac{\partial H}{\partial p_i} = \dot{q}_i \tag{2-6}$$

where a dot over a character signifies the first derivative of the quantity with respect to time. For each direction, the partial derivative of the Hamiltonian with respect to a position coordinate is equal to the negative of the time derivative of the corresponding (or *conjugate*) momentum coordinate. And the partial derivative of the Hamiltonian with respect to a momentum coordinate equals the time derivative of the conjugate position coordinate. Simultaneous solution of these differential equations yields the description of the mechanics of the system.

## 2.2 THE HARMONIC OSCILLATOR

A first example of how to use Hamilton's classical equations of motion is in the application to the mechanics of the *harmonic oscillator*. The harmonic oscillator is a special model problem that will be used again and again. It is the problem of a mass able to move in one direction and connected by a spring to an infinitely heavy wall, as shown in Fig. 2.1. The spring is special because it is harmonic, which means that the restoring force is linear in the coordinate that gives the displacement from the equilibrium. An equivalent statement is that the potential energy for stretching or compressing the spring varies quadratically with the distance of displacement from the equilibrium length. If $x$ is chosen to be the position coordinate of the mass, and $x_0$ is the equilibrium length of the

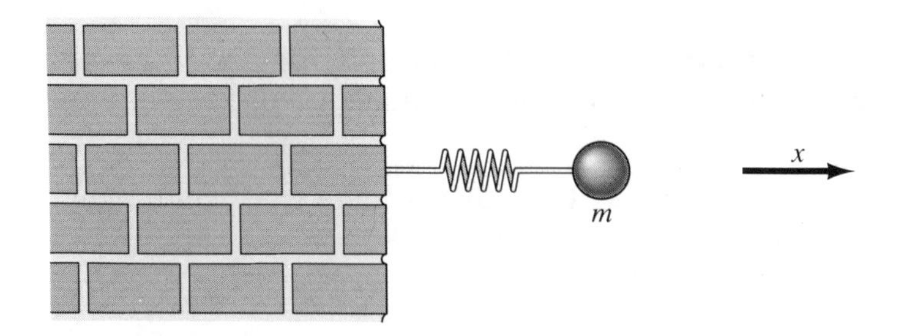


**FIGURE 2.1** The harmonic oscillator is a particle attached by a harmonic spring to an unmovable wall.

spring, then the harmonic potential experienced by the mass in Fig. 2.1 has the following mathematical form.

$$V(x) = \frac{1}{2} k (x - x_o)^2 \tag{2-7}$$

The constant, k, in this expression is called the *spring constant* or the *force constant*.

Since there is but one degree of freedom for the harmonic oscillator, the kinetic energy is simply $p^2/2m$. Thus, the Hamiltonian for this system is

$$H(x,p) = \frac{p^2}{2m} + \frac{1}{2} k (x - x_o)^2$$

Differentiating this Hamiltonian function with respect to x and using Eq. (2-5) yields

$$k (x - x_o) = -\dot{p}$$

Then, differentiating H with respect to p and using Eq. (2-6) yields

$$\frac{p}{m} = \dot{x}$$

The second of these two equations of motion is the definition of the momentum in Newtonian mechanics; that is, multiplying through by m gives p as equal to the mass times the velocity, $p = m\dot{x}$. Solving the equations of motion is done by any standard means of solving differential equations. In this particular example, the first step is to take the time derivative of the second equation (after multiplying through by m):

$$\dot{p} = m\ddot{x}$$

This relates the time derivative of the momentum to the acceleration or the second time derivative of the position function. But the first equation of motion also tells us that the time derivative of the momentum is $-k(x - x_o)$. Therefore, these must be equal to each other:

$$-k(x - x_o) = m\ddot{x}$$

This expression is the differential equation that determines the position function, x(t).

Hamilton's formulation, applied to the harmonic oscillator problem, has yielded a differential equation that relates the position (a function of time) to the second time derivative of the position. This differential equation has a general solution that may be expressed in either of two ways. They are mathematically equivalent:

$$x(t) = x_o + a\sin(\omega t) + b\cos(\omega t) \qquad (2\text{-}8a)$$

$$x(t) = x_o + A e^{-i\omega t} + B e^{i\omega t} \qquad (i \equiv \sqrt{-1}) \qquad (2\text{-}8b)$$

The constants in these functions are a, b, and $\omega$ in the first, and A, B, and $\omega$ in the second. $x_o$ is the position at which the potential is zero; it is the equilibrium length of the spring. The constant $\omega$ must have units of inverse time, and it is a frequency. The other constants must have units of length because x(t) is the position of the particle. Second differentiation of either Eq. (2-8a) or (2-8b) yields

$$\ddot{x}(t) = -\omega^2\left(x(t) - x_o\right) \qquad (2\text{-}9)$$

Comparing this with the differential equation derived for the harmonic oscillator demonstrates that the frequency is simply related to the force constant and the particle's mass.

$$\omega = \sqrt{\frac{k}{m}} \qquad (2\text{-}10)$$

Thus, if the force constant were made larger, i.e., if the spring were stiffer, then Eq. (2-10) says that the frequency would be increased. If the mass were heavier, then the system would be more sluggish, which is to say that the frequency of oscillation would be diminished.

The constants a and b, or A and B, are found from initial conditions. For instance, if at time $t = 0$, it is known that $x(0) = x_o$, then the following is obtained:

$$x(0) \;=\; x_o \;+\; a\sin(0) \;+\; b\cos(0) \;=\; x_o \quad\Rightarrow\quad b = 0 \qquad (2\text{-}11)$$

(The symbol $\Rightarrow$ will be used to mean "implies that" or "leads to.") The remaining unknown constant, a, is given by the maximum displacement from equilibrium that the spring reaches. In other words, a is how far the spring had been stretched or compressed initially before it was released to vibrate freely.

At maximum displacement, all the energy of the oscillator is potential energy. This is when the mass is at one of its *turning points,* either $x_o + a$ or $x_o - a$. Notice that a is in no way restricted to any particular value. It may be zero, meaning the oscillator is not moving in time. It may be 1.0 or 1.0234 or 120,000.1. This statement just reinforces everyday experience that we can stretch a spring to most *any* length desired and release it in the classical world, at least so long as the spring does not break. The significance of this is that we can adjust the energy of the oscillating spring system in a continuous manner. It will turn out that this does not happen in the quantum world.

## 2.3 MOTION THROUGH SEVERAL DEGREES OF FREEDOM

Molecular problems are problems of many particles, and so there are many degrees of freedom. To illustrate how Hamilton's formulation is used with more than one degree of freedom, consider the double oscillator problem depicted in Fig. 2.2. This is a system with two degrees of freedom, the positions along the x-axis of the two masses. The subscripts 1 and 2 are used here for the masses, $m_1$ and $m_2$, and for their positions, $x_1$ and $x_2$. The subscripts a and b are used for the two harmonic springs, with force

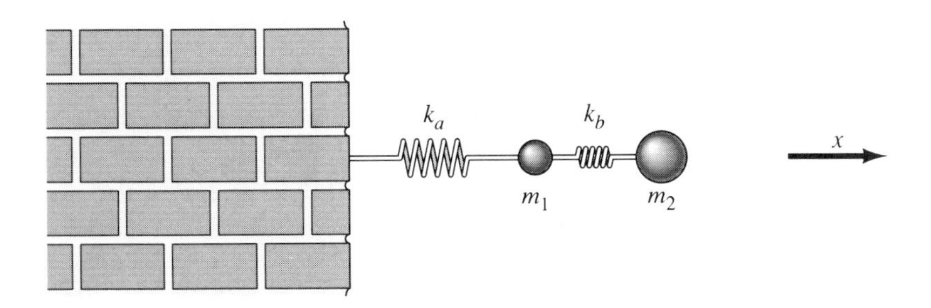


**FIGURE 2.2** A double harmonic oscillator. Both particles 1 and 2 move along the x-axis. Spring a is connected to an unmovable wall and to particle 1. Spring b connects the two particles.

---

constants $k_a$ and $k_b$ and with equilibrium lengths $x_a$ and $x_b$. The classical Hamiltonian for this system is the sum of the kinetic energies of the two particles and the potential energies from stretching the two springs. We analyze spring b by noting that its length is $x_2 - x_1$. Its displacement from equilibrium is this length less $x_b$, assuming that $x_2 > x_1$ always. The potential energy stored in spring b is this displacement squared, times $k_b/2$. Thus, the Hamiltonian is

$$H(x_1,x_2,p_1,p_2) = \frac{p_1^2}{2m_1} + \frac{p_2^2}{2m_2} + \frac{1}{2}k_a(x_1 - x_a)^2 + \frac{1}{2}k_b(x_2 - x_1 - x_b)^2 \tag{2-12}$$

The equations of motion, by Hamilton's formulation, follow from Eqs. (2-5) and (2-6).

$$p_1 / m_1 = \dot{x}_1 \tag{2-13a}$$

$$k_a(x_1 - x_a) - k_b(x_2 - x_1 - x_b) = -\dot{p}_1 \tag{2-13b}$$

$$p_2 / m_2 = \dot{x}_2 \tag{2-14a}$$

$$k_b (x_2 - x_1 - x_b) = -\dot{p}_2 \tag{2-14b}$$

These are *coupled* differential equations that reflect the fact that the two springs are physically coupled: They are connected at mass 1. Actually solving this type of problem or this set of coupled differential equations is considered at the end of this chapter. For now, the important feature is that systems of several degrees of freedom will, in general, have coupled equations of motion.

There are circumstances where the equations of motion in several degrees of freedom are not coupled. In these cases, the Hamiltonian has a special form which shall be referred to as a *separable* form. Separability arises when a particular Hamiltonian may be written as additive, independent functions, e.g.,

$$H(q_1,q_2,q_3,\ldots,p_1,p_2,p_3,\ldots) = H'(q_1,p_1) + H''(q_2,q_3,\ldots,p_2,p_3,\ldots) \tag{2-15}$$

In this instance, H' is a function only of the first position and momentum coordinates, while H" is a separate function involving only the other coordinates. If we apply Eqs. (2-5) and (2-6) to H in order to find the equations of motion for this system, the equations for $q_1$ and $p_1$ are found to be distinct, or mathematically unrelated to any of the equations that involve any of the other coordinates. In fact, the equations for $q_1$ and $p_1$ could just as well have been obtained by taking H' to be the Hamiltonian for the motion in the $q_1$-direction, while the other equations of motion could have been obtained directly from the H" Hamiltonian. Thus, the additive, independent terms in the Hamiltonian may be separated in order to deduce the equations of motion. Furthermore, since the equations of motion are separable, then the actual physical motions of the H' system and the H" system are unrelated and independent. It is possible that a Hamiltonian may be separable in all variables, and then all the motions are independent.

In the double-oscillator system of Fig. 2.2, the last potential energy term involved both position coordinates together, and so that problem was not separable in $x_1$ and $x_2$. To contrast with that situation, we will next consider a problem that ends up displaying separability. It is the oscillator problem in Fig. 2.3. The Hamiltonian is

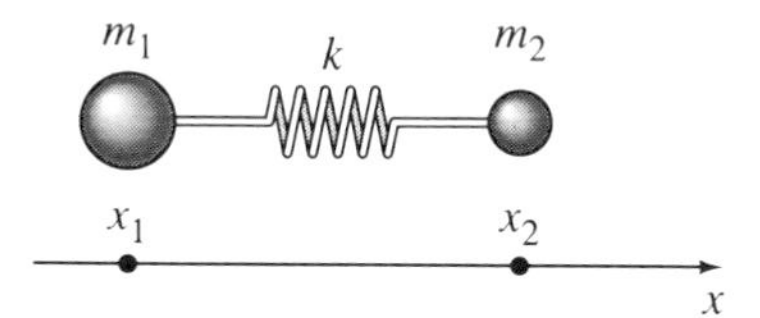


**FIGURE 2.3** An oscillator system where two particles, each constrained to move along the x-axis, are attached by a harmonic spring with force constant k. The positions of the two particles along the x-axis are designated $x_1$ and $x_2$. The equilibrium length of the spring is the constant, $x_o$.

---

$$H(x_1,x_2,p_1,p_2) = \frac{p_1^2}{2m_1} + \frac{p_2^2}{2m_2} + \frac{1}{2}k(x_2 - x_1 - x_o)^2$$

Because of the potential energy term, this appears to be inseparable. However, it turns out that a certain change of coordinates, or more precisely a *coordinate transformation*, does lead to a separable form. This coordinate transformation comes about by defining two new variables. First, we define r to be a coordinate that gives the displacement from the equilibrium length of the spring, and second, we define s to be the position of the center of mass of the whole system:

$$r \equiv x_2 - x_1 - x_o$$

$$s \equiv \frac{m_1x_1 + m_2x_2}{m_1 + m_2}$$

By inspection, the potential energy now has the simple form $kr^2/2$. That is, in the {r,s} coordinate system, the potential energy is independent of s, the position of the system's mass center.

The kinetic energy of this system can be expressed in terms of momentum coordinates that are conjugate to the position coordinates, r and s. First, Eq. (2-5) is used in the $\{x_1,x_2\}$ system to write

$$p_1 = m_1\dot{x}_1 \qquad p_2 = m_2\dot{x}_2$$

Next, the kinetic energy is expressed in terms of the time derivatives of the position coordinates by using these expressions for the momenta.

$$T = \frac{1}{2}(m_1 \dot{x}_1^2 + m_2 \dot{x}_2^2)$$

Next, the time derivatives of r and s are substituted for the time derivatives of $x_1$ and $x_2$. The relation between velocities in the two coordinate systems comes about from taking the time derivative of the equations that relate r and s with $x_1$ and $x_2$:

$$\dot{r} = \dot{x}_2 - \dot{x}_1$$

$$\dot{s} = (m_1 \dot{x}_1 + m_2 \dot{x}_2) / (m_1 + m_2)$$

By rearrangement, we may obtain from these two equations that

$$\dot{x}_1 = \dot{s} - \frac{m_2}{(m_1 + m_2)} \dot{r}$$

$$\dot{x}_2 = \dot{s} + \frac{m_1}{(m_1 + m_2)} \dot{r}$$

Thus,

$$T = \frac{1}{2}\left[ (m_1 + m_2)\dot{s}^2 + \frac{m_1 m_2}{(m_1 + m_2)} \dot{r}^2 \right]$$

At this point, M will be used to designate the total mass of the system, i.e., $M = m_1 + m_2$, and $\mu$ will be used for the quantity $m_1 m_2 / (m_1 + m_2)$. $\mu$ is called the *reduced mass* of a two-body system. And so,

$$T = \frac{1}{2}(\mu \dot{r}^2 + M \dot{s}^2)$$

Notice that this is similar in form to the kinetic energy expressed in the original coordinates $x_1$ and $x_2$. The similarity suggests a definition of the conjugate momenta in the {r,s} coordinate system:

$$p_r = \mu \dot{r}$$

$$p_s = M \dot{s}$$

In fact, it is Hamilton's equations of motion [Eq. (2-5)] that ensures that these are correct definitions.

The Hamiltonian for the oscillator of Fig. 2.3 can now be written fully transformed to the {r,s} coordinate system.

$$H(r, s, p_r, p_s) = \frac{p_s^2}{2M} + \frac{p_r^2}{2\mu} + \frac{1}{2} k r^2$$

This is recognized to be a separable Hamiltonian because the first term is independent of the remaining two terms. The first term is interpreted as the kinetic energy of translation of the whole system. The second term is the kinetic energy for the *internal* motion of vibration, while the last term is the potential energy for vibration. Separability means that the translational motion of the whole system is independent and unrelated to the internal motion.

As discussed above, separability of a Hamiltonian means that the equations of motion could just as well have been developed from the independent terms if they had been treated as individual Hamiltonians. For the example of Fig. 2.3, this says that the equations of motion obtained from $H(r,s,p_r,p_s)$ will be the same as those obtained for the separate problems of internal motion, where the Hamiltonian is

$$H'(r, p_r) = \frac{p_r^2}{2\mu} + \frac{1}{2} k r^2$$

and of translational motion, for which the Hamiltonian is

$$H''(s, p_s) = \frac{p_s^2}{2M}$$

In other words, the four equations of motion we would obtain for the entire system using H turn out to be the same as the two equations of motion obtained from the H' function and the two equations of motion obtained from the H" function. This comes about because the vibration of the system, which is motion in the r-direction, does not affect and is not affected by the translation of the system, which is motion in the s-direction. So, H' alone is sufficient if only internal motion mechanics are of interest; translational motion of the system may be ignored because it is separable.

Comparison of the internal vibrational Hamiltonian for the two-body oscillator of Fig. 2.3 with the Hamiltonian for the one-body oscillator of Fig. 2.1 shows that the differences are only in the "names" of the masses, m versus $\mu$, and the names of the displacement coordinates, x versus r. On this basis, we may say that vibration of the two-body system is mechanically equivalent to vibration of a single body of mass $\mu$ attached to an unmovable wall. *Mechanical equivalence* means two systems have equations of motion (or Hamiltonians) of the same form.

## 2.4 PARTICLE MOTION IN THREE DIMENSIONS

The examples so far have been with particles that have been restricted to move only in a straight line, along the x-axis. A more general analysis is necessary, of course, since atoms and molecules do not have to experience such constraints. A free atom may move in the x-direction or the y-direction or the z-direction. Should any other type of coordinate system be used to give the location of the atom, there will still be three and only three independent coordinates. For a system composed of N atoms, or N particles, whatever the type, there will be a total of 3N independent coordinates required to specify the positions of all N particles. There will be 3N *degrees of freedom* for the system.

In the previous section, the translational motion of the system of two particles along the x-axis was found to be separable from the vibrational motion. In a three-dimensional picture of the system, translational motion is also separable, but the coordinate transformation is different. In three Cartesian dimensions, the positions of the two particles may be specified as $(x_1,y_1,z_1)$ and $(x_2,y_2,z_2)$. The center of mass of the system is also given by three coordinate values (X,Y,Z), and they are

$$X = \frac{1}{M}(m_1 x_1 + m_2 x_2)$$

$$Y = \frac{1}{M}(m_1 y_1 + m_2 y_2) \qquad (2\text{-}16)$$

$$Z = \frac{1}{M}(m_1 z_1 + m_2 z_2)$$

The separation distance between the two particles, r, is the following.

$$r = \sqrt{(x_2 - x_1)^2 + (y_2 - y_1)^2 + (z_2 - z_1)^2} \qquad (2\text{-}17)$$

Two other new coordinates, $\theta$ and $\phi$, complete the transformation. They give the orientation of the two-particle system about the center of mass. As depicted in Fig. 2.4, r, $\theta$ and $\phi$ are *spherical polar coordinates.*

$$x_2 - x_1 = r \sin\theta \cos\phi$$

$$y_2 - y_1 = r \sin\theta \sin\phi \qquad (2\text{-}18)$$

$$z_2 - z_1 = r \cos\theta$$

The overall coordinate transformation is $\{x_1,y_1,z_1,x_2,y_2,z_2\} \rightarrow \{X,Y,Z,r,\theta,\phi\}$. Notice that the number of independent coordinates, six, remains the same.

The transformation to spherical polar coordinates makes the potential energy dependent only on the variable, r, rather than on all six coordinates. The kinetic energy has a piece associated with translational motion of the whole system,

$$T^{\text{translation}} = \frac{1}{2M}(p_X^2 + p_Y^2 + p_Z^2)$$

It also has a piece associated with the vibrational motion and a piece associated with rotation of the two bodies about the center of mass. Thus, the Hamiltonian will consist of independent terms associated with motion in the X-direction (i.e., translation of the system along the x-axis), in the Y-direction, and in the Z-direction, and with vibrational and

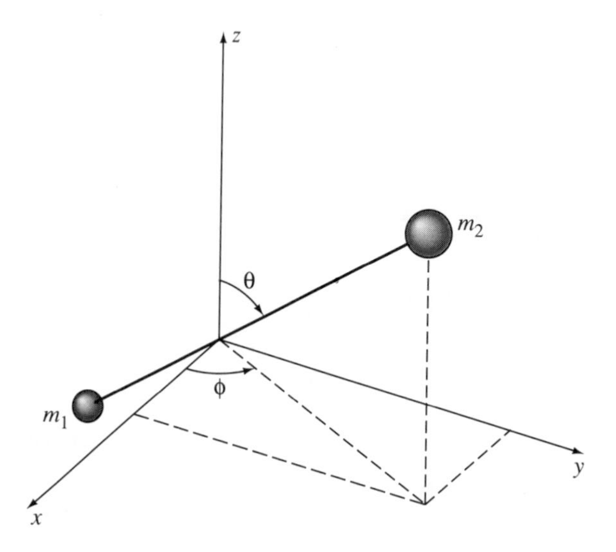


**FIGURE 2.4** Spherical polar coordinates for a two-particle system. r is the distance between $m_1$ and $m_2$. The center of mass is positioned at the origin of the (x,y,z) coordinate system. $\theta$ is the angle between the z-axis and the line connecting the two masses. $\theta$ ranges from 0 to $\pi$. $\phi$ measures the rotation *about* the z-axis of the line connecting the masses. It ranges from 0 to $2\pi$.

---

rotational motion. This describes the separability of the problem. There are three translational degrees of freedom that are separable from the rest.

Had we analyzed a system of N particles with $N > 2$, there would still be only three translational degrees of freedom for the system as a whole. And this motion would be separable from the other motions. Thus, there would be $3N - 3$ degrees of freedom for other-than-translational motion. In the case of two point-mass particles, there were two rotational degrees of freedom corresponding to the two angles $\theta$ and $\phi$. This will be the case for any linear system, and so there will be $(3N - 3) - 2 = 3N - 5$ vibrational degrees of freedom. For nonlinear collections of particles, there are three independent rotations about the mass center since rotations about the x-axis, about the y-axis, and about the z-axis are all different. Thus, for nonlinear systems, $3N - 6$ is the number of degrees of freedom remaining for vibrational motion.

It may not seem clear why there should be a different number of rotational degrees of freedom for linear and for nonlinear systems. The subtle difference has to do with the fact that we are considering particles to be point-masses, entities without volume or size. If some number of these masses were arranged along the x-axis, we could certainly visualize turning the system about the y-axis or turning it about the z-axis. On the other hand, turning it about the x-axis is to do nothing. It is an operation or motion that does not exist. Only if one or more particles were off-axis would such motion be defined, but then that would be a nonlinear arrangement.

## 2.5 HARMONIC VIBRATION OF MANY PARTICLES

The oscillator problem of Fig. 2.2 yielded a nonseparable Hamiltonian because the two springs were connected to one of the masses and this coupled the stretching and contracting of the springs. The corresponding mathematical feature is a cross-term in the potential energy expression, a term that involves both of the position coordinates, $x_1$ and $x_2$. This problem is representative of the complexity of the problems of molecular vibration since we could imagine the chemical bonds of a molecule behaving much like springs connecting atoms. The vibrational potentials of some molecule considered in this way (Fig. 2.5) would have plenty of cross-terms of the atomic displacement coordinates. Pulling a bit on one atom would obviously affect the other atoms. All the motions are coupled through the network of springs, and were the Hamiltonian to be written in terms of atomic position coordinates, it would not be separable. However, the oscillator problem of Fig. 2.3 also had a cross-term involving the particle position coordinates, and yet that problem was finally separated into internal vibration and the translation of the whole system. A coordinate transformation was the key to obtaining that Hamiltonian in separable form. So, we may ask, could a coordinate transformation lead to separability for the general type of problem in Fig. 2.5? In fact, there is a

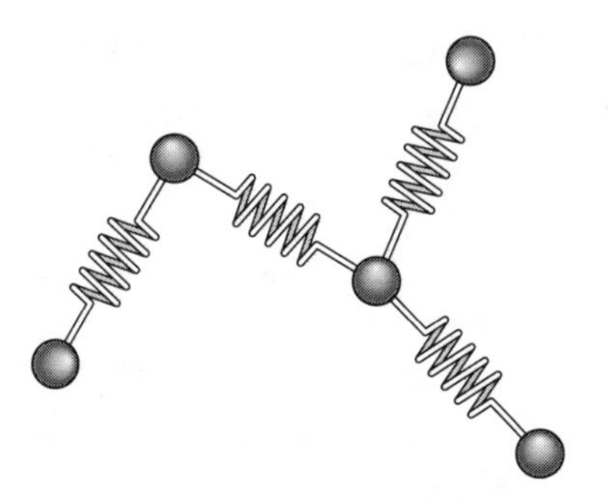

**FIGURE 2.5** A vibrating system of particles connected by springs.

specific coordinate transformation that can accomplish just that. However, the process of making that transformation assumes that all the springs are harmonic, and so when we later apply this process to molecules, it will be as an approximation of the behavior of chemical bonds. The potentials for chemical bonds are always at least somewhat anharmonic, meaning that there is a cubic, quartic, or higher order dependence of the potential energy on the extent of extension or compression of the "spring" (bond).

If separability of the harmonic vibrations is achieved, the Hamiltonian in the final coordinate system must have the following general form.

$$H = \sum_i \left( \frac{1}{2} \alpha_i q_i^2 + \frac{p_i^2}{2\beta_i} \right) \tag{2-19}$$

In other words, complete separability means that each degree of freedom has an independent term in the Hamiltonian. Just the form of this Hamiltonian tells much about the vibrations of the system, and this can be seen by thinking about a system that is mechanically equivalent. That system would consist of a set of different harmonic oscillators, each of the type shown in Fig. 2.1 and none connected to another. For the $i^{th}$ oscillator of this set, there would be a displacement coordinate, $q_i$, a spring constant which could be called $\alpha_i$, and the particle's mass which could be called $\beta_i$. One could initiate vibration by stretching and releasing any one of these oscillators, and there would be no effect on the others. By mechanical equivalence, the same is true for the coupled harmonic oscillator problem (Fig. 2.5); however, the $q_i$-coordinates are more complicated because they turn out to be linear combinations of the many different particle displacement coordinates. A given $q_i$ could correspond to a displacement of all the particles at once.

The coordinates $\{q_i\}$ for which the vibrational Hamiltonian takes on the separable form of Eq. (2-19) are referred to as *normal coordinates*. Motion that follows along the direction of these coordinates is referred to as *normal mode* vibration. For each normal mode, there is a vibrational frequency and, by mechanical equivalence arguments, its value is

$$\omega_i = \sqrt{\frac{\alpha_i}{\beta_i}} \tag{2-20}$$

The important quantities $\alpha_i$ and $\beta_i$ depend on the masses of the particles and the force constants of the springs, and they do so in a way that will have to be worked out through coordinate transformations.

Fig. 2.6 shows another oscillator problem that involves two particles. If the particles are of the same mass and if the three springs happen to

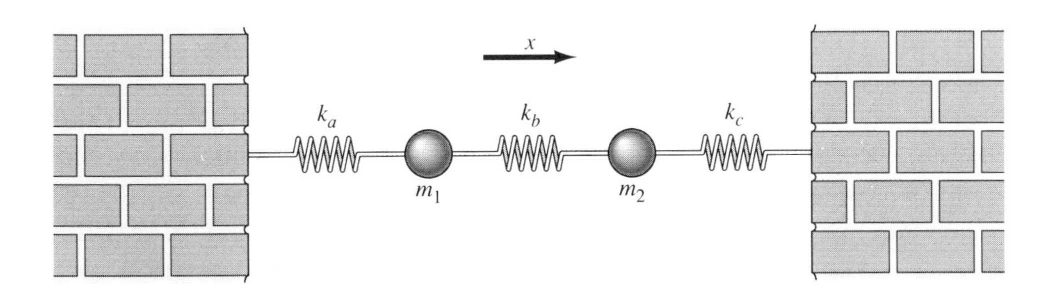


**FIGURE 2.6** A vibrating system of two particles constrained to move in the x-direction and connected by springs to each other and to two unmovable walls.

have the same force constant, the two normal coordinates have a simple qualitative form. One corresponds to the two particles moving together; at the same time, both are moving to the left or else both are moving to the right. The other normal coordinate corresponds to the two particles moving oppositely; if one is moving to the right, the other is moving to the left. These two normal modes of vibration have different frequencies of oscillation. This fact is physically reasonable because in the parallel motion the length of the central spring does not change. Only two springs are stretched or contracted. In the opposed motion, when the two outer springs are squeezed, the inner one is stretched. The system is "stiffer" with respect to the latter type of motion, and it is correct to expect the frequency of that vibrational mode to be greater than the frequency of the parallel motion mode.

With the example of Fig. 2.6 in mind, we can understand several features of a normal mode description. If the system is at rest at its equilibrium structure initially, and then the particles are displaced by an amount that is proportional to one normal coordinate, releasing the particles will cause the system to vibrate only in that mode. In the Fig. 2.6 example, with equal masses and like springs, displacement from equilibrium of one particle to the left and the other to the right by the same amount is displacement proportional to one of the normal coordinates. Displacement of both particles to the right (or left) by the same amount would be a displacement along the other normal coordinate.

When a system vibrates in a pure mode, all the particles in the system will reach their point of maximum displacement at the same instant. They will all pass through their equilibrium positions at the same instant. Simply, they will all vibrate with the same frequency and phase. In carbon dioxide, for instance, the pure symmetric stretching vibrational mode is one that equivalently stretches both carbon-oxygen bonds simultaneously. In this vibrational mode, the carbon atom remains at the midpoint of the line between the oxygen atoms as they oscillate back and forth with the same frequency.

A system such as that in Fig. 2.6 will normally vibrate in a more complicated manner than simple normal mode motion even if the springs are perfectly harmonic. However, any such motion can be represented as a superposition of the different normal mode motions.

## 2.6 CLASSICAL NORMAL MODE ANALYSIS *

Normal mode analysis is the process of finding the normal mode frequencies and normal coordinates for a vibrating system. This means finding a coordinate transformation so as to put the Hamiltonian in the form of Eq. (2-19). Before approaching this generally, consider the problem of Fig. 2.6. With $x_1$ and $x_2$ as the position coordinates and the left wall at x = 0, the kinetic and potential energy expressions are

$$T = \frac{p_1^2}{2m_1} + \frac{p_2^2}{2m_2}$$

$$V = \frac{k_a}{2}(x_1 - x_a)^2 + \frac{k_b}{2}(|x_2 - x_1| - x_b)^2 + \frac{k_c}{2}(x_c - x_2)^2$$

where the constant $x_a$ is the length of spring a when the system is at equilibrium, and $x_b$ is the length of spring b when the system is at equilibrium. Since the walls are at a fixed distance, the constant $x_c$ must be equal to the sum of $x_a$ and $x_b$. (The equilibrium length of spring c is $x_w$, the position of the right wall, less $x_c$.) The kinetic energy term is already in the desired form, but the potential energy has cross-terms. These cross-terms need to be eliminated by some coordinate transformation.

Because the momentum coordinates are conjugate to the position coordinates, any transformation among the position coordinates implies making a transformation of the momentum coordinates as well. One particular type of transformation, a *linear transformation*, can be applied in the same way to position and momentum. For a set of coordinates $\{r_1, r_2, r_3, \ldots\}$, a linear transformation to another set of coordinates $\{s_1, s_2, s_3, \ldots\}$ is one that can be expressed in terms of a special set of coefficients, $c_{ij}$, as

$$\begin{aligned} s_1 &= c_{11} r_1 + c_{12} r_2 + c_{13} r_3 + \ldots \\ s_2 &= c_{21} r_1 + c_{22} r_2 + c_{23} r_3 + \ldots \\ &\ldots \end{aligned} \qquad \text{(2-21a)}$$

or as a matrix equation,

$$\begin{pmatrix} s_1 \\ s_2 \\ s_3 \\ \cdots \end{pmatrix} = \begin{pmatrix} c_{11} & c_{12} & c_{13} & \cdots \\ c_{21} & c_{22} & c_{23} & \cdots \\ c_{31} & c_{32} & c_{33} & \cdots \\ & \cdots & & \end{pmatrix} \begin{pmatrix} r_1 \\ r_2 \\ r_3 \\ \cdots \end{pmatrix} \tag{2-21b}$$

or simply, $\vec{s} = \mathbf{c}\,\vec{r}$ (see Appendix I for notation conventions). This says that the r-coordinates are being combined linearly to form the s-coordinates. A *non*linear transformation might have a term that is quadratic in the r-coordinates. Taking the time derivative of the transformations in Eq. (2-21) reveals that a velocity expressed in the s-coordinate system is related by the same linear coefficients to velocities expressed in the r-coordinate system. The same must be true for momenta, and so, for example,

$$p_{s_1} = c_{11}\, p_{r_1} + c_{12}\, p_{r_2} + c_{13}\, p_{r_3} + \cdots$$

Thus, the momentum and position coordinates will remain as properly conjugate variables as long as any linear transformation is applied to both. A complication to be overcome is that a particular transformation that removes the cross-terms in the potential might introduce cross-terms in the kinetic energy.

The general method of solution to the normal modes problem is best presented in matrix form. (A review of some elements of matrices and linear algebra is in Appendix I.) Two square matrices (or second-rank tensors) are defined by their individual elements as

$$\mathbf{K}_{ij} \equiv \left.\frac{\partial^2 V}{\partial q_i\, \partial q_j}\right|_{eq} \tag{2-22}$$

$$\mathbf{M}_{ij} \equiv \left.\frac{\partial^2 T}{\partial \dot{q}_i\, \partial \dot{q}_j}\right|_{\text{at rest}} \tag{2-23}$$

The coordinates $\{q_i\}$ are whatever the coordinates that have been used to express V. The matrix **K** will be called the force matrix; its elements, the second derivatives of the potential energy evaluated when the system is at its equilibrium position, are force constants. The matrix **M** will be called the mass matrix. Its elements are the second derivatives of the kinetic energy with respect to velocities, evaluated when the system is not moving or at rest. If these definitions are applied to the Hamiltonian of Eq. (2-19), the resulting **K** matrix is diagonal, which means all the off-diagonal elements are zero. The diagonal elements are $\alpha_1$, $\alpha_2$, $\alpha_3$, and so on. Also, the resulting **M** matrix is diagonal, and its diagonal elements are $\beta_1$, $\beta_2$, $\beta_3$, and so on. The **K** matrix for the problem in Fig. 2.6 would not be diagonal because of the cross-terms mentioned above. Clearly, the goal of a coordinate transformation is to bring the **K** and **M** matrices into diagonal form, that is, to simultaneously *diagonalize* these two matrices.

It is generally possible to use linear transformations to simultaneously diagonalize the **K** and **M** matrices for a vibrating system. Note that the coordinate system used at the outset to construct the two matrices can be anything convenient. The final point of the process, the normal coordinates, will be achieved regardless of the starting point. A linear transformation from some initial coordinate system $\{x_i\}$ to an intermediate coordinate system $\{r_i\}$ via coefficients $U_{ij}$ (i.e., $\vec{r} = \mathbf{U}\,\vec{x}$) implies the following transformation of the force and mass matrices.

$$\mathbf{K}^{\{r\}} = \mathbf{U}\,\mathbf{K}\,\mathbf{U}^{-1} \quad \text{and} \quad \mathbf{M}^{\{r\}} = \mathbf{U}\,\mathbf{M}\,\mathbf{U}^{-1} \tag{2-24}$$

This is easily established from the definitions of the two matrices. The matrix designated $\mathbf{U}^{-1}$ is the inverse of **U**, which means that $\mathbf{U}\mathbf{U}^{-1} = \mathbf{U}^{-1}\mathbf{U} = \mathbf{1}$, where **1** is the unit or identity matrix (see Appendix I). A special type of linear transformation, called a *unitary* transformation, is required in the simultaneous diagonalization procedure, and this has been assumed for **U** in Eq. (2-24). The condition of unitarity means that the inverse of the transformation matrix is equal to the transpose of the matrix. Several algorithms exist for finding a unitary transformation matrix that diagonalizes a given matrix (see Appendix II), and the first step in simultaneous diagonalization will be to employ any of those algorithms to diagonalize the mass matrix. (Of course, it may happen to be diagonal to start out with, as would be the case if the initial coordinate system were Cartesian displacement coordinates for each particle.) We will use **U** to designate the transformation matrix that diagonalizes **M**.

The second step in the simultaneous diagonalization procedure is another linear transformation, but it is not unitary. While the first step diagonalized **M**, this transformation changes the diagonal elements so that all are equal to one, and the transformed **M** matrix is the unit matrix, **1**.

Giving the name **D** to the diagonal matrix produced from the first step, i.e., $\mathbf{D} = \mathbf{U\,M\,U}^{-1}$, the second step's transformation matrix is a diagonal matrix, called **d**, whose nonzero elements are

$$d_{ii} \equiv \frac{1}{\sqrt{D_{ii}}}$$

It is its own transpose, and it is applied in the same way on both the left and right sides of a matrix being transformed by **d**. Applying it to **D** yields $\mathbf{d\,D\,d} = \mathbf{d\,U\,M\,U}^{-1}\,\mathbf{d} = \mathbf{1}$, and so this second transformation leads to a coordinate system where the mass matrix is the unit matrix. A subtle aspect is that this second transformation has implicitly scaled the position coordinates by the masses; these are said to be *mass-weighted coordinates*.

The transformations of the first two steps have to be applied to the force matrix, and the resulting matrix will be designated **X**:

$$\mathbf{X} \equiv \mathbf{d\,U\,K\,U}^{-1}\,\mathbf{d}$$

The last step of the simultaneous diagonalization procedure is the diagonalization of **X** by a unitary linear transformation **V**:

$$\alpha \equiv \mathbf{V\,X\,V}^{-1}$$

The nonzero elements of the matrix $\alpha$ are along its diagonal and are the $\alpha$'s for Eq. (2-19)! Of course, this transformation must be applied to the mass matrix, but since that matrix has become equal to **1**, there is no change: $\mathbf{V\,1\,V}^{-1} = \mathbf{V\,V}^{-1} = \mathbf{1}$. This means that the $\beta$'s for Eq. (2-19), which are the diagonal elements of the mass matrix, are all equal to one.

To summarize the whole process, we shall write the vibrational Hamiltonian in terms of each of the various coordinate systems in matrix form. $\vec{x}$ designates the original displacement coordinate set; $\vec{r}$ will designate the coordinate system that results from Step One; and $\vec{s}$ will designate the coordinate system that results from Step Two. $\vec{q}$, the normal coordinate system, results from Step Three. Since the Hamiltonian is the total energy of the system, it must be the same from one coordinate system to the next. Thus, every transformation is applied right next to its inverse in the Hamiltonian; that is, we may freely insert multiplication by the unit matrix anywhere and use the fact that for a unitary matrix $\mathbf{U}$, $\mathbf{1} = \mathbf{U}^{-1}\mathbf{U}$. The three transformation steps may be thought of as grouping matrix products together.

$$
\begin{aligned}
2H &= \vec{x}^T \mathbf{K} \vec{x} + \dot{\vec{x}}^T \mathbf{M} \dot{\vec{x}} \\
&= (\vec{x}^T \mathbf{U}^{-1})(\mathbf{U}\mathbf{K}\mathbf{U}^{-1})(\mathbf{U}\vec{x}) + (\dot{\vec{x}}^T \mathbf{U}^{-1})(\mathbf{U}\mathbf{M}\mathbf{U}^{-1})(\mathbf{U}\dot{\vec{x}}) \\
&= \vec{r}^T(\mathbf{U}\mathbf{K}\mathbf{U}^{-1})\vec{r} + \dot{\vec{r}}^T \mathbf{D} \dot{\vec{r}} \\
&= (\vec{r}^T \mathbf{d}^{-1})(\mathbf{d}\mathbf{U}\mathbf{K}\mathbf{U}^{-1}\mathbf{d})(\mathbf{d}^{-1}\vec{r}) + (\dot{\vec{r}}^T \mathbf{d}^{-1})(\mathbf{d}\mathbf{D}\mathbf{d})(\mathbf{d}^{-1}\dot{\vec{r}}) \\
&= \vec{s}^T \mathbf{X} \vec{s} + \dot{\vec{s}}^T \mathbf{1} \dot{\vec{s}} \\
&= (\vec{s}^T \mathbf{V}^{-1})(\mathbf{V}\mathbf{X}\mathbf{V}^{-1})(\mathbf{V}\vec{s}) + (\dot{\vec{s}}^T \mathbf{V}^{-1})\mathbf{V}\mathbf{V}^{-1}(\mathbf{V}\dot{\vec{s}}) \\
&= \vec{q}^T \alpha \vec{q} + \dot{\vec{q}}^T \dot{\vec{q}}
\end{aligned}
$$

Since the mass matrix in the final coordinate system is the unit matrix, the momentum coordinates are equal to the velocities: $p_i = \dot{q}_i$. And since $\alpha$ is diagonal, the Hamiltonian does have the desired separable form. Then, from Eq. (2-20), the normal mode frequencies are the square roots of the diagonal elements of $\alpha$. The analysis is completed.

If **K** and **M** matrices were constructed for the system in Fig. 2.5 using Cartesian displacement coordinates for each particle, the number of rows and columns of the matrices would be 3N = 12. However, the number of normal modes and the number of normal coordinates is only 3N – 6 = 6. The simultaneous diagonalization procedure will produce six normal coordinates plus six coordinates that correspond to the degrees of freedom that are not vibration, i.e., the translation and rotation of the system as a whole. These six are distinguished from the six normal coordinates by having zero-valued frequencies. That is, since there is no potential for translational or rotational motion, the diagonal values of the force matrix ($\alpha$) in the final coordinate set are zero. Should there be a value of the diagonal force matrix that is less than zero, it would mean that the system is not at an equilibrium structure.

The definition for elements of the **K** matrix in Eq. (2-22) has no restriction that the potentials be harmonic. Elements can be calculated for any potential function. However, using the **K** matrix to represent the potential, which is implicit in the normal mode analysis, excludes all anharmonic components. The excluded components will depend on the displacement coordinates to higher orders than quadratic, and neglecting them amounts to an approximation. The approximation is best for vibrational motions that take the system through small excursions away from the equilibrium. If the maximum displacement in some coordinate happens to be $\delta$, and $\delta$ is small, then a Hamiltonian term quadratic in $\delta$ should be more important than terms that are cubic or quartic in $\delta$. The quality of the approximation will worsen as one attempts to describe larger amplitude displacements. Realistic systems are not strictly harmonic, and so the applicability of normal mode analysis is as a *small amplitude approximation*.

## Exercises

1. Set up the classical Hamiltonian for a particle of mass m that may move in the three directions, x, y, and z, and which experiences a gravitational potential of the form $V(x,y,z) = mgz$. Then, using Eqs. (2-5) and (2-6), write the equations of motion for this particle.
2. Show that a potential of the general form $V(x) = a + bx + cx^2$ is the same as that of a harmonic oscillator because it can be written as $V(x) = V_o + \frac{1}{2} k (x - x_o)^2$. (Find k, $V_o$, and $x_o$ in terms of a, b, and c.)
3. Verify Eq. (2-9) with both Eqs. (2-8a) and (2-8b).
4. From the relation that for any real number $\alpha$, $e^{i\alpha} = \cos(\alpha) + i \sin(\alpha)$, find the constants A and B in Eq. (2-8b) so that this x(t) is the same as the x(t) in Eq. (2-8a) for the special case when b = 0 and a = 1. Next, generalize your result for the constant a left unspecified. This amounts to writing A and B as a function of a.
5. What force constant for a harmonic spring would give rise to a vibrational frequency, $\omega$, of 10 $sec^{-1}$ when a mass of 1 kg is attached? What would be the frequency if such a spring were attached to a particle with the mass of (i) an electron, (ii) a hydrogen atom, (iii) the moon?

6. Write the potential energy of the following system of point-mass particles and harmonic springs as a function of the position coordinates of the particles. Assume that they are constrained to move only in the xy plane.

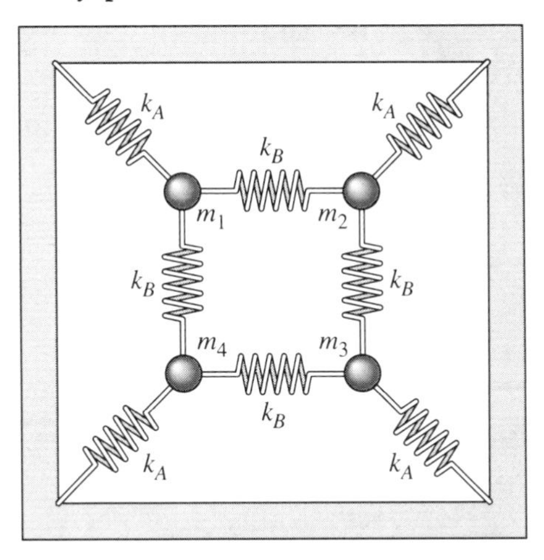


Let $l$ be the length of each side of the square, let $s_a$ be the equilibrium length of the springs with force constant $k_a$, and let $s_b$ be the equilibrium length of the springs with force constant $k_b$.

7. Develop Hamilton's equations of motion for the system in Prob. 6.

8. Set up the Hamiltonian for a particle of mass m that is connected to a harmonic spring with force constant k. The other end of the spring is connected to a pivot that allows the spring-and-mass assembly to turn about freely in any direction. The location of the pivot is fixed in space and does not move. Show that this system is mechanically equivalent in vibrational and rotational motion to a system of two particles connected by the same spring, assuming the particles may move in all three dimensions.

9. In what qualitative manner would a symmetric stretch mode of the system in Prob. 6 differ in having $m_1 = m_2 = m_3 = m_4$ and in having $m_1 = 2m_2 = m_3 = 2m_4$?

10. What would be the qualitative forms of the normal modes of vibration of the oscillator in Fig. 2.6 if there were not two but three equally massive particles connected one to another and to the walls by harmonic springs with the same force constants? And if there were four particles?

*11.* If the two particles in Fig. 2.6 were considered more realistically, there would be a y- and a z-degree of freedom for each particle to move. This would mean four more vibrational degrees of freedom, and they would be associated with bending modes of the system. Qualitatively, what would they look like? Should any of them have the same normal mode frequency?

★*12.* Find the normal mode frequencies for an oscillator system with three degrees of freedom if the **K** and **M** matrices were the following.

$$\mathbf{K} = \begin{pmatrix} 1 & -1 & 0 \\ -1 & 2 & -1 \\ 0 & -1 & 1 \end{pmatrix} \qquad \mathbf{M} = \begin{pmatrix} 1 & 0 & 0 \\ 0 & 2 & 0 \\ 0 & 0 & 1 \end{pmatrix}$$

★*13.* Devise a linear system of springs and particles that would have the **K** and **M** matrices of Prob. 9.

★*14.* Set up the **K** matrix for the system in Prob. 6.

## Bibliography


1. K. R. Symon, *Mechanics* (Addison-Wesley, Reading, Massachusetts, 1971). This is an intermediate level text in classical mechanics, and its coverage goes beyond our immediate interests. The sections on Hamilton's formulation of classical mechanics and the sections on normal modes of vibration, though, may be quite helpful.

2. E. B. Wilson, Jr., J. C. Decius, and P. C. Cross, *Molecular Vibrations. The Theory of Infrared and Raman Vibrational Spectra* (Dover Publications, New York, 1980). The second chapter of this text, which is on the classical mechanics of vibrations of systems of particles, is very relevant.

*Chapter 3*

# Quantum Mechanics - I

*The classical mechanics of the last chapter proves deficient for describing systems of very small particles such as atoms and electrons. Instead, a very different mathematical apparatus is needed, and this is embodied in quantum mechanics. This chapter explores some of basic elements of quantum mechanics and also considers the idealized problem of a harmonic oscillator.*

## 3.1 QUANTUM PHENOMENA

The first of what are now called quantum phenomena were noticed or detected around the beginning of the twentieth century. Numerous experiments could not be explained by the electromagnetic and mechanical laws known or accepted at the time. For example, it was found that a beam of monoenergetic electrons could be diffracted in the same way as a monochromatic beam of light impinging on two slits. The diffraction of light was known to be a manifestation of its wave character, and so to explain the diffraction of electrons, an abstract type of wave character had to be attributed to small particles. Of course, this was entirely outside the paradigm of classical mechanics, which holds that particles are exactly what we sense and observe in our everyday world: They are distinct objects that exist at a specific place at any instant of time, unlike electromagnetic radiation which is distributed in space and time.

It was no doubt unsatisfying to many scientists during the early part of the century that basic physical laws worked out over almost two centuries did not hold for the tiny world of electrons and atoms. Today, we see the classical picture as a special case of broader theories of matter and radiation. It took studies of the tiny-particle world and the introduction of several new concepts to develop the more encompassing theories. This was because key features of the quantum world are manifested in ways not directly perceptible in our everyday world.*

Particle diffraction was an important observation. The particles are free particles meaning they are not subject to a potential. Electrons, for instance, would first have been stripped away from some atoms or molecules. Observing diffraction requires that all the particles in the sample move at the same velocity or momentum, and this can be accomplished by the manner in which the particle beam is formed. The quantitative result of the diffraction experiment is a wavelength, $\lambda$. Measured for different particle momenta, p, the wavelength is found to vary inversely with p. This is called the *de Broglie relation*:

$$p = \frac{h}{\lambda} \tag{3-1}$$

$\lambda$ is the de Broglie wavelength. The proportionality constant of this relation, h, turns out to be a fundamental constant of nature, *Planck's constant*. The value of h is $6.6256 \times 10^{-27}$ erg-sec.

Let us find the de Broglie wavelength for an object from our day-to-day world, a rider and motorcycle traveling at 36 km/hr. If the mass of the rider and cycle is 200 kg, then the momentum of this "particle" is,

* A question that often comes to mind upon first learning of the quantum mehanical revolution is whether a new theory could someday replace quantum mechanics. In this day and age, this may be more a matter for the philosophy of science and how science accepts or rejects new theories, for the successes of quantum ideas appear endless and have stood the test of time. There is good reason for confidence that any major scientific revolution to come will not replace quantum theory but will build upon or encompass it.

$$(200 \text{ kg}) \times (36 \text{ km/hr}) = (2 \times 10^5 \text{ g}) \times (10^3 \text{ cm/sec})$$
$$= 2 \times 10^8 \text{ g-cm/sec}$$

Dividing this into Planck's constant yields the wavelength.

$$\lambda = \frac{6.63 \times 10^{-27} \text{ g-cm}^2\text{/sec}}{2 \times 10^8 \text{ g-cm/sec}} = 3.3 \times 10^{-35} \text{ cm}$$

This is an incredibly tiny wavelength, and it is not surprising that effects associated with something this size are not within the perception of the motorcycle rider. However, the de Broglie wavelength of a free electron moving at the same velocity as the rider and cycle is much longer:

$$\lambda = \frac{6.63 \times 10^{-27} \text{ g-cm}^2\text{/sec}}{(9.1 \times 10^{-28} \text{ g})\,(10^3 \text{ cm/sec})} = 7.3 \times 10^{-3} \text{ cm}$$

The wavelength of the electron, because of its small mass, is significant relative to the dimension of its atomic world, whereas the wavelength of the rider and cycle is negligible relative to the dimensions of their world. One may consider it a valid approximation to treat the mechanics of the rider and cycle classically, whereas the wave character of the free electron is not so ignorable.

Other observations made at the beginning of the twentieth century not only showed unexpected phenomena but also seemed to involve the new constant, h, in many ways. The photoelectric effect, which is the emission of electrons from a metal surface upon irradiation with light, was explained by Einstein with a nonclassical picture of electromagnetic radiation. He argued that the energy delivered by radiation comes in discrete amounts, called quanta, and that the energy is proportional to the frequency of the radiation, $\nu$:

$$E^{\text{radiation}} = h\nu \tag{3-2}$$

Notice that the proportionality constant is Planck's constant. As more problems were explained by the unusual quantum and wave notions, it became certain that classical physics was not complete and could not be applied at the subatomic level. The laws of quantum mechanics that then emerged were worked out by clever and sometimes indirect lines of thought.

## 3.2 SIMPLE WAVE CHARACTER

If it is necessary for a quantum-level description that a free particle have an associated wavelength, then there should be some simple oscillating function associated with the particle's description. This would be a sine or cosine function (or equivalently, a complex exponential), e.g.,

$$A(x) = A_o \cos\left(\frac{2\pi x}{\lambda}\right) \tag{3-3}$$

The constant, $A_o$, is the maximum amplitude of A(x). It is a feature of this type of function that the wavelength may be extracted from the function as an *eigenvalue*. Notice that when A(x) is differentiated twice, the result is a constant, the eigenvalue C, times A(x).

$$\frac{d^2}{dx^2} A(x) = -\left(\frac{2\pi}{\lambda}\right)^2 A_o \cos\left(\frac{2\pi x}{\lambda}\right) = C\,A(x)$$

In such a situation this function is called an *eigenfunction* of the operation of second differentiation. A translation of the German word *eigen* is "of one's own," or loosely, "same." It is one's own function, or the same function, that is produced upon the operation of second differentiation. The constant that shows up multiplying the original function is the eigenvalue. In contrast, A(x) is not an eigenfunction of the operation d/dx since first order differentiation of the cosine function produces a different function, the sine function.

At this point we might believe that somehow the function A(x) or something similar is going to be the basis for a description of the mechanics of a moving free particle. However, we do not have an interpretation of A(x); its meaning is not apparent. One thing, though, is that A(x) has a wavelength embedded in it, and that wavelength shows up in a particular eigenvalue of the function. It was deduced by Schrödinger, Heisenberg, and others that eigenvalues have special significance, and in particular that they give the measurable information about a system, such as the de Broglie wavelength of a free particle. The eigenfunctions themselves came to be known as *wavefunctions* partly because it was the apparent wave character of particles that demanded a new mechanics.

If there are particular operations, such as second differentiation, for which all proper wavefunctions must be eigenfunctions, then there exists a general basis for finding wavefunctions for any system. It is to solve an eigenequation – rather than guess at something like A(x). This focuses attention on the operations that may enter the eigenequation, and so, specific mechanical elements come to be associated with specific mathematical operations. One of these can be demonstrated using the de Broglie relation and the free particle example. If the operation $(d/dx)^2$ is scaled (multiplied) by the constant* $-(h/2\pi)^2$, then by Eq. (3-1) the resulting operator yields the square of the momentum when applied to A(x):

$$-\left(\frac{h}{2\pi}\right)^2 \frac{d^2}{dx^2} A(x) = \frac{h^2}{\lambda^2} A(x)$$

This suggests an operation that in quantum mechanics corresponds to the square of the dynamical variable of momentum. We might anticipate that this is useful, since the classical energy, at least, is a function of momentum, and energy is of key importance in a particle world where energy is quantized.

These several notions about the quantum world form part of a theoretical basis that is usually referred to as the set of *quantum mechanical postulates*. These postulates are ideas or statements that in themselves may or may not be true. However, if predictions that follow naturally from these ideas are proven true, then the ideas are accepted as true even if there is no direct proof.

Postulate I says that *for every quantum mechanical system, there exists a wavefunction that contains a full mechanical description of the system*. It is a function of position coordinates and possibly time. This is quite different from classical mechanics where the full mechanical description would be a function that gives the position coordinates for any instant in time, e.g., $\vec{x}(t)$.

Postulate II explains, in part, that we may glean a picture of a quantum mechanical system from its wavefunction. *For every dynamical variable, there is an associated mathematical operation. Furthermore, if*

* Planck's constant, h, often shows up in equations divided by $2\pi$. Thus, a special symbol, $\hbar$, is used to replace the ratio $h/2\pi$. $\hbar$ (h-bar) is also called Planck's constant, and so it is always important to indicate whether h or $\hbar$ is meant.

*there is an eigenvalue for such an operation and a particular wavefunction, then that eigenvalue is the result that would be obtained from measuring that dynamical variable for the particular system.* An example, so far, is that the operation associated with the square of the momentum has an eigenvalue of $(h/\lambda)^2$ for the free particle wavefunction, A(x).

The dynamical variables in the classical Hamiltonian are position and momentum coordinates. As long as wavefunctions are developed* as functions of position coordinates, as was the case for A(x), the mathematical operation associated with a position coordinate is just multiplication by the coordinate. The operation associated with the momentum variable $p_i$ that is conjugate to the position variable $q_i$ is

$$-i\hbar\frac{\partial}{\partial q_i}$$

where $i \equiv \sqrt{-1}$. Notice that the square of this operation (i.e., applying it twice) is the same as that used above for the square of the momentum. The various mathematical operations are best referred to as *operators*, and operators may be distinguished from variables by a "hat" over the character. Thus, the statement $\hat{x} = x$ means that the operator associated with x is the same as the variable x, while $\hat{p} = -i\hbar\,\partial/\partial x$ means that the operator associated with the momentum is $-i\hbar$ times the operation of differentiation with respect to x. Operators can be combined and applied repeatedly to form new operators. Thus, if the position and momentum operators are known, an operator can be constructed that is associated with the classical Hamiltonian. It is called the Hamiltonian operator, or just the Hamiltonian if the quantum mechanical context is clear.

The Schrödinger equation, which can be taken as Postulate III, is that *the wavefunction of a system must be an eigenfunction of the Hamiltonian operator*. The eigenvalue is the energy of the system, since by association, energy is the quantity given by the classical Hamiltonian. Energy is measurable or observable, and so this statement is similar to

* Quantum mechanics may also be developed so that the functions describing the system are functions of momentum coordinates, not position coordinates. This is termed a momentum representation, and in this representation, position and momentum operators take on a different form. The picture of a quantum system, though, is equivalent, and the choice between representations is largely one of convenience.

the second postulate that tells what to make of eigenvalues of wavefunctions. But the third postulate is a stronger statement than the second postulate in that it *requires* the wavefunction to be an eigenfunction of one specific operator, the Hamiltonian. The most common symbol used for an unspecified quantum mechanical wavefunction is $\psi$, and so the Schrödinger equation is written as

$$\hat{H}\psi = E\psi \tag{3-4}$$

E is the energy eigenvalue. Solution of the Schrödinger equation is central to all quantum mechanical problems.

## 3.3 OPERATORS

Because of the important mathematical role of operators in quantum mechanics, it is useful to understand the algebra of operators. As a general definition, we may say that an operator is a mathematical device that relates two functions. An operator equation is any equation that includes an operator. Thus, $g(x) = 3f(x)$ may be regarded as an operator equation, since the functions $g(x)$ and $f(x)$ are related by the operation of multiplication by 3. The equation could be given more formally as $g(x) = \hat{O}\, f(x)$ where $\hat{O} = 3$.

Though operators may very well be as simple as multiplication by a constant such as 3, they may be multiplication by a variable, such as the operator $\hat{x}$. Or they may involve differentiation with respect to one or more variables. Anything that can relate one function to another function can be labelled an operator.

Two operators may be combined by addition or by multiplication to create another operator. The process of operator addition and the process of operator multiplication, though, require definition.

$\hat{c} = \hat{a} + \hat{b}$ means the result of operating with $\hat{c}$ on an arbitrary function, f, yields the same result as operating with $\hat{a}$ on f and adding that to the result of operating with $\hat{b}$ on f.

$\hat{d} = \hat{a}\,\hat{b}$ means that the result of operating with $\hat{d}$ on an arbitrary function, f, yields the same result as operating on f with $\hat{b}$ and then operating on that result with $\hat{a}$.

With these definitions, operators can be added and multiplied, and so algebraic expressions may be written involving just operators.

Multiplication of operators is successive application of two or more operators. However, this is not the same as multiplication of numbers; the order of the operators being multiplied may make a difference. Consider the example where one operator is multiplication by the variable x, $\hat{a} = x$, and the other operator is differentiation with respect to x, $\hat{b} = d/dx$. To invoke the definition of operator multiplication, let f(x) be an arbitrary function; that is, let it stand for any function whatsoever that we may choose. The product of the operator multiplication of $\hat{a}$ times $\hat{b}$ is found from applying $\hat{b}$ and then $\hat{a}$ to the arbitrary function. But this result is seen to be different from multiplication of $\hat{b}$ times $\hat{a}$.

$$\hat{a}\,\hat{b}\,f(x) = x\,\frac{df(x)}{dx} \qquad \Rightarrow \qquad \hat{d} = x\frac{d}{dx}$$

$$\hat{b}\,\hat{a}\,f(x) = \frac{d}{dx}\big(x\,f(x)\big) = x\,\frac{df(x)}{dx} + f(x) \qquad \Rightarrow \qquad \hat{d} = \left(x\frac{d}{dx} + 1\right)$$

This illustrates that operator multiplication is not commutative. Multiplication of simple numbers, of course, is commutative; 2 times 3 is the same as 3 times 2.

A *commutator* is a special operator constructed from two operators as the difference between the two ways they may be multiplied. The commutator is designated by placing the two operators being multiplied inside square brackets. The definition is

$$[\hat{A},\hat{B}] \equiv \hat{A}\,\hat{B} - \hat{B}\,\hat{A} \tag{3-5}$$

If the two operators do commute, which means that the same result is achieved from applying them in either order, then the commutator is equal to zero. On the basis of the multiplication results for the $\hat{a}$ and $\hat{b}$ operators just considered, their commutator is: $[\hat{a},\hat{b}] = -1$.

A special type of operator equation is the eigenvalue equation, which may be expressed generally as

$$\hat{O} f = c f \tag{3-6}$$

This equation says that for some particular function f and some operator $\hat{O}$, operating on f with $\hat{O}$ yields a constant, c, times f. The Schrödinger equation is, of course, an important example of an eigenvalue equation.

## 3.4 THE HARMONIC OSCILLATOR

The problem of the quantum mechanical description of a harmonic oscillator is our first example of applying the postulates of quantum mechanics. The picture of the system is the same as that in Fig. 2.1, a mass m connected by a harmonic spring with force constant k to an unmovable wall. The steps for treating this problem quantum mechanically, as well as any other problem, are the following.

i. Find the classical Hamiltonian.
ii. By replacing variables in the classical Hamiltonian with corresponding quantum mechanical operators, develop the quantum mechanical Hamiltonian operator.
iii. Find the wavefunction from the Schrödinger equation, which means finding the eigenfunction or eigenfunctions of the Hamiltonian operator.

From the development in Chap. 2, the classical Hamiltonian of the harmonic oscillator is

$$H(x,p) = \frac{p^2}{2m} + \frac{1}{2} k (x - x_o)^2$$

For convenience, let us define the x-coordinate origin to be the location of the mass when the system is at equilibrium. This means choosing the origin such that $x_o = 0$. Thus,

$$H(x,p) = \frac{p^2}{2m} + \frac{1}{2} k x^2$$

The quantum mechanical Hamiltonian is trivially constructed by formally making x, p, and H operators.

$$\hat{H} = \frac{\hat{p}^2}{2m} + \frac{1}{2}k\hat{x}^2$$

Of course, the explicit form of the momentum operator is that used earlier, and the position coordinate operator is just multiplication by x. So, a more useful way of writing the quantum mechanical Hamiltonian is

$$\hat{H} = -\frac{\hbar^2}{2m}\frac{d^2}{dx^2} + \frac{1}{2}kx^2 \qquad (3\text{-}7)$$

The solution of the Schrödinger equation, $\hat{H}\,\psi(x) = E\,\psi(x)$, is the next step.

The Schrödinger equation for the harmonic oscillator happens to be a well-studied differential equation which mathematicians had solved long before the quantum mechanical problem was formulated. There are an infinite number of functions of x that turn out to be valid eigenfunctions of the Hamiltonian operator. A subscript n is used, therefore, to distinguish one solution from another. Each eigenfunction has its own energy eigenvalue, and so the differential equation at hand is

$$-\frac{\hbar^2}{2m}\frac{d^2\psi_n(x)}{dx^2} + \frac{1}{2}kx^2\,\psi_n(x) = E_n\,\psi_n(x) \qquad (3\text{-}8)$$

All the eigenfunctions, or solutions, may be expressed as

$$\psi_n(x) = \psi_n(\tfrac{z}{\beta}) = \frac{N}{\sqrt{2^n\, n!}}\, h_n(z)\, e^{-z^2/2} \quad \text{where } z \equiv \beta x \qquad (3\text{-}9)$$

These wavefunctions have been written as a function of z for conciseness, but z is just x scaled by the constant β. The constant β is related to the force constant and the particle mass:

$$\beta^2 \equiv \sqrt{km}\,/\,\hbar \quad \text{and} \qquad (3\text{-}10)$$

$$N = \left(\frac{\beta^2}{\pi}\right)^{1/4} \qquad (3\text{-}11)$$

The n subscript has zero as its smallest allowed value and may be continued to infinity. The functions $h_n(z)$ are simple polynomials of z (or of $\beta x$) that are called Hermite polynomials.

The harmonic oscillator wavefunctions of Eq. (3-9) are all equal to a constant times a polynomial times a particular exponential function, called a Gaussian function. At z = 0 (i.e., at x = 0), the Gaussian function is at its maximum value, one. As z increases or decreases, the Gaussian function diminshes quickly. For z >> 0 or z << 0, the Gaussian function becomes vanishingly small, and so the asymptotic limit for each of the wavefunctions is zero. Restated in terms of the position variable x instead of z, the Gaussian function is

$$e^{-z^2/2} = e^{-(\beta x)^2/2}$$

Because $\beta$ increases with the mass of the particle and with the spring constant, then a heavier mass or a stiffer spring will make the value of the Gaussian function smaller for any given displacement in x.

The Hermite polynomials are very simple polynomials. It turns out that they may be generated in several ways, such as by the following formula:

$$h_n(z) = (-1)^n e^{z^2} \frac{d^n}{dz^n} e^{-z^2} \qquad (3\text{-}12)$$

Or, they may be generated by a recursive procedure.* The first several are

$$h_0(z) = 1 \qquad h_0(\beta x) = 1$$

$$h_1(z) = 2z \qquad h_1(\beta x) = 2\beta x$$

$$h_2(z) = 4z^2 - 2 \qquad h_2(\beta x) = 4\beta^2 x^2 - 2$$

$$h_3(z) = 8z^3 - 12z \qquad h_3(\beta x) = 8\beta^3 x^3 - 12\beta x$$

* The recursive relation tells how to generate the n + 1 polynomial from the $n^{th}$ order polynomial, given that $h_0(z) = 1$:

$$h_{n+1}(z) = 2z\,h_n(z) - \frac{dh_n(z)}{dz}$$

$$h_4(z) = 16z^4 - 48z^2 + 12 \qquad h_4(\beta x) = 16\beta^4x^4 - 48\beta^2x^2 + 12$$

$$h_5(z) = 32z^5 - 160z^3 + 120z \quad h_5(\beta x) = 32\beta^5x^5 - 160\beta^3x^3 + 120\beta x$$

Notice that $h_0$ is a constant. $h_1$ has one node, meaning that it has one point where it changes sign, namely, z = 0. Generally, $h_n$ has n nodes, and odd functions (i.e., n=1,3,5, . . . ) have one node at the origin (x = z = 0).

A crucial point is to see that these wavefunctions are eigenfunctions of the harmonic oscillator Hamiltonian and to determine the eigenvalues associated with each. For $\psi_0$,

$$\hat{H}\psi_0 = -\frac{\hbar^2}{2m}\frac{d^2\psi_0}{dx^2} + \frac{1}{2}kx^2\psi_0$$

$$= -\frac{\hbar^2}{2m}\beta^2\left(\beta^2x^2 - 1\right)\psi_0 + \frac{1}{2}kx^2\psi_0$$

$$= \frac{\hbar^2}{2m}\beta^2\psi_0 + \left(\frac{k}{2} - \frac{\beta^4\hbar^2}{2m}\right)x^2\psi_0$$

A rearrangement of terms has been made between the second and third lines. This rearrangement leads not only to one term that is just a constant times the wavefunction, but also to a second term that is a different function entirely, $x^2\psi_0$. However, if the expression for $\beta$ from Eq. (3-10) is used, then the factor that multiplies $x^2\psi_0$ is found to be zero. Thus, $\psi_0$ is indeed an eigenfunction of the Hamiltonian. The eigenvalue, $E_0$, of the Schrödinger equation $\hat{H}\psi_0 = E_0\psi_0$ is

$$E_0 = \frac{\hbar^2\beta^2}{2m}$$

$$= \frac{\hbar^2}{2m}\frac{\sqrt{km}}{\hbar}$$

$$= \frac{1}{2}\hbar\sqrt{\frac{k}{m}} = \frac{1}{2}\hbar\omega$$

where $\omega$ is the intrinsic frequency of the harmonic oscillator, the same frequency as in the classical description. If the Hamiltonian is applied to the $\psi_1$ wavefunction from Eq. (3-9), the following results:

$$\hat{H}\psi_1 = -\frac{\hbar^2}{2m}\frac{d^2\psi_1}{dx^2} + \frac{1}{2}kx^2\psi_1$$

$$= -\frac{\hbar^2}{2m}\beta^2\left(\beta^2x^2 - 3\right)\psi_1 + \frac{1}{2}kx^2\psi_1$$

$$= 3\frac{\hbar^2}{2m}\beta^2\,\psi_1 + \left(\frac{k}{2} - \frac{\beta^4\hbar^2}{2m}\right)x^2\,\psi_1$$

$$= 3\frac{\hbar^2}{2m}\beta^2\,\psi_1 = \frac{3}{2}\hbar\omega\,\psi_1$$

The energy eigenvalue for this state is precisely $\hbar\omega$ more than the energy of $\psi_0$. And if the next state, $\psi_2$, were so tested, its eigenenergy would be $\hbar\omega$ more than the energy of $\psi_1$. In fact, the energy eigenvalues of the harmonic oscillator may be expressed generally by using the index n that distinguishes the different states.

$$E_n = (n + \frac{1}{2})\,\hbar\omega \tag{3-13}$$

n will be referred to as a *quantum number* from now on. The state with the lowest energy is always called the *ground state*, and for the harmonic oscillator, this is the $n = 0$ state.

There are two very important features of Eq. (3-13) that are typical of many quantum systems. First, the allowed energies of the system are not continuous. Whereas the classical harmonic oscillator could be made to have any energy at all just by stretching it to a suitable length and releasing it, the quantum mechanical oscillator can have only the particular energies specified by Eq. (3-13). Any energy added to the oscillator or removed from it should come in chunks (quanta) of the amount that separates the allowed energies, i.e., integer multiples of $\hbar\omega$. For an oscillator in our everyday world, $\hbar$ is so small that the allowed energies are essentially continuous; this is the correspondence between the classical and quantum descriptions. The second important feature is that the

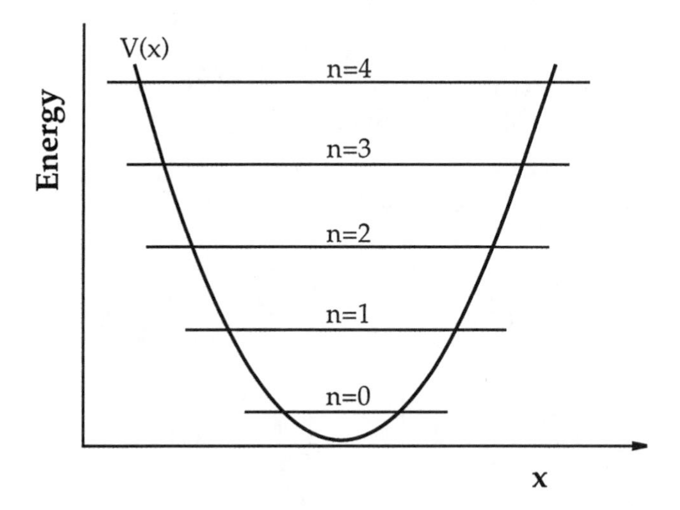


**FIGURE 3.1** The energy levels of the quantum mechanical harmonic oscillator.

ground state energy is not zero. There exists no state of the system for which the energy is zero, and so the quantum mechanical oscillator can never be completely at rest!

There is a helpful way of displaying the quantum mechanical information about the harmonic oscillator, or about other simple problems. The position coordinate, x, labels the horizontal coordinate axis of a two-coordinate graph. The vertical axis is in units of energy, and the first item to be drawn is the potential, V(x). As shown in Fig. 3.1, V(x) for the harmonic oscillator is a parabola that opens upward. Next, straight, horizontal lines are drawn at energies that correspond to the eigenenergies of the system. Each is labelled with the quantum number, n. Notice the regular spacing of these lines that is dictated by Eq. (3-13). The horizontal lines designate the *energy levels* of the system. The particular points where these lines cross the parabola, V(x), are points at which the potential energy is the same as the energy of the oscillator. Were this a classical problem, these points would be those at which the particle has no kinetic energy; it has stopped and is ready to turn back. The points are called

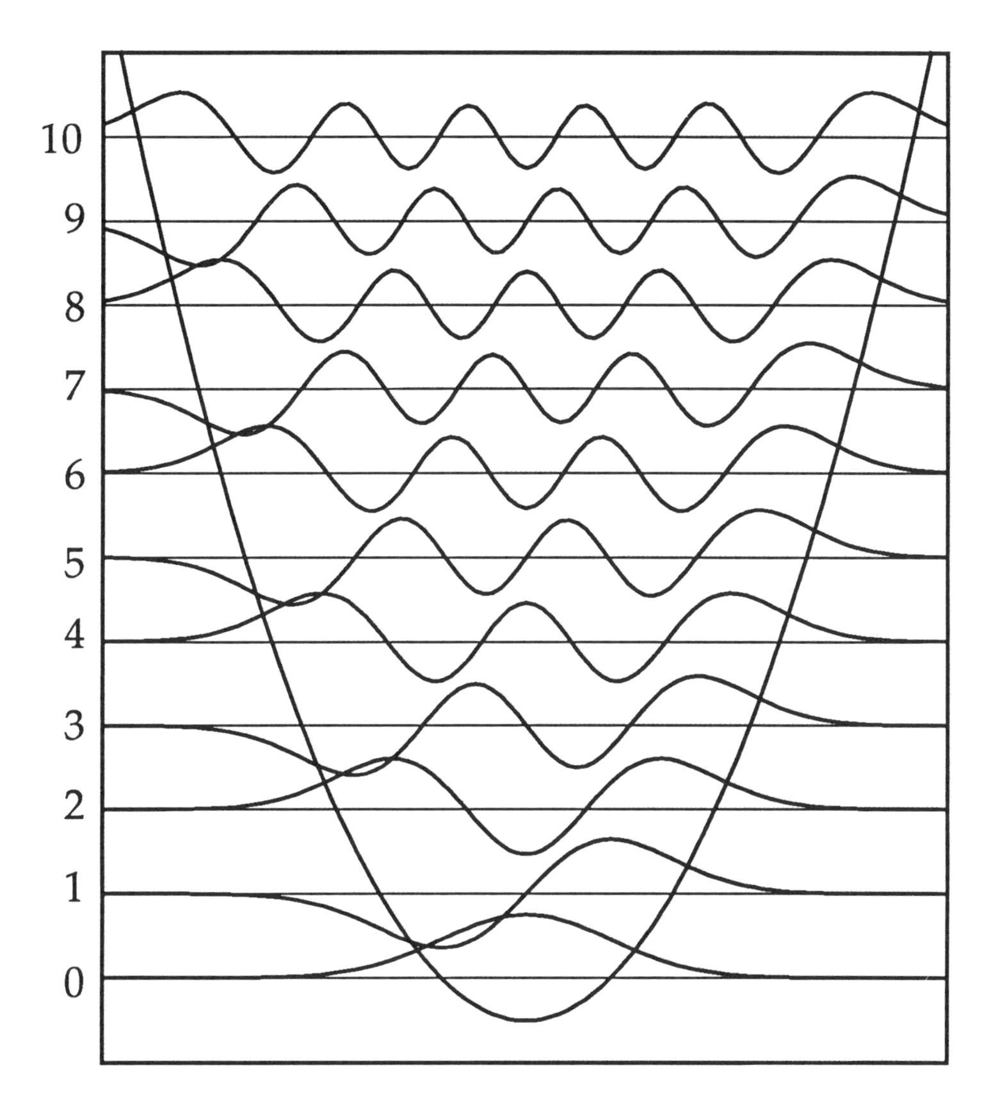


**FIGURE 3.2** The wavefunctions of low-lying levels of the quantum mechanical harmonic oscillator. The functions are drawn so that the baseline is the energy level line as in Fig. 3.1, and that means that the function is zero-valued or has a node at any point where it crosses the energy level line. Since it is the qualitative form that is usually of interest, the vertical axis is in arbitrary units. Notice that for the highest levels drawn, the wavefunction has maximum amplitude close to the classical turning points. A truly classical oscillator is moving at its slowest speed as it changes direction at a turning point, and so it is more likely to be found around the turning points than in the middle.

*classical turning points*, though the quantum mechanical particle does not have to turn back at them.

The horizontal lines in Fig. 3.1 that identify the energy levels serve as baselines for drawing the wavefunctions. Typically, the vertical scale for the wavefunctions is arbitrary since it is the qualitative form of each that is of most interest. Fig. 3.2 is a drawing of this sort, and one of the things we can see is that the number of nodes equals the quantum number.

## 3.5 THE PROBABILITY DENSITY

In a classical picture of electromagnetic radiation, it is the square of the wave amplitude that gives the energy density. Early notions about quantum mechanical wavefunctions of particles borrowed this idea, and it worked. Postulate IV says that *the square of the wavefunction is a function that gives the probability per unit volume, or the probability density, of finding the particles of a system to be at specific locations in space.* Designating the probability density as $\rho$, this says,

$$\rho(q_1, q_2, \ldots) \equiv \psi^*(q_1, q_2, \ldots)\, \psi(q_1, q_2, \ldots) \tag{3-14}$$

Notice that the "square of the wavefunction" is $\psi^*\psi$, where $\psi^*$ is the *complex conjugate* of $\psi$. The complex conjugate of a number or function is that number or function with $-i$ replacing $i$ (the square root of $-1$) wherever it shows up. While the square of a real number is the number times itself, the square of a complex number is the number times its complex conjugate. So, to take the square of a function that may be complex, we must multiply together the function and its complex conjugate.

The probability density function is as close as one can get to attaching physical significance to a wavefunction. Consider the ground state wavefunction of the harmonic oscillator. The probability density function is

$$\rho_0 = \frac{\beta}{\sqrt{\pi}} e^{-\beta^2 x^2}$$

This function has its maximum value at x = 0, which is the position at equilibrium. Postulate IV says that the probability per unit length (length instead of volume since this is a one-dimensional system) of finding the particle is greatest at x = 0. Loosely, the most likely spot to find the particle as it moves about is in the vicinity of x = 0.

The probability density for a number of the lowest energy states of the harmonic oscillator is shown in Fig. 3.3. These functions are sketched using the associated energy level lines as the baseline in the same way that the wavefunctions have been displayed in Fig. 3.2. At a point where the wavefunction has a node, the probability density is zero, and these points are easy to see. Notice that as the quantum number increases, the probability density becomes relatively bigger in the turning point regions than around x = 0. In this way, the quantum mechanical description is beginning to resemble the classical description. (Classically, the particle moves slowest at the turning points and spends more time in those regions.)

Should it be that the probability density function for a particle is constant in some particular region of space, then the probability of finding the particle in that region of space must result from multiplying the probability density by the volume of that region. If the probability density is not constant, then it is necessary to multiply it by an infinitesimal volume element and integrate over the region. Thus, the net probability of finding the harmonic oscillator particle between x = 0.0 and x = 0.1 is

$$\int_{0.0}^{0.1} \rho_0 \, dx$$

The probability of finding the particle *somewhere* or *anywhere* means that the limits of integration span all regions of space. For the ground state of the harmonic oscillator, this is

$$\int_{-\infty}^{\infty} \rho_0 \, dx$$

Evaluation of this integral quantity shows that it has a value of exactly one. This is understandable, since a probability of one means total

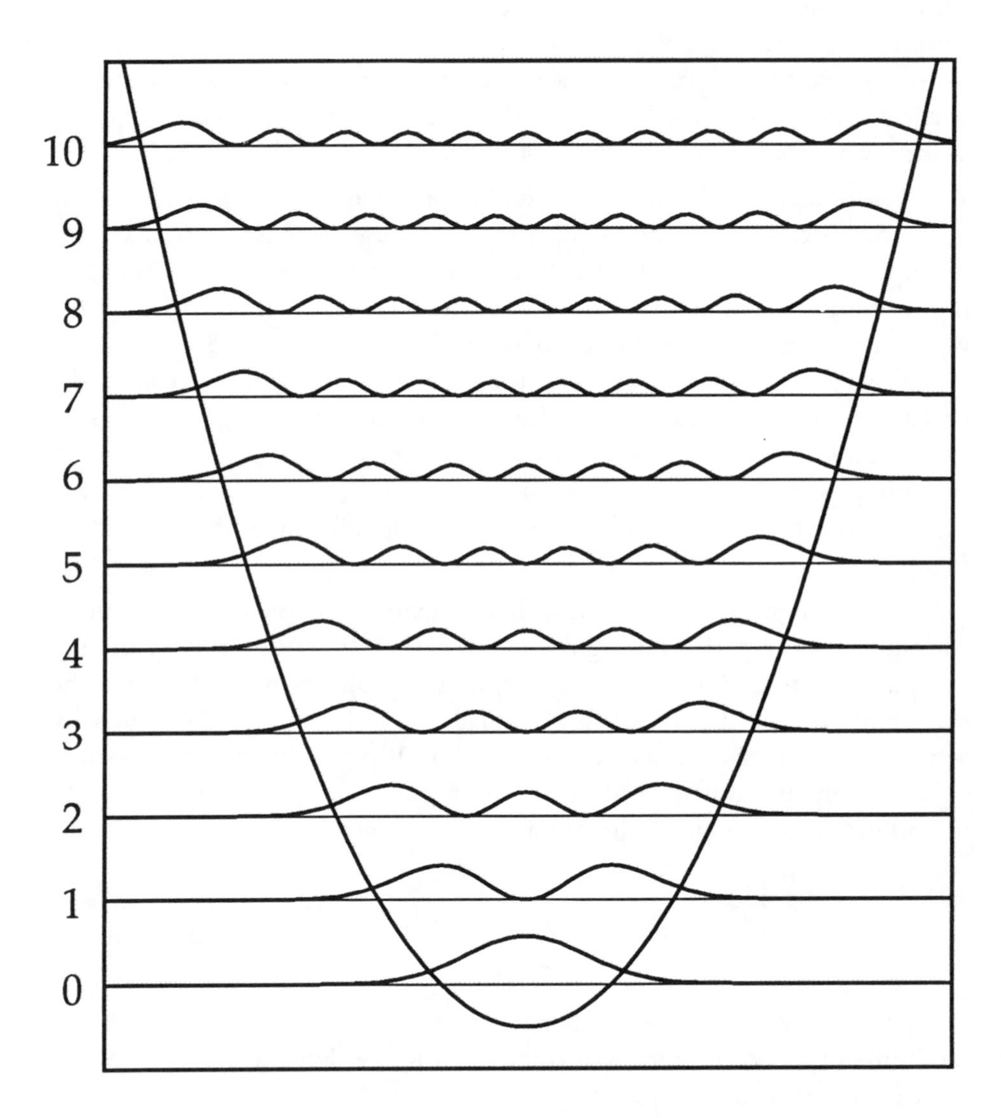


**FIGURE 3.3** The probability density functions of low-lying levels of the quantum mechanical harmonic oscillator. These functions are the squares of the wavefunctions. They are drawn so that the baseline, or point of zero probability density, is the energy level line as in Fig. 3.1. Notice that as n increases, the probability density at $x = 0$ diminishes. Furthermore, at levels such as $n = 10$ the points of maximum probability density are close to the classical turning points, unlike the lower states.

certainty, and yes, we can be certain that the particle is to be found somewhere between $x = -\infty$ and $x = \infty$.

The condition that the probability of finding the harmonic oscillator particle somewhere is one is not implicit in the Schrödinger equation. Notice that the function $\Gamma = 3\psi_0$, for example, is just as much an eigenfunction of the Hamiltonian as $\psi_0$, and it has the same eigenvalue:

$$\hat{H}\Gamma = \hat{H}(3\psi_0) = 3\hat{H}\psi_0 = 3E_0\psi_0 = E_0(3\psi_0) = E_0\Gamma$$

But the integral of $\Gamma^*\Gamma$ over all space is 9, not 1. So, when the condition is satisfied that the integral of the probability density over all space is one, the wavefunction is in the proper form for interpretation by Postulate IV. Such a wavefunction is said to *normalized*. A solution of the Schrödinger equation is not automatically a normalized wavefunction but may be made so by multiplying it by a constant. That constant, called a normalization factor, is obtained from the integral of the square of the unnormalized wavefunction. That is, if $\Gamma$ were an unnormalized wavefunction, then $\Gamma' = N\Gamma$ would be a normalized wavefunction given the following definition of the normalization constant, N.

$$N \equiv \frac{1}{\sqrt{\int_{-\infty}^{\infty} \Gamma^*\Gamma \, dx}} \tag{3-15}$$

If $\Gamma$ happens to be normalized already, this expression will give $N = 1$. (If the wavefunction, $\Gamma$, were a function of several coordinates, $q_1$, $q_2$, $q_3$, and so on, then $dq_1\, dq_2\, dq_3 \ldots$ would be in place of dx in the integral.)

Postulate II says that a measurement of a dynamical variable equals an eigenvalue *if* the wavefunction is an eigenfunction of the operator associated with that variable. In the case of the harmonic oscillator, it is reasonable to ask what would result from measuring the position of the particle, x, even though none of the wavefunctions are eigenfunctions of the operator x. Postulate V says that in such a case only the average of many measurements can be determined from the wavefunction. Individual measurements may differ from one to the next. The average is obtained by integrating the variable's probability density, which is just $\psi^*\psi$ with the associated operator sandwiched between. According to this postulate *the average or mean value of measuring the quantity associated* *with some operator* $\hat{T}$ *for a system described by a normalized wavefunction* $\Psi$ *is*

$$\int \Psi^* \hat{T} \Psi \, d\tau$$

$d\tau$ is the volume element for the coordinate system in which $\Psi$ is given. This average value is also called the *expectation value* of T and is usually designated as $\langle T \rangle$. If subscripts are attached to this symbol, they are used to indicate the quantum number(s) of the particular wavefunction used in the integral.

For the harmonic oscillator, it is directly established* that for all n,

$$\langle x \rangle_n = \int_{-\infty}^{\infty} \psi_n^* \, x \, \psi_n \, dx = 0$$

For all states of the harmonic oscillator, the expectation value of the position of the particle is zero, which is the location of the potential minimum or the middle of the potential well. This is a consequence of the fact that for every state, the proability density at some x is exactly the same as at –x. The probability density function is symmetric with respect to the middle of the well, and there is the same chance of being a certain amount to the left of the middle as being to the right by that amount. An average of many measurements gives zero, although any one measurement need not give zero. In fact, a series of measurements may give many different values that are distributed in some range.

The *standard deviation* is a statistical estimator of the range of most measurements in some series of trials. It is the root mean square (rms) deviation from the average value. Let's say a particular set of measurements of some quantity yielded these five values, 1.2, 2.4, 2.0, 1.8, and 1.6, the average of these being 1.8. The deviations of each measurement from this average are –0.6, 0.6, 0.2, 0.0, and –0.2, respectively. The deviations squared are 0.36, 0.36, 0.04, 0.0, and 0.04, and the mean of these values is 0.16. The rms deviation is $\sqrt{0.16} = 0.4$. Three of the five

* The function in the integrand, $f(x) = \psi_n(x)^* \times \psi_n(x)$, must be an odd function because it is the product of an odd function, x, and an even function, $\psi_n(x)^* \psi_n(x)$. This means that for any value of the variable x, $f(x) = -f(-x)$. When an odd function is integrated from $-\infty$ to $+\infty$, the result must be zero.

measurements happen to be in the range 1.8 ± 0.4.

The same type of deviation in measurement from an average value may be evaluated from the wavefunction in a quantum mechanical description of a system. First, there is an operator that corresponds to "measuring the deviation from the average" of some dynamical variable. If that variable is T this operator is simply $\hat{T} - \langle T \rangle$. It is the T operator less the numerical value that is the expectation value of T. The square of this operator, $(\hat{T} - \langle T \rangle)^2$, corresponds to measuring the square of the deviation from the average. Thus, the mean square of the deviation is the expectation value of this operator,

$$(\Delta T)^2 \equiv \int \Psi^* (\hat{T} - \langle T \rangle)^2 \, \Psi \, d\tau \tag{3-16}$$

and the root mean square deviation is the square root of this number, $\Delta T$. It is called the *uncertainty* in the measurement of the variable.

From Eq. (3-16), another expression may be developed for $\Delta T$.

$$\begin{aligned}
\int \Psi^* (\hat{T} - \langle T \rangle)^2 \, \Psi \, d\tau &= \int \Psi^* (\hat{T}^2 - 2\,\hat{T} \langle T \rangle + \langle T \rangle^2)\, \Psi \, d\tau \\
&= \int \Psi^* \hat{T}^2 \, \Psi \, d\tau - 2 \langle T \rangle \int \Psi^* \hat{T} \, \Psi \, d\tau + \\
&\qquad \langle T \rangle^2 \int \Psi^* \, \Psi \, d\tau \\
&= \langle T^2 \rangle - 2 \langle T \rangle^2 + \langle T \rangle^2 = \langle T^2 \rangle - \langle T \rangle^2
\end{aligned}$$

Therefore,

$$\Delta T = \sqrt{\langle T^2 \rangle - \langle T \rangle^2} \tag{3-17}$$

The uncertainty in T is the square root of the difference between the expectation value of the operator "T squared" and the square of the expectation value of T.

*Heisenberg's uncertainty relation* states that the product of the measurement uncertainties of two conjugate variables (see Chap. 2) is a number on the order of Planck's constant or larger. (Of course, $\hbar$ is a very tiny value in our macroscopic everyday world.) Position and momentum are conjugate variables, and so, if one may be measured with relatively little uncertainty, then the other cannot.

If a wavefunction is an eigenfunction of an operator of interest, the uncertainty is identically zero, which is consistent with Postulate II. For instance, if the operator $\hat{T}$ applied to some function $\Gamma$ yields the eigenvalue t times $\Gamma$, then the following result is obtained.

$$\hat{T}^2 \Gamma = \hat{T}(\hat{T}\Gamma) = \hat{T} t \Gamma = t^2 \Gamma \quad \Rightarrow \quad \langle T^2 \rangle = t^2 = \langle T \rangle^2$$

This means $\Delta T$ is zero. There is no uncertainty in the measurement of T in this case. Every measurement will yield the same value, namely, t. This also means that the uncertainty in the variable conjugate to T must then be infinite according to Heisenberg's uncertainty relation. In later sections, we shall see the circumstances for energy and time to be taken as conjugate variables, and then if a system's energy has no uncertainty, its lifetime (the uncertainty in time) will be infinite.

## Exercises

1. Ignoring relativistic effects calculate the de Broglie wavelength for a free electron with a velocity 10% of the speed of light. How slow would a 10.0 g marble be moving if it had the same de Broglie wavelength?
2. From the definition of operator multiplication and the definition of a commutator, find a single operator that is the same as the following operators.

   *a.* $[\hat{B}, \hat{A}^2]$ *b.* $[\hat{B}^2, \hat{A}]$ *c.* $[\hat{B}^2, \hat{A}^2]$ *d.* $[\hat{A}^2, \hat{B}^2]$

   where $\hat{A} = x$ and $\hat{B} = d/dx$.
3. Evaluate $[e^{-3x}, d^2/dx^2]$ and $[e^{-x}, d^2/dx^2]$.
4. Consider a 10.0 g mass attached to a spring with a force constant such that the oscillation frequency is 1 cycle/sec. This is a system that we could encounter in our macroscopic world, but treating it quantum mechanically, what is the energy separation between different

allowed energy states and what is the zero-point energy? Is the quantization of energy consistent with our perceptions of continuous allowed energies for classical (macroscopic) oscillators? What would be the n quantum number of this oscillator if it had an energy of about 0.001 ergs?

5. Starting from Eq. (3-9), write out the first five harmonic oscillator wavefunctions explicitly in terms of x.

6. Apply the harmonic oscillator Hamiltonian to $\psi_2$ and to $\psi_3$ and verify that they are eigenfunctions.

7. Establish that $\psi_0$ of the harmonic oscillator is an eigenfunction of the following operator, and deduce why it must be.

$$\hat{A} = \frac{\hbar^2}{4m^2}\frac{d^4}{dx^4} - 2\hbar^2\omega^2\left(1 + 2x\frac{d}{dx} + \frac{d^2}{dx^2}\right) + \frac{k^2x^4}{4}$$

8. Find the expectation value of the square of the momentum for the $n = 0$ and $n = 1$ states of the harmonic oscillator.

9. At what value or values of x is the probability density function of the $n = 1$ wavefunction of the harmonic oscillator at maximum value?

10. Find $\Delta p$ for the $n = 0$ and $n = 1$ states of the harmonic oscillator.

11. Find the expectation value of $x^2$ for the $n = 0$, $n = 1$, $n = 2$ and $n = 3$ harmonic oscillator wavefunctions. How does this quantity depend on the quantum number n?

12. Given that the expectation value $< x >$ is zero for all harmonic oscillator states, use the result of Prob. 11 to develop an expression that gives $\Delta x$ in terms of the n quantum number (i.e., that gives $\Delta x$ as a function of n).

13. For a hypothetical harmonic oscillator with a mass equal to the mass of a hydrogen atom and a force constant such that the frequency of vibration is $1.0 \times 10^{14}$ cycles/sec, find the classical turning points for the $n = 0$, $n = 1$, $n = 2$ and $n = 3$ levels. Then calculate the probability density for each wavefunction at one of its turning points. On the basis of these values, what would you expect for levels with very large n quantum numbers?

14. Use the recursion relation for Hermite polynomials to find $h_7(z)$.

15. Evaluate $\psi_n(0)$ (i.e., find the numerical value of the $n^{th}$ wavefunction at $x = 0$) for the n=1, 3, 5, and 7 (normalized) states of the harmonic oscillator. Is there an apparent trend in these values?

## Bibliography

1. W. Kauzman, *Quantum Chemistry* (Academic Press, New York, 1957). The correspondence between classical vibration and the wave nature of quantum mechanical systems was very important in the early understanding of quantum phenomena. The first chapters of this text highlight that.

*INTRODUCTORY TEXTS:*

2. J. C. Davis, *Advanced Physical Chemistry* (Ronald Press, New York, 1965).
3. M. Karplus and R. N. Porter, *Atoms and Molecules* (Benjamin, Menlo Park, California, 1970).
4. F. L. Pilar, *Elementary Quantum Chemistry* (McGraw-Hill, New York, 1990).
5. D. A. McQuarrie, *Quantum Chemistry* (University Science Books, Mill Valley, California, 1983).

## Chapter 4

# Quantum Mechanics - II

*With the basic postulates of quantum mechanics in hand, we now consider some of the formal elements of quantum mechanical problems and the mathematics for solving them. Variation theory and perturbation theory, introduced in this chapter, are two very important tools for solving or for approximately solving the Schrödinger equations that arise for many different types of problems.*

## 4.1 HERMITIAN OPERATORS

The manipulation of wavefunctions and operators can be presented with more concise notation than has been used so far. One way, due to Dirac, is called "bra-ket" notation. First, the integration symbols, $\int \ldots d\tau$, are replaced with angle brackets, < ... >. So, anything within these brackets is to be integrated over all spatial coordinates of the system. Between the brackets there is a vertical bar that separates functions. It is implicit that the complex conjugate is to be used for anything between the left bracket and this vertical bar. An operator acting on a function on the left side of the vertical bar does not act on anything to the right of the vertical bar. An operator to the right of the vertical bar acts only on functions to its right, as would be natural from operator algebra. Thus, an expectation value for the wavefunction $\Psi$ and the operator $\hat{O}$ is written† $<\Psi \mid \hat{O}\Psi>$, but this means exactly the same as $\int \Psi^* \hat{O} \Psi \, d\tau$.

An operator, $\hat{O}$, is said to have the important property of being *Hermitian* for some class of functions if for any pair of functions, $\psi_i$ and $\psi_j$, from that class the following equivalence exists.

$$\int \psi_i^* \hat{O} \psi_j \, d\tau = \int \psi_j [\hat{O}^* \psi_i^*] \, d\tau \qquad (4\text{-}1a)$$

The difference between the left and right integrals is simply that instead of the operator being applied to $\psi_j$, the complex conjugate of the operator is applied to the other function, $\psi_i^*$. Eq. (4-1a) is also written as

$$<\psi_i \mid \hat{O} \psi_j> = <\hat{O} \psi_i \mid \psi_j> \qquad (4\text{-}1b)$$

in Dirac notation.

There are three useful theorems about Hermitian operators that should be established at this point.

***Theorem 4.1*** The expectation value of a Hermitian operator is a real number.

Given a wavefunction $\psi$ and an operator $\hat{G}$ that is Hermitian, then

$<G> = <\psi \mid \hat{G} \psi>$ — by the definition of an expectation value

$= <\hat{G} \psi \mid \psi>$ — since $\hat{G}$ is Hermitian

$= <\hat{G}^* \psi^* \mid \psi^*>^*$ — because (i) any number is equal to the complex conjugate of its complex conjugate, $x = x^{**}$, and (ii) the complex conjugate of an integral of a function is the

† Sometimes a second vertical bar is placed to the right of the operator, thereby sandwiching it between two bars, e.g., $<\psi \mid \hat{O} \mid \psi>$, with no change in meaning.

| | |
|---|---|
| | integral of the complex conjugate of the function |
| $= \langle \psi \mid \hat{G}^{**} \psi^{**} \rangle^*$ | from rewriting the integral, i.e., $\langle f \mid g \rangle = \langle g^* \mid f^* \rangle$ or $\int f^*(x)\, g(x)\, dx = \int g(x)\, f^*(x)\, dx$ |
| $= \langle \psi \mid \hat{G} \psi \rangle^*$ | since $x = x^{**}$ |
| $= \langle G \rangle^*$ | by the definition of an expectation value |

The result is that the expectation value of the operator is equal to its own complex conjugate. This can only be true for real numbers, since they have no imaginary part. So, the expectation value of a Hermitian operator must be real. Since observables or physical quantities that may be measured should have nonimaginary expectation values, then it must be that the operators associated with such quantities are Hermitian. The Hamiltonian, for instance, is a Hermitian operator.

*Theorem 4.2* Eigenvalues of a Hermitian operator are real numbers.

Given a normalized wavefunction $\psi$ that is an eigenfunction of a Hermitian operator $\hat{G}$ with eigenvalue g, then

| | |
|---|---|
| $\hat{G} \psi = g \psi$ | given |
| $\psi^* \hat{G} \psi = \psi^* g \psi = g\, \psi^* \psi$ | multiplying the equation by $\psi^*$ |
| $\langle \psi \mid \hat{G} \psi \rangle = g \langle \psi \mid \psi \rangle$ | from integrating both sides of the equation |
| $= g$ | since $\psi$ is normalized |

Therefore, the eigenvalue is the same as the expectation value, which must be real by the previous theorem.

*Theorem 4.3* Eigenfunctions of a Hermitian operator with different associated eigenvalues are orthogonal functions.

Given a set of normalized wavefunctions $\{\psi_i\}$ that are each eigenfunctions of a Hermitian operator $\hat{G}$ with associated eigenvalues $g_i$, none of which are equal, then

$$<\psi_i|\hat{G}\psi_j> = <\psi_i|g_j\psi_j> = g_j<\psi_i|\psi_j> \quad \text{from given information}$$

$$<\psi_i|\hat{G}\psi_j> = <\hat{G}\psi_i|\psi_j> \quad \text{since } \hat{G} \text{ is Hermitian}$$

$$= <g_i\psi_i|\psi_j> = g_i<\psi_i|\psi_j> \quad \text{from given information}$$

Since $g_i \neq g_j$, then $<\psi_i|\psi_j> = 0$.

The result that $<\psi_i|\psi_j> = 0$ is a statement of *orthogonality;* two functions are said to be orthogonal if the integral of the product of one and the complex conjugate of the other is zero. An analgous use of the term is that two geometrical vectors are orthogonal if their dot product is zero. The angle between two such vectors would be 90°, and in a like, though abstract, sense orthogonal functions may be thought of as being in completely different "directions" in a space of "functions." This theorem shows that every eigenfunction of a Hermitian operator is orthogonal to every other eigenfunction with a different eigenvalue. Furthermore, it is possible to prove the following.

***Corollary 4.3.1*** Eigenfunctions of a Hermitian operator with the same eigenvalues may be transformed into orthogonal functions while remaining eigenfunctions of the operator.

Since it has been stated that the Hamiltonian operator must be Hermitian, then this corollary theorem provides for all the solutions of the Schrödinger equation to be orthogonal functions.

## 4.2 SIMULTANEOUS EIGENFUNCTIONS

It is possible that there may be a set of functions that are simultaneously eigenfunctions of two different operators; however, if this happens, it means the operators commute.

*Theorem 4.4* If a set of functions $\{f_i\}$ are eigenfunctions of two different operators, $\hat{A}$ and $\hat{B}$, then the operators commute.

$\hat{A} f_i = a_i f_i$ and $\hat{B} f_i = b_i f_i$ given

$\hat{A}\hat{B} f_i = \hat{A} (b_i f_i) = b_i \hat{A} f_i = b_i a_i f_i$ applying the operators

$\hat{B}\hat{A} f_i = \hat{B} (a_i f_i) = a_i \hat{B} f_i = a_i b_i f_i$ applying the operators

$\hat{A}\hat{B} f_i - \hat{B}\hat{A} f_i = a_i b_i f_i - b_i a_i f_i$ by subtraction

$= 0$ since the eigenvalues are numbers

Therefore, $\hat{A}\hat{B} - \hat{B}\hat{A} = 0$.

An important corollary, which will not be proved here, is

*Corollary 4.4.1* If two operators commute, it is possible to find a set of functions that are simultaneously eigenfunctions of both.

When a wavefunction is an eigenfunction of an operator corresponding to some observable, the quantum mechanical uncertainty for a measurement of that observable is zero. Each and every measurement should yield the eigenvalue and nothing else. A consequence of the theorem about simultaneous eigenfunctions is that it may be possible to measure an observable of some system with no uncertainty provided that the corresponding operator commutes with the Hamiltonian.

## 4.3 MULTIDIMENSIONAL PROBLEMS AND DEGENERACY

An example of a multidimensional quantum mechanical problem is the two dimensional oscillator shown in Fig. 4.1. The Hamiltonian for this problem is

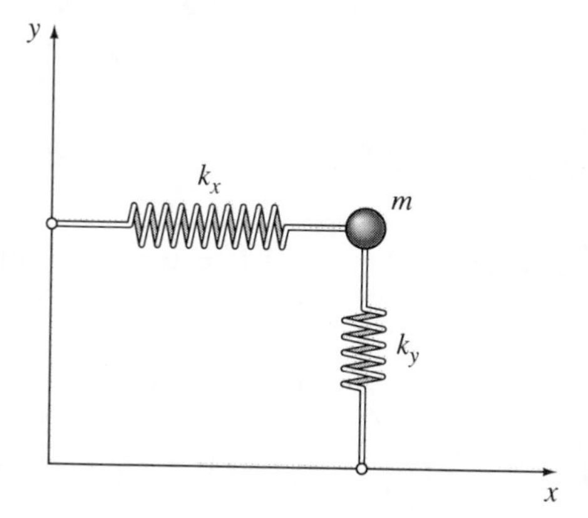


**FIGURE 4.1** A two-dimensional oscillator where a particle of mass m is attached to two harmonic springs with force constants $k_x$ and $k_y$. Each spring is contacted to an unmovable rod, but the connection points may slide freely along the rods (i.e., along the x-axis or along the y-axis). Thus, the potential energy for the $k_x$ spring depends only on the x-coordinate of the particle, while the potential energy of the $k_y$ spring depends only on the y-coordinate of the particle. The equilibrium lengths of the springs are taken to be zero for this example.

$$\hat{H} = \frac{\hat{p}_x^2}{2m} + \frac{k_x x^2}{2} + \frac{\hat{p}_y^2}{2m} + \frac{k_y y^2}{2}$$

This is a separable Hamiltonian, and it has been written so as to make that apparent. The first two terms involve only the x-coordinate and the last two terms involve only the y-coordinate. As was done for separable, classical Hamiltonians, the two sets of terms may be freely designated as independent Hamiltonians:

$$\hat{H}_x \equiv \frac{\hat{p}_x^2}{2m} + \frac{k_x x^2}{2}$$

$$\hat{H}_y \equiv \frac{\hat{p}_y^2}{2m} + \frac{k_y y^2}{2}$$

$$\hat{H} = \hat{H}_x + \hat{H}_y$$

This merely highlights the fact that this particular Hamiltonian is separable.

The wavefunction for this system must be a function of the x- and y-position coordinates, $\psi = \psi(x,y)$. The Schrödinger equation,

$$(\hat{H}_x + \hat{H}_y)\,\psi(x,y) = E\,\psi(x,y)$$

is a separable differential equation. Separability is demonstrated by assuming that the wavefunction, $\psi$, takes on a product form; that is, the wavefunction is assumed to be a product of independent functions of x- and y-coordinates. These independent functions will be designated $X(x)$ and $Y(y)$, and so, the assumption is made that

$$\psi(x,y) = X(x)\,Y(y)$$

Now, if the Hamiltonian is applied to this assumed wavefunction, the component Hamiltonians act only on the respective coordinate functions.

$$\hat{H}_x X(x)\,Y(y) = Y(y)\,\hat{H}_x X(x)$$

$$\hat{H}_y X(x)\,Y(y) = X(x)\,\hat{H}_y Y(y)$$

Thus,

$$Y(y)\,\hat{H}_x X(x) + X(x)\,\hat{H}_y Y(y) = E\,X(x)\,Y(y)$$

and upon multiplying through by $1/(XY)$,

$$\frac{\hat{H}_x X(x)}{X(x)} + \frac{\hat{H}_y Y(y)}{Y(y)} = E \tag{4-2}$$

This has the same form as the equation $f(x) + g(y) = c$. It is a sum of a function of the x-coordinate (only) and a function of the y-coordinate (only), and that sum is equal to a constant. Since x and y are independent variables, the only way for this to be true is if the function of x is constant and the function of y is also constant. Therefore, upon naming the two constants $E_x$ and $E_y$, we find that Eq. (4-2) implies the following.

$$\frac{\hat{H}_x X(x)}{X(x)} = E_x \quad \Rightarrow \quad \hat{H}_x X(x) = E_x X(x) \tag{4-3a}$$

$$\frac{\hat{H}_y Y(y)}{Y(y)} = E_y \quad \Rightarrow \quad \hat{H}_y Y(y) = E_y Y(y) \tag{4-3b}$$

$$\text{where} \quad E_x + E_y = E$$

There are now two Schrödinger equations. Recall that in a classical analysis of a mechanically separable problem, separate equations of motion are obtained too.

The procedure for working with the Schrödinger equation is general for all multidimensional problems that are separable. Separability implies two things:

i. The wavefunction is a product of independent coordinate functions of the separable variables.

ii. The total energy eigenvalue is a sum of energy terms associated with each separable coordinate.

If separability in a problem is not immediately apparent, a product form of the wavefunction may be tested just the same. If application of the Hamiltonian to the product form [e.g., $X(x)Y(y)$] yields an equation of the same type as Eq. (4-2), then the problem is separable. If the problem is not separable, the Schrödinger equation ends up as a coupled differential equation, and solution is usually more difficult.

Separability is as advantageous in quantum mechanics as in classical mechanics. For instance, the Schrödinger equations in Eq. (4-3) may be recognized as simple, one-dimensional, harmonic oscillator Schrödinger equations. Thus, the harmonic oscillator solutions given in Chap. 3 may be used directly. First, though, the independence of the x- and y-motions, which made for the separability, must be taken into account. In particular, the intrinsic vibrational frequencies of these one-dimensional oscillators are not necessarily the same since they depend on the respective spring constants. That is, there are two frequencies for this system, and they will be named $\omega_x$ and $\omega_y$.

$$\omega_x = \sqrt{\frac{k_x}{m}} \quad \text{and} \quad \omega_y = \sqrt{\frac{k_y}{m}} \tag{4-4}$$

Furthermore, there must be two independent quantum numbers since the extent of vibrational motion in the x-direction is independent of that in the y-direction. We may name these quantum numbers $n_x$ and $n_y$.

From the one-dimensional Schrödinger equation solutions, the energy eigenvalues are

$$E_x = (n_x + \frac{1}{2})\hbar\omega_x \quad \text{and} \quad E_y = (n_y + \frac{1}{2})\hbar\omega_y$$

On this basis, the energy of the system as a whole depends on both of the quantum numbers, since it is the sum of the x- and y-motion energies. It is appropriate to subscript the total energy, E, with the quantum numbers, and then the following may be said about the energy levels of the system.

$$E_{n_x n_y} = (n_x + \frac{1}{2})\hbar\omega_x + (n_y + \frac{1}{2})\hbar\omega_y \tag{4-5}$$

The quantum numbers, being those of a harmonic oscillator problem, may be any positive integer or zero.

The wavefunction, $\psi$, for the system as a whole is a product, and it must be labelled by the two quantum numbers in order to distinguish the different Schrödinger equation solutions.

$$\psi_{n_x n_y}(x,y) = X_{n_x}(x)\, Y_{n_y}(y) \tag{4-6}$$

The functions $X_{n_x}$ and $Y_{n_y}$ are simply one-dimensional harmonic oscillator wavefunctions. These product functions describe the *states* of the system as a whole.

An interesting situation would arise if this two-dimensional oscillator were *isotropic*, that is, if it were the same in all directions, or specifically if $k_x = k_y$. In that case, $\omega_x = \omega_y$, and the energy expression, Eq. (4-5), would become

$$E_{n_x n_y} = (n_x + n_y + 1)\hbar\omega \quad \text{with } \omega = \omega_x = \omega_y$$

From this expression, the energies of various states may be tabulated by going through the sequence of all the allowed values of the two quantum numbers:

| $n_x$ | $n_y$ | $E_{n_x n_y} / \hbar\omega$ |
|---|---|---|
| 0 | 0 | 1 |
| 1 | 0 | 2 |
| 0 | 1 | 2 |
| 1 | 1 | 3 |
| 2 | 0 | 3 |
| 0 | 2 | 3 |
| 2 | 1 | 4 |
| 1 | 2 | 4 |
| 3 | 0 | 4 |
| 0 | 3 | 4 |

Notice that there are states, such as those with quantum numbers (1,0) and (0,1), that have exactly the same energy. This is termed *degeneracy*; states with the same energy are said to be degenerate with one another. The lowest energy state is the *ground state*, and in this case it is non-degenerate. It happens to be the only nondegenerate state for this problem. Also notice that at each higher energy level in this problem, the degeneracy, meaning the number of degenerate states, increases. Quantum mechanical degeneracy often has interesting manifestations in molecular spectroscopy, and examples will be given later.

## 4.4 VARIATION THEORY

For most chemical problems, the Schrödinger equation is not strictly separable and the differential equation is not easily solved by analytical means. Or more to the point, the problems are not simple or ideal. The techniques that are best used to find wavefunctions for complicated problems often turn out to be indirect, or at least they appear so. The techniques may also involve approximation as well. Variation theory and perturbation theory are the most powerful techniques and have been very important for understanding many quantum chemical systems.

The *variational principle* is the basis for the variational determination of a wavefunction. This principle tells us that for a given operator the eigenvalue of an eigenfunction is at a minimum value with respect to small adjustments that might be made to the function. As a consequence, the expectation value for some arbitrary function and the operator cannot be less than that minimum. In other words, for an arbitrary function, the expectation value is greater than or equal to the lowest eigenvalue of that operator. Now, the Schrödinger equation calls for finding eigenfunctions of a very specific operator, the Hamiltonian. So, the variational principle helps by saying that the expectation value of the Hamiltonian for any wavefunction we may guess will be greater than or equal to the true ground state energy, the lowest eigenvalue of the Hamiltonian. The expectation value of a guess wavefunction amounts to a guess of a system's energy eigenvalue, and the variation principle indicates that such a guess will err on the high side. This has an important consequence: A guess wavefunction can be improved – made more like the true Hamiltonian's eigenfunction – through any adjustment that lowers the expectation value of the energy. This is because the energy cannot be lowered "too much"; the variational principle dictates that the expectation value of the energy will never be less than the lowest eigenvalue.

For the variational principle to hold, the guess wavefunction must satisfy certain conditions that the true wavefunction satisfies. First, the function and its first derivative must be continuous in all spatial variables of the system and over the entire range of those variables. This requirement says the function does not ever stop and start up somewhere else and that it changes smoothly. Second, the function must be single-valued, which means that at any point in space, the function has only one unique value. Third, if the square of the function is integrated over all space, the result must be a finite value. These conditions follow from the postulates and specifically from the interpretation of the square of a wavefunction as a probability density. The first condition ensures that the integral of the probability density over any particular region of space is well-defined. The second condition ensures that the probability density at every point is unique. The third condition ensures that the function may be normalized. True wavefunctions, guess wavefunctions, and approximate wavefunctions must satisfy these conditions.

The process of adjusting guess wavefunctions, or trial wavefunctions, to minimize the expectation value of the energy is the *variational method*. A simple example, at this point, is the harmonic oscillator problem. The Hamiltonian is that of Eq. (3-7). The guess wavefunction, $\Gamma$, will be taken to be a Gaussian function with a parameter, $\alpha$, that will be adjustable.

$$\Gamma = \left(\frac{\alpha^2}{\pi}\right)^{1/4} e^{-(\alpha x)^2/2}$$

The constant in front of the exponential makes $\Gamma$ normalized for any choice of $\alpha$. Other functional forms might be used for the guess wavefunction; the Gaussian form is the example at hand. The Gaussian function satisfies the conditions mentioned above: It is continuous, has a continuous first derivative, is single-valued, and yields a finite value if it is squared and integrated from $-\infty$ to $\infty$.

The expectation value, designated W, is obtained by integration with the trial function.

$$W = \langle\Gamma \mid \hat{H}\,\Gamma\rangle$$

$$= \langle\Gamma \mid -\frac{\hbar^2}{2m}\frac{d^2}{dx^2} + \frac{1}{2}kx^2 \mid \Gamma\rangle$$

$$= \frac{\hbar^2\alpha^2}{2m} + \left(\frac{k}{4} - \frac{\hbar^2\alpha^4}{4m}\right)\frac{1}{\alpha^2} = \frac{\hbar^2\alpha^2}{4m} + \frac{k}{4\alpha^2}$$

The problem is to adjust $\alpha$ so as to minimize W. W is implicitly a function of the parameter $\alpha$, and so minimization may be accomplished by taking the first derivative.

$$\frac{dW}{d\alpha} = \frac{\hbar^2\alpha}{2m} - \frac{k}{2\alpha^3}$$

At any point where the first derivative of a function of one variable is zero, the value of the function itself is either a minimum or a maximum. [When the first derivative of f(x) is zero, f(x) is just starting to turn upward or else to turn downward, meaning it has reached a minimum or a maximum. From the second derivative, it can be determined if the point is a minimum or maximum.] For $W(\alpha)$, let that point be designated $\alpha_{min}$.

$$\left.\frac{dW}{d\alpha}\right|_{\alpha=\alpha_{min}} = 0 \qquad \Rightarrow \qquad \frac{\hbar^2 \alpha_{min}}{2m} - \frac{k}{2\alpha_{min}^3} = 0$$

The expression for the first-derivative function is set equal to zero when the variable is equal to that of the minimum (or maximum) point. That produces an equation for $\alpha_{min}$, and solving it yields

$$\alpha_{min}^2 = \sqrt{km}\,/\,\hbar$$

This represents the best possible choice for adjusting $\alpha$ to improve the trial wavefunction, since the expectation value of the energy cannot be made any less with any other choice of $\alpha$. Comparison of this result with Eq. (3-10) shows that $\alpha_{min} = \beta$, where $\beta$ is the constant used in the harmonic oscillator eigenfunctions. (We might have expected that would work out.)

The expectation value of the energy is obtained by using the expression for $\alpha_{min}$ in $W(\alpha)$.

$$W(\alpha_{min}) = \frac{\hbar^2 \alpha_{min}^2}{4m} + \frac{k}{4\alpha_{min}^2}$$

$$= \frac{\hbar}{4}\frac{\sqrt{km}}{m} + \frac{\hbar}{4}\frac{k}{\sqrt{km}} = \frac{\hbar}{2}\sqrt{\frac{k}{m}}$$

This, as we know, is the true ground state energy of the harmonic oscillator as worked out in Chap. 3. Thus, by guessing the form of the harmonic oscillator wavefunction to be a Gaussian function, the variational method has determined *which* Gaussian has the lowest expectation value for the energy. And since the true ground state wavefunction is a Gaussian, this best Gaussian *is* the exact wavefunction. The power of this approach is evident upon realizing that the exact wavefunction and eigenenergy were obtained by relatively simple mathematics; there was no formal differential equation solving.

The variational method will not necessarily lead to the exact wavefunction, even though it happened to do so in the prior example. The difficulty is that it is not generally possible to guess the exact form of a wavefunction. The variational method yields the best of whatever functional form is chosen, and that form may be only an approximation of the true wavefunction. This leads to the idea of increasing the degree of mathematical flexibility in the trial wavefunction, and that can be accomplished by using more parameters. For instance, in the harmonic oscillator problem, a more flexible trial function is

$$\Phi = N e^{-\alpha x^{\gamma}}$$

With $\alpha$ and $\gamma$ being adjustable, $\Phi$ could be a Gaussian (when $\gamma = 2$) just as well as something else ($\gamma \neq 2$). The flexibility has been accomplished by using two parameters. As a result, the expectation value of the energy of $\Phi$ is a function of more than one variable (parameter), and the minimization must be carried out with respect to both. Judicious choice of functional forms along with embedding many adjustable parameters is the key to applying variation theory to difficult problems. Using advanced computing systems, calculations have been reported in the research literature wherein wavefunctions for molecules have been found variationally with more than 10 million adjustable parameters!

To understand the variational principle itself, consider a problem with a known set of normalized Hamiltonian eigenfunctions, $\{\psi_i\}$, with corresponding eigenenergies, $\{E_i\}$. It can be proved that any function in the geometrical space of the problem can be represented as a superposition or linear combination of a set of Hamiltonian eigenfunctions. This is much like the idea of a Fourier expansion where a function can be represented by a sum of sine and/or cosine functions. In this sense, an arbitrary function *is* an arbitrary linear combination of the eigenfunctions. That is, anything that may serve as a guess function is the same thing as some linear combination of the $\psi_i$'s. If the coefficients, $\{a_i\}$, in this linear combination are considered arbitrary, then the guess function, $\Gamma$, is an arbitrary function.

$$\Gamma = N \sum a_i \psi_i$$

So, $\Gamma$ stands for any guess wavefunction whatsoever. N is the usual normalization constant. The expectation value of $\Gamma$ with the Hamiltonian is easily simplified because the $\psi$ functions are eigenfunctions of the Hamiltonian, and therefore are orthogonal.

$$W = <\Gamma \mid \hat{H} \, \Gamma>$$

$$= N^2 < \sum_i a_i \psi_i \mid \hat{H} \sum_j a_j \psi_j >$$

$$= N^2 \sum_i \sum_j a_i^* a_j < \psi_i \mid \hat{H} \psi_j >$$

$$= N^2 \sum_i \sum_j a_i^* a_j E_j < \psi_i \mid \psi_j >$$

$$= N^2 \sum_i \sum_j a_i^* a_j E_j \, \delta_{ij} = \sum_j (N^2 a_j^2) E_j \equiv \sum_j a_j'^2 E_j$$

The normalization factor, which is

$$N^2 = \left( \sum_i a_i^2 \right)^{-1}$$

has been absorbed into the primed coefficients in the last line. The primed coefficients squared must each be less than or equal to one because of the following.

$$<\Gamma \mid \Gamma> = 1 = N^2 \sum_i \sum_j a_i^* a_j < \psi_i \mid \psi_j >$$

$$= N^2 \sum_i a_i^2 = \sum_i a_i'^2$$

The sum of their squares is exactly 1.0, since $\Gamma$ is normalized, and so it would be impossible for any one coefficient to be greater than one.

The expectation value W must be greater than or equal to the lowest eigenenergy from the set $\{E_i\}$ because W is a sum of the eigenenergies, each multiplied by a number (e.g., $a_i'^2$) that is less than one. To illustrate this reasoning, consider the eigenenergies of the harmonic oscillator, $(n + 1/2) \hbar\omega$, and a guess wavefunction that is a linear combination of the harmonic oscillator functions, with the normalized expansion coefficient for the $n^{th}$ function being $a_n'$. The expectation value of the energy is

$$W = \frac{\hbar\omega}{2}\left( a_0'^2 + 3\,a_1'^2 + 5\,a_2'^2 + 7\,a_3'^2 + \ldots \right)$$

and the condition on the coefficients is

$$a_0'^2 + a_1'^2 + a_2'^2 + \ldots = 1$$

By inspection, the smallest value for W is $\hbar\omega/2$, which is what we know to be the ground state energy. This will be obtained for the value of W only when $a_0'^2 = 1$ and the other coefficients are zero. Consider that relative to this choice, any change consistent with the constraints will make W a greater number. For instance, with $a_1'^2 = 0.9$, $a_2'^2 = 0.1$, and the other coefficients zero, $W = 3.2\ \hbar\omega$. This is the variational principle in action: The expectation value of the energy for an arbitrary function cannot be less than the lowest eigenvalue.

## 4.5 PERTURBATION THEORY

Perturbation theory offers another method for finding quantum mechanical wavefunctions. It is especially suited to problems that are similar to model or ideal situations differing in only some small way, which is the perturbation. For example, the potential for an oscillator might be harmonic except for a feature such as the small "bump" depicted in Fig. 4.2. Because the bump is a small feature, one expects the system's behavior to be quite similar to that of a harmonic oscillator. Perturbation theory would be a way to correct a description of the system, obtained from treating it as a harmonic oscillator, so as to account for the effects of the bump in the potential. Perturbation theory can yield exact wavefunctions and eigenenergies, but it can also be employed in an approximate way with much savings in effort.

In perturbation theory, the Hamiltonian for any problem is partitioned into two or more pieces. The first piece is one for which the

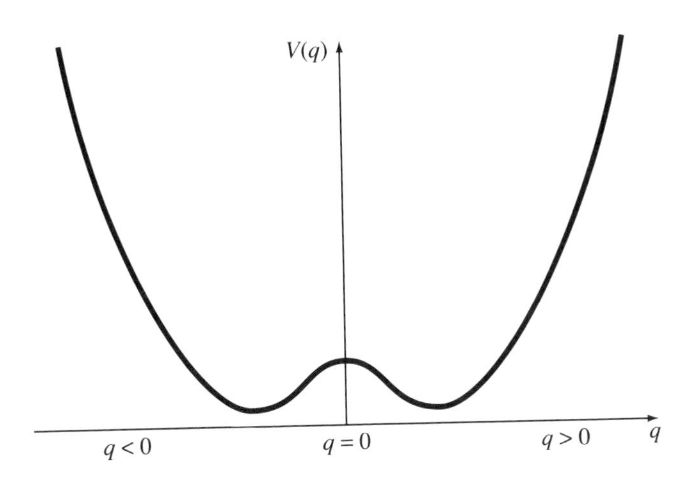


**FIGURE 4.2** A harmonic oscillator potential, a parabola, with a small perturbing potential, a bump in the middle. The composite potential is called a double-minimum potential or a *double-well potential*. An example of this type of potential is found for the water molecule. Water is a bent molecule, which means that the minimum of its potential energy as a function of the atomic positions is at a structure that is bent. We may envision a water molecule being straightened into a linear arrangement of the atoms, but this is necessarily a higher energy arrangement. If we continue, the molecule will be bent again but in the opposite sense:

$$\mathrm{H}\backslash_{\mathrm{O}}/^{\mathrm{H}} \rightarrow \mathrm{HOH} \rightarrow {}_{\mathrm{H}}/^{\mathrm{O}}\backslash_{\mathrm{H}}$$

Eventually it will look like a mirror image of the original equilibrium structure, and so it will then be at a potential minimum. If the bending angle is taken to be a coordinate, then the potential energy as a function of the angle will have the form of a double well potential.

---

eigenfunctions and eigenenergies are known, while everything else represents the perturbation. This first piece and the associated eigenfunctions and eigenenergies are distinguished in notation by a zero superscript. Thus, at the outset, one has a problem where the following Schrödinger equation has been solved.

$$\hat{H}^{(0)} \psi_i^{(0)} = E_i^{(0)} \psi_i^{(0)} \tag{4-7}$$

This is the zero-order perturbation equation. The complete Hamiltonian for the problem may be written with the introduction of a parameter, $\lambda$, which multiplies or scales the remaining piece of the Hamiltonian.

$$\hat{H} = \hat{H}^{(0)} + \lambda \hat{H}^{(1)} \tag{4-8}$$

If $\lambda$ is set to one, this expression gives the true Hamiltonian for the problem. If $\lambda$ is set to zero, it gives the Hamiltonian for the zero-order or unperturbed situation; there is, then, no contribution from $\hat{H}^{(1)}$. Thus, "tuning" $\lambda$ from zero to one lets us go smoothly from the situation we understand [Eq. (4-7)] to the problem we seek to solve. With this embedded parameter, the Schrödinger equation with $\hat{H}$ is really a family of equations covering all the choices of $\lambda$ values. Perturbation theory is a means of dealing with the entire family of equations, even though in the end, one is interested only in the case where $\lambda=1$.

Eq. (4-8) is an expansion of the Hamiltonian operator in a power series in $\lambda$. In fact, a general development of perturbation theory would allow for the possibility that some way of choosing the parameter might give rise to pieces of the Hamiltonian that depend on $\lambda$ quadratically, and to higher powers.

$$\hat{H} = \hat{H}^{(0)} + \sum_{n=1} \lambda^n \hat{H}^{(n)} \tag{4-9}$$

$\hat{H}^{(n)}$ is the $n^{th}$ order Hamiltonian. Writing the Hamiltonian as a power series implies writing the entire Schrödinger equation as an expansion in $\lambda$.

$$\left( \sum_{i=0} \lambda^i \hat{H}^{(i)} \right) \left( \sum_{j=0} \lambda^j \psi_n^{(j)} \right) = \left( \sum_{k=0} \lambda^k E_n^{(k)} \right) \left( \sum_{m=0} \lambda^m \psi_n^{(m)} \right) \tag{4-10}$$

The subscripts on the wavefunctions and on the energies identify a particular state as in Eq. (4-7).

When a particular order of Hamiltonian in Eq. (4-10) is applied to a particular order of wavefunction, the result, $\hat{H}^{(i)}\psi_n^{(j)}$, occurs in the Schrödinger equation with a factor of $\lambda^{(i+j)}$. On the right hand side of Eq. (4-10), energies multiply wavefunctions, and these have factors of $\lambda^{(k+m)}$. Clearly, terms will be found on the left hand side and on the right hand side that enter the equation with a factor of $\lambda^0, \lambda^1, \lambda^2$, and so on. The idea of perturbation theory is to approach a solution that is valid for the entire family of equations that arise from allowing $\lambda$ to be anything from zero to one. This implies a mathematical constraint that Eq. (4-10) remain valid for *all* choices of $\lambda$. At first that might seem impossible since some terms have a linear dependence on the parameter, some a quadratic dependence, and so on. The difference between the case of $\lambda = 1$ and $\lambda = 1/2$ is that a term linear in $\lambda$ would be diminished by one-half for the latter case, but a quadratic term would be diminished by one-fourth. The various terms would enter with a different weighting for every different choice of $\lambda$. However, that does not preclude a general solution. Instead, it dictates that the terms for any particular power of $\lambda$ on the left hand side of Eq. (4-10) must equal those on the right hand side. Below is a list of the first few equations which then result. They are found by multiplying out Eq. (4-10) and collecting terms according to the power of the $\lambda$ factor.

$$\lambda^0 \text{ terms:} \quad \hat{H}^{(0)}\psi_n^{(0)} = E_n^{(0)}\psi_n^{(0)}$$

$$\lambda^1 \text{ terms:} \quad \hat{H}^{(1)}\psi_n^{(0)} + \hat{H}^{(0)}\psi_n^{(1)} = E_n^{(1)}\psi_n^{(0)} + E_n^{(0)}\psi_n^{(1)}$$

$$\lambda^2 \text{ terms:} \quad \hat{H}^{(2)}\psi_n^{(0)} + \hat{H}^{(1)}\psi_n^{(1)} + \hat{H}^{(0)}\psi_n^{(2)}$$

$$= E_n^{(2)}\psi_n^{(0)} + E_n^{(1)}\psi_n^{(1)} + E_n^{(0)}\psi_n^{(2)}$$

Instead of one equation, the original Schrödinger equation, there are now an infinite number of equations because the $\lambda$ power series expansion is infinite. This is reasonable, though, since the ultimate solution is one that applies to the whole family of Schrödinger equations (i.e., the infinite number of choices of $\lambda$).

The $\lambda^0$ equation is the same as Eq. (4-7), and its solutions are known because of how the Hamiltonian was partitioned in the first place. Inspection of the $\lambda^1$ equation shows that it involves zero-order elements in addition to the first-order elements. The unknowns in this equation are the first-order correction to the wavefunction, $\psi_n^{(1)}$, and the first-order correction to the energy, $E_n^{(1)}$. In the $\lambda^2$ equation, zero-order and first-order elements are used in addition to the second-order corrections. Because of this pattern, the process for solving these equations must proceed from the zeroth equation to the first-order equation to the second-order equation, and so on. In this way, at any order, all the required lower order corrections will already be known.

Though more and more terms are involved as the order of $\lambda$, or the perturbation order, goes up, there are common steps in working with the equations. At each and every order, the energy correction is obtained by multiplying that $\lambda$-order equation by $\psi_n^{(0)*}$ and integrating over all space. There is a simplification that removes the two terms with $\psi_n^{(\lambda)}$ upon carrying out this integration. Notice that on the left hand side, this term will be $\hat{H}^{(0)}\psi_n^{(\lambda)}$, while on the right hand side it will be $E_n^{(0)}\psi_n^{(\lambda)}$. These can be grouped together on the left hand side as,

$$(\hat{H}^{(0)} - E_n^{(0)})\,\psi_n^{(\lambda)}$$

When this is integrated with the complex conjugate of the zero-order wavefunction, the result is zero.

$$<\psi_n^{(0)}\,|(\hat{H}^{(0)} - E_n^{(0)})\,\psi_n^{(\lambda)}> \;=\; <(\hat{H}^{(0)} - E_n^{(0)})\psi_n^{(0)}\,|\,\psi_n^{(\lambda)}>$$

$$= \;<(E_n^{(0)}\psi_n^{(0)} - E_n^{(0)}\psi_n^{(0)}\,|\,\psi_n^{(\lambda)}> \;=\; 0$$

The Hamiltonian operator is Hermitian, and the operation of multiplication by a constant is Hermitian; and that made it possible to exchange where the operator was being applied. With this simplification, the first-order equation yields an expression for the first-order correction to the energy.

$$<\psi_n^{(0)}\,|\,\hat{H}^{(1)}\,\psi_n^{(0)}> \;=\; <\psi_n^{(0)}\,|\,E_n^{(1)}\,\psi_n^{(0)}>$$

$$= \;E_n^{(1)}<\psi_n^{(0)}\,|\,\psi_n^{(0)}> \;=\; E_n^{(1)} \qquad (4\text{-}11)$$

This reveals that the first-order correction to the energy is merely the expectation value of the first-order perturbing Hamiltonian with the zero-order wavefunction. The second-order equation yields an expression for the second-order correction to the energy.

$$< \psi_n^{(0)} \mid \hat{H}^{(2)} \psi_n^{(0)} > + < \psi_n^{(0)} \mid \hat{H}^{(1)} \psi_n^{(1)} > =$$

$$< \psi_n^{(0)} \mid E_n^{(2)} \psi_n^{(0)} > + < \psi_n^{(0)} \mid E_n^{(1)} \psi_n^{(1)} >$$

$$E_n^{(2)} = < \psi_n^{(0)} \mid \hat{H}^{(2)} \psi_n^{(0)} > + < \psi_n^{(0)} \mid (\hat{H}^{(1)} - E_n^{(1)}) \psi_n^{(1)} > \quad (4\text{-}12)$$

Using this equation, though, requires knowing the first-order correction to the wavefunction.

To find the corrections to the wavefunction, an idea from the previous section is used that an arbitrary function in some quantum mechanical space can be expressed as a linear combination of eigenfunctions of any Hamiltonian for that space. In perturbation theory, the zero-order equation is presumed to have been solved, and so the prerequisite, a set of eigenfunctions must exist. Thus, the arbitrary functions, first- or second-order corrections to the wavefunction, and so on, can each be expressed as a linear combination of the zero-order functions. The coefficients in that linear expansion constitute the information that has to be obtained. Let us apply this to the first-order equation and let $c_i$ be the expansion coefficient for the $i^{th}$ zero-order function.

$$\psi_n^{(1)} \equiv \sum_i c_i \psi_i^{(0)} \quad (4\text{-}13)$$

If we group Hamiltonians and energies as was done above, then Eq. (4-13) may be used with the first order Schrödinger equation to give

$$(\hat{H}^{(1)} - E_n^{(1)}) \psi_n^{(0)} = -(\hat{H}^{(0)} - E_n^{(0)}) \psi_n^{(1)}$$

$$= -(\hat{H}^{(0)} - E_n^{(0)}) \sum_i c_i \psi_i^{(0)}$$

$$= -\sum_i c_i (E_i^{(0)} - E_n^{(0)}) \psi_i^{(0)} \quad (4\text{-}14)$$

This result can be used to extract one of the desired coefficients by multiplying by the complex conjugate of one of the zero-order wavefunctions and integrating over all space. Let that one function be $\psi_j^{(0)}$, with $j \neq n$ for now.

$$< \psi_j^{(0)} \mid (\hat{H}^{(1)} - E_n^{(1)}) \psi_n^{(0)} > \; = \; -\sum_i c_i (E_i^{(0)} - E_n^{(0)}) < \psi_j^{(0)} \mid \psi_i^{(0)} >$$

$$< \psi_j^{(0)} \mid \hat{H}^{(1)} \psi_n^{(0)} > - E_n^{(1)} < \psi_j^{(0)} \mid \psi_n^{(0)} > \; = \; -\sum_i c_i (E_i^{(0)} - E_n^{(0)}) \delta_{ij}$$

$$< \psi_j^{(0)} \mid \hat{H}^{(1)} \psi_n^{(0)} > \; = \; - c_j (E_j^{(0)} - E_n^{(0)})$$

$$\therefore \quad c_j \; = \; < \psi_j^{(0)} \mid \hat{H}^{(1)} \psi_n^{(0)} > / \; (E_n^{(0)} - E_j^{(0)}) \qquad (4\text{-}15)$$

By letting j take on all values except n, we may use this equation to find all the coefficients except $c_n$. [We can not use Eq. (4-15) with $j = n$ because the denominator is zero then.]

The result presented in Eq. (4-15) can be discussed in several ways. We see that a particular $c_j$ will be zero if the integral quantity involving the zero-order wavefunctions for the $n^{th}$ and $j^{th}$ states with the perturbing Hamiltonian is zero. Thus, if there is no *coupling* or *interaction* between two states brought about by a perturbation, then there will be no *mixing* of their wavefunctions to first order. Also, if the difference in zero-order energies is large, the extent of mixing will be small because this difference is in the numerator in Eq. (4-15).

The procedure that has been developed for first-order corrections to the wavefunction can be used at second and higher orders as well. The resulting expressions will be more complicated. The unanswered question at this point is the value of the coefficient $c_n$. This comes from another procedure, and it too can be used for all orders. This procedure develops from the normalization condition, and specifically from requiring that the perturbed wavefunction be normalized for any choice of the perturbation parameter $\lambda$.

$$1 \; = \; < \psi_n \mid \psi_n > \; = \; < \psi_n^{(0)} + \lambda \psi_n^{(1)} + \ldots \mid \psi_n^{(0)} + \lambda \psi_n^{(1)} + \ldots >$$

$$= <\psi_n^{(0)} \mid \psi_n^{(0)}> + \lambda\left(<\psi_n^{(1)} \mid \psi_n^{(0)}> + <\psi_n^{(0)} \mid \psi_n^{(1)}>\right)$$

$$+ \ldots$$

From the power series expansion, the normalization condition is now written as a sum of terms that depend on different orders of $\lambda$. The first right-hand-side term is, of course, equal to one, and so all the remaining terms must add up to zero. Such a result will be true for any choice of $\lambda$ only if each term is itself zero. The term that has $\lambda$ to the first power consists of an integral with its complex conjugate. If that integral is zero, then so is its complex conjugate. Therefore,

$$0 = <\psi_n^{(0)} \mid \psi_n^{(1)}> = <\psi_n^{(0)} \mid \sum_i c_i \psi_i^{(0)}>$$

$$= \sum_i c_i <\psi_n^{(0)} \mid \psi_i^{(0)}> = \sum_i c_i \delta_{ni} = c_n$$

This establishes that $c_n$ is zero. For higher order corrections to the wavefunction, it will not necessarily be that the $n^{th}$ coefficient is zero; this is a special result at first order.

The perturbation derivations can be carried out to any desired order, but the idea of perturbation theory is that the perturbation, whatever it is, makes a small correction to the zero-order picture. In the typical application, the lowest few orders provide nearly all of the correction needed to make the energies and wavefunctions exact. Of course, *nearly* all is not all; to stop at some low order of perturbation theory (i.e., to truncate the expansion in $\lambda$) is to make an approximation. The quality of the approximation that we make, or the order at which we truncate, is our choice when we use this approach.

An illustration of the efficacy of low-order perturbation theory, in fact, of the first order energy corrections, is given in Fig. 4.3. The model problem is that of a slightly altered harmonic vibrational potential, the type seen in Fig. 4.2. The result of an extensive variational treatment that closely approaches the exact energies and wavefunctions for the lowest several states is shown along with the energy levels obtained from first-order perturbation theory, taking the bump in the potential to be the perturbation of the otherwise harmonic system. The energies of the variational and perturbational treatments may be compared with the

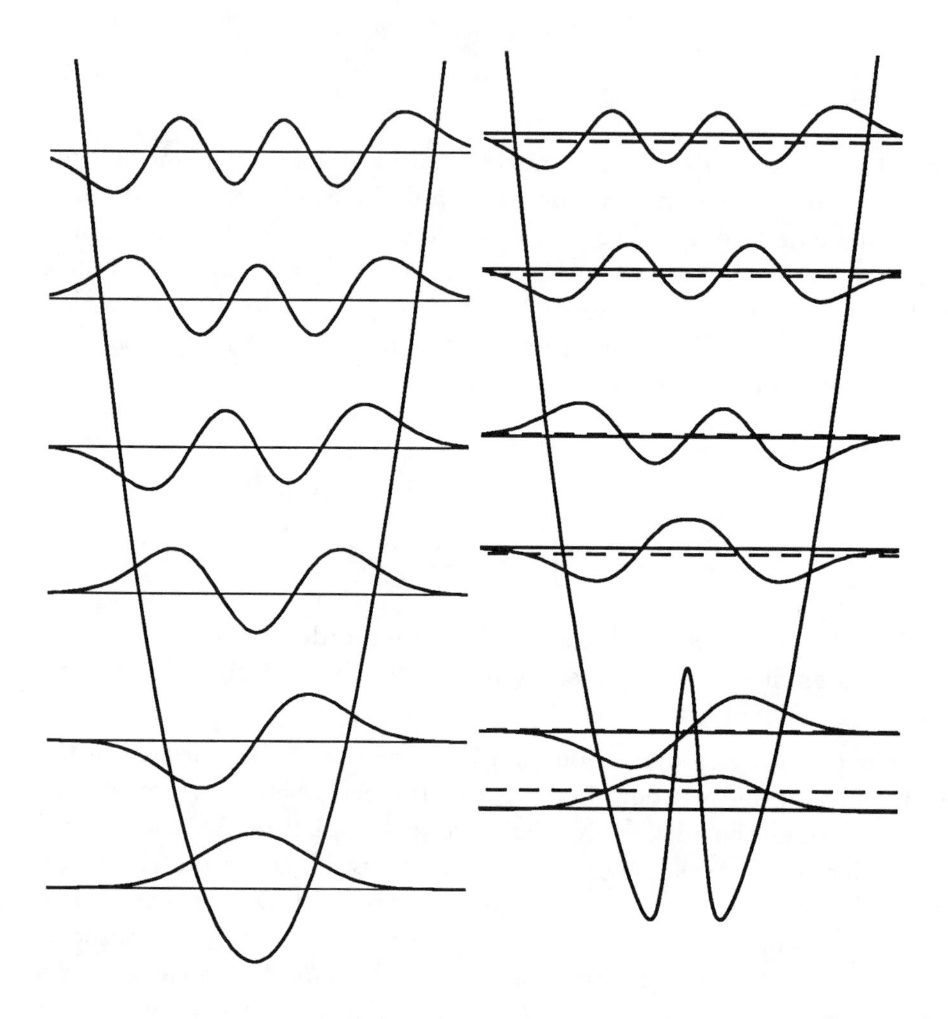

**FIGURE 4.3** On the left is a harmonic vibrational potential with the exact energy levels and corresponding wavefunctions drawn for the first several states. On the right is the same potential augmented, or perturbed, by a bump in the middle. The solid horizontal lines drawn with this potential are the energy levels obtained from an extensive variational treatment of this problem, whereas the dashed lines are the energy levels obtained from first-order perturbation theory, using the system on the left as the zero-order picture and the bump as the perturbation. The wavefunctions drawn for the perturbed system are those obtained from the variational treatment, and they demonstrate the detailed changes in the wavefunctions from the perturbing potential.

energies of the unperturbed (harmonic) oscillator. From that, we see that the first-order corrections are similar to the energy changes obtained from a variational treatment. Also, the correspondence between the variational energy and the first-order perturbation theory energy improves as one goes to higher energy levels. This is because for the higher energy states, the relative effect of the bump in the potential is diminished, and so, perturbation theory at low order is even more appropriate.

## 4.6 TIME DEPENDENCE AND TRANSITIONS

To this point, attention has been limited to wavefunctions that do not evolve in time. However, Hamiltonians may include time-dependent elements that cause the system to change, and an important case of this sort is the effect of electromagnetic radiation. The character of light is that of electric and magnetic fields that oscillate in space and time. When light impinges on a molecule, the oscillating fields may give rise to an interaction, i.e., an element of the complete Hamiltonian, and so the time dependence of the fields then enters the Hamiltonian. We shall see that the interaction with light can induce transitions among the states of a system, and the mechanisms for this are the basis for molecular spectroscopy experiments.

When time, t, becomes a variable in a quantum mechanical problem, an associated operator is needed. Like a position coordinate, time as an operator is just multiplication by t. Also like a position coordinate, time has a conjugate variable with its own associated operator. Now, two operators are conjugate if their commutator is $i\hbar$. This was the case for the position operator x and the momentum operator $p_x$. To find the operator that is conjugate with time, an operator equation is employed. Using f to designate an arbitrary function, and G to be the operator we wish to find, then,

$$i\hbar = [\hat{G},t] \quad \Rightarrow \quad \hat{G}\,t\,f - t\,\hat{G}\,f = i\hbar\,f$$

$$\Rightarrow \quad \hat{G} = i\hbar \frac{\partial}{\partial t}$$

Dimensional analysis of this operator shows that it will produce an energy quantity (namely, erg-sec/sec = erg) upon applying it to a wavefunction. This operator, which will be designated as $\hat{E}$ from now on, plays an important role in the time dependent wavefunctions. In the time dependent generalization of Postulate III (the Schrödinger equation), this operator replaces E, the energy eigenvalue. The *time dependent Schrödinger equation* (TDSE) is

$$\hat{H}\Psi = \hat{E}\Psi \tag{4-16}$$

$\Psi$ is an explicit function of time in this equation, and the Hamiltonian may have an explicit time dependence also. The generalization of Postulate III is that the solutions of the TDSE are the wavefunctions that describe a system in both space and time.

The time-dependent Schrödinger equation is a generalization of the time-independent Schrödinger equation (TISE) but does not invalidate the TISE. This is because the TDSE is always separable in space and time in those cases where the Hamiltonian has no explicit time dependence (which are the only kinds of cases considered until now). We may demonstrate this separability with the following steps, where only one spatial coordinate, x, is used for simplicity. The Hamiltonian operator is taken to have no explicit time dependence.

$$\hat{H}(x)\,\Psi(x,t) = i\hbar\frac{\partial}{\partial t}\Psi(x,t)$$ the TDSE

$$\Psi(x,t) = \psi(x)\,\phi(t)$$ trial product form of $\Psi$

$$\frac{\hat{H}(x)\,\psi(x)}{\psi(x)} = \frac{i\hbar}{\phi(t)}\frac{\partial\phi(t)}{\partial t}$$ substitution into TDSE and division by $\psi(x)\,\phi(t)$

$$\equiv E$$ the left and right hand sides are functions of independent variables and so must each equal a constant, E

$$\therefore \quad \hat{H}(x)\,\psi(x) = E\,\psi(x)$$ separated equation in x

$$\therefore \quad i\hbar \frac{\partial \phi(t)}{\partial t} = E\,\phi(t) \qquad \text{separated equation in t}$$

Thus, with no explicit time dependence in the Hamiltonian, the TDSE is separable into a spatial differential equation and a differential equation in time. The differential equation in the spatial coordinate(s) is seen to be the familiar time-independent Schrödinger equation.

The general solution of the differential equation in time that was obtained for the time-independent Hamiltonian case is

$$\phi(t) = \exp(-i\,Et/\hbar) \tag{4-17}$$

This function is sometimes referred to as a *phase* or said to give the phase of the wavefunction. If the TISE is solved for a particular problem, the result is a spatial wavefunction, but multiplying that function by the $\phi(t)$ in Eq. (4-17) gives a product function that is necessarily a solution of the TDSE. Such a product wavefunction is called a *stationary state* wavefunction because the probability density function is independent of time: $\phi^*\phi = 1$. That is, if $\Psi(x,t) = \psi(x)\,\phi(t)$, then $\Psi(x,t)^*\Psi(x,t) = \psi^*(x)\,\psi(x)$.

It is an important result that even for time-independent Hamiltonian problems, stationary state wavefunctions are not the only possible solutions of the TDSE. In fact, any arbitrary superposition (or linear combination) of different stationary state wavefunctions will be a solution of the TDSE. To illustrate this point, let $\hat{H}_o$ be some particular Hamiltonian with no explicit time dependence and let $\{\psi_i\}$ be the set of its eigenfunctions with associated eigenvalues $\{E_i\}$.

$$\hat{H}_o\,\psi_i = E_i\,\psi_i \tag{4-18}$$

The stationary state wavefunctions for this problem are

$$\Psi_i = \psi_i\,e^{-i E_i t/\hbar} \tag{4-19}$$

An arbitrary superposition of these states is a linear combination of these functions with unspecified or arbitrary coefficients.

$$\Gamma = \sum_i c_i\,\Psi_i$$

We can show that $\Gamma$ is, in fact, a solution of the TDSE by operating on it with $\hat{H}_o$ and with the energy operator and comparing the results:

$$\hat{H}_o \Gamma = \sum_i c_i \hat{H}_o \psi_i e^{-iE_i t/\hbar} = \sum_i c_i E_i \psi_i e^{-iE_i t/\hbar}$$

$$i\hbar \frac{\partial}{\partial t} \Gamma = \sum_i c_i \psi_i (i\hbar) \frac{\partial}{\partial t} e^{-iE_i t/\hbar} = \sum_i c_i E_i \psi_i e^{-iE_i t/\hbar}$$

The results are identical. So, even though $\Gamma$ is not an eigenfunction of the Hamiltonian, it is a solution of the TDSE.

When the Hamiltonian has explicit time dependence, separability of the TDSE into time and spatial differential equations is usually not possible. Analytical solution of the TDSE may be a horrible task. We will take a narrower view of time-dependent Hamiltonian problems, and we will treat the time dependence in the Hamiltonian as a perturbation of a system with a time-independent Hamiltonian. In other words, we will consider any complete Hamiltonian as consisting of two pieces, a time-dependent piece, H', and a time-independent piece, $H_o$; H' is a perturbation of the $H_o$ system.

As in the prior two sections, a suitable linear combination of the eigenfunctions or of Schrödinger equation solutions can form any valid wavefunction for the system. Use of this fact converts the task of solving the differential Schrödinger equation into a task of finding the linear expansion coefficients. The same thing is to be done now, except that the expansion coefficients must be allowed to be functions of time so that the TDSE wavefunction can evolve in time. The linear combination will be made from the stationary states of the TDSE involving just $H_o$ [Eq. (4-18)]. Thus, for the Schrödinger equation,

$$(\hat{H}_o + \hat{H}') \Phi = i\hbar \frac{\partial \Phi}{\partial t} \tag{4-20}$$

we will use the following general expansion for the wavefunction $\Phi$ where the stationary state wavefunctions, $\Psi_i$, are as given in Eq. (4-19).

$$\Phi = \sum_i a_i(t) \Psi_i \tag{4-21}$$

Fully solving Eq. (4-20), then, means finding all the $a_i(t)$ functions. The first step is to substitute Eq. (4-21) for $\Phi$ into Eq. (4-20). At the same time, it is helpful to write out the stationary state wavefunctions as the products of spatial and time functions via Eq. (4-19).

$$\sum_i a_i(t)\, e^{-iE_it/\hbar}\, \hat{H}_o \psi_i \;+\; \sum_i \hat{H}'\, a_i(t)\, e^{-iE_it/\hbar}\, \psi_i \;=\; i\hbar \sum_i \psi_i \frac{\partial}{\partial t} a_i(t)\, e^{-iE_it/\hbar}$$

Notice that on the left hand side of this expression, $\hat{H}_o\psi_i$ can be replaced with $E_i\psi_i$. On the right hand side, the partial differentiation will yield exactly the same thing from the differentiation of the exponential, plus an additional term. Thus, this expression simplifies to

$$\sum_i \hat{H}'\, a_i(t)\, \psi_i\, e^{-iE_it/\hbar} \;=\; i\hbar \sum_i \dot{a}_i(t)\, \psi_i\, e^{-iE_it/\hbar} \qquad (4\text{-}22)$$

where $\dot{a}_i(t)$ means the time derivative of that function.

Since the stationary state wavefunctions used in the expansion for $\Phi$ are orthonormal, a useful thing happens if Eq. (4-22) is multiplied by any stationary state wavefunction, such as $\Psi_k$, and then integrated over all spatial coordinates. Orthonormality simplifies the right hand summation.

$$\sum_i \langle \Psi_k \mid \hat{H}'\, a_i(t)\, \Psi_i \rangle \;=\; i\hbar \sum_i \dot{a}_i(t) \langle \Psi_k \mid \Psi_i \rangle$$

$$= \; i\hbar\, \dot{a}_k(t) \qquad (4\text{-}23)$$

Since k could be any index, Eq. (4-23) represents a whole set of coupled differential equations. If the integral quantities happen to be known, it might be possible to solve the set of equations and obtain the desired $a_i(t)$ functions. Short of doing that, we restrict attention to the short time behavior of $\Phi$ starting from a point, defined as $t = 0$, where the wavefunction $\Phi$ is one and only one stationary state, i.e., where for some state designated "initial,"

$$a_{initial}(0) = 1 \qquad \text{and} \qquad a_{i \neq initial}(0) = 0$$

Eq. (4-23) is valid for any time including the specific time $t = 0$. So, substituting in the $t = 0$ values for the a functions yields

$$\langle \Psi_k \mid \hat{H}'\, \Psi_{initial} \rangle \;=\; i\hbar\, \dot{a}_k(0) \qquad (4\text{-}24)$$

The significance of this result is that for some choices of k, the left-hand side integral might be zero. For those choices, the first time derivative of the $a_k(t)$ function at t = 0 would have to be zero. Since $a_k(t)$ gives the weighting of the k stationary state in the $\Phi$ function, the zero value for the first derivative means that for short times at least, the weighting of the k stationary state will be unchanging. And by the conditions we chose for t = 0, this means that when k is any index other than "initial" the weighting will be zero. The system will not be evolving into the k stationary state (at least in the short time limit). So, whether the $\Phi$ function can evolve from some initial stationary state into another because of the effect of a time-dependent Hamiltonian H' all hinges on whether the integral in Eq. (4-24) is zero or not zero (at least in the short time limit).

When H' corresponds to the interaction with electromagnetic radiation in some particular experimental arrangement, then the qualitative distinction between a zero and a nonzero integral in Eq. (4-24) becomes the basis for spectroscopic *selection rules*. When the integral is nonzero, the *transition* from the initial state to the k state or "final" state is said to be *allowed*. When the integral is identically zero, the transition is said to be *forbidden* or not allowed. As we shall see in later chapters, allowed transitions lead to characteristic features of molecular spectra. Forbidden transitions are often not observed.

## 4.7 DERIVATIVES OF THE SCHRÖDINGER EQUATION *

There are a number of properties of molecules that are defined as some particular derivative of a molecular state energy. In general, derivatives of a state's energy tell how the energy will vary with respect to the variable or variables of differentiation. Consider a Taylor expansion of a function of x, f(x), about the point $x_o$.

$$f(x) = f(x_o) + (x - x_o) f'(x_o) + \frac{1}{2}(x - x_o)^2 f''(x_o) + \frac{1}{3!}(x - x_o)^3 f'''(x_o) + \ldots$$

The first derivative function, f'(x), evaluated for the specific choice of $x = x_o$, is simply a number. It gives the linear dependence of f(x) in the vicinity of $x = x_o$, or in other words, it tells how f(x) will vary linearly with a variation in x from the point $x = x_o$. In an entirely similar way, an eigenenergy may be formally expressed as a Taylor expansion in terms of some parameter or variable that is in the Hamiltonian. With E as an eigenenergy and $\alpha$ as a parameter, we may write

$$E(\alpha) = \sum_{n=0}^{\infty} \frac{\alpha^n}{n!} \left. \frac{\partial^n E(\alpha)}{\partial \alpha^n} \right|_{\alpha=0} \tag{4-25}$$

where the expansion has been taken to be about $\alpha = 0$. It is often of interest to know the values of certain of the $n^{th}$ derivatives of a molecular state energy. For instance, if a molecule were placed between two electrically charged parallel plates, it would experience a uniform electric field. The molecular energy would change linearly with the strength of that field proportional to the *dipole moment* of the molecule, and it would change quadratically in proportion to a quantity called the *polarizability*. The dipole moment is a first derivative value like $f'(x_o)$, and the polarizability is a second derivative.

The time-independent Schrödinger equation says that the function produced by operating with the Hamiltonian on $\psi$ is equal to $\psi$ times the energy eigenvalue; the left hand and right hand sides are functions. Therefore, the Schrödinger equation may be freely differentiated with respect to any variable or any parameter treated as a differentiable variable. The result of differentiating the Schrödinger equation will be a new equation which may be solved to yield energy derivatives.

To illustrate the manipulations of formal differentiation of the TDSE, let us consider the oscillator problem given by the following Hamiltonian.

$$H = \frac{p^2}{2m} + \frac{1}{2} k x^2 + \alpha x^3$$

We now declare $\alpha$ to be a parameter, which means the TDSE may be differentiated:

$$\frac{\partial}{\partial \alpha}(H\psi) = \frac{\partial}{\partial \alpha}(E\psi) \tag{4-26}$$

From this point, a more concise notation for a partial derivative is a superscript on any differentiated quantity; e.g., $f^{\alpha} \equiv \partial f / \partial \alpha$. Then, applying the product rule of differentiation, Eq. (4-26) leads to

$$H^{\alpha}\psi + H\psi^{\alpha} = E^{\alpha}\psi + E\psi^{\alpha} \tag{4-27}$$

The derivative of the energy is the main thing sought, but the derivative of the wavefunction is also unknown. The derivative of the example Hamiltonian, though, is apparent:

$$H^{\alpha} = x^3$$

The differentiation of Eq. (4-26) can be done repeatedly to generate higher order *derivative Schrödinger equations*. Eq. (4-27) is a first-order derivative Schrödinger equation.

In a Taylor series, the expansion of a function is about a point or value for the variable or variables. In Eq. (4-25), it is the set of energy derivatives evaluated at $\alpha = 0$ that are used. Thus, after differentiating the Schrödinger equation, the parameter (in this case $\alpha$) is set to zero, and then a solution is sought for the equation that results. If we start with the zero-order (not differentiated) Schrödinger equation of the oscillator example, setting $\alpha$ to zero yields a familiar problem, the harmonic oscillator. With a subscript o to mean that the parameter $\alpha$ has been set to zero,

$$H_o \equiv H(\alpha{=}0) = \frac{p^2}{2m} + \frac{1}{2}kx^2$$

and, as usual, the corresponding wavefunctions are those of the harmonic oscillator. For this development, a subscript o will be carried to indicate that these are the solutions when $\alpha = 0$, as in

$$H_o\psi_{n\text{-}o} = E_{n\text{-}o}\psi_{n\text{-}o} \tag{4-28}$$

Differentiation and then evaluation at zero parameter value (setting $\alpha$ to zero) leads to a sequence of derivative Schrödinger equations. The first is Eq. (4-28), the zero-order equation, and the next follows from Eq. (4-27).

$$H_o^{\alpha}\psi_{n\text{-}o} + H_o\psi_{n\text{-}o}^{\alpha} = E_{n\text{-}o}^{\alpha}\psi_{n\text{-}o} + E_{n\text{-}o}\psi_{n\text{-}o}^{\alpha} \tag{4-29}$$

The second-order equation is

$$H_o^{\alpha\alpha}\psi_{n\text{-}o} + 2H_o^{\alpha}\psi_{n\text{-}o}^{\alpha} + H_o\psi_{n\text{-}o}^{\alpha\alpha} = E_{n\text{-}o}^{\alpha\alpha}\psi_{n\text{-}o} + 2E_{n\text{-}o}^{\alpha}\psi_{n\text{-}o}^{\alpha} + E_{n\text{-}o}\psi_{n\text{-}o}^{\alpha\alpha}$$

This formal process may be continued as far as desired. Also, it may be carried out with respect to several parameters.

Examination of Eq. (4-29) shows the only unknown elements to be $E^{\alpha}_{n-o}$ and $\psi^{\alpha}_{n-o}$, assuming the zero-order equation has been solved for the $n^{th}$ state. And if these two elements are obtained by solving the first-order equation, then the unknowns in the second-order equation are seen to be $E^{\alpha\alpha}_{n-o}$ and $\psi^{\alpha\alpha}_{n-o}$. In fact, it continues to be that each higher derivative equation depends on the lower order results, and all orders of derivative Schrödinger equations take on the form

$$(H_o - E_{n-o})\psi^{(\alpha...)}_{n-o} + (H^{(\alpha...)}_o - E^{(\alpha...)}_{n-o})\psi_{n-o} = \{\text{ lower order elements }\}$$

where $(\alpha...)$ designates a particular derivative of arbitrary order. This implies a uniformity from one order of differentiation to the next, and it is possible to exploit this uniformity in solving these equations.

Solving the derivative Schrödinger equations proceeds exactly as solving the perturbation theory equations. Inspection of the two sets of equations shows the only formal difference to be the n! term in the series expansion of the derivative approach. [Compare Eq. (4-25) with the expansion of the energy in Eq. (4-10).] An $n^{th}$-order perturbation theory correction to an energy will differ from the $n^{th}$-order derivative of the energy by this factor; otherwise the two approaches are equivalent. So, differentiation of the Schrödinger equation really amounts to a different way of developing perturbation theory, or vice versa. It is a useful alternative because instead of identifying a perturbation in a problem, we identify a parameter or variable in the Hamiltonian and then carry out straightforward differentiation. When there are a number of parameters, it may be easier to select parameters from the Hamiltonian rather than extract different sources of perturbation. The derivatives of the energy and derivatives of the wavefunction are interpreted and used in the same way as any common functional derivative; the concept of a perturbation being "on" or "off" is unnecessary.

Certain molecular properties are defined as derivatives. As already mentioned, the electric dipole moment of a molecule is the first derivative of the molecular energy with respect to the strength of an applied electric field. The magnetizability of a molecule is a second derivative with respect to the strength of a magnetic field. The mathematics of differentiation offer a direct and analytical means for evaluating such properties.

## 4.8 MATRIX METHODS FOR LINEAR VARIATION THEORY*

A special and powerful use of variation theory is with linear variational parameters. That means that the trial wavefunction is taken to be a linear combination of functions in some chosen set. The adjustable parameters are the expansion coefficients of each of these functions. This is, of course, a specialization of the way in which variation theory may be used, but it is powerful because the resulting equations take the form of matrix expressions. Solving the Schrödinger equation becomes a problem in linear algebra, and such problems are ideally suited to computer solution.

For any problem to be treated by linear variational methods, a *basis set* must be selected at the outset. This is simply some set of functions of the coordinates for the problem under study. In the best circumstances, the functions are chosen to be part of the basis set because they are close to the anticipated form of the true wavefunction. The number of functions is not restricted. A few may be used, or many may be used. The basis set may even have an infinite number, though the discussion here will assume a finite basis set.

The linear variational wavefunction, given as a *basis set expansion,* is a linear combination of the functions in the basis. With $\Gamma$ as the wavefunction and $\{\phi_i\}$ designating the set of basis functions, the expansion is

$$\Gamma = \sum_i^N c_i \phi_i \tag{4-30}$$

N is the number of functions in the basis set. The functions in the set are distinguished by the subscript i. The coefficients, $c_i$, are expansion coefficients; they are the adjustable parameters. Notice that basis set expansions have already been used: The first-order perturbation theory corrections to the wavefunction were obtained as an expansion in a basis of the zero-order functions.

The expectation value of the energy of $\Gamma$ is the quantity to be minimized. This is given by the following expression, subject to a normalization constraint on $\Gamma$.

$$<\Gamma \mid H \Gamma> \;=\; <\sum_{i}^{N} c_i \phi_i \mid H \sum_{j}^{N} c_j \phi_j>$$

$$= \sum_{i}^{N} \sum_{j}^{N} c_i^* <\phi_i \mid H \phi_j> c_j \qquad (4\text{-}31)$$

This double-summation expression has a form that can be expressed with matrices (see Appendix I). It happens to be of the form of a row of values (coefficients) times a matrix of (integral) values times a column of coefficients, with the result being a scalar. The rank-one matrices of coefficients will be called column or row vectors, designated as (see Appendix I)

$$\vec{c} = \begin{pmatrix} c_1 \\ c_2 \\ c_3 \\ \cdots \\ c_N \end{pmatrix} \quad \text{and} \quad \vec{c}^{\,T} = (c_1 \;\; c_2 \;\; c_3 \;\; \cdots \;\; c_N)$$

It is also helpful to have a special symbol for the complex conjugate of the row vector: $\vec{c}^{\,\dagger} = \vec{c}^{\,T*}$. The integral values are arranged into a rank-two N-by-N matrix designated **H**.

$$\mathbf{H} = \begin{pmatrix} <\phi_1 \mid H\phi_1> & <\phi_1 \mid H\phi_2> & <\phi_1 \mid H\phi_3> & \cdots & <\phi_1 \mid H\phi_N> \\ <\phi_2 \mid H\phi_1> & <\phi_2 \mid H\phi_2> & <\phi_2 \mid H\phi_3> & \cdots & <\phi_2 \mid H\phi_N> \\ <\phi_3 \mid H\phi_1> & <\phi_3 \mid H\phi_2> & <\phi_3 \mid H\phi_3> & \cdots & <\phi_3 \mid H\phi_N> \\ & & \cdots & & \end{pmatrix}$$

**H** is called the *matrix representation* of the Hamiltonian operator. A matrix representation of an operator is a matrix of integral values arranged into rows and columns according to the basis functions. Clearly, the values in a matrix representation are dependent on the functions that were selected for the basis set. A different basis set implies a different matrix representation. Wherever it is important to keep track of the basis used in the representation, a superscript is added to the designation of the matrix, e.g., $\mathbf{H}^{\phi}$, and it identifies the particular function set. Now, the quantity in Eq. (4-31) can be written with the coefficient vectors and the Hamiltonian matrix in a very simple form.

$$<\Gamma \mid H\Gamma> \;=\; \vec{c}^{\dagger}\,\mathbf{H}\,\vec{c} \;=\; \sum_{i,j} c_i^{*}\,\mathbf{H}_{ij}\,c_j$$

Again, this is the form of an N-long row of numbers times an N-by-N square array times an N-long column of numbers, and that produces just one value (not a vector or matrix).

The overlap of the Γ function with itself is the dot product of the coefficient vector and its complex conjugate transpose if, as we assume here, the basis functions are orthonormal.

$$<\Gamma \mid \Gamma> \;=\; \vec{c}^{\dagger}\vec{c} \;=\; \sum_{j} c_j^{*}\,c_j$$

If this is divided into $<\Gamma \mid H\Gamma>$, the resulting value is the energy expectation whatever the normalization of Γ. This is a way of imposing the normalization constraint.

$$W \;=\; \frac{<\Gamma \mid H\Gamma>}{<\Gamma \mid \Gamma>}$$

W is a function of the coefficients in $\vec{c}$. To apply variation theory, W must be minimized with respect to each of the elements of $\vec{c}$. This means taking the first derivative and setting that function to zero at the point where the coefficients lead to the minimum value of W.

$$\left.\frac{\partial W}{\partial c_i}\right|_{\vec{c}=\vec{c}_{min}} = 0 \qquad (4\text{-}32)$$

There will be one equation for each coefficient.

The partial differentiation of W with respect to one of the coefficients yields the following.

$$\frac{\partial W}{\partial c_i^{*}} \;=\; \frac{\sum_j \mathbf{H}_{ij}\,c_j}{<\Gamma \mid \Gamma>} \;-\; \frac{c_i}{<\Gamma \mid \Gamma>}\left\{\frac{<\Gamma \mid H\Gamma>}{<\Gamma \mid \Gamma>}\right\}$$

(We may also differentiate† with respect to $c_i$ instead of with respect to $c_i^*$.) Notice that the quantity in brackets is W. So, when this first-derivative function is evaluated at the coefficient values that minimize W, i.e. $\vec{\mathbf{c}}_{min}$, then this bracketed quantity must equal E, the variational energy. With this replacement for the bracketed quantity, we have that

$$\frac{\partial W}{\partial c_i^*}\bigg|_{\vec{\mathbf{c}}=\vec{\mathbf{c}}_{min}} = \frac{1}{<\Gamma \mid \Gamma>}\left(\sum_j \mathbf{H}_{ij}\, c_{j\text{-}min} - c_{i\text{-}min}\, E\right)$$

Setting this to zero following the condition for minimization of W yields

$$\sum_j \mathbf{H}_{ij}\, c_{j\text{-}min} - c_{i\text{-}min}\, E = 0$$

Clearly, there will be one such equation for each choice of the index i. The whole set of these equations can be arranged into a single matrix expression.

$$\mathbf{H}\,\vec{\mathbf{c}}_{min} = \vec{\mathbf{c}}_{min}\, E \tag{4-33}$$

This expression is an important and common expression called a matrix eigenvalue equation. It says that the matrix **H** operating on (or multiplying) the vector of coefficients that minimizes W is equal to that vector times a constant, which is the energy eigenvalue. This result means that in general applying linear variation theory will amount to finding an eigenvector of the matrix representation of the Hamiltonian.

There are numerous means for solving the matrix equation given in Eq. (4-33) and Appendix II discusses certain methods. Regardless of which algorithm is applied, the process of finding the eigenvalues and the coefficient vectors is referred to as a *diagonalization* of **H** because it amounts to a transformation of **H** from the original basis into a basis where it is diagonal. Diagonalization is extremely well-suited to

† The differentiation of W, as shown, is with respect to a complex conjugate of one of the coefficients and is carried out by treating each coefficient as independent of its complex conjugate. By carrying out the differentiation with respect to complex conjugates, equations are developed for the simple coefficients. Should the coefficients be taken to be real, which would mean the coefficient and its complex conjugate are the same, then differentiation yields the same expression, but times 2. The factor of 2, though, makes no difference in the coupled equations because of setting the derivative to zero.

computers, and there are many well-crafted computer programs in use for diagonalizing matrices. Some are general routines, while others are specialized for the forms that certain matrices take on. For a matrix with N rows and N columns, it is usual for a diagonalization routine to require a number of multiplications of pairs of real numbers on the order of $N^3$. This means that the computational cost grows with the matrix dimension as $N^3$.

## 4.9 FIRST-ORDER DEGENERATE PERTURBATION THEORY *

Degeneracy complicates perturbation theory. If there is a state, $\psi_j$, that is degenerate with the state of interest, $\psi_n$, at zero-order, then the first-order correction to the wavefunction cannot be found with Eq. (4-15) since the energy denominator is zero: $E_n^{(0)} = E_j^{(0)}$. Of course there is way around this complication.

If a set of states, which may be a subset of the states of some system, happens to be degenerate with respect to some operator, then any normalized linear combination of those states will also be an eigenfunction of that operator with the same eigenvalue. For instance, consider a set of functions $\{\phi_i, i = 1, ..., N\}$ that are degenerate, normalized eigenfunctions of a zero-order Hamiltonian. That is,

$$\hat{H}^{(0)} \phi_i = E_n^{(0)} \phi_i$$

Any linear combination of these functions will be an eigenfunction of this Hamiltonian, as the following steps show.

$$\hat{H}^{(0)} \sum_{j=1}^{N} c_j \phi_j = \sum_{j=1}^{N} c_j \hat{H}^{(0)} \phi_j$$

$$= \sum_{j=1}^{N} c_j E_n^{(0)} \phi_j = E_n^{(0)} \sum_{j=1}^{N} c_j \phi_j$$

Consequently, the zero-order wavefunction in the perturbation treatment for the $n^{th}$ state cannot be presumed to be any particular one of the $\phi_i$ functions. Instead, it must be regarded as an undetermined linear combination of the functions in the degenerate subset.

$$\psi_n^{(0)} = \sum_{i=1}^{N} c_j \phi_j \tag{4-34}$$

This is substituted into the first-order perturbation theory Schrödinger equation to give

$$\hat{H}^{(1)} \sum_{j=1}^{N} c_j \phi_j + \hat{H}^{(0)} \psi_n^{(1)} = E_n^{(1)} \sum_{j=1}^{N} c_j \phi_j + E_n^{(0)} \psi_n^{(1)} \tag{4-35}$$

Now, solution of this equation must yield not only the first-order correction to the energy and the first-order correction to the wavefunction, but also the value of the expansion coefficients in the zero-order wavefunction, the $c_j$'s.

In nondegenerate perturbation theory, the procedure for obtaining the first-order correction to the energy was to multiply by the zero-order wavefunction for the state of interest and integrate. The result, Eq. (4-11), was that the first-order correction to the energy was the expectation value of the perturbing Hamiltonian with the zero-order wavefunction. In the case of degeneracy, it is necessary to carry out this process with each of the degenerate functions. So, we may take one function from the set and multiply Eq. (4-35) by it and then integrate.

$$< \phi_k \,|\, \hat{H}^{(1)} \sum_{j=1}^{N} c_j \phi_j > + < \phi_k \,|\, \hat{H}^{(0)} \psi_n^{(1)} >$$

$$= E_n^{(1)} \sum_{j=1}^{N} c_j < \phi_k \,|\, \phi_j > + E_n^{(0)} < \phi_k \,|\, \psi_n^{(1)} >$$

Just as in the nondegenerate development, the second left-hand-side term is the same as the second right-hand-side term, and so they cancel. Because of orthogonality of the degenerate functions, the summation on the right hand side that is multiplied by the first-order correction to the energy simplifies to $c_k$.

$$\sum_{j=1}^{N} \langle \phi_k | \hat{H}^{(1)} \phi_j \rangle c_j = c_k E_n^{(1)}$$

This expression may be rewritten by using the matrix repesentation of the perturbing Hamiltonian in the $\phi$ basis.

$$\sum_{j=1}^{N} \mathbf{H}_{kj}^{(1)\text{-}\phi} c_j = c_k E_n^{(1)}$$

And when all choices of k are considered, there are N equations of this form. They may all be collected into one matrix equation by arranging the c coefficients into a column vector (see Appendix I).

$$\mathbf{H}^{(1)\text{-}\phi} \vec{c} = \vec{c} E_n^{(1)} \tag{4-36}$$

This happens to be the standard form of the matrix eigenvalue equation (see Appendix II).

The interpretation of Eq. (4-36) is that the first-order correction to the energy of the state of interest (n) is an eigenvalue of the matrix representation of the perturbing Hamiltonian in the basis of the degenerate functions $\{\phi_i\}$. The eigenvector in Eq. (4-36) is the set of coefficients that tells which linear combination of degenerate functions must represent the zero-order state function, that is, what values end up being used in Eq. (4-34). So, in the case of degenerate functions, a diagonalization procedure is used, just as in linear variation theory, except that the matrix representation is limited to just the functions that are degenerate with the state of interest. The diagonalization implies a transformation (see Appendix II) of the $\{\phi_i\}$ set of functions; they will be mixed with each other. If the eigenenergies obtained via Eq. (4-36) are all different, then the perturbation is said to have removed the degeneracy. After solving Eq. (4-36), the first-order correction to the wavefunction and all the higher order corrections may be obtained as in the nondegenerate case, if the degeneracy has been removed. Or if the degeneracy has not been removed, the expressions for higher order corrections are developed by allowing the corrections to the wavefunctions to be linear combinations of the degenerate functions, analogous to Eq. (4-34).

## Exercises

1. Show that the n = 0, 1, and 2 wavefunctions of the harmonic oscillator are orthogonal to each other.
2. Determine whether or not the wavefunctions of the harmonic oscillator could also be eigenfunctions of the operators $\hat{x}$ and $\hat{p}_x$.
3. For the harmonic oscillator, which of the following are Hermitian operators, $\hat{x}$, $\hat{x}^2$, $\hat{p}$, and $\hat{p}^2$?
4. Find the number of degenerate states for the lowest five energy levels of a two-dimensional oscillator of the type shown in Fig. 4.1 given that $k_x = 4\, k_y$.
5. Find the degeneracy of the lowest four energy levels of an isotropic three-dimensional harmonic oscillator ($k_x = k_y = k_z$).
6. Given that for a certain problem it has been determined that the energy levels depend on two quantum numbers, n and m, according to the expression

$$E_{n,m} = -\frac{\hbar a}{n^2} + \frac{\hbar a m^2}{4}$$

where a is a constant. If n can have only the value 1, 2, or 3, and if m can be zero or any integer from –(n+1) to n+1, find the degeneracies of all the energy levels.

7. Apply the variational method to the harmonic oscillator problem using the following as the trial wavefunction.

$$\Gamma(x) = N e^{-\alpha x^4}$$

$\alpha$ is the adjustable parameter, and N is the normalization constant. N has a dependence on the value of $\alpha$ that must be found in order to work out this problem.

8. Repeat Prob. 7 with the following trial function.

$$\Gamma(x) = N x e^{-\alpha x^2}$$

How much higher is this energy than that of the true ground state? Explain this result.

*9.* Find the first order perturbation theory correction to the energies of the ground and first two excited states of the harmonic oscillator in terms of the constant g in the perturbing Hamiltonian,

$$\hat{H}^{(1)} = gx^4$$

*10.* Find the first- and second-order perturbation theory corrections to the energy of the ground state of the harmonic oscillator in terms of the constant g in the perturbing Hamiltonian

$$\hat{H}^{(1)} = g\,x^3$$

*11.* Derive a general expression for the third-order perturbation theory correction to the energy of a quantum state.

*12.* Write the expression for the probability density function for a harmonic oscillator that has been prepared in the time-varying superposition of the ground and first excited time-independent states. That is,

$$\Psi(x,t) = \frac{1}{\sqrt{2}}\,\psi_{n=0}(x)\,e^{-i\,\omega t/2} + \frac{1}{\sqrt{2}}\,\psi_{n=1}(x)\,e^{-3\,i\,\omega t/2}$$

From this, determine the behavior of the probability density as a function of time at the specific point, $x = 0$.

★*13.* Find an expression for the first derivative of the energy, $dE/d\alpha$, evaluated at $\alpha = 0$ for the oscillator with the Hamiltonian

$$H = -\frac{p^2}{2m} + \frac{1}{2}kx^2 + \alpha x^3 + \frac{1}{2}\alpha^2 x^6$$

Also, write out the second- and third-derivative Schrödinger equations for this problem.

★*14.* Apply the variational method to the harmonic oscillator problem using the trial wavefunction

$$\phi = \left\{c_1\,\beta x + c_2\,(\beta^2 x^2 - 1/2)\right\}\,e^{-\beta^2 x^2/2}$$

where the variational parameters are $c_1$ and $c_2$. From a comparison with the exact wavefunctions for the harmonic oscillator, account for the result.

★*15.* For an oscillator with the following Hamiltonian,

$$H = -\frac{p^2}{2} + \frac{x^2}{2} + \frac{x^3}{4}$$

find the variational energies of three states using a linear expansion in a basis set of three functions. The three functions are the $n = 0$, $n = 1$, and $n = 2$ states of the harmonic oscillator with $k = 1$ and $m = 1$. (This requires setting up the matrix representation of the Hamiltonian in the harmonic oscillator basis and then diagonalizing.)

★*16.* For an oscillator with the following Hamiltonian,

$$H = -\frac{p^2}{2} + \frac{x^2}{2} + \frac{x^4}{4}$$

find the variational energies of three states using a linear expansion in a basis set of three functions, as in Prob. 15. The three functions are the lowest three wavefunctions of the harmonic oscillator. Compare the form of the Hamiltonian matrix in this problem, where the perturbation to the harmonic potential (i.e., $x^4/4$) is symmetric, with the form of the Hamiltonian matrix in Prob. 15, where the perturbation $x^3/4$ is antisymmetric.

## Bibliography

(See list of *INTRODUCTORY TEXTS* in Chap. 3)

1. E. Kreyszig, *Advanced Engineering Mathematics* (John Wiley and Sons, New York, 1967).

2. H. G. Hecht, *Mathematics in Chemistry: An Introduction to Modern Methods* (Prentice-Hall, Englewood Cliffs, New Jersey, 1990).

3. H. Margeneau and G. M. Murphy, *The Mathematics of Physics and Chemistry* (Van Nostrand and Co., Princeton, New Jersey, 1956).

*INTERMEDIATE LEVEL TEXTS:*

4. P. W. Atkins, *Molecular Quantum Mechanics*, 2nd ed. (Oxford University Press, New York, 1983).

5. I. Levine, *Quantum Chemistry* (Allyn and Bacon, Boston, 1970).

6. L. I. Schiff, *Quantum Mechanics*, 3rd ed. (McGraw-Hill, New York, 1968).

7. E. Merzbacher, *Quantum Mechanics*, 2nd ed. (John Wiley and Sons, New York, 1970).

# Chapter 5

# Angular Momentum

*There are many problems in quantum chemistry and physics that involve the angular momentum of a system. Often, the angular momentum features can be analyzed apart from a complete solution of the Schrödinger equation. Consequently, techniques for analyzing angular momenta prove to be quite powerful on their own. This chapter is focused entirely on the quantum mechanical treatment of angular momentum. The development is somewhat formal, but that is due to its generality: We shall see in later chapters that analysis of angular momentum is the key to a diverse set of spectroscopic experiments.*

## 5.1 ANGULAR MOMENTUM OPERATORS

The importance of angular momentum in many quantum mechanical systems is its quantization. It is often the case that the angular momentum of a system will be restricted to certain values and not continuously variable. In the earliest days of quantum theories, the Bohr model of the hydrogen atom had some success in explaining atomic spectra by imposing a quantization condition on the electron's orbital angular momentum. This was a presumption, but the ultimate explanation of the mechanics of the hydrogen atom showed that indeed the electron's angular momentum is quantized. Today, many features of molecular spectra are understood in terms of angular momentum properties, and even reaction probabilities have been found that are dependent on the angular momenta of the colliding atoms or molecules.

The fundamental ideas of angular momentum in quantum systems draw on classical mechanical theory, and there one finds a notable correspondence between rectilinear and rotational motions. Rectilinear position and momentum are as basic as angular position and angular momentum. Rectilinear force has an analog in the torque about an axis, and in the appropriate equations of motion, the mass of a straight-moving body plays the same role as a rotating body's moment of inertia. In all respects, the analysis of the mechanics of rotating systems is not so much special as it is a generalization of the types of coordinates so as to include angular coordinates.

Angular momentum, like rectilinear momentum, is a vector quantity, and for a moving point-mass it may be defined as the vector cross-product of the position vector (from the axis of rotation to the mass) and the linear momentum vector.

$$\vec{L} = \vec{r} \otimes \vec{p} \quad \Rightarrow \quad \begin{aligned} L_x &= y\,p_z - z\,p_y \\ L_y &= z\,p_x - x\,p_z \\ L_z &= x\,p_y - y\,p_x \end{aligned} \tag{5-1}$$

For a system of several point-mass particles, the total angular momentum is the vector sum of the angular momenta of each of the particles.

Quantum mechanical operators corresponding to the components of an angular momentum vector may be found directly from the familiar rectilinear position and momentum operators, e.g.,

$$\hat{L}_x = \hat{y}\,\hat{p}_z - \hat{z}\,\hat{p}_y$$

Therefore,

$$\hat{L}_x = -i\hbar\left(y\frac{\partial}{\partial z} - z\frac{\partial}{\partial y}\right)$$

$$\hat{L}_y = -i\hbar\left(z\frac{\partial}{\partial x} - x\frac{\partial}{\partial z}\right) \tag{5-2}$$

$$\hat{L}_z = -i\hbar\left(x\frac{\partial}{\partial y} - y\frac{\partial}{\partial x}\right)$$

With these explicit forms for the operators, it is straightforward to show that the commutator of any pair of them produces the third one:

$$[\hat{L}_x, \hat{L}_y] = i\hbar\hat{L}_z$$

$$[\hat{L}_y, \hat{L}_z] = i\hbar\hat{L}_x \tag{5-3}$$

$$[\hat{L}_z, \hat{L}_x] = i\hbar\hat{L}_y$$

In practice, it is better to work with angular momentum in an angular coordinate system rather than a rectilinear system. This means using the spherical polar coordinate system defined as in Fig. 2.4 and transforming the operators in Eq. (5-2).

The coordinate transformation to and from spherical polar coordinates is

$$x = r\sin\theta\cos\phi \qquad r^2 = x^2 + y^2 + z^2$$

$$y = r\sin\theta\sin\phi \qquad \theta = \arccos(z/r)$$

$$z = r\cos\theta \qquad \phi = \arctan(y/x)$$

This transformation enables us to write the angular momentum component operators in either system. Chain rule differentiation provides the substitution for a differential operator in Eq. (5-2). For instance,

$$\frac{\partial}{\partial x} = \frac{\partial r}{\partial x}\frac{\partial}{\partial r} + \frac{\partial\theta}{\partial x}\frac{\partial}{\partial\theta} + \frac{\partial\phi}{\partial x}\frac{\partial}{\partial\phi}$$

and from the transformation equations,

$$\frac{\partial r}{\partial x} = \frac{x}{r} = \sin\theta\cos\phi, \qquad \text{and so on.}$$

If this is worked through completely, the following operator expressions are obtained.

$$\hat{L}_x = -i\hbar \left(-\sin\phi \frac{\partial}{\partial\theta} - \frac{\cos\phi}{\tan\theta}\frac{\partial}{\partial\phi}\right)$$

$$\hat{L}_y = -i\hbar \left(\cos\phi \frac{\partial}{\partial\theta} - \frac{\sin\phi}{\tan\theta}\frac{\partial}{\partial\phi}\right) \tag{5-4}$$

$$\hat{L}_z = -i\hbar \frac{\partial}{\partial\phi}$$

These can be applied to wavefunctions given in spherical polar coordinates to extract information about the angular momentum of a system.

Another useful operator may be found from the angular momentum component operators. The square of the angular momentum of a classical or a quantum mechanical system is a scalar quantity, since it is simply the dot product of $\vec{L}$ and itself. The dot product of any vector and itself equals the square of the length of the vector. Thus, $\vec{L} \bullet \vec{L}$ tells us the magnitude of the angular momentum because it corresponds to the square of that vector's length. The quantum mechanical operator that corresponds to this dot product is designated $\hat{L}^2$, and it can be applied to wavefunctions to find out about the magnitude of a system's angular momentum.

The explicit form of the $\hat{L}^2$ operator in any coordinate system may be derived from the component expression for a dot product; that is,

$$\hat{L}^2 = \hat{L}_x^2 + \hat{L}_y^2 + \hat{L}_z^2$$

In spherical polar coordinates, the following is obtained with Eq. (5-4).

$$\hat{L}^2 = -\hbar^2 \left( \frac{1}{\sin\theta}\frac{\partial}{\partial\theta} \sin\theta \frac{\partial}{\partial\theta} + \frac{1}{\sin^2\theta}\frac{\partial^2}{\partial\phi^2} \right) \tag{5-5}$$

This operator is related to the *Laplacian operator*, $\nabla^2$ (del squared), in spherical polar coordinates. From the definition of the Laplacian in Cartesian coordinates, and the coordinate transformation given above, we have that

$$\nabla^2 = \frac{\partial^2}{\partial x^2} + \frac{\partial^2}{\partial y^2} + \frac{\partial^2}{\partial z^2} \tag{5-6a}$$

$$= \frac{2}{r}\frac{\partial}{\partial r} + \frac{\partial^2}{\partial r^2} - \frac{\hat{L}^2/\hbar^2}{r^2} \tag{5-6b}$$

The Laplacian is used in the kinetic energy operator for a particle, because $\hat{p}^2/2m = -\hbar^2\nabla^2/2m$. Thus, the square of the angular momentum is intimately connected with the amount of kinetic energy of a moving particle.

## 5.2 SPHERICAL HARMONIC FUNCTIONS

The name *spherical harmonic functions* refers to the special set of functions of the two spherical polar coordinate angles, $\theta$ and $\phi$, that are eigenfunctions of the two operators $\hat{L}^2$ and $\hat{L}_z$. These functions are usually designated $Y(\theta,\phi)$, and the whole set is obtained by solving the two differential eigenequations,

$$\hat{L}^2 Y = \alpha Y$$

$$\hat{L}_z Y = \beta Y$$

There are an infinite number of solutions to these equations, and two integers are needed to label and distinguish the different spherical harmonic functions. The conventional choices for these two integers are $l$ and m. The $\theta$ dependence of these functions may be expressed with a special set of polynomials called the *associated Legendre polynomials.*

Associated Legendre polynomials for a variable z, designated $P_l^{|m|}(z)$, may be generated with the following two formulas.

$$P_l^0(z) = \frac{1}{2^l l!}\frac{d^l}{dz^l}\left[(z^2-1)^l\right] \tag{5-7}$$

$$P_l^m(z) = (1 - z^2)^{m/2} \frac{d^m}{dz^m} P_l^0(z) \tag{5-8}$$

Notice that if $m > l$, these formulas lead to a function that is zero. Certain of these polynomials* are given explicitly in Table 5.1.

In the spherical harmonic functions, the "variable" for the Legendre polynomials is really a function, cos θ. This means that after establishing the polynomials explicitly for z using Eqs. (5-7) and (5-8), the desired equations in terms of θ are obtained by substituting cos θ for z throughout. Trigonometric identities are used to simplify the expressions, and these forms are also presented in Table 5.1. The associated Legendre polynomials are orthogonal functions for the range $z = -1$ to $z = 1$. This corresponds to the range where $\cos\theta = -1$ to $\cos\theta = 1$, which is $\theta = 0$ to $\theta = \pi$. Over this range the functions are normalized by a simple factor that uses $l$ and m:

$$\sqrt{\frac{(2l+1)\,(l-|m|)!}{2\,(l+|m|)!}}$$

The symbol Θ is used to designate a normalized associated Legendre polynomial in θ; that is,

$$\Theta_{lm}(\theta) = \sqrt{\frac{(2l+1)\,(l-|m|)!}{2\,(l+|m|)!}}\; P_l^{|m|}(\cos\theta) \tag{5-9}$$

Notice that the absolute value of m is used in Eq. (5-9), and so the Θ functions are the same for m and –m.

Spherical harmonics have a simple exponential dependence on the angle $\phi$ via functions designated $\Phi_m$.

$$\Phi_m(\phi) = e^{im\phi} / \sqrt{2\pi} \tag{5-10}$$

The spherical harmonics are products of Φ and Θ functions.

* The polynomials for m = 0, that is, those generated with only Eq. (5-7), are generally referred to as the Legendre polynomials, and then the superscript is suppressed. The polynomials that follow from Eq. (5-8) are the *associated* Legendre polynomials.

TABLE 5.1 Associated Legendre Polynomials Through $l = 4$.

| $P_l^0(z)$ | $P_l^{\|m\|}(z)$ | $P_l^{\|m\|}(\cos\theta)$ |
|---|---|---|
| $l = 0$ | $P_0^0 = 1$ | $P_0^0 = 1$ |
| $l = 1$ | $P_1^0 = z$ | $P_1^0 = \cos\theta$ |
| | $P_1^1 = \sqrt{1-z^2}$ | $P_1^1 = \sin\theta$ |
| $l = 2$ | $P_2^0 = (3z^2 - 1)/2$ | $P_2^0 = (3\cos^2\theta - 1)/2$ |
| | $P_2^1 = 3z\sqrt{1-z^2}$ | $P_2^1 = 3\sin\theta\cos\theta$ |
| | $P_2^2 = 3(1 - z^2)$ | $P_2^2 = 3\sin^2\theta$ |
| $l = 3$ | $P_3^0 = (5z^3 - 3z)/2$ | $P_3^0 = \cos\theta\,(5\cos^2\theta - 3)/2$ |
| | $P_3^1 = 3(5z^2 - 1)\sqrt{1-z^2}/2$ | $P_3^1 = 3\sin\theta\,(5\cos^2\theta - 1)/2$ |
| | $P_3^2 = 15z(1 - z^2)$ | $P_3^2 = 15\cos\theta\sin^2\theta$ |
| | $P_3^3 = 15(1 - z^2)^{3/2}$ | $P_3^3 = 15\sin^3\theta$ |
| $l = 4$ | $P_4^0 = (35z^4 - 30z^2 + 3)/8$ | $P_4^0 = (35\cos^4\theta - 30\cos^2\theta + 3)/8$ |
| | $P_4^1 = \sqrt{1-z^2}\,(35z^3 - 15z)/2$ | $P_4^1 = \sin\theta\cos\theta\,(35\cos^2\theta - 15)/2$ |
| | $P_4^2 = (1-z^2)(105z^2 - 15)/2$ | $P_4^2 = \sin^2\theta\,(105\cos^2\theta - 15)/2$ |
| | $P_4^3 = 105z(1-z^2)^{3/2}$ | $P_4^3 = 105\sin^3\theta\cos\theta$ |
| | $P_4^4 = 105(1-z^2)^2$ | $P_4^4 = 105\sin^4\theta$ |

$$Y_{lm}(\theta,\phi) = \Theta_{lm}(\theta)\ \Phi_m(\phi)\ (-1)^{[m+|m|]/2} \tag{5-11}$$

The factor of $(-1)^{[m+|m|]/2}$ is an arbitrary phase factor that is introduced to conform to common conventions; its value is always 1 or –1. Notice that the $\Phi$ functions are not dependent on the $l$ integer. They are easily shown to be orthogonal and to be normalized for the range $\phi = 0$ to $\phi = 2\pi$. Thus, the spherical harmonic functions are a set of orthonormal functions over the usual ranges of angles in the spherical polar coordinate system. As a formula, this is

$$\int_0^{2\pi}\int_0^{\pi} Y^*_{lm}(\theta,\phi)\ Y_{l'm'}(\theta,\phi)\ \sin\theta\ d\theta\ d\phi = \delta_{ll'}\ \delta_{mm'} \tag{5-12}$$

This can be a very helpful expression for working with wavefunctions given in terms of $\theta$ and $\phi$.

With explicit forms of the spherical harmonic functions, it is possible to show that each is an eigenfunction of the operators $\hat{L}^2$ and $\hat{L}_z$ and that the eigenvalues come from $l$ and m.

$$\hat{L}^2\, Y_{lm} = l(l+1)\hbar^2\ Y_{lm} \tag{5-13}$$

$$\hat{L}_z\, Y_{lm} = m\hbar\, Y_{lm} \tag{5-14}$$

$l$ must be zero or a positive integer; solutions do not exist for other values. m is restricted because if its absolute value were greater than $l$, then the associated Legendre polynomial would be zero. This would give rise to a zero-valued spherical harmonic function which would not serve as a wavefunction since it would correspond to zero probability density everywhere. Therefore, $|m| \le l$, or $m = -l, -l+1, \ldots, l-1, l$.

## 5.3 THE RIGID ROTATOR

An important model problem involving angular momentum is the rigid rotator. Two masses, $m_1$ and $m_2$, are taken to be connected by a massless rod that is absolutely rigid. Thus, the separation distance between the masses is fixed at some value, R, the length of the rod. Fig. 2.4 serves to depict this problem, provided we take the radial spherical polar coordinate, r, to have the fixed value R (i.e., r = R). After separating the motion corresponding to translation of the center of mass of the system, the quantum mechanical kinetic energy operator for the system in Fig. 2.4 is

$$\hat{T} = -\frac{\hbar^2}{2\mu}\nabla^2$$

where $\mu$ is the reduced mass and is given in the internal coordinates, r, $\theta$, $\phi$. Using Eq. (5-6b) but taking r to be fixed at the value R, we have

$$\hat{T} = \frac{\hat{L}^2}{2\mu R^2}$$

In this model situation, there is no potential of any sort, and so the kinetic energy operator and the Hamiltonian are one and the same.

The Schrödinger equation for the rigid rotator must be

$$\frac{\hat{L}^2}{2\mu R^2}\psi = E\psi$$

Since we know that the spherical harmonics are eigenfunctions of the $\hat{L}^2$ operator, then they are also eigenfunctions of the rigid rotator's Hamiltonian since it is proportional to $\hat{L}^2$. That is, the functions $Y_{lm}$ are the wavefunctions of the rigid rotator. The eigenenergies are obtained from Eq. (5-13):

$$\frac{\hat{L}^2}{2\mu R^2}Y_{lm} = l(l+1)\frac{\hbar^2}{2\mu R^2}Y_{lm}$$

$$\therefore \quad E_{lm} = l(l+1)\frac{\hbar^2}{2\mu R^2} \tag{5-15}$$

Notice that the energies increase quadratically with the quantum number *l*. Also, each energy level has a degeneracy of (2*l* + 1), because there are (2*l* +

1) choices of the m quantum number for any particular $l$; for all such choices, the eigenenergies are the same.

The physical picture of the rigid rotator is that of a dumbell whirling about its center of mass. Its allowed states are those given by the allowed values of the two quantum numbers:

$$l = 0, 1, 2, 3, \ldots$$

$$m = -l, \ldots, l$$

The wavefunctions are eigenfunctions of the Hamiltonian, and also of $\hat{L}^2$ and of $\hat{L}_z$. The eigenvalues of $\hat{L}^2$ are as given in Eq. (5-13). On the basis of the postulates, we should expect that a measurement of the square of the angular momentum will produce one of these values. The square root of such a measurement is interpreted as the magnitude or length of the angular momentum vector. For a given $Y_{lm}$ state, this quantity is $\sqrt{l(l+1)}\,\hbar$. Thus, the only allowed lengths of the angular momentum vector are 0, $\sqrt{2}\,\hbar$, $\sqrt{6}\hbar$, $\sqrt{12}\hbar$, $\sqrt{20}\hbar$, and so on.

The eigenvalue of $\hat{L}_z$ for a $Y_{lm}$ function is $m\hbar$. This is the value of the z-component of the angular momentum vector. So, for each state of the rigid rotator, we know the length of the angular momentum vector and its z-component. However, we do not know the x- or y-component precisely since the wavefunctions are not eigenfunctions of the operators $\hat{L}_x$ and $\hat{L}_y$. The information we do have can be depicted by vectors indicating the possible orientations of the angular momentum vector with respect to the z-axis for the different possible lengths of the vector; this is shown in Fig. 5.1. So, our whirling quantum dumbell can "spin" only in fixed or discrete amounts and then only with the angular momentum vector at certain fixed angles.

The three operators, $\hat{L}^2$, $\hat{L}_z$, $\hat{H}$, commute in this model problem, because according to Theorem 4.4, if a set of functions exists that are simultaneously eigenfunctions of any two operators, then those operators must commute. The spherical harmonic functions are eigenfunctions of each of these three operators.

The Hamiltonian must commute with $\hat{L}^2$ since it is proportional to $\hat{L}^2$. This may be seen by considering the commutator of some arbitrary operator $\hat{G}$ and that operator scaled by a constant g:

$$[\hat{G}, g\hat{G}] \equiv \hat{G}(g\hat{G}) - g\hat{G}\hat{G} = g(\hat{G}^2 - \hat{G}^2) = 0$$

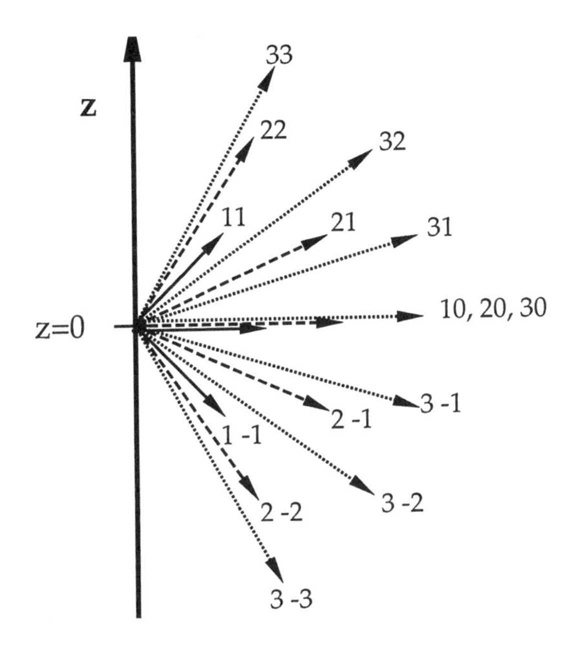


**FIGURE 5.1** The angular momentum vectors of the $Y_{lm}$ states of the rigid rotator for $l = 1$ (solid lines), $l = 2$ (dashed lines), and $l = 3$ (dotted lines). The arrows represent the orientation with respect to the z-axis for the particular $(l,m)$ state. The orientations with respect to the x-axis and the y-axis are not determined, and so this picture represents a planar slice through three-dimensional space such that the slice includes the z-axis and is perpendicular to the x-y plane. The angle between this planar slice and the x-axis or the y-axis, though, is arbitrary.

The commutator of $\hat{L}^2$ with any of the angular momentum component operators happens to be zero, which means that $\hat{L}^2$ also commutes with $\hat{L}_x$, $\hat{L}_y$ and $\hat{L}_z$. This can be shown using the explicit forms of these operators given in Eqs. (5-2) and (5-5), and it is a general result for any type of angular momentum. Another result that may also be obtained from using the explicit forms of the component operators is that they do not commute with each other. This is shown in Eq. (5-3). Therefore, the largest set of mutually commuting operators for the rigid rotator problem consists of three, the Hamiltonian, the $\hat{L}^2$ operator, and any one of the component operators.

It is significant that all of the component operators cannot be part of the mutually commuting set. If the component operators, the Hamiltonian, and $\hat{L}^2$ were all mutually commuting, then the corollary to Theorem 4.4 would say that a set of functions can be found that are simultaneously eigenfunctions of the Hamiltonian, of $\hat{L}^2$, and of all three components. In turn, that would say that we can measure each component of the angular momentum with no uncertainty, which would mean that for any state of the rigid rotator we can know exactly where the angular momentum vector points. That the component operators are *not* mutually commuting means that the wavefunctions of the system can be eigenfunctions of only one component operator, and that measurements with no uncertainty can be accomplished for only that one component; the other two will have nonzero uncertainty.

It is by convention that we choose the spherical harmonic functions such that they are eigenfunctions of the z-component operator and not the x- or y-component operator. But, there is nothing at all special about the z-direction in space. Rather, since we know that there will be one direction in which the angular momentum component is quantized (i.e., the wavefunctions will be eigenfunctions of *one* component operator), then the convention is that that direction, whatever it happens to be, becomes the z-direction.

## 5.4 COUPLING OF ANGULAR MOMENTA

In many chemical problems, there are multiple sources of angular momenta. For instance, electrons orbiting about a molecular axis may give rise to angular momentum at the same time that the rotation of the molecule as a whole about its mass center gives rise to angular momentum. Classically and quantum mechanically, angular momenta add vectorially, and the total angular momentum of a closed system is conserved. In the classical world, we may know the orientation and length of a given angular momentum vector at any instant in time. And so we may add together all such vectors to know the total angular momentum vector. In the quantum world, there is uncertainty in the orientation of an angular momentum vector, and so the addition or coupling of momenta calls for a different sort of analysis.

In a situation where some source of angular momentum is quantized, the associated angular momentum operator commutes with the Hamiltonian. The wavefunctions, then, are simultaneous eigenfunctions. Because of this, we may approach problems involving angular momentum, and we may develop general rules, without considering a specific problem or a specific Hamiltonian. In other words, much can be learned from considering the angular momentum features of quantum states on their own. An extension of the bra-ket notation convention is an aid in this endeavor. Instead of placing a function designation (e.g., $\psi$) in a bra or ket, we will place angular momentum quantum numbers for all angular momentum operators for which the wavefunctions must be eigenfunctions (e.g., $| \, l \, m >$ for a spherical harmonic function instead of $|Y_{lm}>$). Of course in a real problem, there may be a number of states with the same angular momentum quantum numbers but different energies, and so we will not be distinguishing among them in this analysis. Another common practice is to use J in the same way that L has already been used, except that J may represent *any* type of angular momentum, whereas L is usually reserved for orbital angular momentum.

Rules for adding angular momentum need only be developed for two sources at a time. The addition of three sources can be accomplished by adding two together and then adding that result to the third angular momentum vector. So, consider two independent sources, $\vec{J}_1$ and $\vec{J}_2$. In the absence of a physical interaction between the two sources, the wavefunctions are eigenfunctions of the total angular momentum operators for each source and of the z-component operators for each source. The designation of an angular momentum state in this situation requires four quantum numbers,

$$| \, j_1 \; m_1 \; j_2 \; m_2 >$$

$j_1$ and $j_2$ are the total angular momentum quantum numbers for the two independent sources, while $m_1$ and $m_2$ are the z-component quantum numbers. Thus, we may evaluate the effect of total source and z-component operators on the wavefunctions.

$$\hat{J}_1^2 \mid j_1\ m_1\ j_2\ m_2 > \ = \ j_1 (j_1 + 1) \hbar^2 \mid j_1\ m_1\ j_2\ m_2 >$$

$$\hat{J}_2^2 \mid j_1\ m_1\ j_2\ m_2 > \ = \ j_2 (j_2 + 1) \hbar^2 \mid j_1\ m_1\ j_2\ m_2 >$$

$$\hat{J}_{1z} \mid j_1\ m_1\ j_2\ m_2 > \ = \ m_1 \hbar \mid j_1\ m_1\ j_2\ m_2 >$$

$$\hat{J}_{2z} \mid j_1\ m_1\ j_2\ m_2 > \ = \ m_2 \hbar \mid j_1\ m_1\ j_2 m_2 >$$

The possible values for the quantum number $m_1$ are $-j_1, \ldots, j_1$, and by counting those possibilities, we find there are $(2j_1 + 1)$ of them. This is the multiplicity of states from the $j_1$ angular momentum. The total number of states for the whole system is a product of multiplicities, $(2j_1 + 1)(2j_2 + 1)$.

Coupling of angular momenta comes about through an interaction that is brought into the Hamiltonian of the problem. An interaction that couples momenta must involve both angular momentum vectors. So, the form is generally that of a dot product of vectors,* e.g. $\vec{J}_1 \bullet \vec{J}_2$. Such an interaction implies an energetic preference in how the two angular momentum vectors combine in forming a resultant vector. The possibilities, though, are restricted by the quantization of the total angular momentum and its z-component. For instance, the maximum possible z-component of the total angular momentum cannot be any more than the sum of the maximum possible z-components of the two vectors being combined. From Fig. 5.1, the maximum z-component for any angular momentum vector occurs when the associated m quantum number is equal to the j quantum number. Therefore, we may state a rule:

$$\begin{aligned} \text{Maximum resultant z-component} &= [\text{maximum value of } m_1 + \\ &\qquad \text{maximum value of } m_2]\hbar \\ &= (j_1 + j_2)\hbar \end{aligned}$$

This must also equal the maximum z-component quantum number of the resultant vector, designated M, times $\hbar$. That is,

* The interaction between two vectors may require an intervening second-rank tensor, as in $\vec{J}_1 \mathbf{G} \vec{J}_2$, but this adds no complication to this part of the discussion.

$$\text{Maximum value of } M = j_1 + j_2$$

Since any z-component quantum number can take on only certain values because of the length of the associated vector (i.e., $M = -J, \ldots, J$), then knowing the maximum value of M means knowing J. Thus, we conclude that one possible value for the resultant J quantum number is $j_1 + j_2$.

There are other possible resultant J quantum numbers. Notice that there are two ways in which a resultant vector may be composed for which $M = j_1 + j_2 - 1$. One way is with the first vector arranged so that $m_1 = j_1 - 1$, while for the second, $m_2 = j_2$. This is illustrated in Fig. 5.2. The second way is with $m_1 = j_1$, but $m_2 = j_2 - 1$. One $\hbar$ less in the resultant z-component comes about through addition of one or the other component vectors at an orientation with one $\hbar$ less in the z-component. These two arrangements of the source vectors do not distinctly correspond to resultant angular momentum states, but they do indicate that there will exist two states for which $M = j_1 + j_2 - 1$. One of those is expected, since it represents one of the possible choices of M with $J = j_1 + j_2$ (the possible J value already identified). The fact that there is another state with the same z-component means that there must be a state with a different J value, such that its maximum z-component is $(j_1 + j_2 - 1)\hbar$. That is, the next possible J value is $j_1 + j_2 - 1$.

If the process were continued to the next step down in the z-component of the resultant vector, it would turn out that there has to be another possible J value; this one being $j_1 + j_2 - 2$. And this pattern would continue up until $|j_1 - j_2| = 0$. Thus, the rule for adding angular momenta is that the possible values for the resultant or total J quantum number span a range given by the quantum numbers associated with the source vectors:

$$J = j_1 + j_2, j_1 + j_2 - 1, j_1 + j_2 - 2, \ldots, |j_1 - j_2| \tag{5-16}$$

The resultant states are eigenfunctions of the operators $\vec{J}^2$, $\vec{J}_1^{\,2}$, $\vec{J}_2^{\,2}$, and $\hat{J}_z$ and may be designated $|J\,M\,j_1\,j_2\rangle$. They are not necessarily eigenfunctions of the z-component operators of the sources.

Incorporating a new interaction into a system's Hamiltonian, such as a coupling of angular momenta, should never change the number of states. As mentioned above, the number of states in the absence of any interaction is a product of multiplicities, $(2j_1 + 1)(2j_2 + 1)$. The number of states corresponding to the possible resultant J values in Eq. (5-16) is a sum of the

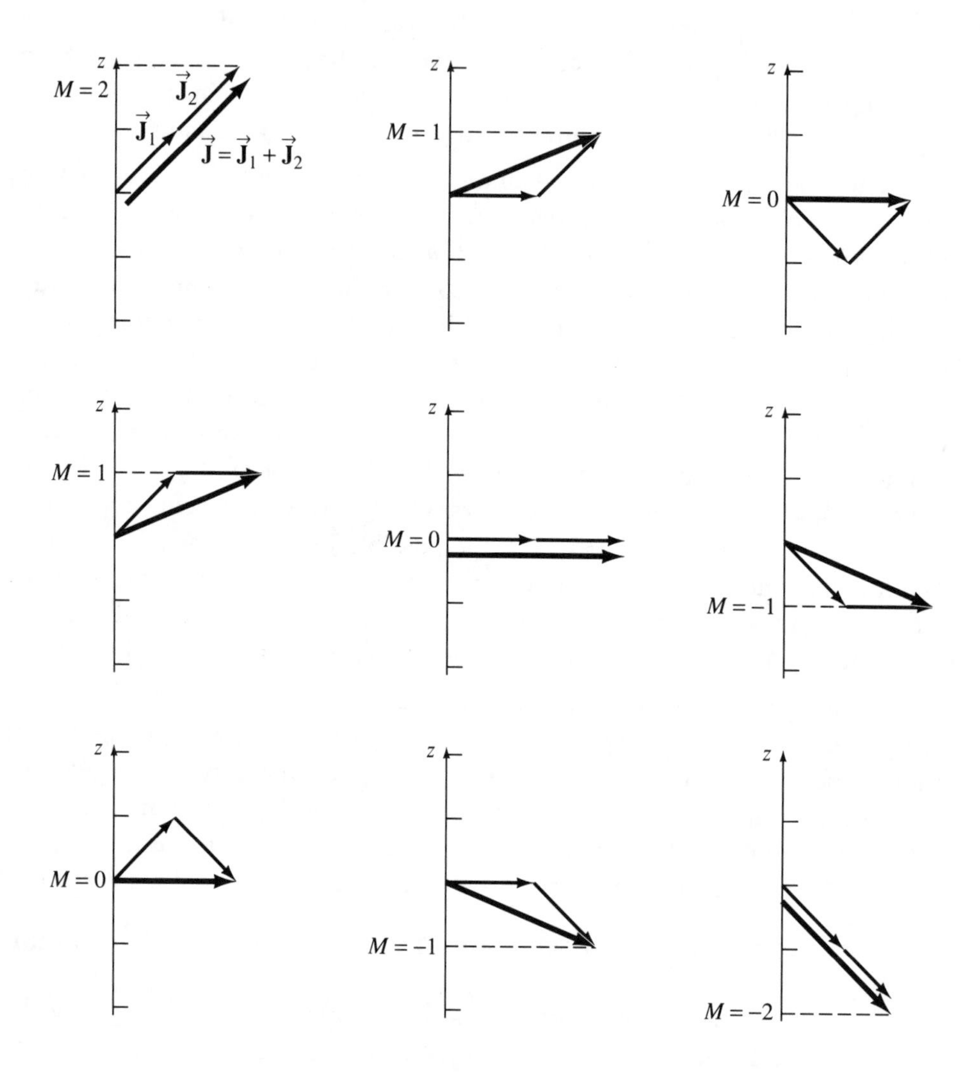


**FIGURE 5.2** For two angular momentum vectors, $\vec{J}_1$ and $\vec{J}_2$, with associated quantum numbers $j_1$ and $j_2$ both equal to 1, there will be $(2j_1 + 1)(2j_2 + 1) = 9$ different combinations of their possible orientations with respect to the z-axis. These nine ways are shown here, and for each a resultant vector from the sum of $\vec{J}_1$ and $\vec{J}_2$ has been drawn. In each case, the z-component of the resultant vector may be obtained from the sum of the z-component quantum numbers $m_1$ and $m_2$.

multiplicities, (2J + 1), for each possible value of J. It can be proven that this is the same number of states, i.e.,

$$(2j_1 + 1)(2j_2 + 1) = \sum_{J=|j_1-j_2|}^{j_1+j_2} (2J + 1)$$

This is demonstrated for a few examples. We specify values for the source momenta, $j_1$ and $j_2$, and find the resultant J's according to Eq. (5-16). The sum of the multiplicities for the possible J values is compared with the product of the $j_1$ and $j_2$ multiplicities.

| $j_1$ | $j_2$ | $(2j_1+1)(2j_2+1)$ | J | (2J+1) | $\Sigma$ (2J+1) |
|---|---|---|---|---|---|
| 0 | 0 | **1** | 0 | 1 | **1** |
| 1 | 0 | **3** | 1 | 3 | **3** |
| 0 | 2 | **5** | 2 | 5 | **5** |
| 1 | 1 | **9** | 2 | 5 | |
| | | | 1 | 3 | |
| | | | 0 | 1 | **9** |
| 2 | 1 | **15** | 3 | 7 | |
| | | | 2 | 5 | |
| | | | 1 | 3 | **15** |
| 2 | 2 | **25** | 4 | 9 | |
| | | | 3 | 7 | |
| | | | 2 | 5 | |
| | | | 1 | 3 | |
| | | | 0 | 1 | **25** |

The J's in this list are the possible values according to Eq. (5-16).

An interesting and necessary feature of the coupled angular momentum states is that they are eigenfunctions of the coupling interaction operator, $\vec{J}_1 \bullet \vec{J}_2$. This can be shown by starting from an expression for the dot product of $\vec{J} = \vec{J}_1 + \vec{J}_2$ and itself.

$$\vec{J} \bullet \vec{J} = (\vec{J}_1 + \vec{J}_2) \bullet (\vec{J}_1 + \vec{J}_2)$$

$$= \vec{J}_1 \bullet \vec{J}_1 + 2\vec{J}_1 \bullet \vec{J}_2 + \vec{J}_2 \bullet \vec{J}_2$$

Rearranging terms yields

$$\vec{J}_1 \bullet \vec{J}_2 = [\vec{J} \bullet \vec{J} - \vec{J}_1 \bullet \vec{J}_1 - \vec{J}_2 \bullet \vec{J}_2] / 2$$

$$= [\hat{J}^2 - \hat{J}_1^2 - \hat{J}_2^2] / 2 \qquad (5\text{-}17)$$

This says that the operator corresponding to the dot product of the two source vectors is the same as a combination of the three operators that give the lengths of the angular momenta. The coupled states, $| J \ M \ j_1 \ j_2 >$, are eigenfunctions of these three operators, and so they must be eigenfunctions of the combination in Eq. (5-17).

$$(\vec{J}_1 \bullet \vec{J}_2) \, | J M j_1 \, j_2 > \; = \; [\hat{J}^2 - \hat{J}_1^2 - \hat{J}_2^2] / 2 \; | J M j_1 \, j_2 >$$

$$= [J(J+1) - j_1 (j_1+1) - j_2 (j_2+1)] \hbar^2 / 2 \; | J M j_1 \, j_2 > \quad (5\text{-}18)$$

Eq. (5-18) will be used later to evaluate the effect of Hamiltonian terms that take the form of a dot product of two angular momentum vectors. The states that are appropriate in the absence of a coupling interaction, $| \, j_1 \; m_1 \; j_2 \; m_2 >$, are not eigenfunctions of $\vec{J}_1 \bullet \vec{J}_2$.

## 5.5 RAISING AND LOWERING OPERATORS *

There are a number of operator relations that are quite powerful in working out angular monetum problems. Certain of these involve operators called *raising operators* and *lowering operators*. As will be seen, these operators act on an angular momentum eigenfunction and increase the z-component of the vector by one step (raising it) or decrease it by one step (lowering it). The length of the angular momentum vector is unchanged. Raising and lowering operators may be generally defined for any angular momentum source, and they are

$$J_+ \equiv J_x + iJ_y \tag{5-19a}$$

$$J_- \equiv J_x - iJ_y \tag{5-19b}$$

(As the context should be clear, at this point, the operator designation ^ will be neglected in this and the next section.) The + subscript identifies the raising operator, and the – subscript identifies the lowering operator.

The first property of the raising and lowering operators is that they commute with the operator corresponding to the square of the angular momentum. This is seen by evaluating the commutator. (Many of the raising and lowering operator expressions can be written concisely as a single equation, as is done below. These equations use the $\pm$ or $\mp$ symbol and are deciphered by either using the top of each symbol throughout to be one equation, or using the bottom of each symbol throughout to be the other equation.)

$$[J^2, J_\pm] = [J^2, (J_x \pm iJ_y)]$$

$$= [J^2, J_x] \pm i[J^2, J_y] = 0 \tag{5-20}$$

Because the $J^2$ operator commutes with each component operator, it must commute with the raising operator and with the lowering operator.

Next, we can work out the effect of the raising and lowering operators by applying them to an angular momentum function. Using $\Gamma$ to designate any particular eigenfunction of $J^2$ and $J_z$, we must have that

$$J^2\Gamma = \mu\hbar^2\Gamma$$

$$J_z\Gamma = \nu\hbar\Gamma$$

$\mu$ and $\nu$ (times Planck's constant) are the eigenvalues for $\Gamma$. To understand the effect of a raising (or lowering) operator, we investigate what happens when it is applied to $\Gamma$:

$$J_+\Gamma = \Omega \tag{5-21}$$

A new function, $\Omega$, is whatever is produced. Applying $J^2$ to $\Omega$ and using the commutation property of Eq. (5-20) shows that $\Omega$'s eigenvalue is the same as $\Gamma$'s.

$$J^2\Omega = J^2(J_+\Gamma) = J_+(J^2\Gamma) = J_+(\mu\hbar^2\Gamma) = \mu\hbar^2(J_+\Gamma) = \mu\hbar^2\Omega$$

This means that $\Omega$ corresponds to an angular momentum vector with the same length as the vector that corresponds to the $\Gamma$ function.

Applying $J_z$ to $\Omega$ will determine the z-component of the angular momentum vector associated with this function.

$$J_z\Omega = J_zJ_+\Gamma \qquad \text{by Eq. (5-21)}$$

$$= J_z(J_x + iJ_y)\Gamma \qquad \text{by Eq. (5-19a)}$$

$$= J_zJ_x\Gamma + iJ_zJ_y\Gamma$$

$$= (i\hbar J_y + J_xJ_z)\Gamma + i(-i\hbar J_x + J_yJ_z)\Gamma \qquad \text{by Eq. (5-3)}$$

$$= \hbar(J_x + iJ_y)\Gamma + (J_x + iJ_y)J_z\Gamma \qquad \text{rearranged terms}$$

$$= \hbar J_+\Gamma + J_+(\nu\hbar)\Gamma \qquad \text{by Eq. (5-19a)}$$

$$= \hbar(\nu + 1)J_+\Gamma = \hbar(\nu + 1)\Omega$$

This shows that $\Omega$ is an eigenfunction of $J_z$, but its eigenvalue is bigger than the eigenvalue of $\Gamma$ by $\hbar$. In other words, the z-component is one step up. And so the net effect of the operator $J_+$ is to change an angular momentum function into one with the M quantum number increased by one. Likewise, the $J_-$ operator will produce a function whose M quantum number is one less than that of the function the operator was applied to.

If the raising operator is applied to an angular momentum function where the z-component is already at the maximum, (i.e., where M = J), then the result is zero. For instance,

$$L_+Y_{ll} = 0$$

$$J_+ \mid J\ M{=}J > \ = 0$$

Likewise, applying the lowering operator to a function with the lowest possible z-component (i.e., where M = –J) would produce zero.

The application of a raising or a lowering operator changes the M quantum number of the angular momentum state, thereby changing one function into another. However, that statement does not completely describe the effect of the operator; it does not tell if the resulting function ends up multiplied by some constant. [To say that $\hat{O}$ acts on f(x) to yield a function with the properties of g(x), does not say $\hat{O}f(x) = g(x)$, only that $\hat{O}f(x)$ is proportional to g(x).] It happens that a raising or lowering operation does introduce a constant, but its value can be obtained using the normalization condition on angular momentum eigenfunctions. With the designation $\psi_{JM}$ for normalized angular momentum eigenfunctions, the effect of the raising operator may be expressed in terms of an as yet unknown constant, c.

$$J_+ \psi_{JM} = c\,\psi_{JM+1} \qquad \Rightarrow$$

$$< J_+ \psi_{JM} \mid J_+ \psi_{JM} > \ = \ < c\,\psi_{JM+1} \mid c\,\psi_{JM+1} >$$

$$= c^2 < \psi_{JM+1} \mid \psi_{JM+1} > \ = c^2 \qquad (5\text{-}22)$$

Thus, because $\psi_{J\,M+1}$ is normalized, $c^2$ can be found from evaluating the integral on the left-hand side of Eq. (5-22).

Several steps are required to find the integral that is equal to $c^2$. From the definition of the raising and lowering operators, each is seen to be the complex conjugate of the other. And because these are Hermitian operators, they can be rearranged in the integral we are seeking to evaluate.

$$< J_+ \psi_{JM} \mid J_+ \psi_{JM} > \ = \ < \psi_{JM} \mid J_- J_+ \psi_{JM} > \qquad (5\text{-}23)$$

The operator product on the right hand side of Eq. (5-23) can be put in a different form by going back to the definition of the raising and lowering operators.

$$J_- J_+ = (J_x - iJ_y)\,(J_x + iJ_y)$$

$$= J_x^2 + i(J_x J_y - J_y J_x) + J_y^2$$

$$= (J_x^2 + J_y^2 + J_z^2) - J_z^2 + i[J_x, J_y]$$

$$= J^2 - J_z^2 - \hbar J_z \tag{5-24}$$

The last step above requires substituting for the commutator according to Eq. (5-3). With this expression for the product of the lowering and raising operators, the integral in Eq. (5-23) can be evaluated because the function being acted on is an eigenfunction of the operators. This is shown in the following steps.

$$< \psi_{JM} \mid J_- J_+ \psi_{JM} > = < \psi_{JM} \mid (J^2 - J_z^2 - \hbar J_z) \psi_{JM} >$$

$$= \hbar^2 [J(J+1) - M^2 - M] < \psi_{JM} \mid \psi_{JM} >$$

$$= \hbar^2 [J(J+1) - M^2 - M] \tag{5-25}$$

Therefore,

$$c^2 = \hbar^2 [J(J+1) - M^2 - M]$$

$$c = \hbar \sqrt{J(J+1) - M^2 - M}$$

The choice of the positive root for c is by convention. (The negative root could be used so long as the choice is made consistently.) If the same analysis is carried through for the lowering operator, its c factor is obtained. Then, the complete effect of the raising and lowering operators on an arbitrary $|J\,M>$ function is established. Written together, we have that

$$J_\pm \mid J\,M> = \hbar \sqrt{J(J+1) - M^2 \mp M} \; \mid J\,M \pm 1> \tag{5-26}$$

## 5.6 COUPLED ANGULAR MOMENTUM EIGENFUNCTIONS *

The angular momentum functions designated in Sec. 5.4 as $| j_1 \; m_1 \; j_2 \; m_2 >$ are referred to as *uncoupled states,* while the functions designated $| J \; M \; j_1 \; j_2 >$ are *coupled states.* A set of uncoupled state functions is called an *uncoupled basis,* and a set of coupled state functions is a *coupled basis.* There will always exist a linear transformation between the two bases, and the functions in both are eigenfunctions of $\hat{J}_1^2$ and $\hat{J}_2^2$. The uncoupled functions, though, are eigenfunctions of the two operators $\hat{J}_{1\text{-}z}$ and $\hat{J}_{2\text{-}z}$, whereas the coupled functions are eigenfunctions of $\hat{J}^2$ and $\hat{J}_z$. This implies that we may transform from the coupled to the uncoupled basis by finding a transformation that makes the matrix representations of $\hat{J}_{1\text{-}z}$ and $\hat{J}_{2\text{-}z}$ diagonal. Or we can transform to the coupled basis by finding a transformation that makes the representations of $\hat{J}^2$ and $\hat{J}_z$ diagonal.

One way of working out the transformation between coupled and uncoupled bases uses the raising and lowering operators of the last section, starting from the *parallel coupling* case, or the maximum M case discussed in Sec. 5.4. This is the one possible arrangement where $J = M = j_1 + j_2$, and for this case, the coupled state is exactly the same as the uncoupled state:

$$| J=(j_1+j_2) \; M=(j_1+j_2) \; j_1 \; j_2 > \; = \; | j_1 \; m_1=j_1 \; j_2 \; m_2=j_2 >$$

To this, we may apply raising or lowering operators in order to generate the relationships among the other possible states.

Since a lowering operator is formed from x- and y-component operators, a lowering operator for the total angular momentum can be obtained in terms of lowering operators of the two angular momentum sources.

$$J_x = J_{1\text{-}x} + J_{2\text{-}x}$$

$$J_y = J_{1\text{-}y} + J_{2\text{-}y}$$

$$J_+ = J_{1+} + J_{2+} \quad \text{and} \quad J_- = J_{1-} + J_{2-}$$

Applying the lowering operator to the parallel coupling case produces a new state.

$$J_- \mid J\,J\,j_1\,j_2>_c \quad = \quad (J_{1-} + J_{2-}) \mid j_1\,j_1\,j_2\,j_2>_u \tag{5-27}$$

The left hand side of Eq. (5-27) has the coupled state with maximum z-component, which means that M has the value J. The subscript c on the ket vector is added for clarity in this discussion. The operator $J_-$ will act on this function to lower the M value from J to J – 1. That is, with Eq. (5-26), we have that

$$J_- \mid J\,J\,j_1\,j_2>_c \quad = \quad \hbar\sqrt{J(J+1) - J^2 + J} \mid J\;J{-}1\;j_1\;j_2>_c \tag{5-28}$$

The right hand side of Eq. (5-27) has the uncoupled state (u subscript) where the z-components of the two sources of angular momentum are at their maximum, which means $m_1$ has the value $j_1$ and $m_2$ has the value $j_2$. One of the lowering operators being applied, the $J_{1-}$ operator, will change the function from that with $m_1 = j_1$ to that with $m_1 = j_1 - 1$. This operator will not affect $m_2$.

$$J_{1-} \mid j_1\,j_1\,j_2\,j_2>_u \quad = \quad \hbar\sqrt{j_1(j_1+1) - j_1^2 + j_1} \mid j_1\;j_1{-}1\;j_2\;j_2>_u \tag{5-29}$$

A similar result is obtained from applying $J_{2-}$.

$$J_{2-} \mid j_1\,j_1\,j_2\,j_2>_u \quad = \quad \hbar\sqrt{j_2(j_2+1) - j_2^2 + j_2} \mid j_1\;j_1\;j_2\;j_2{-}1>_u \tag{5-30}$$

The results from Eqs. (5-28,29,30) may be substituted into Eq. (5-27) to give the coupled state for M = J – 1 (one down from the maximum z-component case) in terms of uncoupled functions. The process can be repeated to find the next lower M case, and so on, with the following overall result.

$$\begin{aligned}\sqrt{J(J+1) - M^2 + M} \mid J\;M{-}1\;j_1\;j_2>_c \\ = \sqrt{j_1(j_1+1) - m_1^2 + m_1} \mid j_1\;m_1{-}1\;j_2\;m_2>_u \\ + \sqrt{j_2(j_2+1) - m_2^2 + m_2} \mid j_1\;m_1\;j_2\;m_2{-}1>_u\end{aligned} \tag{5-31}$$

Eq. (5-31) is a linear transformation between the coupled and uncoupled angular momentum functions.

As a detailed example of this scheme, let us consider the case with $j_1 = 1$ and $j_2 = 1$. From the rule given as Eq. (5-16), the possible values for the total angular momentum quantum number upon coupling the two sources will be J = 2, J = 1, and J = 0. The maximum z-component situation corresponds to J = 2 and M = 2 in the coupled states picture, and to $m_1 = 1$ and $m_2 = 1$ in the uncoupled states picture.

$$| 2\,2\,1\,1 >_c \;=\; | 1\,1\,1\,1 >_u \tag{5-32a}$$

Notice that since $j_1$ and $j_2$ are fixed values – they correspond to the lengths of the angular momentum vectors of the sources – it is not necessary to list them in the state designations. Thus, Eq. (5-32a) is written more concisely with $| \, J \, M >$ for the coupled state and $| \, m_1 \, m_2 >$ for the uncoupled state:

$$| 2\,2 >_c \;=\; | 1\,1 >_u \tag{5-32b}$$

Applying appropriate lowering operators to both sides of Eq. (5-32b) leads to the following steps that invoke the results of Eqs. (5-28,29,30).

$$J_- \, | 2\,2 >_c \;=\; (J_{1-} + J_{2-}) \, | 1\,1 >_u$$

$$\hbar\sqrt{2(2+1) - 2^2 + 2} \;\; | 2\,1 >_c \;=\; \hbar\sqrt{1(1+1) - 1^2 + 1} \;\; | 0\,1 >_u$$

$$+ \; \hbar\sqrt{1(1+1) - 1^2 + 1} \;\; | 1\,0 >_u$$

$$\therefore \quad | 2\,1 >_c \;=\; \frac{1}{\sqrt{2}} \, | 0\,1 >_u \;+\; \frac{1}{\sqrt{2}} \, | 1\,0 >_u \tag{5-33}$$

This says that the coupled state where J = 2 and M = 1 is a linear combination of two uncoupled states, one where $m_1 = 0$ and $m_2 = 1$, and the other where $m_1 = 1$ and $m_2 = 0$.

Applying lowering operations to Eq. (5-33) produces another coupled state in terms of the uncoupled states.

$$J_- \, | 2\,1 >_c \;=\; \frac{1}{\sqrt{2}} \left[ (J_{1-} + J_{2-}) \, | 0\,1 >_u \;+\; (J_{1-} + J_{2-}) \, | 1\,0 >_u \right]$$

$$\hbar\sqrt{2(2+1)-1^2+1}\;|\,2\;0>_c \;=\; \frac{\hbar}{\sqrt{2}}\Big[\Big(\sqrt{1(1+1)-0^2+0}\;|\,-1\;1>_u$$

$$+\sqrt{1(1+1)-1^2+1}\;|\,0\;0>_u\Big)$$

$$+\Big(\sqrt{1(1+1)-1^2+1}\;|\,0\;0>_u$$

$$+\sqrt{1(1+1)-0^2+0}\;|\,1\;-1>_u\Big)\Big]$$

$$\therefore\quad |\,2\;0>_c \;=\; \frac{1}{\sqrt{6}}\big[\;|\,-1\;1>_u + 2\;|\,0\;0>_u + \;|\,1\;-1>_u\big] \tag{5-34}$$

This says that the coupled state where J = 2 and M = 0 is a linear combination of three uncoupled states. And this process can be continued to find the coupled state J = 2 and M = –1, and then J = 2 and M = –2. Of course, this last situation is analogous to how we started, and so we must arrive at

$$|\,2\;-2>_c \;=\; |\,-1\;-1>_u$$

An alternative to the process described is to start from this lowest M case and apply raising operators, but the results would be identical.

The application of the lowering operator to the original, maximum z-component state produced all the J = 2 states, but there remain J = 1 and J = 0 states. These are found, in sequence, by the condition of orthogonality. It is always true that the M quantum number of the coupled state equals the sum of the m quantum numbers of the uncoupled states that combine to form a coupled state. That is, $M = m_1 + m_2$, and this can be verified for the examples in Eqs. (5-33) and (5-34). Consequently, one of the remaining coupled states, that corresponding to J = 1 and M = 1, must be a linear combination only of the uncoupled states $|\,1\;0>_u$ and $|\,0\;1>_u$. There is only one linear combination of these which will be orthogonal to the state in Eq. (5-33), and it is the opposite combination. Therefore, the following must be true.

$$|\,1\;1>_c \;=\; \frac{1}{\sqrt{2}}\;|\,0\;1>_u \;-\; \frac{1}{\sqrt{2}}\;|\,1\;0>_u \tag{5-35}$$

From this relation, lowering operations can be performed to find the J = 1 and M = 0 coupled state function, and then the J = 1 and M = –1 function. Orthogonality is used again to find the last unknown coupled state function, that for which J = 0 and M = 0.

In this example, there are nine uncoupled states and, of course, nine coupled states. The coefficients that have been worked out are linear expansion coefficients, and so they can be arranged into a 9-by-9 matrix. This will be the transformation (matrix) from the uncoupled basis to the coupled basis. This matrix and its inverse contain the coefficients to make up one kind of basis function from the other set. These coefficients are used in many problems in quantum physics, and they are often called *vector coupling coefficients*. Since they can be worked out once and for all for specific choices of $j_1$ and $j_2$, there are many tabulations of the values, and many alternate procedures have been devised for finding them. They are sometimes scaled by different factors for convenience in different circumstances. Names that are given to vector coupling coefficients under various prescriptions include Clebsch-Gordon coefficients, Racah coefficients, 3-j symbols, and 6-j symbols.

## Exercises

1. Verify the derivation of the expressions in Eq. (5-3).
2. Beginning with Eq. (5-2), carry out the explicit change of variables to arrive at Eq. (5-4). Use these operators to verify Eq. (5-5).
3. Derive Eq. (5-6b) from Eq. (5-6a).
4. Show that $\hat{L}^2$ commutes with $\hat{L}_x$ and $\hat{L}_y$ by using the explicit spherical polar coordinate forms.
5. Show that for any integer m, the function $\sin^m\theta \cos\theta\, e^{im\phi}$ is an eigenfunction of $\hat{L}^2$. Find the eigenvalue.
6. Find the commutator of $\hat{L}_x$ and $\hat{L}_y$ in spherical polar coordinates.
7. Write out the explicit form of $Y_{54}(\theta,\phi)$.

8. Evaluate $< Y_{20} \mid \hat{L}_x Y_{20} >$ and $< Y_{20} \mid \hat{L}_x^2 Y_{20} >$.

9. Verify Eq. (5-12) for $l = 2$, $m = 1$ and $l' = 1$, $m' = 1$.

10. Find the operator, $L_+$, as an explicit differential operator in $\theta$ and $\phi$ by using the explicit forms of the $L_x$ and $L_y$ operators. Apply $L_+$ to $Y_{21}$ to generate $Y_{22}$. Then apply it to $Y_{22}$.

11. Express the product $J_+J_-$ in terms of $J^2$ and $J_z$.

12. What are the values of the total angular momentum quantum number J for a problem with three coupled sources of angular momentum, $j_1 = 2$, $j_2 = 1$, and $j_3 = 1$?

13. If it were possible to have an angular momentum source with an angular momentum quantum number of 1/2, what would be the values of the total angular momentum quantum number J for a system of two such sources coupled together? And of three, and four?

★14. Apply the lowering operation to Eq. (5-35) in order to find the coupled state $\mid 1\ 0 >_c$ in terms of uncoupled states.

★15. Find the linear transformation matrix between the sets of coupled and uncoupled angular momentum functions for two angular momentum sources, $j_1 = 2$ and $j_2 = 1$.

★16. If there were an angular momentum source with $j_1 = 1/2$, what would be the length of the corresponding angular momentum vector? Next, find the linear transformation matrix between the sets of coupled and uncoupled angular momentum functions for two such angular momentum sources, i.e., $j_1 = 1/2$ and $j_2 = 1/2$.

## Bibliography

1. K. R. Symon, *Mechanics* (Addison-Wesley, Reading, Massachusetts, 1971). This is a useful source for the classical mechanics of angular momentum.

2. M. E. Rose, *Elementary Theory of Angular Momentum* (John Wiley and Sons, New York, 1957).

3. M. Tinkham, *Group Theory and Quantum Mechanics* (McGraw-Hill, New York, 1964). This is an advanced text that provides a general treatment of angular momentum coupling via the mathematics of group theory. The development is formal and high level, but it is extremely powerful.
4. L. I. Schiff, *Quantum Mechanics* (McGraw-Hill, New York, 1968). This advanced text has a particularly useful section on angular momentum coupling coefficients and raising and lowering operators.
5. W. H. Flygare, *Molecular Structure and Dynamics* (Prentice-Hall, Englewood Cliffs, New Jersey, 1978). The use of coupled and uncoupled bases for spectroscopic angular momentum problems is demonstrated.
6. C. S. Johnson, Jr., and L. G. Pedersen, *Problems and Solutions in Quantum Chemistry and Physics* (Addison-Wesley, Reading, Massachusetts, 1974).
7. R. N. Zare, *Angular Momentum* (John Wiley and Sons, New York, 1988). This advanced text covers important applications in quantum chemistry.

*Chapter 6*

# Vibrational-Rotational Spectroscopy of a Diatomic Molecule

*The quantum mechanical fundamentals that have been developed so far are applied to the vibrational and rotational motions of diatomic molecules in this chapter. The analysis begins with an idealization of a typical diatomic, and it continues to a realistic treatment. The resulting picture provides the basis for interpeting infrared and microwave spectra of diatomics obtained in laboratory experiments, and this chapter explores the way in which structural and energetic information is deduced from these spectra.*

## 6.1 MOLECULAR SPECTROSCOPY

Molecular spectroscopy refers to a wide category of experiments that involve probing individual molecules often by means of electromagnetic radiation. Radiation is used to induce transitions between molecular eigenstates. Monitoring the radiation in a spectroscopic experiment, directly or indirectly, is a means of investigating the energies of the states of molecules. Quantum theory relates the measurements of spectra to molecular features such as the stiffness of different chemical bonds and the molecular structure. Spectroscopic investigation of molecules was a revolution in chemistry that took place mostly after World War II. It has provided the detailed picture of molecular structure, molecular dynamics and molecular properties upon which numerous principles of synthesis, molecular biology, and materials science have been established. New spectroscopic experiments are continuing to be devised, aided partly by technological advances in electronics, and very subtle molecular features are being characterized.

A basic spectroscopic experiment involves a light source that can be focused on and passed through a sample. Detectors are required to monitor the light intensity before and after the sample. If the sample absorbs some of the impinging radiation, then the light intensity drops upon passing through the sample. Since the energy of a quantum of radiation is proportional to its frequency, then the frequency at which light is absorbed by the sample is of significance; it indicates the energy required to take an individual molecule or atom from one eigenstate to another. So, the basic experiment involves monitoring the light intensity as the frequency of the light source is tuned through some range of frequencies. The information that is obtained is a set of frequencies at which the sample absorbs. These correspond to differences among the energies of the states of the species in the sample. Unfortunately, there is nothing that directly tells which states are associated with a given energy difference. Thus, the task of analyzing a spectrum usually starts with assigning frequencies of absorption to specific pairs of states. The job of making these assignments involves the use of quantum mechanics so as to find out what types of states and energies are possible.

## 6.2 BASIC TREATMENT OF VIBRATION AND ROTATION

One of the most widely used types of spectroscopy probes the vibrations and rotations of simple molecules. The first such case to consider is that of a diatomic because the mechanics are just those of a two-body system. After separating out translational motion, the Hamiltonian is the kinetic energy operator, conveniently expressed in spherical polar coordinates (Fig. 2.4), plus a potential that depends only on the separation of the atoms. The potential, V(r), is something that develops because of the electrons and the chemical bonding in which they participate. Later chapters will examine the quantum mechanics behind bonding and the origin of the internuclear potentials, V(r), for diatomics. For now, we approach the quantum mechanics of a diatomic molecule by leaving the potential unspecified.

The Schrödinger equation for a vibrating and rotating diatomic, or generally for a two-body problem, with $\mu$ being the reduced mass, is

$$-\frac{\hbar^2}{2\mu}\nabla^2\psi(r,\theta,\phi) + V(r)\,\psi(r,\theta,\phi) = E\,\psi(r,\theta,\phi) \tag{6-1}$$

With Eq. (5-6b), the operator $\nabla^2$ may be expressed in spherical polar coordinates and in terms of the angular momentum operator, $\hat{L}^2$.

$$-\frac{\hbar^2}{2\mu}\left(\frac{2}{r}\frac{\partial\psi}{\partial r} + \frac{\partial^2\psi}{\partial r^2}\right) + \frac{\hat{L}^2}{2\mu r^2}\psi + V(r)\,\psi = E\,\psi \tag{6-2}$$

This differential equation is separable into radial and angular parts, where the angular wavefunctions must be the spherical harmonics. That is, if we assume a product form for the eigenfunction, $\psi$,

$$\psi(r,\theta,\phi) = Y_{lm}(\theta,\phi)\,R(r) \tag{6-3}$$

then applying the operator $\hat{L}^2$ (and the factor $2\mu r^2$) produces

$$\frac{\hat{L}^2}{2\mu r^2}\psi = \frac{l(l+1)\hbar^2}{2\mu r^2}Y_{lm}R \tag{6-4}$$

Therefore, with the product form of $\psi$ of Eq. (6-3) substituted into the Schrödinger equation, Eq. (6-2), we have

$$-\frac{\hbar^2 Y_{lm}}{2\mu}\left(\frac{2}{r}\frac{\partial R}{\partial r} + \frac{\partial^2 R}{\partial r^2}\right) + \frac{l(l+1)\hbar^2}{2\mu r^2}Y_{lm}R + V(r)\,Y_{lm}R = E\,Y_{lm}R$$

Dividing this equation by the spherical harmonic function, $Y_{lm}$, establishes the separability of the Schrödinger equation because the resulting differential equation is only in terms of the variable r.

$$-\frac{\hbar^2}{2\mu}\left(\frac{2}{r}\frac{\partial R}{\partial r} + \frac{\partial^2 R}{\partial r^2} - \frac{l(l+1)}{r^2}R\right) + V(r)\,R = E\,R \tag{6-5}$$

This is called the radial Schrödinger equation of the two-body problem.

Though the two-body Schrödinger equation is separable, that is, $\psi(r,\theta,\phi) = R(r)Y_{lm}(\theta,\phi)$, it is important that the radial equation, Eq. (6-5), is still connected with the angular part of the problem. It incorporates the angular momentum quantum number *l*, and so Eq. (6-5) really represents an infinite number of differential equations corresponding to the infinite possible choices of *l* (namely, 0, 1, 2, . . . ). Thus, any radial function, R(r), found from solving Eq. (6-5) will have to be labelled with the *l* value.

At this point, it becomes convenient to rewrite Eq. (6-5) in terms of a new variable, s, which is the displacement from some fixed radial separation.

$$s \equiv r - r_e \tag{6-6}$$

$r_e$ is a fixed value, and it might be chosen to be the equilibrium value of r or the separation of the two bodies at the potential minimum. Then, s = 0 corresponds to $r = r_e$, the potential minimum. On a graph of V versus s, the point s = 0 is the bottom of the potential well. It is also convenient to introduce a new function, *S*, defined in terms of the still undetermined function R.

$$S(s) = S(r - r_e) \equiv r\,R(r) \tag{6-7}$$

Since the differential in s equals the differential in r [i.e., ds = dr from Eq. (6-6)], then differentiating *S*(s) with respect to s must be the same as differentiating rR(r) with respect to r. Doing this twice produces the following relation between *S* and R.

$$\frac{\partial^2 S}{\partial s^2} = r\left(\frac{2}{r}\frac{\partial R}{\partial r} + \frac{\partial^2 R}{\partial r^2}\right) \tag{6-8}$$

The right hand side is a part of Eq. (6-5), and so with the appropriate substitutions, and then with multiplication by r, Eq. (6-5) is now written in the following way.

$$-\frac{\hbar^2}{2\mu}\left[\frac{\partial^2 S(s)}{\partial s^2} - \frac{l(l+1)}{(s+r_e)^2} S(s)\right] + V(s+r_e)\, S(s) = E\, S(s) \qquad (6\text{-}9)$$

This is the radial Schrödinger equation expressed in terms of $S(s)$ instead of R(r).

We may immediately consider the special case of Eq. (6-9) when the angular momentum is zero, that is, $l = 0$. The Schrödinger equation is then

$$-\frac{\hbar^2}{2\mu}\frac{\partial^2 S(s)}{\partial s^2} + V(s+r_e)\, S(s) = E\, S(s) \qquad (6\text{-}10)$$

This equation is to be solved, but only upon specification of the potential function, V. One possibility is for the potential to be harmonic, $V(s+r_e) = ks^2/2$. Notice that then Eq. (6-10) becomes identical with the standard harmonic oscillator problem of Eq. (3-8), except that here the variable has been named s. A harmonic potential is the simplest, somewhat realistic choice for the stretching potential of a chemical bond. However, the harmonic form must be recognized as an approximation or as an idealization of the true molecular potential because, unlike a harmonic function, the potential for lengthening a chemical bond must reach a plateau or a constant potential energy once the chemical bond has completely broken and the atoms are not interacting. The harmonic function $kx^2/2$, on the other hand, goes to infinity as x gets larger; this "harmonic chemical bond" never breaks. Furthermore, even in the equilibrium region, there is no fundamental reason that the potential function should have precisely the shape of a parabola. But in many cases, a harmonic approximation is a fairly good approximation for the lower energy states. To be more realistic about the potential requires a more general functional form, such as a higher order polynomial. This is discussed further on.

The cases when $l > 0$ in Eq. (6-9) present only a small complication of the basic treatment because the possibly troublesome $l(l+1)$ term can be combined with the potential. That is, we may regard the potential as a combination of the $l(l+1)$ term with the original potential.

$$V_l^{eff}(s + r_e) \equiv V(s + r_e) + \frac{l(l+1)\hbar^2}{2\mu(s + r_e)^2} \tag{6-11}$$

Of course, instead of one potential, there is now a potential for each choice of $l$. Any one of these new potentials is termed an *effective potential* because the effect of the angular momentum term has been built into it.

Next, consider a power series expansion of the s dependence of the effective potential in Eq. (6-11), i.e., $(s + r_e)^{-2}$.

$$\frac{1}{(s + r_e)^2} = \frac{1}{r_e^2} - \frac{2s}{r_e^3} + \frac{6s^2}{r_e^4} - \frac{24 s^3}{r_e^5} + \ldots \tag{6-12}$$

This, of course, is an infinite-order polynomial in s, and if the fundamental potential, V, is a polynomial as well, then the pieces on the right hand side of Eq. (6-11) can be combined term by term. An approximation of the expansion in Eq. (6-12) may be made by truncating it. The validity of the approximation depends on how large s may get relative to $r_e$. If s is likely to be less than 10% of $r_e$, for instance, then the third term will be only three-tenths of the second term, the fourth term will be only four-tenths of the third term, and so on. The terms will quickly become tiny, and truncation after a few should serve as a high-quality approximation. For typical chemical bonds in diatomic molecules, the classical turning point, which characterizes the extent of displacement from equilibrium, is often on the order of or less than 10% of the equilibrium bond length: A truncated power series is a good approximation, normally.

The most drastic truncation of the expansion in Eq. (6-12) is to keep only the first term. Using this approximation in the diatomic's vibration-rotation Hamiltonian corresponds to a neglect of the coupling of the vibrational motion with the rotational motion. It would be as if the system were effectively a rigid rotator, even though its ongoing vibration makes it anything but rigid. With this drastic approximation, Eq. (6-9) becomes

$$-\frac{\hbar^2}{2\mu}\left[\frac{\partial^2 S(s)}{\partial s^2} - \frac{l(l+1)}{r_e^2} S(s)\right] + V(s + r_e)\, S(s) = E\, S(s)$$

It is not necessary to rewrite this with $V^{eff}$ since the constant operator term may be brought to the right hand side and combined with E.

$$-\frac{\hbar^2}{2\mu}\frac{\partial^2 S(s)}{\partial s^2} + V(s + r_e)\, S(s) = E'\, S(s)$$

$$\text{where} \quad E' \equiv E - \frac{\hbar^2 l(l+1)}{2\mu r_e^2}$$

If V were harmonic, this would again be the Schrödinger equation of the harmonic oscillator, and E' would be found to be the harmonic oscillator energies.

From knowing the solutions of the harmonic oscillator problem, we arrive at the simplest idea of the energies, E, of the rotating-vibrating diatomic system. Using J instead of $l$ for the angular momentum quantum number, because that is the convention for molecular rotation, the possible state energies for the diatomic whose potential has been approximated as harmonic and for which vibration-rotation interaction is neglected are given as

$$E_{nJ} = (n + \frac{1}{2})\hbar\omega + \frac{\hbar^2 J(J+1)}{2\mu r_e^2} \tag{6-13}$$

The energy levels are given by the vibrational quantum number, n, and the rotational quantum number, J. The approximations made to arrive at this result are checked, in the end, by comparing energies predicted from Eq. (6-13) with those measured in a spectroscopic experiment, and we shall consider certain specific data later.

The typical sizes of the constants in Eq. (6-13) are such that a change in the vibrational quantum number, n, will affect the energy more than a change in the rotational quantum number. For instance, for the molecule hydrogen fluoride, the quantity $\hbar\omega$ is around 4140 $cm^{-1}$, whereas $\hbar^2/(2\mu r_e^2)$ is only 21 $cm^{-1}$. Thus, the energy will be 4140 $cm^{-1}$ greater if n is increased from zero to one, but only 42 $cm^{-1}$ greater if J is increased from zero to one. A schematic representation of the associated energy level pattern is given in Fig. 6.1.

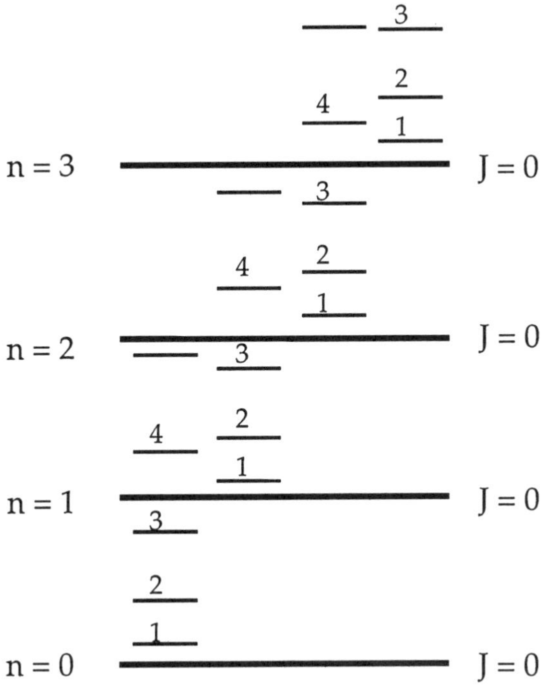

3
2
4
1
n = 3 J = 0
3
4 2
1
n = 2 J = 0
3
4 2
1
n = 1 J = 0
3
2
1
n = 0 J = 0

**FIGURE 6.1** The energy levels of a vibrating-rotating diatomic molecule according to Eq. (6-13). Each horizontal line in this sketch represents an energy level with energy increasing in the vertical direction. The levels are labelled by the vibrational quantum number, n, and the rotational quantum number, J. The long, thick lines are all $J = 0$ levels, and the energy spacing between any pair of these lines is $\hbar\omega$. The diagram has been terminated at $n = 3$, but the energy levels continue on infinitely in the same pattern. Levels for which J is not zero have been drawn as short, thin lines to help organize the drawing, but there is nothing fundamentally different between these and the $J = 0$ levels. These lines appear as stacks or manifolds originating from each $J = 0$ line. The leftmost manifold is for all the states for which $n = 0$, and the rightmost manifold is for all those for which $n = 3$. The numbers above four of these lines show the sequence of the J quantum numbers. Higher J levels exist, but have not been drawn. The degeneracy of each level is $(2J + 1)$.

## 6.3 CENTRIFUGAL DISTORTION

The truncation of Eq. (6-12) can be made to include more terms. With the truncation after two terms, the improved approximation is

$$\frac{1}{(s + r_e)^2} \cong \frac{1}{r_e^2} - \frac{2s}{r_e^3}$$

This is a first order polynomial in s. So, within this approximation, the effective potential of Eq. (6-11) has the form

$$V^{eff}(s + r_e) = V(s + r_e) + a + b\,s$$

And if the original potential is harmonic, or taken to be harmonic, then the effective potential is

$$V^{eff}(s + r_e) = a + b\,s + \frac{1}{2} k\,s^2 \qquad (6\text{-}14)$$

$$\text{where} \quad a = \frac{\hbar^2 J\,(J+1)}{2\mu r_e^2} \quad \text{and} \quad b = -\frac{2\,a}{r_e}$$

Of course, this is still a harmonic potential, since it is a quadratic polynomial in s. As shown in Fig. 6.2, it is a parabola differing from the parabola of $V(s + r_e)$ in the position of its minimum. Instead of the minimum being $V(r_e)=0$, the minimum is at some position we will call $\delta$, and then $V^{eff}(\delta + r_e) = V_o$, where the constant $V_o$ is not necessarily zero.

The effect of shifting the minimum of the parabolic potential of a harmonic oscillator changes the energies of the eigenstates in only one way. The energy amount that the parabola vertically shifts simply adds to each eigenenergy. For instance, wavefunctions that satisfy

$$-\frac{\hbar^2}{2\mu}\frac{d^2\psi_n}{dx^2} + \frac{1}{2} k\,x^2\,\psi_n = (n + \frac{1}{2})\,\hbar\omega\,\psi_n$$

have to satisfy the Schrödinger equation,

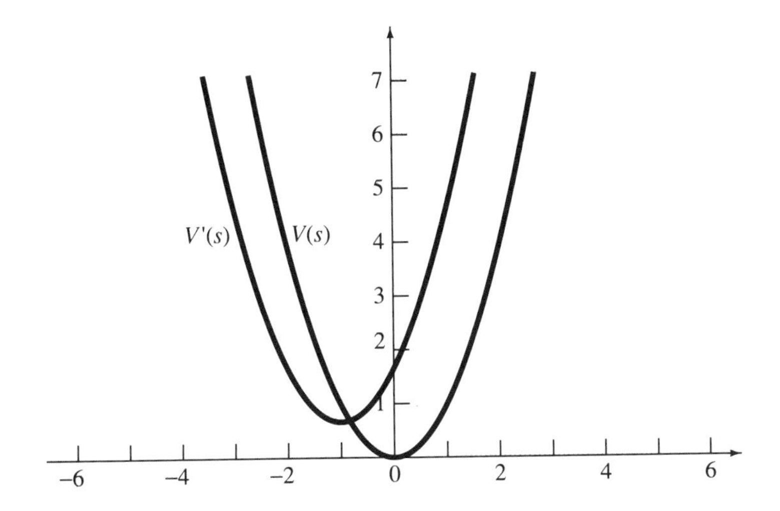


**FIGURE 6.2** A plot of the function $V(s) = s^2/2$ and of the function $V'(s) = 1 + s + s^2/2$. Both functions are parabolas with the same curvature; however, the minimum of $V'(s)$ has been shifted by an amount $\delta = -1$, and the value of the potential at that point is $V'(\delta) = 1/2$ rather than zero.

$$-\frac{\hbar^2}{2\mu}\frac{d^2\psi_n}{dx^2} + \left(V_o + \frac{1}{2}kx^2\right)\psi_n = \left[V_o + \left(n + \frac{1}{2}\right)\hbar\omega\right]\psi_n$$

with $V_o$ being a constant. The eigenenergies here differ only by the constant $V_o$. So, for the harmonic oscillator with the effective potential of Eq. (6-14), all the energy levels are shifted by whatever energy the minimum of the potential has been shifted. (If the x-value of the minimum is shifted, there is no change in the eigenenergies because the force constant is the same; the curvature of the parabola is unaffected by displacement of the minimum to the left or right.) Therefore, the effect of the better approximation of Eq. (6-12) that we are considering, can be determined by finding the location of the minimum of the effective potential in Eq. (6-14). That means finding $\delta$ and $V_o$, as defined above. $\delta$ is the value of s when the first derivative of $V^{eff}$ is zero, and so

$$\delta = -\frac{\hbar^2 J(J+1)}{k\mu r_e^3} \tag{6-15}$$

$V_o$ is the value of $V^{eff}$ when s = δ, and so

$$V_o = -\frac{\hbar^4 [J(J+1)]^2}{2k\mu^2 r_e^6} + \frac{\hbar^2 J(J+1)}{2\mu r_e^2} \quad (6\text{-}16)$$

Thus, the energy levels are now those of the harmonic oscillator but with the constant, $V_o$, added:

$$E_{nJ} = (n + \frac{1}{2})\hbar\omega + \frac{\hbar^2 J(J+1)}{2\mu r_e^2} - \frac{\hbar^4 [J(J+1)]^2}{2k\mu^2 r_e^6} \quad (6\text{-}17)$$

The difference between this and Eq. (6-13) is just the last term. This last term is said to be associated with the effect of centrifugal distortion.

The physical idea of centrifugal distortion is that rotational motion contributes to the stretching of the "spring" between the particles because of centrifugal force. Thus, the separation distance, which is taken to be fixed at $r_e$ in the most drastic approximation of Eq. (6-12), will actually become somewhat longer as J increases. This means that the actual rotational energy will be less than what Eq. (6-13) dictates. The new term in Eq. (6-17) provides the kind of downward correction to the energies that is expected to arise from the centrifugal distortion of the bond. Notice that the J dependence of this term is quartic; it will diminish the energies of high-J states much more than low-J states.

## 6.4 VIBRATION-ROTATION COUPLING

Another improvement that can be made in the vibrational-rotational energy level expression for a diatomic comes from carrying through the next (third) term in the truncation of Eq. (6-12). This term is quadratic in the displacment coordinate, s, and so its inclusion implies a different force constant in the effective potential.

$$V_J^{eff}(s + r_e) = V(s + r_e) + \frac{J(J+1)\hbar^2}{2\mu}\left(\frac{1}{r_e^2} - \frac{2s}{r_e^3} + \frac{6s^2}{r_e^4}\right) \quad (6\text{-}18)$$

If the original potential is taken to be harmonic, a simple way to include this new term is to combine it with the original quadratic potential. The resulting potential has the form

$$V_J^{eff}(s + r_e) = a + bs + \frac{1}{2}(k + 2c)\, s^2 \qquad (6\text{-}19)$$

where a and b are the same as in Eq. (6-14) and $c = -3b/r_e$. The vibrational frequency for this oscillator is now seen to depend on the rotational quantum number J.

$$\omega_J = \sqrt{\frac{(k + 2c)}{m}} = \sqrt{\frac{k}{m} + \frac{6J(J+1)\hbar^2}{(mr_e^2)^2}}$$

$\omega_J$ will be slightly greater than $\omega$, except for $J = 0$, and this will affect the spacing of the energy levels. With $\omega_J$ replacing $\omega$ in Eq. (6-17), the relative effect is to separate adjacent rotational levels by an amount that is almost linear in J. This effect on the energies of the system is referred to as a vibration-rotation coupling effect.

An alternative approach, which is the more standard, is to treat the J(J+1) term of the potential in Eq. (6-18) as a perturbation of the non-rotating harmonic oscillator. The part of the perturbation that goes as $1/r_e^2$ is exactly treated at first order because it is a constant term. It turns out that the second-order energy correction from the part of the perturbation that goes as $-2s/r_e^3$ is the same as the exact result. So, with second order as a chosen point of truncation of the perturbative corrections to the energy, the complete energy level expression [using the entire J(J+1) term in Eq. (6-18) as the perturbation] is,

$$E_{n,J} = \hbar\omega\,(n + \frac{1}{2}) + \frac{\hbar^2 J(J+1)}{2\mu r_e^2} - \frac{\hbar^4 J^2 (J+1)^2}{2k\mu^2 r_e^6} + \frac{3\hbar^2}{\omega\mu^2 r_e^4}(n + \frac{1}{2})\, J(J+1) \qquad (6\text{-}20)$$

The third term is the same as that in Eq. (6-17) and is associated with centrifugal distortion. The last term is associated with vibration-rotation coupling. Notice that it involves both the vibrational quantum number, n, and the rotational quantum number, J.

## 6.5 VIBRATIONAL ANHARMONICITY

*Vibrational anharmonicity* refers to the effects on energy levels of an otherwise harmonic oscillator from those parts of the stretching potential that are anharmonic, that is, those that do not vary as the square of the displacement coordinate. The most realistic vibrational potentials of molecules are not strictly harmonic, and so for a most precise quantum mechanical treatment there must be an incorporation of anharmonicity effects to some degree.

An example of a potential that has the correct qualitative form for the stretching of a diatomic is the Morse potential, shown in Fig. 6.3. It is constructed for a specific diatomic by knowing (or guessing) the dissociation energy, $D_e$, and choosing a parameter, $\alpha$, to yield the desired shape.

$$V^{Morse}(s) = D_e \left(1 - e^{-\alpha s}\right)^2 \tag{6-21}$$

where s is the coordinate corresponding to the displacement from equilibrium separation of the atoms. At $s = 0$, the potential is zero, and asymptotically it approaches $D_e$ as s approaches infinity. For the region $s < 0$, the potential rises very steeply, and so it has the proper qualitative form throughout. This potential can be compared with a harmonic potential by writing its power series expansion in s.

$$V^{Morse}(s) = D_e \left[ \alpha^2 s^2 - \alpha^3 s^3 + \frac{7}{12} \alpha^4 s^4 - \frac{1}{4} \alpha^5 s^5 + \ldots \right]$$

The first term is the harmonic part. For $\alpha s < 1$, the typical situation, the terms diminish in size as the series continues, but even for $\alpha s \cong 1$, the terms diminish in importance, though *after* the cubic term. The cubic and quartic terms are the leading source of anharmonicity for near equilibrium displacements of a Morse oscillator, and in fact, it is very often the case that

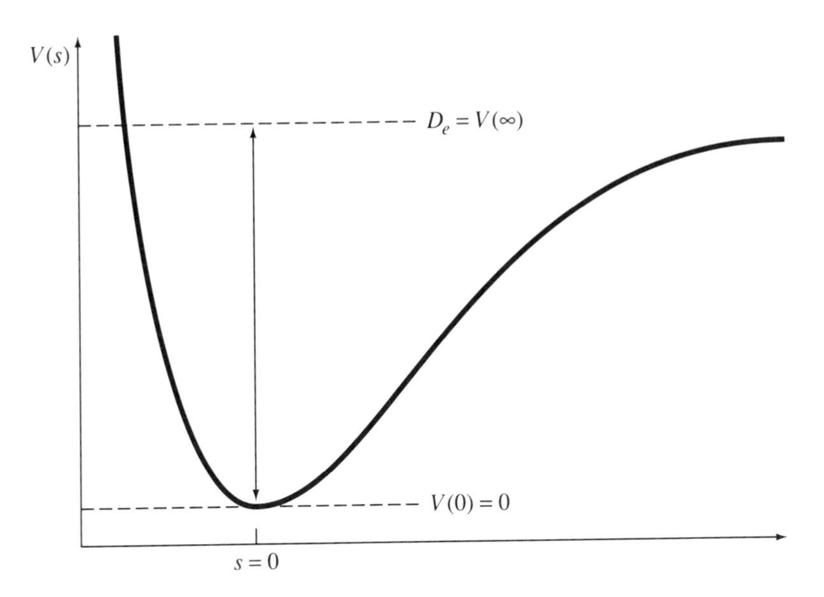


**FIGURE 6.3** The functional form of the Morse potential of Eq. (6-21).

a cubic potential function is the most important contributor to anharmonicity effects of real molecular potentials.

To incorporate anharmonicity effects in a description of a diatomic, a third order polynomial function is a starting point for the potential, and we may express this generally as

$$V(s) = \frac{1}{2}ks^2 + gs^3 \tag{6-22}$$

This is a starting point because higher order terms or other functional forms might be employed for still greater precision in representing a true potential. The cubic term, though, is often the most significant anharmonic element of a potential, as noted for the Morse potential. The cubic term (and any higher order terms) may be well treated as a perturbation of the harmonic oscillator. The first-order corrections for all states turn out to be zero. The second order corrections arising from the cubic term $gs^3$ are

$$E_n^{(2)} = -\frac{7}{16}\frac{g^2\hbar^2}{\mu k^2} - \frac{15}{4}\frac{g^2\hbar^2}{\mu k^2}\left(n+\frac{1}{2}\right)^2 \tag{6-23}$$

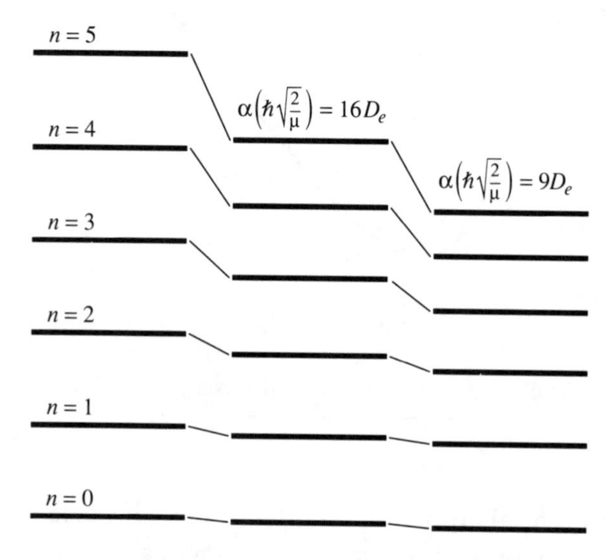


**FIGURE 6.4** A representative correlation diagram of several of the low-lying energy levels of a harmonic oscillator with the energy levels of an anharmonic oscillator. The anharmonic energy levels are those given by Eq. (6-23) assuming that the potential function is the power series expansion of a Morse potential truncated at the cubic term. The closer spacing of the levels with increasing vibrational quantum number is typical.

---

A cubic potential term will usually have a negative sign; that is, $g < 0$ in Eq. (6-22), and this was the case for the Morse oscillator. Regardless of the sign of g, the associated anharmonicity corrections of Eq. (6-23) are a lowering of the energy of each vibrational state relative to the harmonic picture. Furthermore, the extent of lowering increases with n. The energy levels of a typical anharmonic oscillator tend to become more closely spaced with increasing energy as illustrated in Fig. 6.4.

## 6.6 SELECTION RULES

The vibrational-rotational energy levels of a diatomic are only part of the information needed to understand and interpret spectra. The other part is the information about what particular transitions among the states may take place in an experiment. From time-dependent perturbation theory (Chap. 4), the allowed transitions are those among stationary states that are mixed by the perturbing Hamiltonian. If H' is a Hamiltonian corresponding to the interaction of the diatomic with electromagnetic radiation, then the stationary states (n,J) and (n',J') are mixed if the following is true.

$$< \Psi_{nJ} \mid H' \Psi_{n'J'} > \neq 0$$

If this integral is nonzero, the transition from one state to the other is allowed (written nJ $\longleftrightarrow$n'J' ). Otherwise, it is forbidden (written nJ$\not\longleftrightarrow$n'J' ).

The perturbing Hamiltonian in a basic spectroscopic experiment is the interaction with the electric field of the radiation. A uniform electric field interacts via the dipole moment of a charge distribution; the energy of interaction is the negative of the dot product of the field, $\vec{E}$, and the dipole. Before considering the electric field of light, let us consider a static, uniform electric field. One way such a field can be generated is by oppositely charging two parallel plates as shown in Fig. 6.5. A simple charge distribution that would possess or give rise to a dipole moment is an arrangement of a positive point charge and a negative charge, as in Fig. 6.5. This simple illustration shows that the orientation of the charge distribution with respect to the field will affect the interaction energy. At 0°, the interaction is energetically favorable, which means it takes on a negative value. At 180°, the interaction energy should be the same size, but positive in sign. At 90°, the interaction is zero. This type of orientational dependence is mathematically expressed as a dot product of two vectors. One vector points from one point charge to the other (the dipole moment vector), and the other vector points from one plate to the other (the electric field vector).

With radiation oscillating at a frequency $\omega_r$, the interaction Hamiltonian is

$$H' = -\vec{\mu} \bullet \vec{E} \cos(\omega_r t) \tag{6-24}$$

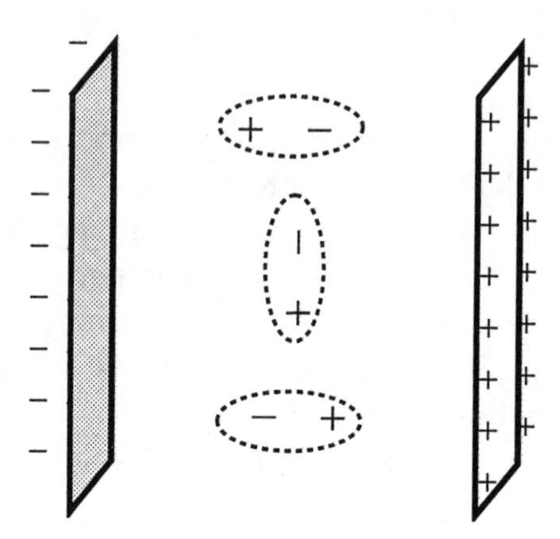

**FIGURE 6.5** An illustration of a dipole placed in a uniform electric field. The field develops between the two oppositely charged, parallel plates. A dipole moment is a property of a separated positive and negative charge, such as within the circled area. At the three orientations shown, there will be three different interaction energies because the positive end of the dipole is attracted toward the negatively charged plate and the negative end of the dipole is attracted toward the positively charged plate.

where $\vec{\mu}$ is the dipole moment vector. (Notice that $\mu$ is used both for dipoles and for the reduced mass of a two-body system, but the context should make it clear which $\mu$ is meant.) To treat electromagnetic radiation encountering molecules randomly oriented in space, we must assume an arbitrary or unspecified orientation. The orientation with respect to the molecular axis is the rotational angle, $\theta$, and since the dipole moment has a nonzero component only along the molecular axis, the dot product in Eq. (6-24) varies with the cosine of $\theta$. For a realistic picture of a molecule, we must allow for the dipole moment to vary with the geometry of the molecule. Were there a fixed distribution of charges, the dipole moment would be unchanging, but in a molecule, the vibrational excursions of the nuclei mean that the charge distribution is not fixed. For a diatomic molecule, then, the dipole moment is a function of the displacement, i.e., $\mu(s)$. Thus, the interaction Hamiltonian may be expressed as

$$H' = -|\vec{E}| \cos(\omega_r t)\, \mu(s) \cos\theta \qquad (6\text{-}25)$$

The selection rules for dipole transitions will be determined by integrals with this operator.

So far, the wavefunctions for a diatomic have been labelled by the quantum numbers n and J. The quantum number M, which gives the z-component of the rotational angular momentum, has been suppressed. At this point, we need to make explicit use of the wavefunctions, and since these are products of radial and spherical harmonic functions, we will now include the M quantum number:

$$\psi_{nJM}(s,\theta,\phi) = S_n(s)\, Y_{JM}(\theta,\phi)$$

The general integral expression needed for determining the selection rules is

$$<\psi_{nJM} | H' \psi_{n'J'M'}> \tag{6-26a}$$

$$= -|\vec{E}|\cos(\omega_r t) < S_n(s)\, Y_{JM}(\theta,\phi) \,|\, \mu(s) \cos\theta\, S_{n'}(s)\, Y_{J'M'}(\theta,\phi) >$$

The integral factors into integrals over the radial and angular coordinates.

$$<\psi_{nJM} | H' \psi_{n'J'M'}> \tag{6-26b}$$

$$= -|\vec{E}|\cos(\omega_r t) < S_n(s) \,|\, \mu(s)\, S_{n'}(s) > \, < Y_{JM}(\theta,\phi) \,|\, \cos\theta\, Y_{J'M'}(\theta,\phi) >$$

It is the integral over the radial coordinate that involves the dipole moment function.

The integral in Eq. (6-26b) over $\theta$ and $\phi$ may be evaluated for specific spherical harmonic functions by explicit integration. It is also possible to obtain the following result that is general for any choice of J, M, J', and M'.

$$< Y_{JM} | \cos\theta\; Y_{J'M'} > = \delta_{MM'} \sqrt{\delta_{J,J'+1} \frac{J^2 - M^2}{4J^2 - 1} + \delta_{J,J'-1} \frac{J'^2 - M^2}{4J'^2 - 1}} \tag{6-27}$$

[Eq. (6-27) is developed conveniently by using the recursion relations of the Legendre polynomials to replace $\cos\theta\, Y_{J'M'}$ by a sum of $Y_{J'+1,M'}$ and $Y_{J'-1,M'}$. The orthonormality of the spherical harmonic functions gives the result.] The immediate information from Eq. (6-27) is the selection rule that the J quantum number must change by one between the initial and final states of a dipole allowed transition. That is, the transition is allowed only if J = J'−1 or if J = J'+1, and that can be stated as $|J - J'| = 1$.

The integral over the radial coordinate in Eq. (6-26) can be evaluated in several ways. One particularly helpful approximation starts with a power series expansion of the dipole moment function, $\mu(s)$. The expansion is then truncated at some low order thought to be appropriate for the system at hand.

$$\mu(s) = \mu_e + s \left.\frac{d\mu}{ds}\right|_{s=0} + \frac{1}{2} s^2 \left.\frac{d^2\mu}{ds^2}\right|_{s=0} + \cdots \tag{6-28}$$

$\mu_e = \mu(0)$ is the dipole moment when the molecule is at its equilibrium length (s = 0). Thus,

$$< S_n(s) \mid \mu(s)\, S_{n'}(s) > \; = \; \mu_e < S_n(s) \mid S_{n'}(s) > + \left.\frac{d\mu}{ds}\right|_{s=0} < S_n(s) \mid s\, S_{n'}(s) >$$

$$+ \frac{1}{2} \left.\frac{d^2\mu}{ds^2}\right|_{s=0} < S_n(s) \mid s^2\, S_{n'}(s) > + \; \cdots$$

If the radial wavefunctions are those of the harmonic oscillator (that is, if the harmonic oscillator approximation of the potential is invoked), then explicit integration can be performed for any choice of n and n'. The result is the following expression (with m being used here for the reduced mass).

$$< S_n(s) \mid \mu(s)\, S_{n'}(s) > \; = \; \mu_e\, \delta_{nn'} + \left.\frac{d\mu}{ds}\right|_{s=0} \sqrt{\frac{\hbar}{2\sqrt{mk}}} \left( \sqrt{n}\, \delta_{n,n'+1} + \sqrt{n'}\, \delta_{n+1,n'} \right)$$

$$+ \frac{1}{2} \left.\frac{d^2\mu}{ds^2}\right|_{s=0} \frac{\hbar}{2\sqrt{mk}} \left( (2n+1)\, \delta_{nn'} + \sqrt{n(n-1)}\, \delta_{n,n'+2} + \sqrt{n'(n'-1)}\, \delta_{n+2,n'} \right) + \cdots \tag{6-29}$$

(In Sec. 6.10, we shall see that this may also be obtained immediately from matrix operations that avoid explicit integration.) The interpretation of Eq. (6-29) is made term by term. The first term indicates that transitions are allowed wherein n and n' are the same, which means where the vibrational quantum number does not change, but only if there is a nonzero equilibrium dipole moment ($\mu_e$). The next term indicates that

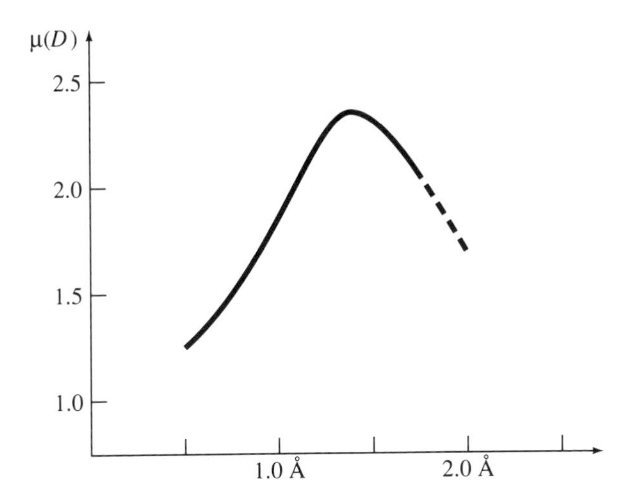


**FIGURE 6.6** The dipole moment function of the hydrogen fluoride molecule as a function of the separation distance between H and F in Å. This curve is based on spectroscopic data [R. N. Sileo and T. A. Cool, J. Chem. Phys. **65**, 117 (1976) and T. E. Gough, R. E. Miller and G. Scoles, Faraday. Discuss. Chem. Soc. **71**, 77 (1981)] with an extrapolation to the asymptotic value of zero at infinite separation. The equilibrium separation distance is at 0.92 Å, and in the vicinity of the equilibrium, the dipole moment curve is very nearly linear.

transitions will be allowed where the difference between n and n' is one, but only if there is a nonzero first derivative of the dipole moment function, $d\mu/ds$. Similarly, the next term depends on the second derivative of the dipole moment function. If it is not zero, transitions will be allowed for $n = n'$ and for a change of 2 in the n quantum number.

Fig. 6.6 shows the dipole moment function of the hydrogen fluoride molecule. It shows a very typical feature of diatomic molecules, which is that the dipole moment function of the molecule is very nearly linear in the displacement coordinate in the vicinity of the equilibrium. Of course, at large hydrogen-fluorine separations, there is significant curvature. Approximating the dipole moment function by truncating the expansion in Eq. (6-28) after the term that is linear in s will, therefore, be a very good approximation for the lower energy vibrational states. When this approximation is made and when the stretching potential is taken to be strictly harmonic, then only the first two right-hand-side terms in Eq. (6-29) are carried. This is called a *doubly harmonic approximation,* and it gives as the selection rule that the vibrational quantum number can be unchanged or can change by one; however, there must be a permanent equilibrium dipole for the first case and a *changing* dipole for the second case. A molecule such as $N_2$ has a zero dipole moment because by symmetry it is nonpolar. That symmetry remains whether the molecule is stretched or compressed, and so it does not have a changing dipole (i.e., $d\mu/ds = 0$).

For a typical polar diatomic molecule, we conclude that electromagnetic radiation will induce changes in the states of the system wherein the vibrational quantum number changes by zero or one and the rotational quantum number changes by one. Though other transitions may be seen, these are always expected. Vibrational transitions are said to be *fundamental* if the vibrational quantum number changes from zero to one. They are said to be *overtones* if the vibrational quantum number changes from 0 to 2, 3, and so on.

## 6.7 ANALYSIS OF INFRARED SPECTRA

At this point, we need to change our thinking somewhat. So far, the vibration and rotation of diatomics has been approached from basic quantum mechanical principles. We have developed energy level expressions for different types of potentials and different approximations, and we have determined which transitions are allowed. In effect, we have learned how to *predict* a spectrum given the molecule's potential, V(s). But the stretching potentials of diatomics are not known *a priori*. In fact, they are the very information that is to be extracted from experiment, not the other way around. Analyzing and interpreting spectra and extracting molecular information require matching predictions, based on quantum mechanics, with laboratory measurements.

The result of a spectroscopic experiment is a set of transition frequencies, as well as the intensities or strengths of the lines. As an example of the information that is available from an experiment, Fig. 6.7 shows an idealization of the high-resolution infrared spectrum of the hydrogen fluoride molecule. With modern instrumental techniques, the frequencies of the lines in this spectrum can be measured with a precision of about 0.01 $cm^{-1}$ and better. Each of the spikes or lines is at a frequency that corresponds to a particular transition between quantum states of HF. Since the frequency of radiation is proportional to the energy of the photons, or the energy absorbed by a molecule, then knowing the frequency of a transition is equivalent to knowing the energy of the transition. In fact, wavenumbers or $cm^{-1}$ serve as units of measure for both frequency and energy (see Appendix V).

The length or height of each spike or transition line is related to the intensity of that transition, which is to say that the depletion of incident radiation at a given frequency depends on the likelihood that the sample will absorb at that frequency. The intensities and frequencies are the primary data from the spectrum. The pattern of the lines appears to be regular or organized, but it needs to be deciphered or analyzed in order to make use of the data. We will first consider the analysis of the frequencies of the lines and then the intensities.

The energy of a transition is the difference between the energies of an initial state and a final state. Therefore, instead of an expression for the energies of the vibrational-rotational states of a diatomic, we need an expression for energy differences. This can be accomplished by combining the information in an energy level expression with that in the selection rules. The result will be a transition energy expression, and it will contain the information about where in the spectrum each of the allowed transitions should be found.

To start with the simplest level of treatment, let us use the energy level expression of Eq. (6-13) and the doubly harmonic selection rules. These rules say that upon going from some initial n,J state to another state by absorption of electromagnetic radiation,

i. J must increase or decrease by one, *and*

ii. n must increase by one.

This presents two situations, one where J increases and one where J decreases. It is easiest to obtain two transition energy expressions, one for each case. For the first situation, an energy difference expression is obtained by subtracting the initial state's energy, expressed according to Eq. (6-13), from the final state's energy:

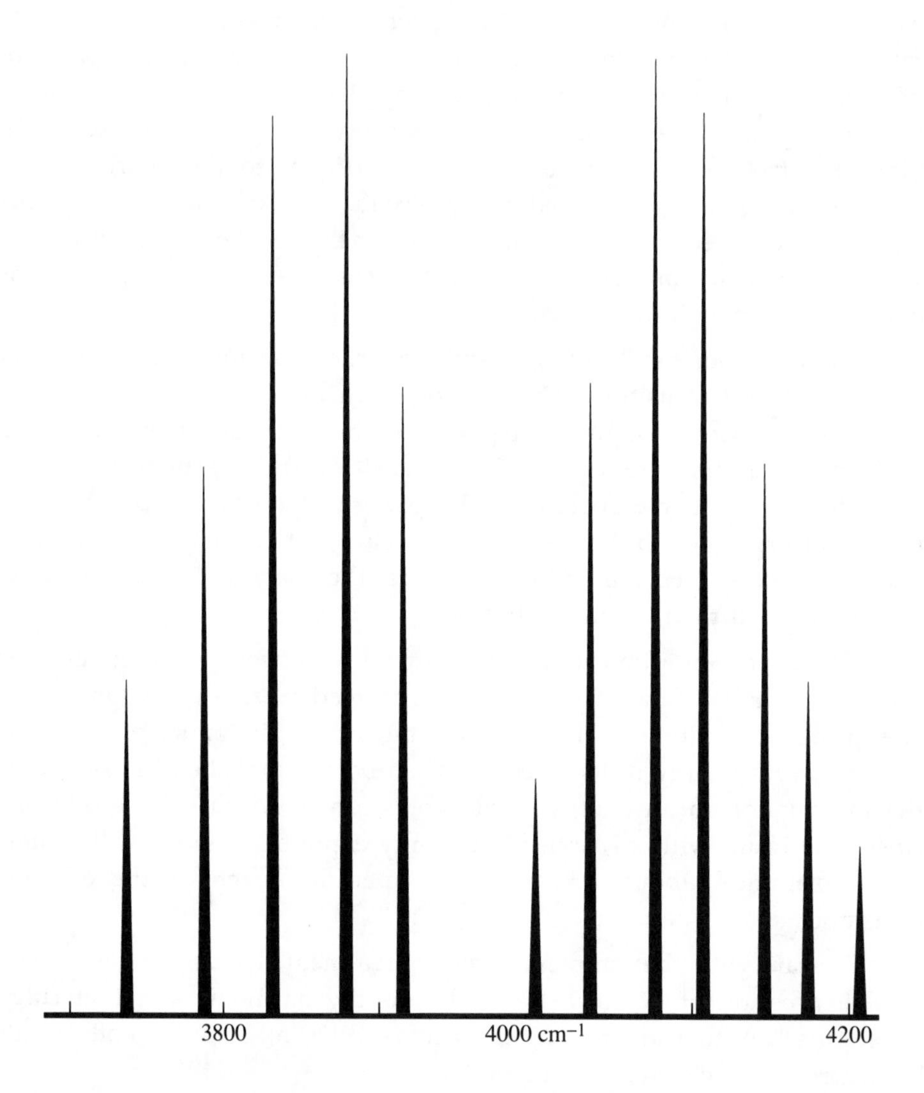


**FIGURE 6.7** An idealized high-resolution spectrum (i.e., a spectrum re-drawn to suppress instrumental noise, etc.) of a low-density gas phase sample of hydrogen fluoride. The frequencies of the peaks are the transition frequencies, and the relative heights of the peaks correspond to the intensities. This drawing shows the characteristic form of an infrared spectrum of a diatomic molecule: two sets of lines, called branches, with diminishing peak height to the left and to the right. The branch on the left, at lower transition frequencies, is the P-branch whereas the branch on the right is the R-branch. The separation between the two branches is roughly twice the separation between lines within the branches.

$$E_{n+1,J+1} = (n + \frac{3}{2})\hbar\omega + \frac{\hbar^2 (J+1)(J+2)}{2\mu r_e^2}$$

minus: $$E_{n,J} = (n + \frac{1}{2})\hbar\omega + \frac{\hbar^2 J(J+1)}{2\mu r_e^2}$$

$$\Delta E_{n,J\rightarrow n+1,J+1} = \hbar\omega + \frac{\hbar^2}{2\mu r_e^2}(2J + 2) \qquad (6\text{-}30)$$

This is an energy difference for the allowed transition where the initial state's J quantum number increases by one.

The other situation is that where the J quantum number decreases by one. This requires a separate energy difference expression:

$$E_{n+1,J-1} = (n + \frac{3}{2})\hbar\omega + \frac{\hbar^2 (J-1)J}{2\mu r_e^2}$$

minus: $$E_{n,J} = (n + \frac{1}{2})\hbar\omega + \frac{\hbar^2 J(J+1)}{2\mu r_e^2}$$

$$\Delta E_{n,J\rightarrow n+1,J-1} = \hbar\omega - \frac{\hbar^2}{2\mu r_e^2}2J \qquad (6\text{-}31)$$

These two $\Delta E$ expressions should be regarded as predictions of the positions of transition lines that may be observed in the spectrum, at least according to the particular level of treatment that was used for the state energies.

In both Eqs. (6-30) and (6-31), the same set of constants shows up in the second term. Collecting them into one constant, B, called the *rotational constant*, makes it simpler to write the expressions. Also, if the energy, E, the vibrational frequency, $\omega$, and the rotational constant, B, are all expressed in $cm^{-1}$, then these expressions become quite concise.

$$\Delta E_{n,J \to n+1,J+1} \text{ (cm}^{-1}) = \omega + 2BJ(J+1) \tag{6-32}$$

$$\Delta E_{n,J \to n+1,J-1} \text{ (cm}^{-1}) = \omega - 2BJ \tag{6-33}$$

B is much smaller than ω, and so the transitions in the spectrum will all be clustered in the vicinity of the frequency ω. That is the first feature to identify in a vibrational spectrum. In the spectrum in Fig. 6.7, the middle of the collection of lines is at about 3961 $cm^{-1}$, and so that gives a fair idea of the value of ω.

The next thing to do with the spectrum is to *assign* individual lines. Assignment of a spectrum means that the transition lines have been associated with particular initial and final states. For a vibrating-rotating diatomic, this means identifying the n,J quantum numbers for the initial state and the n',J' quantum numbers for the final state for each line. An ordered list of transition energies, generated with Eqs. (6-32) and (6-33), reveals how to make these assignments. We start with Eq. (6-33) because it gives frequencies that are all below those given by Eq. (6-32); it must give the leftmost line seen in the spectrum. We pick out a value of J and work up in frequency until reaching J = 1, which is where the highest frequency possible with Eq. (6-33) is reached. Then, we use Eq. (6-32) and start with J = 0, the lowest frequency that this equation can give, and then work up.

| J (initial) → | J' (final) | ΔE | Δ(ΔE) |
|---|---|---|---|
| 4 | 3 | $\omega - 8B$ | |
| | | | 2B |
| 3 | 2 | $\omega - 6B$ | |
| | | | 2B |
| 2 | 1 | $\omega - 4B$ | |
| | | | 2B |
| 1 | 0 | $\omega - 2B$ | |
| | | | **4B** |
| 0 | 1 | $\omega + 2B$ | |
| | | | 2B |
| 1 | 2 | $\omega + 4B$ | |
| | | | 2B |
| 2 | 3 | $\omega + 6B$ | |
| | | | 2B |
| 3 | 4 | $\omega + 8B$ | |

This table shows that there will be a number of transition frequencies in the vicinity of ω, but offset by some multiple of 2B.

Δ(ΔE) are the differences between transition energies. They tell how far apart adjacent spectral lines are. The pattern here is one of lines separated by 2B, except at one crucial point where the separation is twice as much, 4B. Examining Fig. 6.7 shows that in the middle of the cluster of lines, there is a point where the separation is about twice as big as otherwise seen. Thus, the line to the left (lower frequency) of this separation is assigned to be the J = 1 to J' = 0 transition (with n changing by one, as well). The line to the right, then, corresponds to an initial state of J = 0 and a final state where J' = 1. The remaining lines follow in the sequence in the table.

A confirmation of the assignment comes from the relative intensities. Notice that the strengths of the transitions, or the heights of the lines in the spectrum, diminish to the right and to the left. That is, the transitions with the highest initial J quantum number are weakest. The primary reason for this has to do with the *population* of the different quantum states, or the number of molecules in each of the states. The *Maxwell-Boltzmann distribution law,* which is stated without development or proof, is that for a sample of N molecules in thermal equilibrium at a temperature T, where the molecules may exist in stationary states with energies $E_i$ (for each of the $i^{th}$ states), the number of molecules in each state, $N_i$, is

$$N_i = N \frac{e^{-E_i/kT}}{\sum_j e^{-E_j/kT}} \tag{6-34}$$

where k is the *Boltzmann constant.* The denominator in Eq. (6-34) is called the partition function, and in this expression, it serves to ensure that $\Sigma N_i = N$. The Maxwell-Boltzmann law says that the number of molecules in a state will diminish exponentially with the energy of the state for some given temperature.

If we use the vibrational-rotational energy level expression of Eq. (6-13), the population of some state with quantum numbers n, J, and $M_J$, relative to the population of the state with the same vibrational quantum number but with J = 0, is,

$$\frac{N_{n,J,M_J}}{N_{n,0,0}} = \frac{e^{-E_{nJ}/kT}}{e^{-E_{n0}/kT}} = e^{-\hbar^2 J(J+1)/(2\mu r_e^2 kT)} \tag{6-35}$$

So, as J becomes larger, the population diminishes relative to the population of the lowest J state (J = 0). We have already realized that transitions may be observed with any initial J quantum number, but because there are different numbers of molecules in each J state, the number of transitions originating from a given J state must follow proportionately.

One further point is that the transition lines do not distinguish the $M_J$ quantum number. For any initial nJ energy level, transitions may occur from each and every one of the $M_J$ states, and there are (2J+1) of these. This means that a transition strength will depend on the population of an nJ energy level, rather than on the population of a single state. The population of an energy level must be the sum over the populations of all the states of that energy. But according to Eq. (6-34), the populations of states of the same energy are identical. So, the population of the level is simply the population of any one state times the number of degenerate states, which is a factor called the degeneracy. For vibrational-rotational states, this factor, often designated $g_J$, is simply (2J+1). Therefore, the relative intensities, I, of the lines in a high-resolution vibrational absorption band, as in Fig. 6.7, will depend on the initial state's J quantum number,

$$I_J \propto (2J+1)\, e^{-\chi J (J+1)} \quad \text{where } \chi = \hbar^2 / (2 \mu r_e^2 k T) \qquad (6\text{-}36)$$

Fig. 6.8 is a plot of this function for different values of $\chi$, treating J as a continuous variable. Notice that a curve drawn through the tops of the transition lines in Fig. 6.7 would have the form of a curve in Fig. 6.8. Indeed, at low resolution, where the individual lines overlap or are not resolved, the characteristic infrared absorption spectrum looks like the curves in Fig. 6.8.

## 6.8 SPECTROSCOPIC CONSTANTS FROM VIBRATIONAL SPECTRA

The quantum mechanical analysis of a vibrating-rotating diatomic molecule has provided an energy level expression, for each of the various

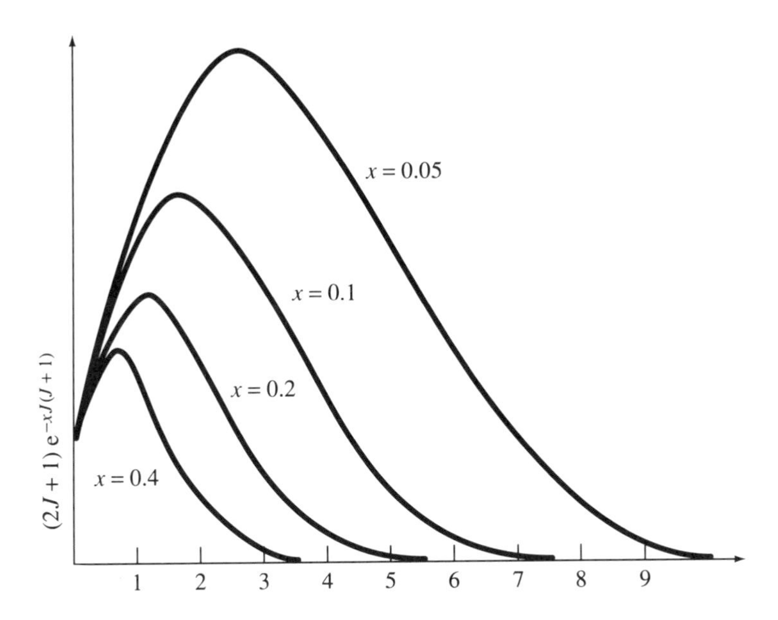


**FIGURE 6.8** A plot of the relative transition intensities of the rotational lines in a diatomic molecule's vibrational absorption band. The curves are plots of the quantity $(2J + 1) \exp[-\chi J(J + 1)]$ with J treated as a continuous variable. The numerical values of $\chi$ are shown next to each curve. Notice that as the temperature of a sample increases, $\chi$ becomes smaller.

levels of approximation, where the energy depends on the quantum number n and on the quantum number J. The J dependence is always expressed as the quantity $J(J+1)$, and the n dependence can always be expressed in terms of $(n+1/2)$. An overview of Eqs. (6-13,17,20,23), the various energy level expressions, reveals that they all take the form of an expansion of the energy in terms of polynomials* in $(n+1/2)$ and $J(J+1)$. That is, the energy of an nJ level may have the following general form.

* It is helpful to regard $(n+1/2)$ as a single variable, e.g., x, and $J(J+1)$ as another independent variable, e.g., y. The polynomial expansion is then just a standard type, $E(x,y) = c_{00} + c_{10}x + c_{01}y + c_{11}xy + c_{20}x^2 + c_{02}y^2 + \ldots$ .

$$E_{nJ} = \sum_{i=0}^{\infty} \sum_{k=0}^{\infty} c_{ik} \left(n + \frac{1}{2}\right)^i [J(J+1)]^k \qquad (6\text{-}37)$$

The various approximations that have been discussed lead to certain truncations of this expansion and to specific values of the constants, $c_{ik}$. For instance, we may say that Eq. (6-13) is of the form of Eq. (6-37) with $c_{10} = \hbar\omega$, $c_{01} = \hbar^2/(2\mu r_e^2)$, and all other $c_{ik}$'s equal to zero.

Transition energy expressions may be developed for any desired truncation of Eq. (6-37) by means of the differencing procedures in the prior sections. These will give $\Delta E$ for the transition lines in terms of the $c_{ik}$'s. Analysis of spectra may then be carried out by first assigning lines, and then by fitting the measured transition frequencies (energies) to the $\Delta E$ expression being employed; that is, the $c_{ik}$'s are adjusted* until the $\Delta E$ expression is correct or very nearly correct for all the observed transition lines. Notice how this has taken the whole approach one step away from the details of the quantum mechanical analysis: The measured transition energies or frequencies must fit a simple polynomial expression, Eq. (6-37), and the main task is to find the right $c_{ik}$'s. At that stage, we look back at the quantum mechanical analysis and relate the $c_{ik}$'s to various kinds of molecular information, such as the bond length and the stretching force constant.

There is well-established terminology and notation for the most important $c_{ik}$'s of Eq. (6-37), and some of this has been introduced already. The $c_{ik}$'s are spectroscopic constants; that is, they are values deduced or extracted from measurement of spectra. The names and designations for the most important ones are the following. The corresponding $c_{ik}$'s are shown in brackets; in some cases, the conventional definition of the spectroscopic constants introduces a negative sign, and so the values may turn out to correspond to $-c_{ik}$'s.

* The usual process for this adjustment and fitting is *least squares fitting*. In this process, values of the expansion coefficients are obtained for which the average for all lines (data points) of the square of the error between the measured value and the function value (from the $\Delta E$ expression) is minimized. Least squares analysis is a routine computational procedure that "fits" a chosen function to a set of data by means of adjusting coefficients or parameters contained in the function.

| | |
|---|---|
| $\omega_e$ | harmonic (equilibrium) vibrational frequency $[c_{10}]$ |
| $\omega_e\chi_e$ | equilibrium anharmonicity; this product is considered to be one value $[-c_{20}]$ |
| $\omega_e y_e$ | equilibrium second anharmonicity constant; this product is considered to be one value $[c_{30}]$ |
| $B_e$ | equilibrium rotational constant $[c_{01}]$ |
| $D_e$ | equilibrium centrifugal distortion constant* $[-c_{02}]$ |
| $H_e$ | equilibrium second centrifugal distortion constant $[c_{03}]$ |
| $\alpha_e$ | vibration-rotation coupling constant $[-c_{11}]$ |
| $\gamma_e$ | $[c_{21}]$ |
| $\beta_e$ | $[c_{12}]$ |

From these, the mean values of the rotational and centrifugal distortion constants are defined to be specific for individual vibrational levels. This means that these new constants depend on the vibrational quantum number n.

$$B_n = B_e - \alpha_e(n + \frac{1}{2}) + \gamma_e(n + \frac{1}{2})^2 + \dots \qquad (6\text{-}38)$$

$$D_n = D_e + \beta_e(n + \frac{1}{2}) + \dots \qquad (6\text{-}39)$$

There are two reasons why one may wish to use the mean values of B and D. First, the transition frequencies are more immediately related, as will be shown, and second, it is a way to regard vibrational motion as affecting the rotational and centrifugal distortion constants.

The most common system of units for reporting infrared spectroscopic constants is that of wavenumbers or $cm^{-1}$ (see Appendix V). A tilde is sometimes used to emphasize that these are the units. So, the vibrational frequency in wavenumbers, $\tilde{\omega}_e$, is related to $\omega_e$ in $sec^{-1}$ by a factor of $2\pi c$. (Of course, as discussed in Appendix V, we may freely state $\omega$ and other constants in $cm^{-1}$, $sec^{-1}$, ergs, and other units, and as long as the units are clear from the context, a special symbol is not really necessary.) Wavenumbers are suitable as an energy scale, and consequently, we may write an energy expression such that the energy and every constant are

* The symbol $D_e$ is often used for dissociation energy. The context in which one finds this symbol will usually show whether it is an energy or a centrifugal distortion constant.

taken to be in wavenumbers. So, with the named spectroscopic constants in place of the $c_{ik}$'s in Eq. (6-37), we have

$$\tilde{E}_{nJ} = \tilde{c}_{00} + \tilde{\omega}_e (n + \frac{1}{2}) - \tilde{\omega}_e\tilde{\chi}_e (n + \frac{1}{2})^2 + \tilde{\omega}_e\tilde{y}_e (n + \frac{1}{2})^3$$

$$+ \tilde{B}_e J(J+1) - \tilde{\alpha}_e J(J+1) (n + \frac{1}{2}) + \tilde{\gamma}_e J(J+1) (n + \frac{1}{2})^2$$

$$- \tilde{D}_e [J(J+1)]^2 - \tilde{\beta}_e [J(J+1)]^2 (n + \frac{1}{2})$$

$$+ \tilde{H}_e [J(J+1)]^3 + \ldots \tag{6-40a}$$

The mean values of Eqs. (6-38) and (6-39) make this expression more concise:

$$\tilde{E}_{nJ} = \tilde{c}_{00} + \tilde{\omega}_e (n + \frac{1}{2}) - \tilde{\omega}_e\tilde{\chi}_e (n + \frac{1}{2})^2 + \tilde{\omega}_e\tilde{y}_e (n + \frac{1}{2})^3$$

$$+ \tilde{B}_n J(J+1) - \tilde{D}_n [J(J+1)]^2$$

$$+ \tilde{H}_e [J(J+1)]^3 + \ldots \tag{6-40b}$$

The value $\tilde{c}_{00}$ cannot be obtained by measurement of transition frequencies because it will not show up in an energy difference ($\Delta E$) equation. It contributes, usually in a small way, to the zero point energy of the system.

In practice, many terms in the energy level expression are small. The ones not explicitly presented in Eq. (6-40) (i.e., those represented by ...) are rarely determined and are very unlikely to be sizable, except in unusual cases. But of the remaining terms, certain approximations prove to have little effect in analyzing and assigning lines in most diatomic spectra. These approximations are

$$\tilde{\omega}_e\tilde{y}_e \cong 0$$

$$\tilde{D}_n \cong \tilde{D}_e$$

$$\tilde{H}_e \cong 0$$

$$\tilde{B}_n \cong \tilde{B}_e - \tilde{\alpha}_e (n + \frac{1}{2})$$

When they are used with Eq. (6-40), the resulting energy level expression is rather simple, and the process of fitting transition frequencies may require only little computation. It is important to realize, though, that the truncation of terms in Eq. (6-40) may have an effect on the spectroscopic constants that are extracted from the transition frequency data. For instance, if one assumes $\tilde{\omega}_e \tilde{y}_e \cong 0$, a different value of $\tilde{\omega}_e \tilde{\chi}_e$ might be obtained than if this assumption were not made. Table 6.1 lists values of constants that have been measured for a number of diatomic molecules. The list shows that vibrational frequencies tend to range from hundreds of wavenumbers to several thousand. Anharmonicity constants show a range of almost two orders of magnitude.

There is an equivalent set of vibrational constants sometimes presented from spectroscopic analysis. They are the following.

$$\omega_0 = \omega_e - \omega_e \chi_e + \frac{3}{4} \omega_e y_e + \ldots \qquad (6\text{-}41a)$$

$$\omega_0 \chi_0 = \omega_e \chi_e - \frac{3}{2} \omega_e y_e + \ldots \qquad (6\text{-}41b)$$

$$\omega_0 y_0 = \omega_e y_e + \ldots \qquad (6\text{-}41c)$$

If these are used in Eq. (6-40b), the following energy level expression results.

$$\tilde{E}_{nJ} = \tilde{E}_{00} + \tilde{\omega}_0 n - \tilde{\omega}_0 \tilde{\chi}_0 n^2 + \tilde{\omega}_0 \tilde{y}_0 n^3$$

$$+ \tilde{B}_n J(J+1) - \tilde{D}_n [J(J+1)]^2 + \ldots \qquad (6\text{-}42)$$

The point of using these forms is to put the energy level expression in a form where the energies are given relative to the energy of the ground state $\tilde{E}_{00}$.

TABLE 6.1 Spectroscopic Constants (in $cm^{-1}$) of Certain Diatomic Molecules. [a]

| | $\omega_e$ | $B_e$ | $\omega_e\chi_e$ | $\alpha_e$ |
|---|---|---|---|---|
| BN | 1514.6 | 1.666 | 12.3 | 0.025 |
| BO | 1885.4 | 1.7803 | 11.77 | 0.0165 |
| BaO | 669.8 | 0.3126 | 2.05 | 0.0014 |
| BeO | 1487.3 | 1.6510 | 11.83 | 0.0190 |
| CH | 2861.6 | 14.457 | 64.3 | 0.534 |
| CD | 2101.0 | 7.808 | 34.7 | 0.212 |
| CN | 2068.7 | 1.8996 | 13.14 | 0.0174 |
| CO | 2170.2 | 1.9313 | 13.46 | 0.0175 |
| CaH | 1299 | 4.2778 | 19.5 | 0.0963 |
| HBr | 2649.7 | 8.473 | 45.21 | 0.226 |
| HCl | 2989.7 | 10.59 | 52.05 | 0.3019 |
| HF | 4138.5 | 20.939 | 90.07 | 0.770 |
| DF | 2998.3 | 11.007 | 45.71 | 0.293 |
| HgH | 1387.1 | 5.549 | 83.01 | 0.312 |
| LiH | 1405.6 | 7.5131 | 23.20 | 0.2132 |
| MgH | 1495.7 | 5.818 | 31.5 | 0.1668 |
| MgO | 785.1 | 0.5743 | 5.1 | 0.0050 |
| NO | 1904 | 1.7046 | 13.97 | 0.0178 |
| OH | 3735.2 | 18.871 | 82.81 | 0.714 |
| OD | 2720.9 | 10.01 | 44.2 | 0.29 |
| SiN | 1151.7 | 0.7310 | 6.560 | 0.0057 |
| SiO | 1242.0 | 0.7263 | 6.047 | 0.0049 |

[a] These values have been collected in Ref. 6 in the bibliography for this chapter. The values are for the most abundant isotopes, except for the deuterated molecules CD, DF, and OD.

## 6.9 ANALYSIS OF MICROWAVE SPECTRA

Analysis of microwave spectra of diatomics may be simpler than the analysis of vibrational spectra. The energy of microwave radiation is so much less than the photon energies of infrared radiation that the vibrational quantum number will not change in the course of a transition. Of course, with no change in n, there is the requirement that the molecule have a permanent dipole moment for any transitions to be allowed. Then, the selection rule is simply that the allowed transitions are those for which J increases by one. Instead of a band of transitions arising from the populated rotational (J) levels, a single transition line is usually observed. The appropriate transition energy expression is obtained by taking differences using the energy level expression for the vibrational-rotational states of a diatomic. However, because the n quantum number is unchanging, and since $\Delta J = +1$, the transition energies are rather simple.

A $J = 0$ to $J = 1$ transition will be observed at a frequency of about 2B, whereas the next transition, $J = 1$ to $J = 2$, will be at 4B, which is twice the frequency. In the microwave and radiofrequency region of the electromagnetic spectrum, a doubling of the frequency may require hardware alterations. So, only one transition is usually seen for a particular equipment setup. (Of course, for heavy diatomic molecules, the rotational constant, B, is small, and then it may turn out that several transitions, such as $J = 4$ to 5, $J = 5$ to 6, and $J = 6$ to 7, are close enough in frequency as to be observable in one scan.)

It is characteristic of the technology of microwave spectroscopy that frequencies are measurable to very high precision. Until the introduction of infrared lasers, microwave spectroscopy far outran vibrational spectroscopy in the precision and accuracy of spectral measurements. The main information obtained from a microwave spectrum is the rotational constant, and so the precision available with this type of experiment means that precise values of the rotational constant are obtained. This, in turn, implies that very precise values of the bond length of a diatomic can be deduced from a microwave spectrum. As discussed in the next chapter, the analysis of microwave spectra of polyatomic molecules tends to be more challenging.

It is an interesting fact that rotational transitions of molecules can be observed for molecules in interstellar space. In this case, the transitions are

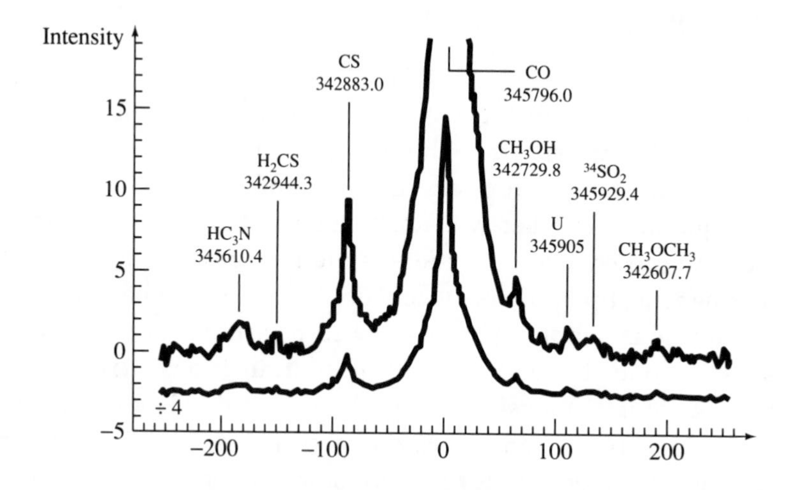


**FIGURE 6.9** Emission line spectrum of Orion A showing the existence in space of CO, CS, and several polyatomic molecules. Transition frequencies are given in MHz, and the spectrum is shown twice, once at full scale and once at 1/4 the vertical scale. [Reprinted with permission from the report by P. R. Jewell, J. M. Hollis, F. J. Lovas and L. E. Snyder, *Astrophys. J. Suppl. Series* **70**, 833 (1989).]

from higher energy states to lower energy states, and so they involve emission of a photon instead of absorption. The selection rules are the same, and rotational emission spectra match absorption spectra. In large interstellar clouds, molecules may be rotationally excited by collisions with other atoms or molecules and by other processes. They may then emit photons as they de-excite to a lower state. Transitions that occur at radiofrequencies are observed on Earth by huge antennas called radiotelescopes. The measurement precision in this region of the spectrum means that molecules are readily "fingerprinted" by their rotational transition frequencies, and so radiotelescopes can be employed to detect the existence of molecules in space. That is, by operating a radiotelescope at the specific transition frequency of a particular molecule, one can determine if those molecules exist in the region of space on which the telescope is focused. In fact, the signal strength can be used to determine the relative abundance of molecules, and maps of various interstellar clouds have been constructed that show the varying relative amounts of certain species. This is not limited to diatomic molecules, though so far many of the molecules that have been found to exist in interstellar clouds are diatomics. Fig. 6.9

shows the telescope signal obtained over a very narrow frequency range, and the emission lines that were detected are identified with certain specific molecules. Radiotelescopes are an application of molecular spectroscopy on a very large scale, and the information from them is important in understanding many astrophysical processes.

## 6.10 DETAILED SECOND ORDER TREATMENT ✯

The treatment of anharmonicity may be continued beyond cubic terms, and the power series expansion of Eq. (6-12) may be continued further, as well. Generally, the higher order terms of either sort have small effects on the energy levels, though there are interesting exceptions. This makes perturbation theory an appropriate tool for interpreting spectroscopic results, and the most workable zero-order picture is the harmonic description. It is quite often the case that just second-order perturbation theory energies are sufficient because of how small the perturbations are.*

As a general way of treating both anharmonicity and rotation-vibration coupling, we start by considering a perturbing potential affecting a harmonic oscillator. A polynomial expansion of a perturbing potential, truncated at the quartic term, is

$$V^{pert}(s) = a + bs + cs^2 + gs^3 + fs^4 \tag{6-43}$$

The first- and second-order perturbation corrections to the energies of the harmonic oscillator can be worked out in terms of the constants, a, b, c, g, and f. Then, we may go back to an original stretching potential and to Eq. (6-12) to decide what these constants are.

The first order-corrections to the energies are

* Sometimes first- or second-order results are given as if they were exact or infinite order. In practice, this may seem so when higher order corrections are tiny, but formally, second and infinite order do not yield identical expressions.

$$E_n^{(1)} = < \psi_n^{(0)} \mid (a + bs + cs^2 + gs^3 + fs^4)\, \psi_n^{(0)} > \qquad (6\text{-}44)$$

$$= a + c < \psi_n^{(0)} \mid s^2\, \psi_n^{(0)} > + f < \psi_n^{(0)} \mid s^4\, \psi_n^{(0)} >$$

The odd powers in the potential yield identically zero integrals, and that leads to the simplification.

The second-order corrections to the energies are,

$$E_n^{(2)} = \sum_{m \neq n} \frac{< \psi_n^{(0)} \mid (a + bs + cs^2 + gs^3 + fs^4)\, \psi_m^{(0)} >^2}{E_n^{(0)} - E_m^{(0)}}$$

The denominator is some integer multiple of $\hbar\omega$ and the constant potential term does not contribute. Thus,

$$E_n^{(2)} = \sum_{m \neq n} - < \psi_n^{(0)} \mid (bs + cs^2 + gs^3 + fs^4)\, \psi_m^{(0)} >^2 / [(m-n)\hbar\omega] \qquad (6\text{-}45)$$

To evaluate the first- and second-order energy corrections, we need to evaluate integrals involving the harmonic oscillator wavefunctions and the operators, s, $s^2$, $s^3$, and $s^4$. These are the integrals that are arranged in the matrix representations of the operators.

The matrix representation of the operator s for the harmonic oscillator has $\mathbf{S}_{ij} = <\psi_i \mid s\, \psi_j>$ as the i-row, j-column element, and so from integrating,

$$\mathbf{S} = \sqrt{\frac{\hbar}{2\sqrt{\mu k}}} \begin{pmatrix} 0 & 1 & 0 & 0 & 0 & \cdots \\ 1 & 0 & \sqrt{2} & 0 & 0 & \cdots \\ 0 & \sqrt{2} & 0 & \sqrt{3} & 0 & \cdots \\ 0 & 0 & \sqrt{3} & 0 & \sqrt{4} & \cdots \\ 0 & 0 & 0 & \sqrt{4} & 0 & \cdots \\ & & \cdots & & & \end{pmatrix} \qquad (6\text{-}46)$$

Because the only nonzero elements are found one position off the diagonal of this matrix, it is possible to provide the same information about **S** by means of a general expression for the individual elements.

$$[\mathbf{S}]_{nm} = \sqrt{\frac{\hbar}{2\sqrt{\mu k}}} \left( \sqrt{n}\, \delta_{n,m+1} + \sqrt{m}\, \delta_{n+1,m} \right) \tag{6-47}$$

Here, the rows and columns are meant to be labelled by the oscillator's quantum numbers, which start at zero, not one.

The matrix representation of the operator $s^2$ may be found by explicit integration, element by element, or by another procedure that is essentially matrix multiplication. The zero-order wavefunctions constitute a complete set, and it is a property of a complete set that a special operator can be formed that acts on any function in the same way as the operation of multiplication by one.

$$\hat{1} = \sum_{k=0}^{\infty} | \psi_k^{(0)} > < \psi_k^{(0)} | \tag{6-48}$$

The operator on the right* may be freely inserted since it is equivalent to multiplication by one. Doing this in the expression for an element of the matrix representation of $s^2$ gives an important result.

$$[\mathbf{S}^2]_{nm} = < \psi_n^{(0)} | s^2 \psi_m^{(0)} > = < \psi_n^{(0)} | s \hat{1} s \psi_m^{(0)} >$$

$$= \sum_k < \psi_n^{(0)} | s \left( | \psi_k^{(0)} > < \psi_k^{(0)} | \right) s \psi_m^{(0)} >$$

$$= \sum_k < \psi_n^{(0)} | s \psi_k^{(0)} > < \psi_k^{(0)} | s \psi_m^{(0)} >$$

$$= \sum_k [\mathbf{S}]_{nk} [\mathbf{S}]_{km}$$

This establishes that the matrix representation of an operator that is a product of operators may be obtained from the corresponding product of the matrix representations of the operators. In this case, multiplication of the matrix **S** by itself produces the matrix representation of $s^2$.

* The product of two functions written as $| f > < g |$ is interpreted as $f(x) \int dx'\, g^*(x')$, so that $< h | f > < g | h >$ means $\int h^*(x) f(x)\, dx \int g^*(x') h(x')\, dx'$.

By matrix multiplication or by explicit integration, the matrix representations of the other operators needed for the first- and second-order corrections are found to be the following.

$$\mathbf{s}^2 = \frac{\hbar}{2\sqrt{\mu k}} \begin{pmatrix} 1 & 0 & \sqrt{2} & 0 & 0 & \ldots \\ 0 & 3 & 0 & \sqrt{6} & 0 & \ldots \\ \sqrt{2} & 0 & 5 & 0 & \sqrt{12} & \ldots \\ 0 & \sqrt{6} & 0 & 7 & 0 & \ldots \\ 0 & 0 & \sqrt{12} & 0 & 9 & \ldots \\ & & \ldots & & & \end{pmatrix} \tag{6-49}$$

Notice that the diagonal elements of this matrix are a simple sequence of the odd integers, 1, 3, 5, and so on. The only other nonzero elements are found to be arranged parallel to the diagonal. They follow the sequence $\sqrt{1 \cdot 2}, \sqrt{2 \cdot 3}, \sqrt{3 \cdot 4}, \sqrt{4 \cdot 5}$, and so on.

Multiplication of the $\mathbf{s}^2$ matrix by $\mathbf{s}$ leads to the next matrix representation.

$$\mathbf{s}^3 = \left(\frac{\hbar}{2\sqrt{\mu k}}\right)^{\frac{3}{2}} \begin{pmatrix} 0 & 3 & 0 & \sqrt{6} & 0 & 0 & \ldots \\ 3 & 0 & 6\sqrt{2} & 0 & \sqrt{24} & 0 & \ldots \\ 0 & 6\sqrt{2} & 0 & 9\sqrt{3} & 0 & \sqrt{60} & \ldots \\ \sqrt{6} & 0 & 9\sqrt{3} & 0 & 12\sqrt{4} & 0 & \ldots \\ 0 & \sqrt{24} & 0 & 12\sqrt{4} & 0 & 15\sqrt{5} & \ldots \\ 0 & 0 & \sqrt{60} & 0 & 15\sqrt{5} & 0 & \ldots \end{pmatrix} \tag{6-50}$$

The diagonal elements of this matrix are zero, which is in keeping with the fact that $s^3$ is an odd function. Parallel to the diagonal are nonzero elements that follow the sequence 3, $6\sqrt{2}, 9\sqrt{3}, 12\sqrt{4}, 15\sqrt{5}$, and so on. And two places over is another sequence, $\sqrt{3 \cdot 2 \cdot 1}, \sqrt{4 \cdot 3 \cdot 2}, \sqrt{5 \cdot 4 \cdot 3}, \sqrt{6 \cdot 5 \cdot 4}$, and so on. Thus, it is easy to continue this matrix representation to any size, and a concise expression analogous to Eq. (6-47) may be set up.

One further matrix multiplication leads to the next representation matrix.

$$s^4 = \frac{\hbar^2}{4\,\mu k} \begin{pmatrix} 3 & 0 & 6\sqrt{2} & 0 & \sqrt{24} & 0 & \cdots \\ 0 & 15 & 0 & 10\sqrt{6} & 0 & \sqrt{120} & \cdots \\ 6\sqrt{2} & 0 & 39 & 0 & 14\sqrt{12} & 0 & \cdots \\ 0 & 10\sqrt{6} & 0 & 75 & 0 & 18\sqrt{20} & \cdots \\ \sqrt{24} & 0 & 14\sqrt{12} & 0 & 123 & 0 & \cdots \\ 0 & \sqrt{120} & 0 & 18\sqrt{20} & 0 & 183 & \cdots \end{pmatrix} \tag{6-51}$$

The sequences of nonzero values on the diagonal and parallel to the diagonal may be expressed in the way that reveals how this matrix may be continued:

(i) $6 \bullet \frac{1}{4} + \frac{3}{2},\ 6 \bullet \frac{9}{4} + \frac{3}{2},\ 6 \bullet \frac{25}{4} + \frac{3}{2},\ 6 \bullet \frac{49}{4} + \frac{3}{2},\ \cdots$

(ii) $6\sqrt{2 \bullet 1},\ 10\sqrt{3 \bullet 2},\ 14\sqrt{4 \bullet 3},\ 18\sqrt{5 \bullet 4},\ 22\sqrt{6 \bullet 5},\ \ldots$

(iii) $\sqrt{4 \bullet 3 \bullet 2 \bullet 1},\ \sqrt{5 \bullet 4 \bullet 3 \bullet 2},\ \sqrt{6 \bullet 5 \bullet 4 \bullet 3},\ \sqrt{7 \bullet 6 \bullet 5 \bullet 4},\ \ldots$

Fig. 6.10 shows the location of nonzero elements in the four matrix representations. A regular pattern is also found for the matrix representations of $s^5$, $s^6$, and so on.

First order perturbation theory corrections are simply the diagonal elements of the matrix representation of the perturbation. Using the matrix representations with Eq. (6-44), this gives

$$E_n^{(1)} = a + c\frac{\hbar}{\sqrt{\mu k}}(n + \frac{1}{2}) + f\frac{\hbar^2}{4\mu k}\left[6\,(n + \frac{1}{2})^2 + \frac{3}{2}\right] \tag{6-52}$$

Because there are only a few off-diagonal elements in any row of the matrix representations that have been considered, the summation in Eq. (6-45) for the second-order energy corrections may be worked out completely. After some simplifications, the resulting expression for the second order energy corrections is

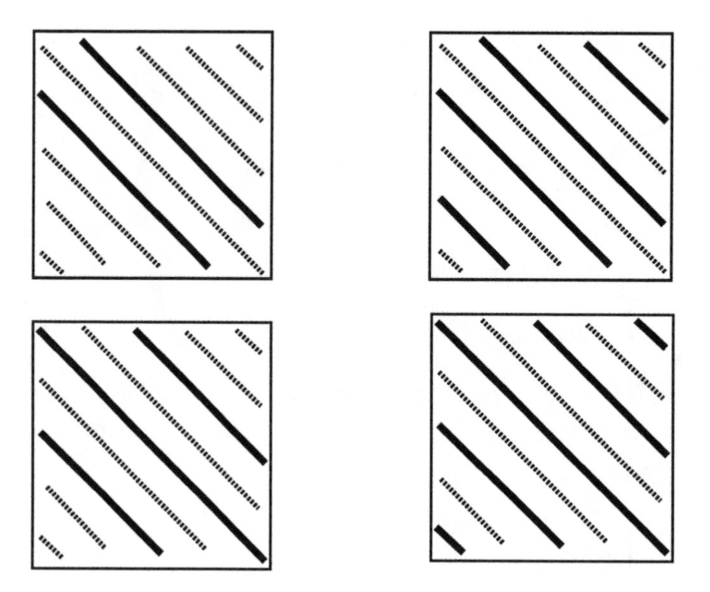

**FIGURE 6.10** A schematic diagram of the matrix representations of the operators s (upper left), $s^2$ (lower left), $s^3$ (upper right), and $s^4$ (lower right) of the harmonic oscillator. The dark, solid lines represent the positions of the only nonzero elements in the matrices.

$$E_n^{(2)} = -\frac{b^2}{2k} - \left(c^2 + \frac{3}{2}bg\right)\frac{\hbar}{2\mu k\omega}(n+\frac{1}{2}) \tag{6-53}$$

$$-\left(\frac{5}{4}g^2 + cf\right)\frac{3\hbar^2}{\mu k^2}(n+\frac{1}{2})^2 + \left(\frac{7}{4}g^2 - 3cf\right)\frac{\hbar^2}{4\mu k^2}$$

$$-\frac{f^2\hbar^2}{4\mu^2 k^2}\left(17(n+\frac{1}{2})^3 - 3(n+\frac{1}{2})^2 + \frac{79}{4}(n+\frac{1}{2}) - \frac{3}{4}\right)$$

The zero-order energies of the harmonic oscillator,

$$E_n^{(0)} = \hbar\omega(n+\frac{1}{2})$$

may be summed with the first- and second-order corrections to yield vibration-rotation state energy levels that are highly accurate, even though still approximate.

The constants a, b, c, f, and g need to be extracted from the effective potential. Thus, a, b, and c are the coefficients in Eq. (6-12).

$$a = \frac{J(J+1)\hbar^2}{2\mu r_e^2}$$

$$b = -\frac{J(J+1)\hbar^2}{\mu r_e^3}$$

$$c = \frac{3J(J+1)\hbar^2}{\mu r_e^4}$$

g and f are the cubic and quartic terms of the simple stretching potential, and they can include the cubic and quartic terms in the expansion of Eq. (6-12). In many situations, we may anticipate the relative sizes of terms and, on that basis, choose to neglect those of least importance. For instance, rarely, if ever, is the $f^2$ term in Eq. (6-43) kept. For similar reasons, it is usually not necessary to go beyond second order corrections.

## 6.11 A SPECIAL NUMERICAL TREATMENT *

The general Schrödinger equation for a vibrating-rotating two-body system can be solved exactly for any specified potential by converting the differential equation to a difference equation. The difference equation can be solved to any precision, and doing this with the aid of a computer is likely to be the most convenient. The particular treatment, which is often called the Numerov-Cooley method, is not general; it is only workable for a one-dimensional differential equation. But since it offers a way around making approximations, it is a powerful means for working with the radial Schrödinger equation of two-body systems. Furthermore, it serves to illustrate that alternative mathematical formulations can be effective in quantum mechanical problems.

Converting a differential equation to a difference equation is a standard mathematical procedure. In the case of the radial Schrödinger equation, the conversion means partitioning or discretizing the displacment coordinate, s, into uniform, small steps. The wavefunction is then found as a set of values for each of these steps rather than in analytical form. A wavefunction that is represented in this way, as a set of function values at specific coordinate values, is called a *numerical representation.* A numerical representation may be found for any function. As an example, consider the following list of coordinate and function values.

| COORDINATE VALUE, x | FUNCTION VALUE, f(x) |
|---|---|
| −3 | 9 |
| −2 | 4 |
| −1 | 1 |
| 0 | 0 |
| 1 | 1 |
| 2 | 4 |

A bit of examination suggests that this may be a representation of the function $f(x) = x^2$. Certainly, each of the points would lie on a graph of that function. However, the representation is not the same thing as the function. Perhaps if there were more points in the representation, there would be one that does not precisely lie on the graph of $f(x) = x^2$. So, a numerical representation only becomes faithful to the function it represents by having many closely spaced points. By making the steps in the discretized displacement coordinate, s, smaller and smaller, the representation of the vibrational wavefunction becomes more complete or more exact. For diatomic molecules, step sizes of a hundredth of an Å will usually yield eigenenergies that are accurate to 0.1 $cm^{-1}$ and better.

Following Cooley's presentation of the difference equation method for a radial Schrödinger equation, the step size will be designated h. Letting $s_0$ be the most negative displacement of interest (a close-in separation where the potential energy is so great that the wavefunction must have a negligibly small amplitude), then the steps are

$$s_i = s_0 + ih \qquad i = 1, 2, \ldots, N \tag{6-54}$$

At each step, we will obtain a value or amplitude for the wavefunction, and these will be designated $Y_i$.

$$Y_i \equiv \Psi(s_i) \tag{6-55}$$

The set of these amplitudes is the representation. The operations performed on the wavefunction in the Schrödinger equation included second differentiation with respect to the displacement coordinate. We will use these values at each of the steps,

$$Y_i^{(2)} \equiv \left.\frac{d^2\psi}{ds^2}\right|_{s_i} \tag{6-56}$$

Other derivatives will be designated similarly,

$$Y_i^{(k)} \equiv \psi^{(k)}(s_i) = \left.\frac{d^k\psi}{ds^k}\right|_{s_i} \tag{6-57}$$

And the potential energy at each step is given as

$$V_i = V(s_i) \tag{6-58}$$

The task is to solve for the values in the set $\{Y_i\}$.

A rather subtle step is required next. The wavefunction or any of its derivatives is written as a power series expansion about one of the step points.

$$\psi^{(k)}(s) = \psi^{(k)}(s_i) + (s - s_i)\,\psi^{(k+1)}(s_i) + \frac{1}{2}(s - s_i)^2\,\psi^{(k+2)}(s_i)$$

$$+ \frac{1}{6}(s - s_i)^3\,\psi^{(k+3)}(s_i) + \ldots \tag{6-59}$$

Next, the power series expansion is used to express the value of the wavefunction (k = 0) or any derivative of the wavefunction (k > 0) at the steps on the left and on the right of the point of the power series expansion, $s_i$. Then, these two values of the wavefunction are added together.

$$\psi^{(k)}(s_i+h) + \psi^{(k)}(s_i-h) = 2\left[\psi^{(k)}(s_i) + \frac{1}{2}h^2\psi^{(k+2)}(s_i) + \frac{1}{24}h^4\psi^{(k+4)}(s_i) + \ldots\right]$$

$$= 2\sum_{n=0}^{\infty}\frac{h^{2n}}{(2n)!}\psi^{(k+2n)}(s_i) \qquad (6\text{-}60)$$

Notice that the odd-power terms in the power series expansion cancel upon the addition of $\psi^{(k)}(s_i+h)$ and $\psi^{(k)}(s_i-h)$.

Eq. (6-60) can be simply written in terms of the Y amplitudes.

$$Y^{(k)}_{i+1} + Y^{(k)}_{i-1} = 2Y^{(k)}_i + h^2 Y^{(k+2)}_i + \frac{h^4}{12}Y^{(k+4)}_i + \ldots \qquad (6\text{-}61)$$

Since the step size, h, is intended to be small and would approach zero if we were to achieve a continuous numerical representation, we can employ truncations of Eq. (6-61) for terms that depend on h. The following two truncations will be used because it turns out the final solution is achieved more rapidly than by certain other truncations.

$$Y_{i+1} + Y_{i-1} = 2Y_i + h^2 Y^{(2)}_i + \frac{h^4}{12}Y^{(4)}_i \qquad (6\text{-}62)$$

$$Y^{(2)}_{i+1} + Y^{(2)}_{i-1} = 2Y^{(2)}_i + h^2 Y^{(4)}_i \qquad (6\text{-}63)$$

Now, the Schrödinger equation written in terms of the amplitudes at the steps is

$$-\frac{\hbar^2}{2\mu}Y^{(2)}_i + V_i Y_i = E Y_i \qquad (6\text{-}64)$$

where $\mu$ is the reduced mass. These last three equations are used together to solve the problem. The goal is to find an expression that yields a Y amplitude from information at one and two steps to the left (or right). Then, with some means of getting the two innermost points, one could obtain the third point, then the fourth, and so on, until all amplitudes are known.

If Eq. (6-63) is multiplied by 1/12 and then subtracted from Eq. (6-62), the terms that depend on $h^4$ are eliminated. The result is

$$Y_{i+1} + Y_{i-1} - 2Y_i = h^2 Y_i^{(2)} + \frac{1}{12}\left( Y_{i+1}^{(2)} + Y_{i-1}^{(2)} - 2\,Y_i^{(2)} \right) \tag{6-65}$$

It is convenient to combine certain amplitudes and designate the combinations as Z:

$$Z_i \equiv Y_i - \frac{1}{12} Y_i^{(2)} \tag{6-66}$$

Then, Eq. (6-65) is

$$Z_{i+1} + Z_{i-1} + 2\,Z_i = h^2 Y_i^{(2)}$$

$$= h^2 \frac{2\mu}{\hbar^2} (V_i - E)\, Y_i \tag{6-67}$$

where the Schrödinger equation, Eq. (6-64), has been used in the second line to substitute for $Y_i^{(2)}$.

Eq. (6-67) can be used to find $Z_{i+1}$ from the values $Z_{i-1}$ and $Z_i$, the amplitudes at two adjacent steps, though only if E, the energy eigenvalue, is already known. Let us assume that E is known, even though it is actually something we seek to find in this whole process. We will take the amplitude at the first point to be zero, which is the asymptotic limiting value.

$$Y_0 = Z_0 = 0$$

The next point will have an amplitude that is small. It can be freely chosen to be any small number. If it is chosen to be too large or too small, this will propagate from point to point in a way that is corrected later by normalizing the amplitudes. The small number chosen as the amplitude of the next point is $\varepsilon$, and so $Z_1 = \varepsilon$. To find Y amplitudes for this point and other points, Eq. (6-66) is used, but with $Y_i^{(2)}$ replaced using the Schrödinger equation, just as was done in arriving at Eq. (6-67).

$$Z_i = Y_i - \frac{h^2}{12}(V_i - E)\,Y_i \quad \text{or}$$

$$Y_i = Z_i \Big/ \left[1 - \frac{h^2}{12}(V_i - E)\right] \tag{6-68}$$

With $Z_0$, $Y_0$, $Z_1$, and $Y_1$, Eq. (6-67) is used to find $Z_2$, and then Eq. (6-68) is used to find $Y_2$. The process is performed again to find $Z_3$ and $Y_3$, and eventually to find the amplitudes at all the steps. After all the amplitudes are found, a normalization factor can be applied to each of them to make the representation correspond to a normalized vibrational wavefunction.

The process described assumes that the energy eigenvalue is known, but of course, this is not likely to be the case. So, instead of a true value for E, the process that has been described is carried out for a guess value of the energy eigenvalue, E. The resulting amplitudes will be approximate and not a true representation. However, it is possible to refine the guess iteratively until it approaches the true energy eigenvalue, and there are a number of ways of refining the guess energy value.

If the application of Eq. (6-67) is done starting from the left side (i.e., at $s_0$) and then repeated starting from the right side (i.e., at $s_N$), the amplitudes and the slope of the wavefunction will match, point for point, only if a true energy eigenvalue is used for E. This provides the test for whether or not the iterative refinement of E has yielded the correct value. Also, the size of the mismatch at some middle point when the value of E is only a guess can be a guide to how much to adjust the energy in the search for the true eigenenergy. By using different guess energies, the iterative process can be made to converge to different vibrational states. This special process is easily implemented for computers, and it is really a viable approach only with computers. It ends up being a procedure that can find the vibrational wavefunctions and energies for most any V(s) that is specified. Fig. 6.11 gives an example of the wavefunctions generated with this method in application to a realistic molecular problem. In this example, the potential is noticeably anharmonic. Even more complicated potentials, including those with multiple minima may be analyzed with this numerical technique.

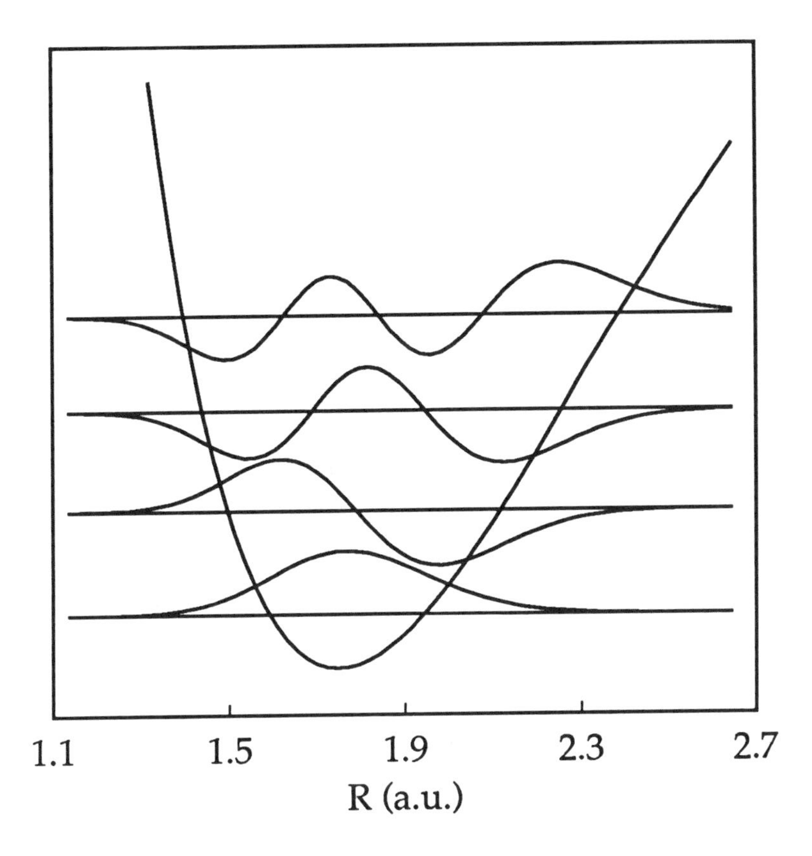


**FIGURE 6.11** Vibrational wavefunctions for the hydrogen fluoride molecule obtained from Numerov-Cooley treatment using a theoretical determination of the realistic (anharmonic) HF vibrational potential, V(s). The horizontal axis is in bohr radii (0.5218 Å).

## Exercises

1. There is a connection between the "exact" analysis of vibration-rotation coupling in a harmonic oscillator and the second-order perturbative treatment that yielded Eq. (6-20). To see this connection, find a power series expansion of $f(x) = \sqrt{a + x}$ about the point $x = 0$. Next use this expansion for the expression for $\omega_J$ in Sec. 6.4, with

$a = k/m$ and x being the entire second term. Substitute this expansion for $\omega$ in Eq. (6-17). At what order of truncation of the expansion of the exact treatment result is there a correspondence with the expression in Eq. (6-20)?

2. Apply perturbation theory in a basis of harmonic oscillator functions to arrive at Eq. (6-23).
3. Verify Eq. (6-27) by writing the product, $\cos\theta\, Y_{J'M'}$, as a linear combination of other spherical harmonic functions and then using the orthogonality of the spherical harmonics to evaluate the integral.
4. With Eq. (6-34), verify that $\Sigma N_i = N$.
5. The equilibrium bond length of LiH is 1.595 Å. Find the equilibrium rotational constant in $cm^{-1}$ for LiH and the isotopic forms, LiD and $^6LiH$.
6. Assume that it is possible to measure a vibrational transition frequency for a diatomic for which the J quantum number remains unchanged at zero. And assume that there is some diatomic for which the following transition frequencies (in $cm^{-1}$) are then obtained: 1600 for n = 0 to n = 1, 3100 for n = 0 to n = 2, and 4450 for n = 0 to n = 3. Find values for the vibrational frequency, the anharmonicity constant, and the second anharmonicity constant on the basis of these frequencies. Next, compare these values with those obtained for the vibrational frequency and the anharmonicity constant if it is assumed that the second anharmonicity constant, $\omega_e y_e$, is zero.
7. Spectroscopic constants of CO are given in Table 6.1. Find the equilibrium bond length. Then, predict the transition frequencies for the first several P- and R-branch lines in the fundamental absorption band of $^{13}C^{17}O$.
8. The fundamental infrared band of HCl is complicated because the natural isotopic abundance of chlorine means that transitions for two isotopes may be readily observed. That is, unless isotopically purified, the spectrum of an HCl sample is actually a superposition of the spectra of two isotopes of HCl. Which are they? For the first several P- and R-branch lines, calculate the separation in $cm^{-1}$ for the corresponding lines in the spectra of the two isotopic forms. Make a sketch of the expected form of the fundamental band spectrum.

9. Assume that the sample temperature for the spectrum in Fig. 6.7 is 100 K. By evaluating the changes in the relative intensities of the lines, make a sketch of the spectrum that would be expected if the sample temperature were 10 K, and then if it were 1000 K.

10. As a function of temperature, from 0 K to 1000 K, make a plot of the intensity ratio of the first P-branch line to the second line for the fundamental vibrational band of HF.

11. Using data in Table 6.1, calculate the band center frequency (i.e., the average of the frequency of the first P-branch and first R-branch lines) for the $n = 0 \to 1$, $n = 1 \to 2$, $n = 2 \to 3$ and $n = 3 \to 4$ vibrational bands of LiH.

12. Repeat the derivation of Eqs. (6-30) and (6-31) with the inclusion of an energy term, $D\,[\,J\,(\,J + 1\,)\,]^2$, associated with centrifugal distortion.

13. From the following general energy level expression for a diatomic,

$$\tilde{E}_{n,J} = (n + \frac{1}{2})\,\tilde{\omega} + \tilde{B}\,J\,(J+1) - \tilde{\alpha}\,(n + \frac{1}{2})\,J\,(J+1) - \tilde{D}\left[J\,(J+1)\right]^2$$

find the branch separation for the fundamental vibrational band in terms of $\tilde{\omega}$, $\tilde{B}$, $\tilde{\alpha}$ and $\tilde{D}$. That is, develop an expression for the difference in transition energies of the two transition lines closest to the band center.

14. Assume that the vibrational spectrum of LiH is well represented by the following energy level expression.

$$\tilde{E}_{n,J} = 1405.65\,(n + \frac{1}{2}) - 23.20\,(n + \frac{1}{2})^2 + 7.513\,J\,(J+1)$$

$$- 0.213\,(n + \frac{1}{2})\,J\,(J+1) - 0.01\left[J\,(J+1)\right]^2$$

Find the equilibrium bond length. Next, find the corresponding energy level expression for LiD, taking the potential energy function to be the same as for LiH.

15. Using the harmonic oscillator wavefunctions given in Chap. 3, confirm the values of the matrix elements, $[\mathbf{S}^2]_{02}$, $[\mathbf{S}^3]_{12}$, and $[\mathbf{S}^4]_{13}$, by explicit integration.

*16. Truncate the matrix representation of **S** in Eq. (6-49) to the upper $6 \times 6$ corner. Multiply this matrix by itself and compare with the matrix representation of $\mathbf{S}^2$.

*17. What is the maximum number of steps from the diagonal of the matrix representation $\mathbf{S}^6$ for the harmonic oscilllator for which elements may be non-zero?

## Bibliography

(See INTRODUCTORY and INTERMEDIATE LEVEL texts listed at the ends of Chaps. 3 and 4.)

1. G. W. King, *Spectroscopy and Molecular Structure* (Holt, Rinehart and Winston, New York, 1964). This provides a very comprehensive introduction to vibrational-rotational spectroscopy as well as other types of molecular spectroscopy.
2. G. M. Barrow, *Introduction to Molecular Spectroscopy* (McGraw-Hill, New York, 1962). This is a very useful text at an introductory to intermediate level.
3. W. S. Struve, *Fundamentals of Molecular Spectroscopy* (John Wiley and Sons, New York, 1989). This is an intermediate to advanced level text that will be helpful for further study.
4. J. M. Hollas, *Modern Spectroscopy* (John Wiley and Sons, New York, 1987). This is a valuable text on both the theory and laboratory techniques of spectroscopy.
5. G. Herzberg, *Molecular Spectra and Molecular Structure. I. Spectra of Diatomic Molecules* (Von Nostrand Reinhold, New York, 1950). This is a research level volume that may be consulted for further study on the spectroscopy of diatomics. It also includes a compilation of spectroscopic data.
6. J. W. Cooley, *Math. Comput.* **15**, 363 (1961). This is the original research report on Numerov-Cooley analysis.

# Chapter 7

# Polyatomic Vibrational and Rotational Spectra

*Interpretation of the vibrational and rotational spectra of polyatomic molecules requires more complicated analysis than for diatomic spectra. There are more vibrational degrees of freedom, and for nonlinear molecules there is one more rotational degree of freedom. Even so, characteristic features of diatomic spectra persist to some extent, and the quantum mechanical analysis is only somewhat more involved. The usual process of working through the mechanics for a model system, determining selection rules, and predicting spectra is followed for polyatomic problems in this chapter.*

## 7.1 ROTATIONAL SPECTROSCOPY OF LINEAR MOLECULES

The rotation of a rigid, linear triatomic or polyatomic molecule is mechanically equivalent to the rotation of a rigid diatomic. All are the rotations of an infinitesimally thin rod with point masses. So, the basic analysis needed to understand the rotational spectroscopy of linear polyatomic molecules follows that of diatomics at the outset. The difference amounts to a generalization of the moment of inertia when there are more than two atoms. From classical mechanics it is known that the moment of inertia, I, about the center of mass of a linear arrangement of point masses is

$$I = \sum_{i<j} m_i m_j r_{ij}^2 \Big/ \sum_i m_i \tag{7-1}$$

where $r_{ij}$ is the distance between the i and j atoms, $m_i$ is the mass of the $i^{th}$ atom, and the sum in the numerator is over all pairs of atoms in the molecule. The denominator, of course, equals the total mass of the molecule. For a diatomic molecule, Eq. (7-1) reduces to $m_1 m_2 r_{12}^2/(m_1 + m_2)$, and this is the same as $\mu r_e^2$, taking $r_e$ to be the same thing as $r_{12}$.

The rotational Schrödinger equation is no different than that for a diatomic because of the mechanical equivalence.

$$\frac{\hat{L}^2}{2I}\psi(\theta,\phi) = E\,\psi(\theta,\phi) \tag{7-2}$$

is the generalization of the two-body rotational Schrödinger equation [Eq. (5-15)] and it comes about merely by replacing $\mu r^2$ by I. Therefore, the wavefunctions for a rotating linear diatomic or polyatomic must be the spherical harmonic functions because they are the eigenfunctions of $\hat{L}^2$. The eigenenergies are

$$E_J = J(J+1)\frac{\hbar^2}{2I} \tag{7-3}$$

where J is the rotational quantum number.

The selection rules are the same as for a diatomic: $\Delta J = +1$ is an allowed transition if the molecule has a nonzero dipole moment. Transitions will be seen for HCCF but not for HCCH since the latter molecule has no dipole. The factor $\hbar^2/2I$ in Eq. (7-3) is defined to be the rotational constant, B, for the linear polyatomic, and so the spacing between the $J = 0$ and $J = 1$ energy levels is simply 2B. The separation between the $J = 1$ and $J = 2$ levels is 4B, and so on as already found for diatomics. A microwave spectrum will measure a transition frequency corresponding to a change of one in the J quantum number, and this value is 2B, 4B, 6B, etc., according to which particular transition is detected. We may expect the J = 0 to J = 1 transition to be the easiest to detect in most circumstances.

The rotational constant is inversely related to the moment of inertia, and the moment of inertia is a function of the bond lengths. For a diatomic molecule, measurement of the rotational constant implies determination of the bond length. But for a linear triatomic molecule, there are two unknowns in the problem, the two bond lengths. One value, the value of B, is not sufficient to find two bond lengths.

For a linear polyatomic of N + 1 atoms, there are N bond lengths to be found in order to know the structure of the molecule (assuming there is a basis for already knowing that the molecule in question is linear). Finding the bond lengths is basically a problem of establishing N equations for the N unknowns. The N equations are established from measured rotational constants for isotopically substituted forms of the molecule. Isotopic substitution is a change in an atomic mass, and with that comes a change in the molecule's moment of inertia. However, we assume that the bond lengths are unchanged because the chemical bonding is surely unaffected by the numbers of neutrons in the nuclei. (Actually, this assumption is valid only in regard to equilibrium structures. For the on-average length of a bond where there is zero-point vibrational motion, isotopic substitution may influence the vibrational averaging slightly and thereby have an effect on the bond length. But this can be taken into account in a more detailed analysis.)

For each isotopic form of a molecule with an experimentally determined B, and hence I, we may write Eq. (7-1) with the bond lengths as the only unknowns. So, with N isotopic forms, we have N equations from which to find the N unknowns. A standard example is HCN. Microwave transition frequencies for the J = 0 to J = 1 transitions were found to be 88,631.6 MHz for $H^{12}C^{14}N$ and 72,414.6 MHz for $D^{12}C^{14}N$ [W. Gordy, *Phys. Rev.* **101**, 599 (1956)]. The B rotational constants are one-half of these values, and if they are converted to moments of inertia, we have that I(HCN) = $18.937 \times 10^{-40}$ g-cm$^2$ and I(DCN) = $23.178 \times 10^{-40}$ g-cm$^2$. [See Ref. 1 (p. 187) in the bibliography for this chapter.] The two equations to solve are

$$I(\text{HCN}) =$$

$$\left( m_H\, m_C\, r_{HC}^2 + m_C\, m_N\, r_{CN}^2 + m_H\, m_N\, (r_{HC} + r_{CN})^2 \right) / \left( m_H + m_C + m_N \right)$$

I(DCN) =

$$\left( m_D\, m_C\, r_{HC}^2 + m_C\, m_N\, r_{CN}^2 + m_D\, m_N\, (r_{HC} + r_{CN})^2 \right) / \left( m_D + m_C + m_N \right)$$

Using the atomic masses from Appendix IV, the first equation becomes

$$\frac{18.937 \times 10^{-40}\ \text{g-cm}^2 \times 27.0109}{1.66057 \times 10^{-24}\ \text{g/amu} \times 10^{-16}\ \text{cm}^2/\text{Å}^2} = 308.0308\ \text{Å}^2$$

$$= 12.0939\, r_{HC}^2 + 168.0369\, r_{CN}^2 + 14.1126 \left( r_{HC}^2 + 2\, r_{HC}\, r_{CN} + r_{CN}^2 \right)$$

27.0109 is the total mass of HCN in amu. The corresponding equation for DCN is

$$391.0607\ \text{Å}^2 = 24.1692\, r_{HC}^2 + 168.0369\, r_{CN}^2$$

$$+ 28.2036 \left( r_{HC}^2 + 2\, r_{HC}\, r_{CN} + r_{CN}^2 \right)$$

Solution of these two simultaneous quadratic equations will produce the H-C and C-N bond lengths in Å. The values are 1.066 Å and 1.156 Å, respectively.* These two values together are termed the *substitution structure* of HCN since they have been obtained on the basis of isotopic substitution. The procedure can be used to determine the substitution structure of any linear polyatomic molecule, though solution of the N quadratic equations can be tedious if not done with a computer.

## 7.2 THE HARMONIC PICTURE OF POLYATOMIC VIBRATIONS

In low-resolution infrared spectroscopy of diatomic molecules, the

* These lengths have been rounded, and so if substituted into the right hand side of the two equations, one will obtain 307.9746 instead of 308.0308 for HCN, and 391.2677 instead of 391.0607 for DCN.

rotational fine structure is lost, and some feature in a spectrum assigned to be a fundamental transition is but a single peak. The frequency of that peak is taken to be the vibrational frequency in the absence of any more precise experiments, and that is the extent of the information obtained. So, this low-resolution information corresponds mostly with a nonrotating picture or else a rotationally averaged picture of the molecule's dynamics; to the extent that we can analyze the data, we need only consider pure vibration. To understand the internal dynamics of polyatomic molecules, it is helpful to start with a "low-resolution" analysis. This means neglecting rotation, or presuming the molecules to be nonrotating. Further on and with heightened sophistication, we may consider a molecule to be both vibrating and rotating, as it would be in a gas phase sample.

The pure vibrations of a polyatomic molecule may be quite complicated. It is convenient to think first about the potential energy for vibration motions in order to understand their nature. The potential energy will be a function of the atomic positions. For a molecule of N atoms, there are 3N atomic degrees of freedom, but only $3N-6$ ($3N-5$ for linear molecules) are left after removing the degrees of freedom for molecular translation and rotation. Thus, there are $3N-6$ (or $3N-5$) coordinates that describe the structure of the molecule but that do not give the molecule's position and orientation in space. These $3N-6$ (or $3N-5$) coordinates are called *internal coordinates*, and most often the bond lengths and bond angles comprise a suitable, though not unique, set. For instance, the internal structure of a water molecule can be specified by the two O-H bond lengths and the H-O-H bond angle. These constitute a set of three internal coordinates, and of course, $3N-6=3$ for water.

A *force field* is any potential for the vibrations of a molecule expressed in terms of some chosen set of internal coordinates. In principle, we arrive at complete understanding of the vibrations of a molecule if we know the force field precisely. Thus, we think of vibrational information in relation to the force field, and this is analogous to thinking about the vibrations of a diatomic in relation to the functional form of the stretching potential, $V(x)$. Approximate force fields may be constructed (sometimes guessed) in several different ways, and they may be used for deducing or computing vibrational information. Laboratory measurement of vibrational frequencies provides the ultimate test of such computed information.

The simplest force field for a molecule is one that is harmonic. This means that the potential energy has only linear and quadratic terms involving the $3N-6$ coordinates. Some of these terms may be cross-terms, e.g., a product of two coordinates $r_1r_2$. As discussed in Chap. 2, a potential that is harmonic in all coordinates may be written so that there are no cross-terms if the original coordinates are transformed to the normal coordinates. We will designate normal coordinates as $\{q_1, q_2, q_3, \ldots\}$, and the classical Hamiltonian in terms of normal coordinates is

$$H = \frac{1}{2} \sum_{i=1}^{3N-6} \left( \dot{q}_i^2 + \omega_i^2 q_i^2 \right) \tag{7-4}$$

This is obviously a separable problem, and as we have already considered in Chap. 2, it is equivalent to a problem of $3N-6$ independent harmonic oscillators. The vibrational frequencies of the oscillators are the $\omega_i$'s. The Schrödinger equation that develops from the quantum mechanical form of this Hamiltonian is also separable. The energy level expression comes from the sum of the eigenenergies of the separated harmonic oscillators or modes. For each, there is a quantum number, $n_i$.

$$E_{n_1 n_2 n_3 \ldots} = \sum_{i=1}^{3N-6} \left( n_i + \frac{1}{2} \right) \hbar \omega_i \tag{7-5}$$

The quantum numbers may take on values of 0, 1, 2, ....

A vibrational state of the polyatomic is specified by a set of values for the $3N-6$ quantum numbers in Eq. (7-5). The lowest energy state is the state that has all the quantum numbers equal to zero. The energy of this state, which is $\hbar \Sigma \omega_i / 2$, is the zero-point vibrational energy. Excited vibrational states are the infinite number of states for which any or all of the quantum numbers are not zero. As shown for the water molecule in Fig. 7.1, the energy levels for even a small polyatomic molecule become increasingly numerous at higher and higher energies above the ground state. That is, with the states arranged by their energies, the number of states found within some small energy increment is in an overall way increasing with energy. With states arranged by their energies, the counting of the number of states per some unit energy value is often referred to as the *density of states*. The density of vibrational states of a polyatomic molecule is a function of the energy that is increasing with energy.

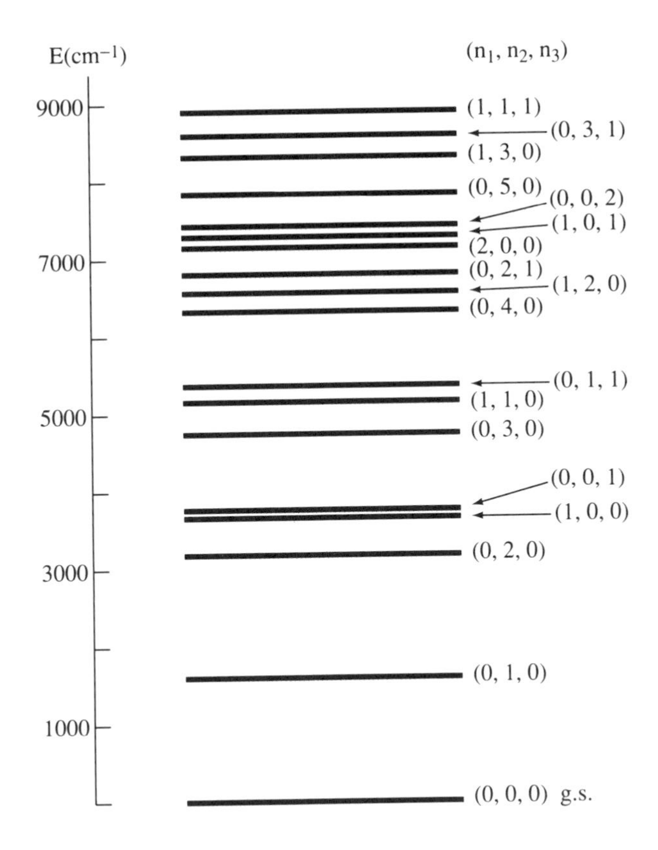


**FIGURE 7.1** The low-lying vibrational state energy levels of the water molecule in a harmonic picture.

From our detailed examination of diatomic vibrational motion, we know that the harmonic picture is an approximation with notable limitations. It is an approximation that is at its best for small amplitude displacements, and that means for low energy states. If we restrict attention to the low-energy states, then there is important qualitative information that goes with the harmonic picture. This qualitative information is the nature of the normal modes of vibration. The coordinate transformation from atomic displacement coordinates to normal coordinates is a transformation that is equally valid for a classical or a quantum mechanical picture, and so we may use classical notions about normal mode vibrations to understand the forms of molecular normal mode vibrations.

Let us use the carbon dioxide molecule as an example. It should have 3N−5 or four modes of vibration. We may represent each mode by the "direction" of the associated normal coordinate, and that means the directions in which each of the atoms move in the course of the vibration. Recall that a normal mode of vibration is the simplest motion of a system of particles, and that for a system vibrating in one mode, all the particles move in phase and at the same frequency. They reach their maximum points of displacment at the same instant, and they pass through their equilibrium positions at the same instant. If we were able to take a "freeze-frame" view of carbon dioxide vibrating purely in one of its normal modes, then the directions the atoms are moving at the instant the particles are at their equilibrium positions would serve to describe the nature of the vibration. We could represent these directions of motion by arrows at each atom, and this is a very common way of representing molecular normal modes.

One of the normal modes of carbon dioxide is a bend along the O-C-O axis. The arrow representation is

$$\overset{\uparrow}{\mathrm{O}} - \overset{\downarrow}{\mathrm{C}} - \overset{\uparrow}{\mathrm{O}}$$

This represents a vibration where the instantaneous direction of motion at the equilibrium point has the oxygen atoms moving up and the carbon atom moving down. Another normal mode is a *breathing motion* or a *symmetric stretch*:

$$\overset{\leftarrow}{\mathrm{O}} - \mathrm{C} - \overset{\rightarrow}{\mathrm{O}}$$

In this motion, the oxygen atoms move away from the carbon atom in phase. If we may use a classical picture for the moment, we would expect the oxygen atoms to continue to their turning points and then reverse directions and eventually pass through the equilibrium positions again. At that instant, their motions might be represented by

$$\overset{\rightarrow}{\mathrm{O}} - \mathrm{C} - \overset{\leftarrow}{\mathrm{O}}$$

This arrow diagram looks different than the previous one, though it is for the same normal mode of vibration. The freeze-frame picture has been taken at a different instant when the particles are at their equilibrium. Either arrow diagram is a correct representation of the symmetric stretching vibration, and it is important not to consider the two drawings to mean different modes.

The mathematics behind the transformation to normal coordinates, which are represented by the arrow diagrams, provides certain rules that may be used to guess the qualitative form of the normal modes of molecules. First, a normal mode is a vibrational motion, and so any motion along the normal coordinate (following the arrows in the diagrams) must not lead to a rotation or translation of the molecule. For carbon dioxide, the following two diagrams are examples of pure translation and pure rotation.

$$\overset{\rightarrow}{O} - \overset{\rightarrow}{C} - \overset{\rightarrow}{O} \qquad\qquad \overset{\uparrow}{O} - C - \overset{\downarrow}{O}$$

If an arrow picture resembles these to any extent, which means that the motion is partly a rotation or translation, then it is not that of a normal mode.

Normal coordinates are independent coordinates. This means that they are orthogonal just as unit vectors along the x-axis, along the y-axis, and along the z-axis of Cartesian space are orthogonal. The normal coordinates are more complicated and abstract. Even so, we may test for orthogonality in about the same way we determine that the x,y,z unit vectors are orthogonal: The dot product of any two different unit vectors is zero. The dot product of a normal coordinate and another normal coordinate is found by adding up the dot products of the corresponding arrows on each atom. As an example, let us use the other stretching mode of carbon dioxide, called the asymmetric stretch. Its arrow diagram is

$$\overset{\rightarrow}{O} - \overset{\leftarrow}{C} - \overset{\rightarrow}{O}$$

This is orthogonal to the symmetric stretch, as seen by taking the dot product of corresponding arrows on the atoms:

$$\overset{\rightarrow 1}{O} - \overset{\leftarrow 2}{C} - \overset{\rightarrow 3}{O}$$

$$\overset{\rightarrow 4}{O} - C - \overset{\leftarrow 5}{O}$$

Taking each of the arrows to be of unit length, the dot product of arrows 1 and 4 is +1. The dot product of the arrows on the right most atom (3 and 5) is −1. For the carbon atom arrows, the dot product is zero, since carbon is not displaced at all in the course of the symmetric stretch. The sum of the three numbers, 1, 0, and −1, is zero. This is the mathematical statement that these two motions, or these two normal coordinates, are distinct or orthogonal.

For all molecules there will be a mode that is best thought of as a breathing mode. All the atoms will move away from the equilibrium. The symmetric stretch mode of carbon dioxide is its breathing mode. To work out arrow diagrams for all the normal modes of a molecule, we may start with a diagram for the breathing mode. Then, we need to identify all other arrow diagrams that are orthogonal, making sure that the diagrams do not include any rotation or translation. If the molecule has symmetry, we can expect the arrow diagrams to reflect that in some way, and that may help systematize our search for the normal coordinate diagrams. Though this process is not necessarily unique, the qualitative information about the nature and types of vibrations will be correct.

As another example, let us find arrow diagrams for the normal modes of the water molecule. The breathing motion must be a simultaneous stretch of the two O-H bonds. If we represent this with an arrow diagram showing the hydrogens moving and the oxygen staying fixed, we will have a representation of a motion where the center of the mass of the molecule moves. Translation will be part of the motion. So, an arrow must be placed on oxygen to make sure that the motion does not include any translation. The representation of the symmetric stretching vibration is

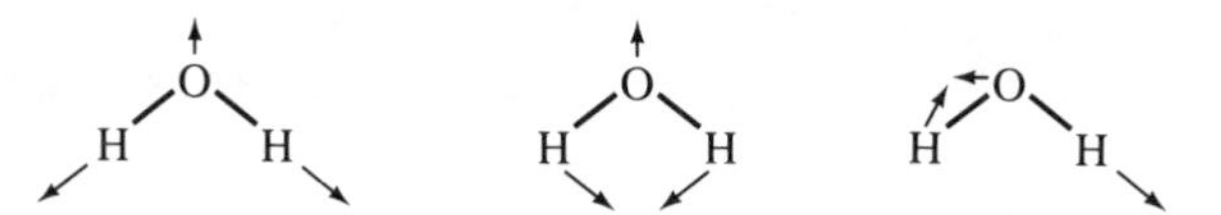

The arrow on the oxygen is smaller than the arrows on the hydrogens because it is so much heavier; the center of mass will remain in place for a displacement of oxygen that is relatively smaller than the hydrogen displacements. We may expect a bending motion, and again, an arrow must be used for oxygen to ensure that the motion does not include any translation or rotation.

An orthogonality test will show that this is an acceptable mode because it is orthogonal to the symmetric stretch. The number of modes is 3N – 6 or 3. So, the remaining mode will be represented by a set of arrows that describe a motion that is orthogonal to both the symmetric stretch and the bend. After some thought, we may realize that this mode is

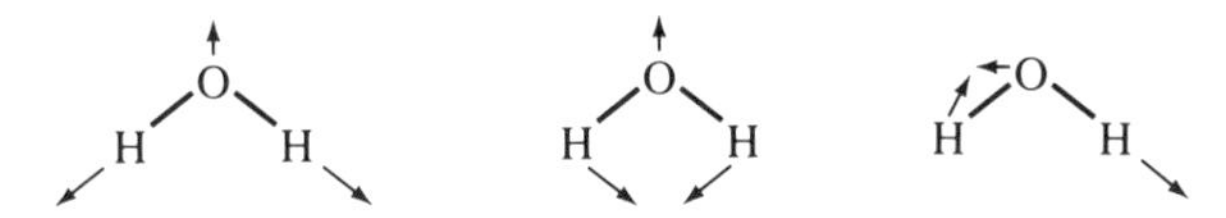


This is an asymmetric stretch.

The normal mode pictures are useful in several ways. One way is to roughly anticipate the relative vibrational frequencies. If we consider a classical analog of carbon dioxide, three balls connected by two springs, then the symmetric stretch is stiffer than the asymmetric stretch. The symmetric stretch requires stretching two springs. The asymmetric stretch is like sliding the center ball back and forth between the two balls on the ends. So, we expect the symmetric stretch to be at a higher frequency than the asymmetric stretch, and usually this is the case.

## 7.3 POLYATOMIC VIBRATIONAL SPECTROSCOPY

Another use of the normal mode picture is in understanding transitions and spectra. The selection rule for a diatomic molecule was found to be that the quantum number changes by one, at least under the harmonic approximation of the potential. Also, the dipole moment has to change in the course of the vibration or the transition is forbidden. Carbon monoxide, for instance, has an allowed fundamental transition, whereas $N_2$ does not. The separation of variables that is accomplished with the normal mode analysis says that each mode may be regarded as an independent one-dimensional oscillator. Thus, we may borrow the results for the simple harmonic oscillator to learn about transitions in polyatomic molecules. First, a transition will be allowed if the vibrational quantum number for a mode changes by one. Second, the vibrational motion must involve a changing dipole moment, or else the transition is forbidden.

The second condition for spectroscopic transitions among polyatomic vibrational levels requires that we determine if a molecule's dipole moment will change in the course of a particular normal mode vibration. The classical definition for the components of the dipole moment vector arising from a fixed distribution of point charges is

$$\mu_x = \sum_i q_i x_i$$

$$\mu_y = \sum_i q_i y_i \qquad (7\text{-}6)$$

$$\mu_z = \sum_i q_i z_i$$

where $q_i$ is the charge of the $i^{th}$ particle. To make a qualitative determination that a dipole moment is either changing or not changing in the course of a normal mode vibration, we can use a very simple scheme along with Eq. (7-6). That scheme is to consider each of the atoms to have a particular partial charge, and the only requirement is that atoms that are chemically equivalent have the same charge. Then we can consider whether a displacement of the atoms in the directions of the arrows that represent the normal modes will produce a change in any of the dipole moment components.

If we return to the example of carbon dioxide, we should assign a partial charge to the oxygen atoms – call it $\delta$ – and a partial charge to the carbon, which must be $-2\delta$ for the molecule to be neutral. The carbon is at the origin of the coordinate system. If the atoms are displaced in the direction of the symmetric stretch mode, the contribution to the dipole moment from one oxygen will cancel the contribution from the other because one will move in the +z-direction and the other in the –z-direction. Thus, in the course of the symmetric stretch, the dipole moment of carbon dioxide is unchanging. This means that transitions where the symmetric stretch quantum number is changed will not be (easily) seen in an infrared absorption spectrum.

The asymmetric stretch will have allowed transitions since the dipole moment changes in the course of this vibration. The two oxygens move in the same direction, and so their contributions will not cancel. The carbon moves oppositely, but it has an opposite partial charge. The bending motion will also give rise to a changing dipole moment.

The diatomic selection rule was found to be $|\Delta n| = 1$ when the potential is strictly harmonic. Other transitions become allowed with an anharmonic potential, but then the $|\Delta n| = 1$ transitions stand out as being stronger. For polyatomic molecules, the selection rule under the assumption of a strictly harmonic potential is $|\Delta n_i| = 1$, while $|\Delta n_j| = 0$ for all the j modes other than the i mode. This means that only one vibrational quantum number can change in any single transition event. Of course, molecules do not have *strictly* harmonic potentials, and just as for diatomics, this selection rule is really telling which transitions will be the strong ones. Other transitions become allowed because of anharmonicity. The transitions that obey this harmonic selection rule and that originate from the ground vibrational state are called *fundamental transitions*, which is the same term used with diatomics. Transitions with $|\Delta n_i| > 1$ that originate in the ground state are called *overtone transitions*. If two or more vibrational quantum numbers change in a transition, it is called a *combination transition*.

The vibrational states of molecules are often designated by a list of the vibrational quantum numbers. The ordering of the list is according to the vibrational frequency. The quantum number for the highest vibrational frequency mode is first in the list. For carbon dioxide, a state would be represented as $(n_1, n_2, n_3)$, where $n_1$ refers to the symmetric stretch, $n_2$ refers to the bend, and $n_3$ refers to the asymmetric stretch. Only three quantum numbers are used for this type of designation because the in-plane bend and the out-of-plane bend are two modes that are degenerate; they have the same frequency, and only one quantum number is used. The ground state is (0,0,0). The following is a representative list of possible transitions and their type, if appropriate.

| | | | |
|---|---|---|---|
| (0,0,0) | → | (0,0,1) | fundamental |
| (0,0,0) | → | (1,0,0) | fundamental |
| (0,0,0) | → | (0,2,0) | overtone |
| (1,0,0) | → | (2,0,0) | |
| (0,0,0) | → | (1,0,1) | combination |

With the many degrees of freedom in a polyatomic molecule, the many different normal mode frequencies, and the many types of transitions that might be seen, it is evident that polyatomic vibrational spectra may be quite

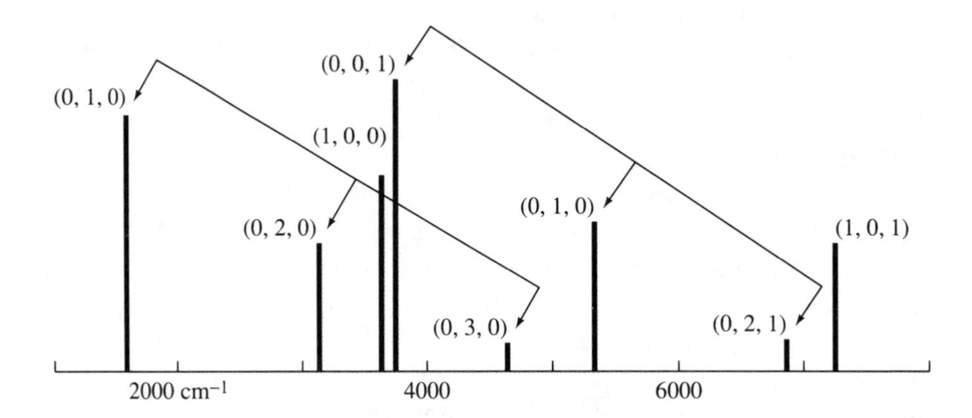


**FIGURE 7.2** A "stick spectrum" of water in the infrared. A stick spectrum is an idealized version of a spectrum obtained in a laboratory experiment where simple vertical lines (sticks) take the place of the peaks actually recorded in the spectrum. The sticks are drawn to about the same height as the corresponding peaks but only for peaks that are assigned. This helps to simplify the spectrum and see regularity in the line positions. The strongest lines in this spectrum are the fundamental transitions.

---

congested and challenging to analyze. An idealized example is given in Fig. 7.2.

The analysis of a full infrared spectrum of a molecule in the gas phase usually starts by finding the fundamental transitions. Most often, these are the strongest transitions. Next, we may look for *progressions*, which are a set of transitions originating from the same initial state and involving the excitation of one mode by successive quantum steps. For instance, the progression built on the fundamental transition (0,0,0,0) $\rightarrow$ (0,1,0,0) of some hypothetical molecule would be the set of transitions from the ground state to (0,2,0,0), and (0,3,0,0), and so on. It is useful to look for these overtone transitions because, to the extent that the harmonic picture holds, we expect them to be found in the spectrum at $2\omega_i$, $3\omega_i$, and so on, where $\omega_i$ is the fundamental transition frequency. Generally, the strongest fundamental transitions will have the strongest overtone progressions. Combination bands, or the low-resolution peaks in the spectrum from combination transitions, may be identified by matching the frequencies with sums and differences of fundamental transition frequencies, and of course, allowing for the fact that the harmonic picture will not hold perfectly for real molecules.

Another type of infrared absorption band is called a *hot band*, and it arises from any transition that does not originate in the ground vibrational state. Recall that the Boltzmann distribution dictates that in a sample of molecules, the population of excited states will grow as the sample is warmed. In a room temperature sample, it is often possible for the population of a low-lying excited vibrational state to be large enough for transitions to originate from this state and to be detectable. Hot band transitions may occur throughout the spectrum. For instance, a hot band corresponding to the transition $(1,0,0) \rightarrow (2,0,0)$ for a linear triatomic molecule would likely be at a frequency very close to that of the fundamental transition $(0,0,0) \rightarrow (1,0,0)$. The transition moments would be almost the same for the two transitions, and so it is only their populations that lead to any difference in the spectrum. That fact can be exploited: If the temperature of the sample is lowered and the spectrum taken again, the hot band's intensity will diminish relative to a band that originates in the ground vibrational state. This is the consequence of diminishing the excited state's population with decreasing temperature. It is a common practice, in fact, to remove hot band congestion from a spectrum by cooling the sample.

## 7.4 CHARACTERISTIC FREQUENCIES

The chemical bond between carbon and oxygen in a carbonyl functional group is known to be only slightly affected by what the carbonyl is attached to. The C–O bonding in formaldehyde is not sharply different than that in acetone. This means that the force constant for stretching the carbon-oxygen bond will be similar in both molecules. In itself, this does not imply that there be similar vibrational frequencies for the two molecules. Even within the harmonic picture, we expect the vibration of the carbonyl to be coupled to the rest of the molecule. In formaldehyde, the stretching and contracting of the C-O bond certainly will not leave the hydrogens in place. Nonetheless, there are similar vibrational frequencies.

Normal modes tend to be delocalized, which means that a single normal mode tends to be a displacement of most all of the atoms in the molecule. That is one reason why a similar force constant for the carbonyl in two different molecules does not in itself ensure a similar vibrational frequency. However, the typical mix of force constants for bends and stretches often results in a particular normal mode appearing to be mainly a localized vibration, such as a carbonyl stretch. For instance, it is possible to identify a normal mode in both formaldehyde and acetone that is mostly like a stretching of the carbonyl. The vibrational frequencies of these modes are similar partly because the coupling with the rest of the molecule, or the delocalization of the mode, ends up having a rather small effect on the frequency. There is reason, then, to think about molecular vibrations as if they were localized.*

The frequencies that are associated with functional groups are called *characteristic frequencies*. These are the frequencies at which to expect vibrational transitions in molecules that contain the given functional group. Table 7.1 lists some characteristic frequencies. There is a range for each characteristic frequency because of the effect of coupling with the rest of the molecule and because there are small variations in the bonding within a functional group because of what it is bonded to. Characteristic frequencies are a useful tool in chemical analysis. A sample of an unknown that exhibits a strong absorption at 1700 $cm^{-1}$, for instance, quite probably contains a carbonyl group in its structure since that is within the range of the characteristic frequency of carbonyl stretching.

* A number of experiments have revealed that highly excited states often have vibrations that are more localized than the ground state, and so it is possible to selectively excite a particular bond stretch by inducing a transition to such a state. That in turn points to the possibility of using electromagnetic radiation to "pump" the excitation of a particular bond and eventually to dissociate a molecule by breaking that bond. Dissociation by heating, on the other hand, will not be selective.

TABLE 7.1 Characteristic Vibrational Stretching Frequencies of Functional Groups. [a]

| | |
|---|---|
| C=O | 1700 $cm^{-1}$ |
| C-O | 1100 $cm^{-1}$ |
| C=C | 1650 $cm^{-1}$ |
| C≡C | 2200 $cm^{-1}$ |
| C≡N | 2200 $cm^{-1}$ |
| C-N | 1200 $cm^{-1}$ |
| C-H | 3000 $cm^{-1}$ |
| N-H | 3400 $cm^{-1}$ |
| O-H | 3500 $cm^{-1}$ |

[a] The characteristic frequencies are ranges around these values, sometimes amounting to a few hundred wavenumbers.

## 7.5 INVERSION AND INTERCONVERSION

A fascinating occurrence in certain molecular systems is the process of *inversion* or *interconversion*. The classic example of inversion is exhibited by the ammonia molecule. It is a pyramid-shaped molecule at its equilibrium. Let us imagine it as three protons in a plane to the left of the nitrogen, so that an umbrella-opening type of vibration moves the plane of the protons to the right and the nitrogen to the left. If this is continued far enough, a point is reached where the protons and the nitrogen are all in the same plane. This is called the inversion point. If the motion continues still further, the protons will be in a plane to the right of the nitrogen. But then, the molecule will look as it did originally, except that it has been reflected. It will have achieved a structure that is equivalent in terms of the vibrational potential energy and internal structural parameters to the original structure. The potential energy will have a unique form along this inversion pathway. It will be at a maximum at the inversion point and will have two equivalent minima corresponding to the ammonia pyramid pointing left and pointing right. This is illustrated in Fig. 7.3.

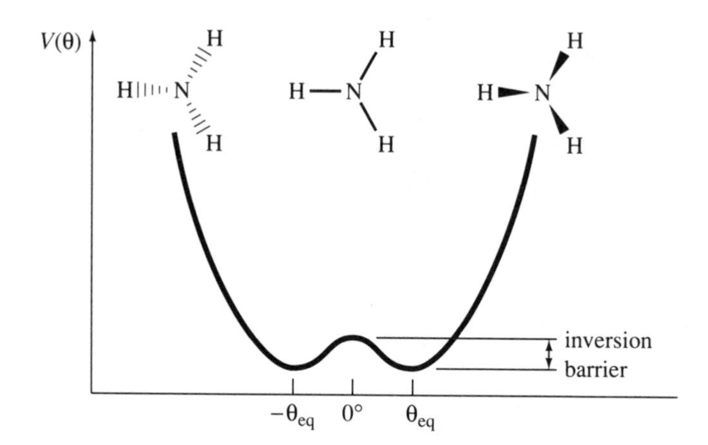


**FIGURE 7.3** The double-minimum potential for the inversion of the ammonia molecule.

An interconversion is any process that takes a molecule into an equivalent form and that cannot be labelled as an inversion (umbrella-opening motion). For example, the weakly bound molecular complex $(HF)_2$ has an equilibrium structure with one of the protons positioned between the fluorines. An interconversion process, illustrated in Fig. 7.4, interchanges the hydrogens and takes the molecule into an equivalent form. There are two points along this path where the potential energy is at its minimum, and so this is a double-well potential. Interconversions can be complicated motions, and also, the number of equivalent structures can be quite numerous.

The energy levels of a one-dimensional double-well problem are interesting, though working them out can be involved. As mentioned in Chap. 4, perturbation theory may be useful if the inversion barrier is low in energy relative to the ground vibrational state. For processes such as the ammonia inversion or the $(HF)_2$ interconversion, the one-dimensional analysis means that motion along the inversion or interconversion pathway has been separated from other vibrational motions, and that is an approximation, of course.

Fig. 7.5 presents exactly calculated energy levels for several different states of a double-well potential. For the states that are high in energy relative to the barrier height, the level spacing is very similar to that of an

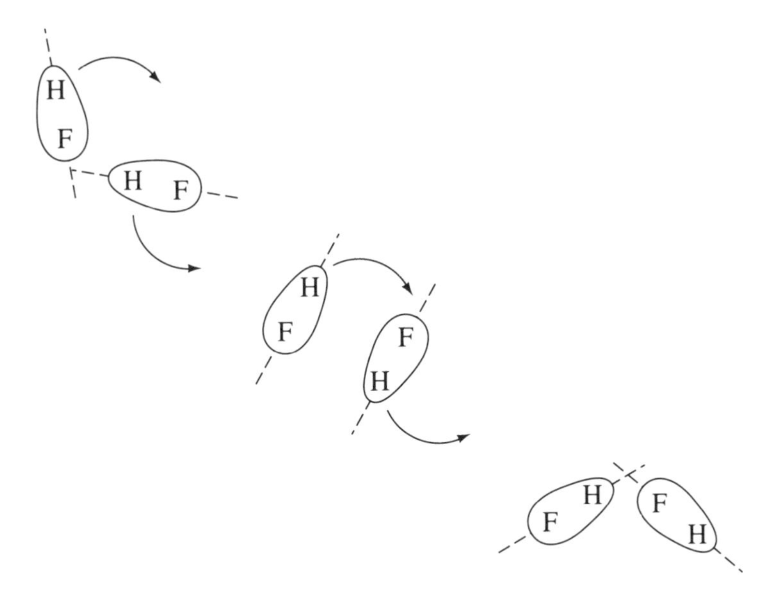


**FIGURE 7.4** An interconversion path for the weak complex of two hydrogen fluoride molecules, $(HF)_2$. The equilibrium arrangement is the nonlinear structure at the top. If the two HF molecules rotate clockwise, the second structure from the top will be reached. And if the molecules are rotated further, the bottom structure will be reached. Notice that the bottom and top structures are equivalent arrangements of the two HF molecules. The middle structure represents the inversion barrier or transition state structure.

unperturbed harmonic oscillator. However, instead of evenly spaced levels at lower energies, the system has levels brought closer together in pairs. In fact, the lowest such pair of levels is a nearly degenerate set of states. These are levels at energies comparable to or less than the height of the barrier. The limiting situation of an immense barrier between the wells is really the situation of two separate wells, and then these levels would be degenerate. That is, there would be a matching set of levels in each distinct potential well.

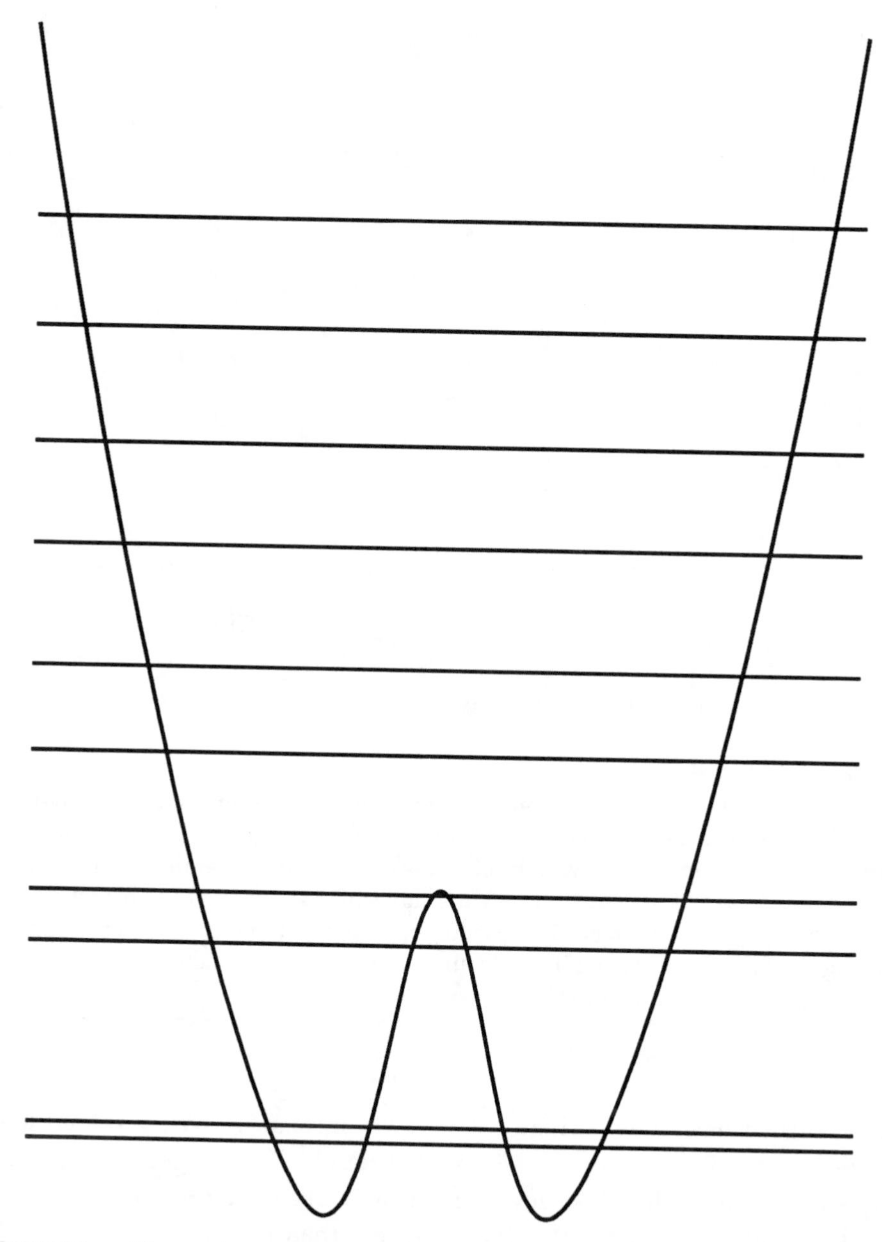

**FIGURE 7.5** The ten lowest-lying vibrational energy levels of a one-dimensional system experiencing a double-minimum potential. The potential is harmonic away from the middle, and the highest levels shown have an energy spacing characteristic of a harmonic oscillator. However, the energies of the first six states are significantly perturbed by the bump in the potential. The perturbation brings the energies of pairs of states closer together.

## 7.6 ROTATIONAL STATES OF NONLINEAR MOLECULES

There are three orthogonal rotations that a rigid, nonlinear arrangement of mass points (i.e., a polyatomic molecule) may undergo. There are three degrees of rotational freedom, and in a classical description, there are three rotational frequencies, $(\omega_1, \omega_2, \omega_3) \equiv \vec{\omega}^T$. These correspond to the rotations about each of the three orthogonal axes of rotation, which can be the x, y, and z axes. The classical kinetic energy of an object rotating about its center of mass is:

$$T = \frac{1}{2} \vec{\omega}^T \mathbf{I} \vec{\omega} \tag{7-7}$$

where **I** is a square array of values called the *inertia tensor*. For a body that consists of point-masses, the inertia tensor elements are

$$\mathbf{I} = \begin{pmatrix} \sum m_i (y_i^2 + z_i^2) & -\sum m_i x_i y_i & -\sum m_i x_i z_i \\ -\sum m_i x_i y_i & \sum m_i (x_i^2 + z_i^2) & -\sum m_i y_i z_i \\ -\sum m_i x_i z_i & -\sum m_i y_i z_i & \sum m_i (x_i^2 + y_i^2) \end{pmatrix} \tag{7-8}$$

This is a real, symmetric matrix, and it may be transformed by a unitary transformation matrix to take it into diagonal form (see Appendix II). The unitary transformation implies a change in the corrdinate system, and the new coordinate axes are designated (a, b, c) instead of (x, y, z). These new axes are called the *principal axes*.

In the principal axis system, there are three nonzero elements of the inertia tensor, and these are the moments of inertia. These are frequently designated $I_A$, $I_B$, and $I_C$ under a convention that $I_A \leq I_B \leq I_C$. The kinetic energy of Eq. (7-7) takes on a simpler form because **I** is diagonal.

$$T = \frac{1}{2} (\omega_A, \omega_B, \omega_C) \begin{pmatrix} I_A & 0 & 0 \\ 0 & I_B & 0 \\ 0 & 0 & I_C \end{pmatrix} \begin{pmatrix} \omega_A \\ \omega_B \\ \omega_C \end{pmatrix}$$

$$= \frac{1}{2} (I_A \omega_A^2 + I_B \omega_B^2 + I_C \omega_C^2)$$

In terms of the angular momenta about each axis, the kinetic energy may be expressed as

$$T = \frac{J_A^2}{2I_A} + \frac{J_B^2}{2I_B} + \frac{J_C^2}{2I_C} \tag{7-9}$$

There is no potential for the free rotation of a molecule, and so this is also the classical Hamiltonian.

If the three moments of inertia, $I_A$, $I_B$, and $I_C$, are equal, the molecule is classified as a *spherical top* molecule. If two of the elements are equal, it is a *symmetrical top* molecule, and there are two types of these. An *oblate top* is one for which $I_A = I_B \leq I_C$. A silver dollar is an oblate top, and so we may think of oblate molecules as flat disks. A *prolate top* is one for which $I_A \leq I_B = I_C$, and this corresponds to a rodlike structure. Finally, a molecule is classified as an *asymmetrical top* molecule if the three moments of inertia are all different.

We may rearrange the kinetic energy, or the rotational Hamiltonian, of Eq. (7-9) to produce a term with the total angular momentum.

$$H = \frac{J_A^2 + J_B^2 + J_C^2}{2I_B} + J_A^2\left(\frac{1}{2I_A} - \frac{1}{2I_B}\right) + J_C^2\left(\frac{1}{2I_C} - \frac{1}{2I_B}\right)$$

$$= \frac{J^2}{2I_B} + J_A^2\left(\frac{1}{2I_A} - \frac{1}{2I_B}\right) + J_C^2\left(\frac{1}{2I_C} - \frac{1}{2I_B}\right) \tag{7-10}$$

For symmetrical top molecules, one or the other of the last two terms in Eq. (17-10) will be zero, since either $I_A = I_B$ or $I_C = I_B$. Of course, for spherical top molecules, both of the last two terms would vanish. Converting to the quantum mechanical operators, we have the Hamiltonians for the two types of symmetrical tops.

$$\hat{H}^{oblate} = \frac{\hat{J}^2}{2I_B} + \hat{J}_C^2\left(\frac{1}{2I_C} - \frac{1}{2I_B}\right) \tag{7-11}$$

$$\hat{H}^{prolate} = \frac{\hat{J}^2}{2\,I_B} + \hat{J}_A^2\left(\frac{1}{2\,I_A} - \frac{1}{2\,I_B}\right) \tag{7-12}$$

From the general rules of angular momentum, we know that the total angular momentum operator, $\hat{J}^2$, commutes with any component operator, and therefore with the square of any component operator. From this we conclude that the Hamiltonian, $\hat{J}^2$, and $\hat{J}_C$ form a mutually commuting set of operators for the oblate top, and that the Hamiltonian, $\hat{J}^2$, and $\hat{J}_A$ form a mutally commuting set for prolate tops. Thus, the wavefunctions for each molecule type will be eigenfunctions of all three operators. The eigenvalues are $J(J+1)\hbar^2$ for $\hat{J}^2$, as usual, and for either $\hat{J}_A$ or $\hat{J}_C$, an integer quantum number, K, times $\hbar$. $K\hbar$ is the component of the rotational angular momentum along the symmetry axis of the disk (oblate top) or rod (prolate top) molecule, and K can range from –J to J in steps of one. The symmetry axis for either type of top is the unique axis.

From the eigenvalues of $\hat{J}^2$ and $\hat{J}_A$ or $\hat{J}_C$, we may immediately obtain the energy eigenvalues for the Hamiltonians of Eqs. (7-11) and (7-12). These energies must be labelled by the two quantum numbers J and K.

$$E_{J,K}^{oblate} = \frac{J(J+1)\hbar^2}{2\,I_B} + K^2\hbar^2\left(\frac{1}{2\,I_C} - \frac{1}{2\,I_B}\right) \tag{7-13}$$

$$E_{J,K}^{prolate} = \frac{J(J+1)\hbar^2}{2\,I_B} + K^2\hbar^2\left(\frac{1}{2\,I_A} - \frac{1}{2\,I_B}\right) \tag{7-14}$$

At this point it is convenient to define three rotational constants analogous to the definition of the B rotational constant used for diatomic molecules.

$$A = \frac{\hbar^2}{2\,I_A} \tag{7-15a}$$

$$B = \frac{\hbar^2}{2\,I_B} \tag{7-15b}$$

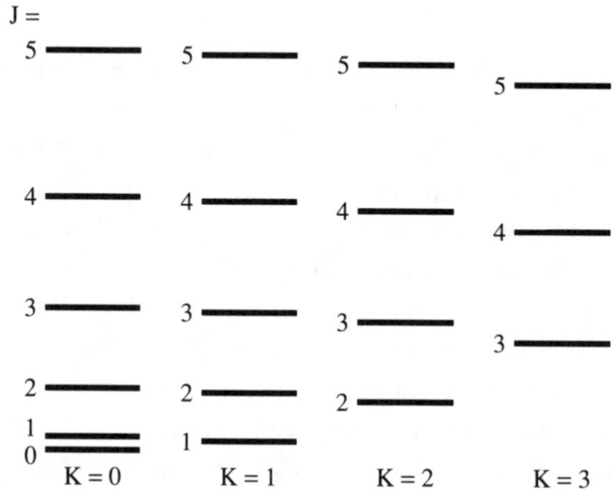

**FIGURE 7.6** An energy level diagram for a oblate top molecule. For clarity, the lines representing the energy levels have been grouped into columns according to the K quantum number. The allowed pure rotational transitions will be between levels in the same column because of the selection rule ΔK=0.

$$C = \frac{\hbar^2}{2 I_C} \tag{7-15c}$$

The energy level expressions become

$$E_{J,K}^{\text{oblate}} = B J (J+1) + K^2 (C - B) \tag{7-16}$$

$$E_{J,K}^{\text{prolate}} = B J (J+1) + K^2 (A - B) \tag{7-17}$$

A typical pattern for the energy levels of the oblate top molecules is shown in Fig. 7.6.

The selection rules for the rotational spectra of symmetric top molecules are

$$|\Delta J| = 1 \text{ and } \Delta K = 0 \tag{7-18}$$

Also, the molecule must possess a permanent dipole moment, and for symmetric top molecules, this will be along the symmetry axis. The selection rules result in a simple transition energy expression:

$$\Delta E_{J,K \rightarrow J+1,K} = 2B(J+1) \tag{7-19}$$

There is no K dependence, and so no A or C dependence, because the K quantum number does not change. This means that the transition frequencies will determine only the B rotational constant. Even so, structural information can be obtained by isotopic substitution, just as for linear polyatomic molecules.

The analysis, so far, has assumed that the molecules are rigid, and of course, this is only a rough approximation. The molecules are vibrating, and so the rotational constants represent a vibrational average of structural parameters. Also, just as in the case of diatomic molecules, molecular rotation may give rise to centrifugal distortion effects. Following the diatomic analysis, these are usually extracted from spectroscopic data by adding corrections to the basic energy expression. Usually, the most sizable of these correction terms enter the energy expressions of Eqs. (7-16) and (7-17) as,

$$-D_J J^2 (J+1)^2 - D_{JK} K^2 J(J+1) - D_K K^4$$

The three constants, $D_J$, $D_{JK}$, and $D_K$, are all referred to as centrifugal distortion constants.

The spectra of asymmetric tops are usually complicated, and the quantum mechanical analysis is beyond the scope of this text. With all three moments of inertia (or rotational constants) different, all terms in the Hamiltonian of Eq. (7-10) remain; there is no simplification. However, it may be that two of the moments of inertia are not too different, and so the molecule may be regarded as nearly prolate or nearly oblate. In that case, the extra piece of the Hamiltonian is treated as perturbation.

Table 7.2 is a collection of the rotational constants that have been extracted from spectroscopic analysis for a number of small molecules.

TABLE 7.2 Rotational Constants in ($cm^{-1}$) of Several Small Polyatomic Molecules.

| | A | B | C |
|---|---|---|---|
| $CH_4$ | 5.2412 | 5.2412 | 5.2412 |
| $NH_3$ | 9.4443 | 9.4443 | 6.196 |
| $H_2O$ | 27.877 | 14.512 | 9.285 |
| $NO_2$ | 8.0012 | 0.4336 | 0.4104 |
| $H_2CO$ | 9.4053 | 1.2953 | 1.1342 |
| $H_2CCO$ | 9.37 | 0.34335 | 0.33076 |

## 7.7 VIBRATIONAL ANHARMONICITY IN POLYATOMICS AND HIGH-RESOLUTION SPECTRA *

Vibrational anharmonicity refers to any part of the molecular force field that has more than a quadratic dependence on the displacement coordinates. If the potential or the force field is expressed as a power series in terms of displacement coordinates, it is normally the cubic and quartic terms that are most important for the lowest lying vibrational states. One way of understanding the effects of vibrational anharmonicity in a polyatomic molecule is to treat the cubic and quartic terms as perturbations of the harmonic picture of the system.

Let us consider the following hypothetical situation. The harmonic part of the force field of a molecule was used to find the normal coordinates $\{q_i\}$ in the usual procedure, and then when the complete force field was expressed in the normal coordinates, there were no cross-terms or no coupling terms (the hypothetical part of this). The potential energy in this situation is

$$V(q_1,q_2,q_3,\dots) = \sum_{i=1}^{3N-6} \left( \frac{1}{2}\omega_i^2 q_i^2 + \alpha_i q_i^3 + \beta_i q_i^4 + \dots \right) \qquad (7\text{-}20)$$

The Hamiltonian in this case is separable by the assumption of no coupling terms. There are no cross terms in this potential function, and of course, there are no cross terms in the kinetic energy when expressed in normal coordinates. The significance is that the treatment of the anharmonicity terms that enter with the $\alpha$ and $\beta$ coefficients in Eq. (7-20) is identical with the treatment of anharmonicity in the one-dimensional oscillator. This is because separability means this is equivalent to a problem of independent one-dimensional oscillators.

To incorporate the effects of the particular type of cubic anharmonicity terms of Eq. (7-20) to second order, we may borrow the one-dimensional result in Eq. (6-23). This means we use the harmonic potential for the zero-order Schrödinger equation and then treat the potential terms that are diagonal in the normal coordinates with second order perturbation theory. This yields the following energy level expression.

$$E^{(2)}(n_1,n_2,n_3,\ldots) = \sum_{i=1}^{3N-6} \left\{ \left(n_i + \frac{1}{2}\right)\hbar\omega_i - \left(\frac{7}{16} + \frac{15}{4}\left(n_i + \frac{1}{2}\right)^2\right) \frac{\alpha_i^2 \hbar^2}{\omega_i^4} \right\} \quad (7\text{-}21)$$

This equation looks rather involved; however, it is simply a sum over energy contributions that depend on one vibrational quantum number only. In each contribution, there is the harmonic zero-order term. Then there is the second-order term that has a contribution that lowers the zero-point energy, plus a term that is negative and quadratic in $(n_i+1/2)$. So, just as for diatomics, cubic anharmonicity may have a greater effect with increasing quantum number, and this means that the levels will get more closely spaced at higher energies. We may also treat quartic terms by perturbation theory and borrow the results through second order that have been developed for a diatomic.

Cubic and quartic anharmonic terms in a molecular force field do not necessarily conform to our hypothetical example of having no cross terms. In general, there could be terms such as $q_iq_j^2$, $q_i^2q_j$, $q_iq_j^3$, $q_i^2q_j^2$, and $q_i^3q_j$ that are as important as the terms in Eq. (7-20). These terms can also be treated as perturbations in the harmonic picture, but there is an important difference between these and the terms in Eq. (7-20). These cross terms remove the separability in the harmonic picture, and as we will see, they couple the states of the system.

To understand the analysis of the cross terms, let us examine a system that has just two vibrational degrees of freedom. For instance, this might be the stretching vibrations of a triatomic molecule. For a perturbative analysis, we use a zero-order Hamiltonian that has a harmonic potential expressed in terms of the two normal coordinates, $q_1$ and $q_2$. The normal coordinates are mass-weighted, and the normal mode frequencies then show up in the potential:

$$H^{(0)} = -\frac{\hbar^2}{2}\frac{\partial^2}{\partial q_1^2} - \frac{\hbar^2}{2}\frac{\partial^2}{\partial q_2^2} + \frac{1}{2}\omega_1^2 q_1^2 + \frac{1}{2}\omega_2^2 q_2^2 \tag{7-22}$$

The perturbing Hamiltonian is the cross-term in the anharmonic part of the potential that we will consider for this example.

$$H^{(1)} = c\, q_1 q_2^2 \tag{7-23}$$

The zero-order energy expression is

$$E^{(0)}_{n_1 n_2} = (n_1 + \frac{1}{2})\hbar\omega_1 + (n_2 + \frac{1}{2})\hbar\omega_2 \tag{7-24}$$

The zero-order wavefunctions are just products of the one-dimensional harmonic oscillator wavefunctions in the two directions:

$$\Psi^{(0)}_{n_1 n_2}(q_1,q_2) = \psi^{\text{har osc}}_{n_1}(q_1)\,\psi^{\text{har osc}}_{n_2}(q_2) \tag{7-25}$$

As before, this follows from the separability of the Hamiltonian in Eq. (7-22).

It is helpful to consider the matrix representation of the perturbing Hamiltonian before applying perturbation theory. The basis for the matrix representation is the set of zero-order functions, as given in Eq. (7-25). The rows and columns of the matrix representation are labelled by a single state, but a single state is specified by the two quantum numbers $n_1$ and $n_2$. So, we need to select an order for the functions in the basis and use it consistently in any matrix operations. The following will be the order used here for the states, and the rows and columns of the matrix representation will assume this order:

$$\Psi_{00}^{(0)} \quad \Psi_{10}^{(0)} \quad \Psi_{01}^{(0)} \quad \Psi_{20}^{(0)} \quad \Psi_{11}^{(0)} \quad \Psi_{02}^{(0)} \quad \Psi_{30}^{(0)} \quad \Psi_{21}^{(0)} \quad \Psi_{12}^{(0)} \quad \Psi_{03}^{(0)} \quad \Psi_{40}^{(0)} \quad \ldots$$

An arbitrary matrix element of the matrix representation of the perturbing Hamiltonian is obtained as,

$$< \Psi_{n_1 n_2}^{(0)} \mid c q_1 q_2^2 \, \Psi_{n_1' n_2'}^{(0)} > \; = \; c < \psi_{n_1}^{\text{har osc}} \mid q_1 \, \psi_{n_1'}^{\text{har osc}} > < \psi_{n_2}^{\text{har osc}} \mid q_2^2 \, \psi_{n_2'}^{\text{har osc}} > \tag{7-26}$$

where the integral over $q_1$ and $q_2$ has been separated into integrals over harmonic oscillator functions. The integrals on the right side of Eq. (7-26) are elements of the matrix representations that were obtained in Chap. 6. So, substituting from Eqs. (6-46) and (6-49) gives

$$\mathbf{H}^{(1)} = \frac{c\hbar}{2\omega_2}\sqrt{\frac{\hbar}{2\omega_1}} \begin{pmatrix} 0 & 1 & 0 & 0 & 0 & 0 \\ 1 & 0 & 0 & \sqrt{2} & 0 & \sqrt{2} \\ 0 & 0 & 0 & 0 & 3 & 0 \\ 0 & \sqrt{2} & 0 & 0 & 0 & 0 \\ 0 & 0 & 3 & 0 & 0 & 0 \\ 0 & \sqrt{2} & 0 & 0 & 0 & 0 \end{pmatrix} \tag{7-27}$$

with the basis truncated to just the first six functions in the list. The diagonal elements of this matrix are all zero, and so the first-order corrections to the state energies are all zero.

The off-diagonal elements in the matrix representation of the perturbing Hamiltonian dictate a mixing of state functions in first-order perturbation theory. For example, the value of one in the first row and second column is the matrix element between the functions $\Psi_{00}^{(0)}$ and $\Psi_{10}^{(0)}$. The wavefunctions correct through first order will be mixtures of these two, and the extent of the mixing is linear in the parameter c. The first-order wavefunctions, then, do not necessarily have the product form of the zero-order wavefunctions, and of course, that makes sense since the perturbing interaction is a coupling of the two vibrations. From the other nonzero elements in Eq. (7-27), other functions would be mixed at first order and at higher orders.

The second-order energy corrections will be the squares of the off-diagonal elements in Eq. (7-27) divided by a difference in zero order energies corresponding to the row and column of the off-diagonal element. For instance, the mixing of $\Psi_{00}^{(0)}$ and $\Psi_{10}^{(0)}$ will lead to a lowering of the lowest energy state, the $\Psi_{00}$ state, by the amount $c^2\hbar^2/8\omega_1^2\omega_2^2$, whereas it will lead to a raising of the higher of the pair of states, the $\Psi_{10}$ state, by that amount.

These anharmonic corrections become very noticeable in a certain circumstance termed *Fermi resonance.* A cubic or quartic anharmonicity in the molecular force field may produce a sizable second-order correction to the energies of two states if their zero-order energies are nearly equal. The separation between two states will be increased by a quantity we shall call Δ. An approximate expression for Δ is from the standard second-order, non-degenerate perturbation theory corrections to the energy.

$$\Delta \approx \frac{2 < \Psi_i^{(0)} \mid H' \Psi_j^{(0)} >^2}{E_i^{(0)} - E_j^{(0)}}$$

where H' is the cubic (or quartic) potential term and $\Psi_j$ is the lower zero order state. The numerator is typically small, but nonetheless, Δ can be large if the denominator is close to zero. (In fact, a surer analysis may be to use variational theory, because the smallness of the energy denominator implies slow convergence in the perturbational expansion.) As a hypothetical example, let us consider a molecule with three modes whose harmonic frequencies are 403, 552, and 799 $cm^{-1}$. A list of the zero order energies of a number of the lowest states, relative to the ground state, with the states labelled by three quantum numbers $n_1$, $n_2$, and $n_3$, for the three modes is,

$E_{000}$ = 0 $cm^{-1}$ (relative to $E_{000}$)
$E_{100}$ = 403
$E_{010}$ = 552
$E_{001}$ = 799 a
$E_{200}$ = 806 a
$E_{110}$ = 955
$E_{020}$ = 1104
$E_{101}$ = 1202 b
$E_{300}$ = 1209 b

$$
\begin{array}{lcll}
E_{011} & = & 1351 & c \\
E_{210} & = & 1358 & c \\
E_{120} & = & 1507 & \\
E_{002} & = & 1598 & d \\
E_{201} & = & 1605 & d \quad \ldots
\end{array}
$$

The pairs of energies with a letter following (e.g., $E_{001}$ and $E_{200}$) are closely spaced relative to the separation between other pairs of states. Even a small anharmonicity involving the normal coordinate for the first mode and the normal coordinate for the third mode could lead to sizable second-order energy corrections that could push these pairs of states apart. Changes of 50 or even 100 $cm^{-1}$ are not uncommon. So, we might actually observe a fundamental transition from the ground state around 749 $cm^{-1}$ instead of 799 $cm^{-1}$, and an overtone transition at 856 $cm^{-1}$ instead of 806 $cm^{-1}$.

Fermi resonance may involve more than two states. If there are three that are closely spaced and if there are anharmonic potential terms that give a nonzero matrix element between the zero-order states, then there may be substantial shifts from the energies expected on the basis of a harmonic picture. Fermi resonance is a complicating factor in analyzing polyatomic vibrational spectra because it displaces bands and obscures the sequencing that might be helpful in making assignments. On the other hand, once assignments are made, the deviation of absorption lines from where they would otherwise be expected is a measure of the Fermi resonance correction, $\Delta$, and this can be used to deduce the anharmonic elements of the molecule's true force field.

Rotational fine structure in infrared spectra is a challenge to resolve and a challenge to interpret. As molecular size increases, rotational constants diminish, and so the spacing between individual rotational lines is diminished. Sometimes only band contours are obtained instead of the individual rotational lines. There is also the problem of congestion of many different vibrational bands because of the number of vibrational modes in polyatomic molecules. Bands may overlap, and this can seriously complicate the spectral analysis even if the rotational fine structure is resolved. It is quite remarkable, then, that contemporary spectroscopic techniques have achieved great success in resolving rotational fine structure, assigning transition lines, extracting rotational constants, and

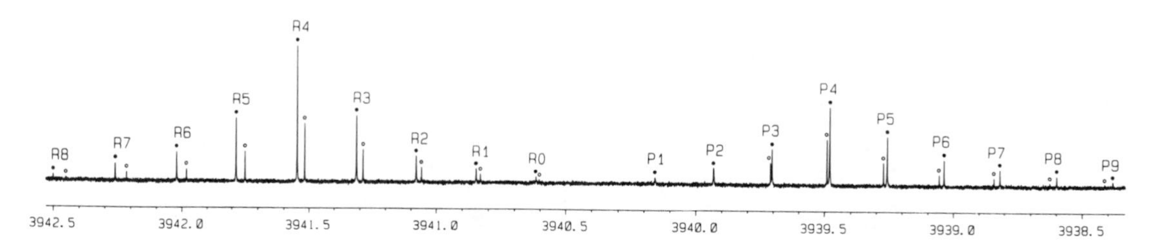


**FIGURE 7.7** An example of a highly-resolved infrared spectrum of a polyatomic molecule and of the assignment of transition lines. This is part of the vibrational spectrum of a molecule formed by the weak attachment of an HF molecule to an HCl molecule. The horizontal axis gives the transition frequency in $cm^{-1}$. This vibrational band is associated with an excitation primarily of the HF submolecule stretching mode, and P- and R-branch lines are clearly seen. The lines marked with a dot are from HF-$H^{35}Cl$ whereas the lines marked with a circle are from HF-$H^{37}Cl$. [Spectrum from G. T. Fraser and A. S. Pine, *J. Chem. Phys.* **91**, 637 (1989).]

determining molecular structure parameters in fairly large molecules and in a number of exotic and fleeting species, such as in the example in Fig. 7.7.

## Exercises

1. From the following measured rotational constants of three isotopic forms of NNO, find the substitution structure (i.e., the two bond lengths).

| | |
|---|---|
| $^{14}N^{14}N^{16}O$ | 12,561.7 MHz |
| $^{15}N^{14}N^{16}O$ | 12,137.3 MHz |
| $^{14}N^{14}N^{18}O$ | 11,859.1 MHz |

2. Using standard bond length values as an estimate of the true bond lengths in HCCH and FCCH, obtain the rotational constants of these two molecules in MHz in order to see the effect of a heavy atom replacement of hydrogen on the transition frequency.
3. Shown here are sketches of the normal modes of acetylene. Which modes are infrared-active?

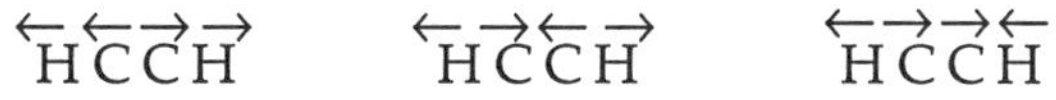

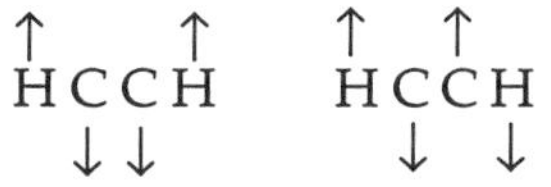

```
←←→→         ←→←→         ←→→←
HCCH         HCCH         HCCH

↑  ↑     ↑ ↑
HCCH     HCCH      (in plane and out of plane)
 ↓↓        ↓ ↓
```

4. Determine if there should be a qualitative difference in the normal modes of vibration, both in the form of the modes and in their infrared activity, of NNO and of $CO_2$.
5. Make a sketch using arrows to show a reasonable guess of the qualitative form of the normal modes of vibration for

   *a*. formaldehyde    *b*. ethylene    *c*. hydrogen peroxide
6. To give an idea of what is typical for small molecules, estimate the zero-point energies of the water molecule, of acetylene, and of formaldehyde either on the basis of actual vibrational frequencies or on the basis of characteristic infrared frequencies.
7. A characteristic H-C stretching frequency is 3000 $cm^{-1}$. Let us assume that this is the frequency for the normal mode of HCN that looks most like a stretch of the H-C bond. This mode might be modelled as a pseudo diatomic vibration if we were to think of the CN as one "atom" and the H as the other. Within this model, we may estimate the effects of certain changes to the molecule as if they were simply changes to the mass of the pseudo diatomic. Doing that, what would be the frequency for this vibration if $^{13}C$ were substituted for $^{12}C$? What would be the frequency if the group C-CN were substituted for the N? (This second case offers an idea of why, and to what extent, frequencies are "characteristic" of the bonding environment.)
8. Obtain structural data and then classify the following molecules as spherical, oblate, prolate, or asymmetrical tops: methane, methyl fluoride, ammonia, benzene and formaldehyde.

9. Assume that the room temperature infrared spectrum of some polyatomic molecule showed two similar bands, one of which was very strong. The band center of the first was at 2000 $cm^{-1}$ and that of the second was at 2100 $cm^{-1}$ and the first was 2.5 times as intense as the second. If the spectrum is to be obtained at a temperature 100° C colder, what change in relative intensities would indicate that the band at 2100 $cm^{-1}$ was probably a fundamental band and the band at 2000 $cm^{-1}$ was a hot band originating from a vibrational state 100 $cm^{-1}$ above the ground state?

10. A molecule whose chemical formula is known to be $CH_4N_2O$ is found to have infrared absorption bands at 3450, 3350, 1690, 1600, and 1160 $cm^{-1}$. On the basis of the characteristic frequencies in Table 7.1, propose a possible structure for this molecule. What additional information might be extracted from the spectrum to test the prediction?

11. Consider a double well potential of the form $V(x) = x^2/2 + 3/2\, e^{-(\beta x)^2/2}$. If a particle of mass m = 1 were to experience this potential, what would be the energies of the first five states from the following treatments?

    i. Zero-order perturbation energies with the exponential part of the potential being the perturbation
    ii. First-order perturbation theory
    iii. Second-order perturbation theory

★12. Repeat Prob. 11 with a linear variational treatment where the basis set is the set of the first five harmonic oscillator wavefunctions [i.e., the wavefunctions for m = 1 and $V_0(x) = x^2/2$].

## Bibliography

(See books listed at the end of Chap. 6.)

1. W. H. Flygare, *Molecular Structure and Dynamics* (Prentice-Hall, Englewood Cliffs, New Jersey, 1978). The sections on microwave spectroscopy in this text offer a very detailed analysis.

2. E. B. Wilson, Jr., J. C. Decius, and P. C. Cross, *Molecular Vibrations. The Theory of Infrared and Raman Vibrational Spectra* (Dover Publications, New York, 1980). This is an advanced level treatment of the topic.
3. G. Herzberg, *Molecular Spectra and Molecular Structure. II. Infrared and Raman Spectra of Polyatomic Molecules* (Van Nostrand Reinhold, New York, 1945). This is a research level volume that is quite comprehensive in the quantum mechanical analysis for polyatomic spectra.

# Chapter 8

# Particles Encountering Potentials

*This chapter concerns the problems of particles, called bound particles, that are trapped in potential wells, and of unbound particles that experience potentials that may give rise to their reflection or scattering. Very simple types of potentials are used in order to demonstrate certain significant quantum phenomena associated with particles encountering potentials. There are intriguing ways in which quantum mechanical particles prove to be quite different from what we would see for particles in our macroscopic world.*

## 8.1 THE PARTICLE-IN-A-BOX PROBLEM

The particle-in-a-box is a special model problem in quantum mechanics. It is the problem of a hypothetical particle of mass m that may move in one dimension, along the x-axis, and that experiences a potential, V(x), with a very simple form. V(x) is zero in some region, $0 < x < l$, and is

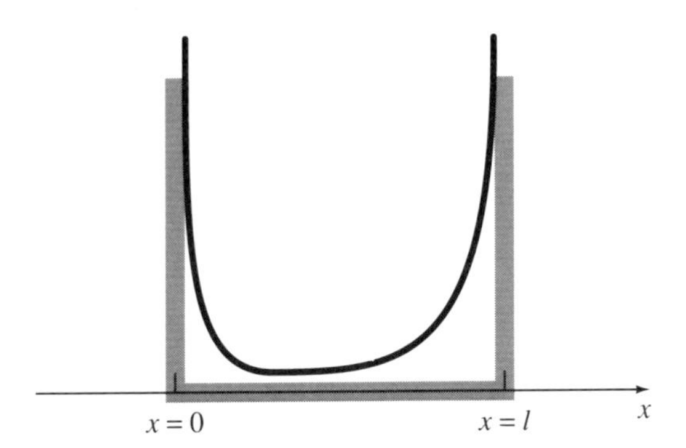


**FIGURE 8.1** The particle-in-a-box potential, which is an impenetrable potential outside x = 0 and x = $l$ and zero inside, might serve as an approximation of the smoothly changing potential represented by the solid line.

infinite elsewhere. This is not a realistic situation by any means because realistic potentials do not become infinite throughout a region and do not change abruptly from zero to infinite. However, certain potential functions may be relatively flat over a region and turn sharply upward at the ends of the region. In such a case, V(x) would serve as an approximation, as shown in Fig. 8.1.

We assume that even quantum particles will not be found in regions where the potential is infinite. Then, V(x) is really a "box"; the particle will be found only in the region $0 < x < l$. Within that region, the potential is zero, and so the time-independent Schrödinger equation is extremely simple,

$$-\frac{\hbar^2}{2m}\frac{d^2\psi(x)}{dx^2} = E\,\psi(x) \tag{8-1}$$

The general solution of this differential equation may be written two ways.

$$\psi(x) = A\,e^{i\sqrt{2mE}\,x/\hbar} + B\,e^{-i\sqrt{2mE}\,x/\hbar} \tag{8-2a}$$

$$\psi(x) \quad = \quad a \sin\left(\sqrt{2mE}\, x/\hbar\right) \quad + \quad b \cos\left(\sqrt{2mE}\, x/\hbar\right) \qquad \text{(8-2b)}$$

The constants, A and B, or a and b, and the energy eigenvalues do not end up being specified in the course of finding the general solution. Of course, the Schrödinger equation that was used is only valid in a certain region, the confines of the box, and so certain information about the problem has been taken into account.

The particle-in-a-box potential is discontinuous at $x = 0$ and at $x = l$. This presents a complication in the way we may approach the problem because we cannot write one Hamiltonian that is correct for the region $x < 0$, and for the region $0 < x < l$, and for the region $x > l$. In general, discontinuous potentials require breaking up the problem into regions. (The boundaries of the regions are wherever the potentials are discontinuous.) Separate Schrödinger equations may be solved for each region, but there are overriding conditions that must be met in order for us to interpret the squares of wavefunctions as probability densities. The conditions are that the wavefunctions are single-valued everywhere, that they are continuous, and that their first derivatives are continuous. Within each region, these conditions are automatically satisfied. At the boundaries, these conditions have to be imposed.

In the particle-in-a-box problem, the wavefunction must be zero-valued at the edges and outside the confines of the box. Thus, the allowed solutions within the box, Eq. (8-2), are only acceptable as wavefunctions if they become zero-valued at the edges of the box. This is expressed as

$$\psi(0) = 0 \quad \text{and} \quad \psi(l) = 0$$

Using the trigonometric form of the solutions in the box, the first of these conditions forces the constant b to be zero. The second condition leads to the following result.

$$\psi(l) \quad = \quad a \sin\left(l\sqrt{2mE}\,/\,\hbar\right) \quad = \quad 0 \qquad \Rightarrow$$

$$l\sqrt{2mE}\,/\,\hbar \quad = \quad n\,\pi \qquad \text{where } |n| = 0, 1, 2, 3, \ldots \qquad \text{(8-3)}$$

Rearranging this result shows it to be a condition on the energy:

$$E = \frac{n^2 \pi^2 \hbar^2}{2m\, l^2} \tag{8-4}$$

The allowed energies, then, are not continuous because of the boundary potential. In fact, particles that are bound or trapped by potentials have discrete energy levels. Notice that the separation of adjacent energy levels increases with n because of the quadratic dependence of the energy eigenvalues on n in Eq. (8-4). In the harmonic oscillator, the bound states were separated by an equal amount; the eigenenergies varied linearly with the quantum number.

Using the relation in Eq. (8-3) allows us to express the wavefunctions as

$$\psi_n(x) = a \sin\left(\frac{n\pi x}{l}\right) \tag{8-5a}$$

At this point, we reject the possibility of $n = 0$ as unphysical. It would correspond to a wavefunction that is zero-valued everywhere and has a zero probability density everywhere. Also, we exclude the negative integers as possible quantum numbers. Even though they are allowed by Eq. (8-3), they do not generate unique wavefunctions. Instead, they correspond to a wavefunction with a sign or phase change since

$$\sin\left(\frac{-n\pi x}{l}\right) = -\sin\left(\frac{n\pi x}{l}\right)$$

Thus, the allowed values of n are 1, 2, 3, and so on. Finally, the normalization condition determines the constant a, which turns out be independent of n:

$$\int_0^l \psi_n^* \psi_n \, dx = \int_0^l a^2 \sin^2(n\pi x / l)\, dx = a^2 l / 2 \Rightarrow 1$$

The wavefunctions for the particle-in-a-box problem are

$$\psi_n(x) = \sqrt{\frac{2}{l}} \sin\left(\frac{n\pi x}{l}\right) \tag{8-5b}$$

with the wavefunction being zero outside the box.

## 8.2 UNBOUND STATES

The Schrödinger equation for a particle free to move in one dimension with no potential is

$$-\frac{\hbar^2}{2m}\frac{d^2\psi(x)}{dx^2} = E\,\psi(x) \tag{8-6}$$

The solutions of this differential equation may be obtained by simple integration.

$$\psi(x) = A\,e^{-i\sqrt{2mE}\,x/\hbar} + B\,e^{i\sqrt{2mE}\,x/\hbar} \tag{8-7}$$

where A and B are constants. Notice that there is no restriction on the energy E. Any value is allowed because the particle is free. Sometimes the constant in the exponential is collected into one constant, k, called the *wave constant* for the particle.

$$k = \sqrt{2mE}\,/\,\hbar \tag{8-8}$$

$$\psi(x) = A\,e^{-ikx} + B\,e^{ikx} \tag{8-9}$$

If the particle is free to move in three dimensions, the Schrödinger is, of course, separable, and the wavefunctions are products of independent functions of x, y, and z that have the form given in Eq. (8-7) or (8-9). However, the wave constants are not necessarily the same in each direction; there will be a $k_x$, a $k_y$, and a $k_z$ in the wavefunction, and these are independent. Collected into a vector, they comprise the *wave vector* for the moving free particle.

$$\vec{k} = (k_x, k_y, k_z) \tag{8-10}$$

The energy is partitioned in the three directions because of the separability of the Schrödinger equation.

$$-\frac{\hbar^2}{2m}\nabla^2\,\Psi(x,y,z) = \frac{\hbar^2}{2m}(k_x^2 + k_y^2 + k_z^2)\,\Psi(x,y,z)$$

$$= (E_x + E_y + E_z)\,\Psi(x,y,z) \tag{8-11}$$

This is equivalent to expressing the kinetic energy of a free classical particle with velocity components as $m(v_x^2 + v_y^2 + v_z^2)/2$.

The wavefunction of the free particle in three dimensions is an eigenfunction of $p_x^2$, $p_y^2$ and $p_z^2$. It may also be an eigenfunction of $p_x$ or $p_y$ or $p_z$. The general solution of the differential equation, or the function that satisfies Eq. (8-11), is the product of functions of the form of Eq. (8-9).

$$\Psi(x,y,z) = \left( A_x e^{-i k_x x} + B_x e^{i k_x x} \right)\left( A_y e^{-i k_y y} + B_y e^{i k_y y} \right) \times \left( A_z e^{-i k_z z} + B_z e^{i k_z z} \right) \qquad (8\text{-}12)$$

If either $A_x$ were zero or $B_x$ were zero, the wavefunction would be an eigenfunction of $p_x$. If either $A_y$ were zero or $B_y$ were zero, the wavefunction would be an eigenfunction of $p_y$, and likewise for the z-direction. This is seen from the following.

$$\hat{p}_x \left( A_x e^{-i k_x x} \right) = -i\hbar A_x \frac{\partial}{\partial x} e^{-i k_x x} = -\hbar k_x \left( A_x e^{-i k_x x} \right)$$

The eigenvalue for the momentum is $-\hbar k_x$. Furthermore, there is no uncertainty in the momentum component because this function is also an eigenfunction of $p_x^2$. If $A_x$ were zero instead, the eigenvalue would be $\hbar k_x$. The positive eigenvalue implies that the momentum is that associated with motion in the +x direction, whereas the negative eigenvalue is associated with motion in the opposite direction. Thus, the complex exponential forms of the wavefunctions of a free particle may be associated with motion to the right or to the left along each coordinate axis. Thus, the set of elements termed the wave vector, $\vec{k}$, in Eq. (8-10) seems appropriately named. We may now see it as a vector that identifies the direction that a free particle is moving in three-dimensional space.

The wavefunctions of the free particle are said to form a *continuum of energy states* because a continuous choice of energies (or wave vectors) is possible. Unlike the harmonic oscillator or the particle in the box, there is no restriction on what energies are allowed. This is a typical feature of a particle that is unbound, where "unbound" means that the particle's kinetic energy is sufficient to surmount any potential energy barrier in the direction of its motion. A classical particle meeting this description would continue in that direction endlessly.

## 8.3 TRANSMISSION, REFLECTION, AND TUNNELING

An interesting feature of the harmonic oscillator wavefunctions (Chap. 3) was that the probability density is not zero beyond the classical turning points, though it is small. Since the classical turning points are the points at which the particle's total energy equals the potential energy, then the nonzero possibility of finding the particle in a region where the potential is still greater is remarkable. This is the quantum phenomenon called *tunneling*, which is a nonzero probability density for finding a particle in a region where the potential is greater than the particle's energy.

Tunneling becomes particularly interesting in the situation where the potential has the form shown in Fig. 8.2. If a particle's energy, E, were in the range $V_{III} < E < V_{II}$, and if it were somehow placed into region I, we might expect some tunneling from region I into region II. And then, as long as the wavefunction did not diminish completely to zero as it continued to the right in region II, there would be some possibility that it would enter region III. There, it would behave as a free particle with kinetic energy equal to $E - V_{III}$. This tunneling occurrence would make a trapped particle free, and that is something that in the classical world would be the same thing as walking through a brick wall. Tunneling through a barrier is an immensely important physical process. It explains spontaneous nuclear fission and beta decay, and in chemistry, it explains why certain reactions take place even when the reactants have insufficient energy to surmount the activation barrier.

The probability that tunneling will take place can be deduced from the wavefunctions. The step potential in Fig. 8.2 serves as a good example of the possibility of tunneling and also of the quantum mechanical complications of potentials that change sharply at some point. So, the next task is to understand something about the wavefunctions for the problem represented by Fig. 8.2.

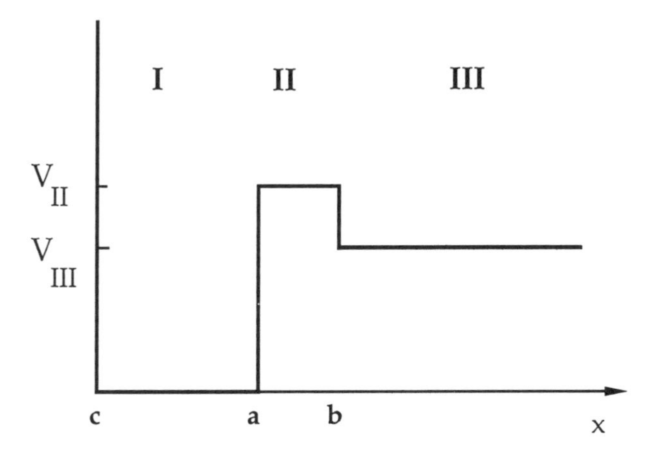


**FIGURE 8.2** A hypothetical step potential for a one-dimensional particle. To understand this system, it is helpful to break it into regions. In the first region (I), there is a potential well that is almost like the potential of the particle-in-a-box problem. The second region (II) has a constant, but not infinite potential. The third region (III) has a flat potential, and this continues to infinity.

The connection or continuity conditions at region boundaries are of central importance in finding the wavefunctions for step potentials. First, we may assume a free-particle type of wavefunction within each region because the potential is unchanging, or is flat. For the problem represented by Fig. 8.2, this means

$$\psi_I(x) = A_I e^{-ik_I x} + B_I e^{ik_I x} \qquad \text{where } k_I = \sqrt{2mE}\,/\hbar \tag{8-13}$$

$$\psi_{II}(x) = A_{II} e^{-ik_{II} x} + B_{II} e^{ik_{II} x} \qquad \text{where } k_{II} = \sqrt{2m(E-V_{II})}\,/\hbar \tag{8-14}$$

$$\psi_{III}(x) = A_{III} e^{-ik_{III} x} + B_{III} e^{ik_{III} x} \qquad \text{where } k_{III} = \sqrt{2m(E-V_{III})}\,/\hbar \tag{8-15}$$

Then, for the wavefunction to be continuous at the boundary regions, the following conditions must be satisfied.

$$\psi_I(a) = \psi_{II}(a) \tag{8-16}$$

$$\psi_{II}(b) = \psi_{III}(b) \tag{8-17}$$

For the wavefunction to have a continuous first derivative everywhere, the following conditions must be satisfied as well.

$$\left.\frac{d\psi_I}{dx}\right|_{x=a} = \left.\frac{d\psi_{II}}{dx}\right|_{x=a} \tag{8-18}$$

$$\left.\frac{d\psi_{II}}{dx}\right|_{x=b} = \left.\frac{d\psi_{III}}{dx}\right|_{x=b} \tag{8-19}$$

Also, the wavefunction must vanish at x = c because the potential is infinite at that point:

$$\psi_I(c) = 0 \tag{8-20}$$

Therefore, we have five connection conditions on the six constants $A_I$, $A_{II}$, $A_{III}$, $B_I$, $B_{II}$, and $B_{III}$. Though these are insufficient to completely determine the six constants, they do imply relationships among them.

Let us take as an initial condition on the system in Fig. 8.2 that each and every particle came streaming in from the far right (i.e., x > b) with an energy, $E > V_{III}$. The quantum mechanical statement of this condition is that in region III, the eigenvalue associated with momentum must be negative, since a negative eigenvalue corresponds to motion in the −x direction which is to the left. The momentum operator is $-i\hbar\, d/dx$, and so $\psi_{III}$ will have a negative eigenvalue with the momentum operator only if $B_{III}$ equals zero. Thus, with the initial condition that a particle encountered the potential in Fig. 8.2 by approaching from the far right, we may say that

$$\psi_{III} = A_{III}\, e^{-i k_{III} x}$$

The first boundary to consider is at x = b, the boundary between regions II and III. Since a coordinate origin has not been specified, we may choose it to be at this boundary; that is, b = 0. The continuity conditions of Eqs. (8-17) and (8-19) are now

$$A_{III} = A_{II} + B_{II}$$

$$-k_{III} A_{III} = -k_{II} A_{II} + k_{II} B_{II}$$

By rearranging, we can obtain $A_{II}$ and $B_{II}$ in terms of $A_{III}$.

$$A_{II} = \frac{A_{III}}{2}\left(1 + \frac{k_{III}}{k_{II}}\right) \tag{8-21}$$

$$B_{II} = \frac{A_{III}}{2}\left(1 - \frac{k_{III}}{k_{II}}\right) \tag{8-22}$$

The ratio, $k_{III}/k_{II}$, is the key factor in the relative size of the amplitudes, $A_{II}$ and $B_{II}$.

The initial condition that we are using is that of particles approaching from the far right. This is implied by $B_{II}$ being zero. In the same way, we may say that if $B_{II}$ were zero, then the wavefunction in region II would correspond to motion to the left; or, if $A_{II}$ were zero, the wavefunction would correspond to motion to the right. When neither coefficient is zero, the postulates provide an interpretation of the probability of motion being to the left or to the right. At any point in region II, the normalized probability that a measurement of the momentum will yield a negative value (motion to the left) is

$$P_{II}^{left} = \frac{A_{II}^* A_{II}}{A_{II}^* A_{II} + B_{II}^* B_{II}} \tag{8-23}$$

The probability of motion in the opposite direction is

$$P_{II}^{right} = \frac{B_{II}^* B_{II}}{A_{II}^* A_{II} + B_{II}^* B_{II}} \tag{8-24}$$

And clearly, the probability of motion in either direction is properly one.

Given the initial condition of the particle approaching from the far right of region III, we may regard the probability in Eq. (8-23) as a measure of the likelihood of *transmission*, T, since it gives the likelihood that the particle will continue in the same direction it was going initially. We may regard the probability in Eq. (8-24) as a measure of the likelihood of *reflection*, R, since it is the probability that the particle has reversed direction or has been reflected back. From Eqs. (8-21) and (8-22), we may substitute to obtain the following expressions.

$$T \equiv P_{II}^{left} = \frac{(1+g)^2}{(1+g)^2 + (1-g)^2} = \frac{(1+g)^2}{2(1+g^2)}$$

$$R \equiv P_{II}^{right} = \frac{(1-g)^2}{2(1+g^2)}$$

where $g = k_{III}/k_{II}$, and where it has been assumed that $E > V_{II}$. Next, using Eqs. (8-14) and (8-15), g can be expressed in terms of E, the particle's energy, $V_{II}$, and $V_{III}$. This leads to the following result.

$$T = \frac{1}{2} + \frac{\sqrt{(E - V_{II})(E - V_{III})}}{2E - V_{II} - V_{III}} \qquad (8\text{-}25)$$

$$R = \frac{1}{2} - \frac{\sqrt{(E - V_{II})(E - V_{III})}}{2E - V_{II} - V_{III}} \qquad (8\text{-}26)$$

From this, we see that the likelihood of reflection and transmission at a step potential boundary can have a simple dependence on the difference between the particle's energy and the heights of the two potential steps.

To understand the implications of Eqs. (8-25) and (8-26), we shall consider certain specific choices of the potentials and the particle's energy.

1. If $V_{II}$ were equal to $V_{III}$, then the rightmost terms in Eqs. (8-25) and (8-26) would reduce to 1/2. Then, we would have T = 1 and R = 0. This is reassuring for the condition that $V_{II}=V_{III}$ means that there would be no potential step (i.e., no boundary) and so there would be complete transmission.

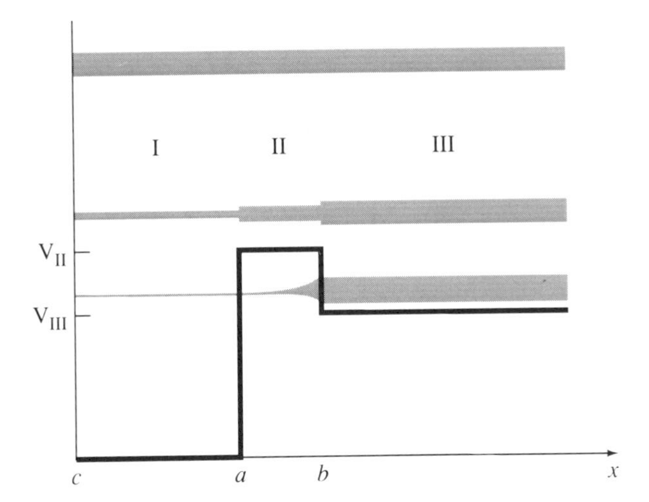


**FIGURE 8.3** The transmission of a beam of particles coming from the right and encountering the potential of Fig. 8.2 is represented by three horizontal lines corresponding to three specific particle energies. The width of each line is a representation of the transmission probability as a function of x. In the highest energy case, there is a negligible diminishment at the boundary between regions II and III, but at lower energies, there is a noticeable diminishment.

2. If the particle's energy, E, were much, much greater than both $V_{II}$ and $V_{III}$, then the rightmost terms would be very nearly equal to 1/2. Again, that would lead to T = 1 and R = 0. Classically, the difference, $E - V_{II}$ (or $E - V_{III}$), is the particle's kinetic energy in the region. So, if it has a lot of kinetic energy, the step in the potential is not too important; mostly the particle is likely to continue ahead. This is illustrated in Fig. 8.3.
3. If E were much, much greater than $V_{III}$, but only 10% bigger than $V_{II}$, then the rightmost terms in Eqs. (8-25) and (8-26) could be approximated:

$$\frac{\sqrt{(E - V_{II})(E - V_{III})}}{2E - V_{II} - V_{III}} \cong \frac{\sqrt{(E - V_{II})E}}{2E - V_{II}} = \frac{\sqrt{(1.1V_{II} - V_{II})\,1.1\,V_{II}}}{2\,(1.1\,V_{II}) - V_{II}}$$

$$= \sqrt{0.11}\;/\;1.2 \;=\; 0.28$$

Then, T = 0.78 and R = 0.22, and in contrast to the prior specific case, there is a significant probability that the incoming particle will be reflected back. This is at odds with the experience in our macroscopic, classical mechanical world where an object encountering a potential will not be stopped or reflected if it has energy in excess of the potential energy. In the quantum world, there is a probability of being reflected even if a particle's energy is more than the potential barrier. That probability grows as the particle's energy diminishes, and were E equal to $V_{II}$, then T and R would both be 1/2. This is illustrated in Fig. 8.3.

The important conclusion from this analysis is that a quantum mechanical particle may be reflected by encountering a potential barrier even if it has sufficient energy to pass over the barrier.

Let us next consider the case where the particle is approaching from the far right, just as before, but that the energy happens to be such that $V_{III} < E < V_{II}$. For this case, $k_{II}$ will be imaginary [see Eq. (8-14)], and so the general solution to the Schrödinger equation for region II is best expressed as

$$\psi_{II}(x) = A_{II} e^{-\alpha_{II} x} + B_{II} e^{\alpha_{II} x} \tag{8-27}$$

$$\text{where } \alpha_{II} = \sqrt{2m(V_{II} - E)} / \hbar$$

The connection conditions lead to relations similar to Eqs. (8-21) and (8-22).

$$A_{II} = \frac{A_{III}}{2}\left(1 + i\frac{k_{III}}{\alpha_{II}}\right) \tag{8-28}$$

$$B_{II} = \frac{A_{III}}{2}\left(1 - i\frac{k_{III}}{\alpha_{II}}\right) \tag{8-29}$$

Notice that $B_{II}$ is the complex conjugate of $A_{II}$. If we substitute this result into Eq. (8-27) and then group the real and imaginary terms, we obtain the following expression for the wavefunction.

$$\psi_{II}(x) = \frac{A_{III}}{2}\left(e^{-\alpha_{II} x} + e^{\alpha_{II} x}\right) + i\frac{A_{III} k_{III}}{2\alpha_{II}}\left(e^{-\alpha_{II} x} - e^{\alpha_{II} x}\right)$$

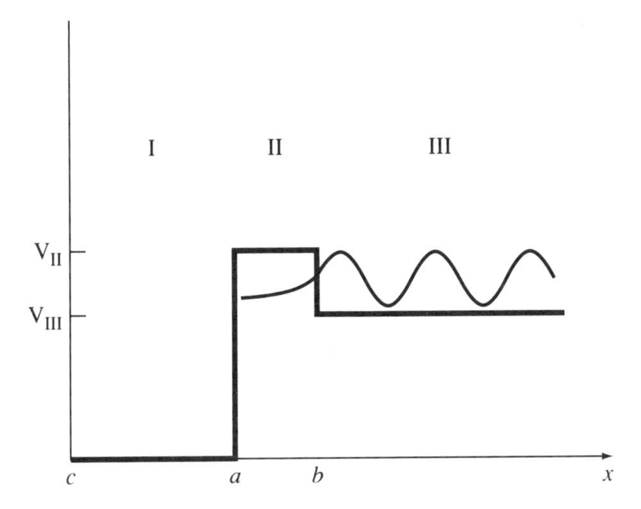


**FIGURE 8.4** The qualitative form of the real part of the wavefunction for a particle encountering the potential of Fig. 8.2 coming from the right with an energy less than $V_{II}$ but greater than $V_{III}$.

$$= A_{III}\left[\cosh(\alpha_{II}x) - i\frac{k_{III}}{\alpha_{II}}\sinh(\alpha_{II}x)\right] \tag{8-30}$$

This wavefunction is complicated, but the important result is that it is not strictly zero. That means the particle penetrates into region II even though it does not have the energy to surmount the potential barrier ($E<V_{II}$). Were it a classical particle, there would be no chance of finding it in region II, and so this quantum feature is referred to as tunneling through a barrier. The real part of the wavefunction is sketched in Fig. 8.4. The particle is said to be in a *classically forbidden* region, somewhat like passing through a solid wall. We do not develop reflection and transmission probabilities in this situation since this is not a free particle and the wavefunction does not consist of simple eigenfunctions of the momentum operator.

The analysis of the problem represented in Fig. 8.2 can be continued to region I. Using the connection conditions, though, becomes tedious, and so we turn to a better example of a potential well problem where there is tunneling. This is the problem represented in Fig. 8.5. This is almost the same as the problem in Fig. 8.2, but the difference is the absence of a barrier, which is the feature in region II of Fig. 8.2.

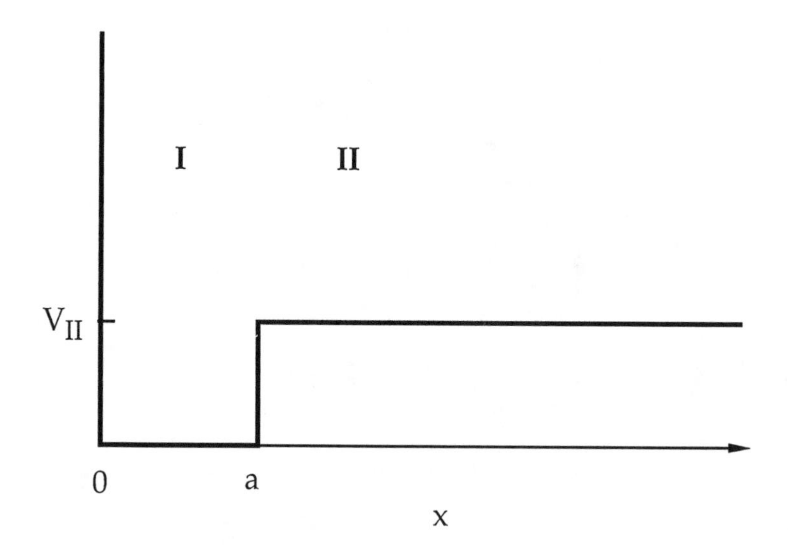


**FIGURE 8.5** A hypothetical step potential for a one-dimensional particle. In region I the potential is zero. In region II the potential is constant. At x = 0, the potential is taken to be infinite.

To analyze the problem in Fig. 8.5, we will start with region I, and as in the problem of Fig. 8.2, the general form for the wavefunction is given in Eq. (8-13). It is useful to apply right away the boundary condition that $\psi(0)= 0$ because the potential is infinite at x = 0. This condition requires that $A_I = -B_I$. Thus,

$$\psi_I(x) = A_I\left(e^{-ik_Ix} - e^{ik_Ix}\right)$$

$$= (2\,i\,A_I)\sin(k_Ix) = A_I'\sin(k_Ix) \tag{8-31}$$

If $E > V_{II}$, the wavefunction in region II is that already written as Eq. (8-14), and so the connection conditions at the boundary between region I and region II are,

$$A_I'\sin(k_Ia) = A_{II}\,e^{-ik_{II}a} + B_{II}\,e^{ik_{II}a}$$

$$A_I' k_I \cos(k_I a) = i k_{II} \left( -A_{II} e^{-i k_{II} a} + B_{II} e^{i k_{II} a} \right)$$

These can be manipulated, as desired, to express the relationship between $A_I'$, $A_{II}$, and $B_{II}$. And using Eqs. (8-23) and (8-24), we may develop expressions for the probability of right- or left-moving particles in region II.

If $E < V_{II}$ for the problem in Fig. 8.5, then the connection conditions are employed differently. In this circumstance, the wavefunction in region II must have the following form.

$$\psi_{II}(x) = A_{II} e^{-\alpha_{II} x} \quad \text{where} \quad \alpha_{II} = \sqrt{2m(V_{II} - E)} / \hbar \tag{8-32}$$

A term we might expect, $B_{II} e^{\alpha_{II} x}$, is missing. $B_{II}$ must equal zero because otherwise the probability density would increase to infinity as x goes to inifinity, and an infinite probability density is unphysical. Now, the connection conditions at the boundary between region I and region II are

$$A_I' \sin(k_I a) = A_{II} e^{-\alpha_{II} a} \tag{8-33}$$

$$A_I' k_I \cos(k_I a) = -\alpha_{II} A_{II} e^{-\alpha_{II} a} \tag{8-34}$$

Either one of these equations is sufficient to relate $A_I'$ and $A_{II}$ for any value of E. The two equations are not redundant, however, and so they must be providing another specification. It comes about by dividing Eq. (8-34) into Eq. (8-33) and then multiplying the resulting equation by $k_I$:

$$\tan(k_I a) = -\frac{k_I}{\alpha_{II}} = -\sqrt{\frac{E}{V_{II} - E}} \tag{8-35}$$

This is a condition on $k_I$, or equivalently, on the energy E. It will turn out from finding the values of E that satisfy this equation that only discrete values are possible. For instance, if the energy is small relative to $V_{II}$, that is, if $E \ll V_{II}$, then $\tan(k_I a)$ will be very nearly zero. The condition that $\tan(k_I a)$ is exactly zero is that

$$k_I a = n\pi \quad \text{where n is an integer}$$

It is important to notice that this condition was obtained for the particle-in-a-box problem. Therefore, the low-lying states of the problem represented by Fig. 8.5 will resemble the states of the particle in a box except that there will be some tunneling into the classically forbidden region II. Also, from Eq. (8-32), we expect the extent of this tunneling to increase as E increases.

The simple step potential problems illustrate several features of quantum systems. (i) A particle trapped in a potential, meaning the potential rises higher than the particle's energy, has discontinuous allowed energies. The basic examples are the harmonic oscillator and the particle in a box. If the particle is not trapped, then there is a continuum of allowed energy states. (ii) A particle may tunnel into a region where it is classically forbidden, meaning where its energy is below the potential energy. The likelihood for tunneling diminishes as the difference in the potential and the particle's energy grows. (iii) The likelihood that a particle will pass through a barrier region (e.g., a step potential) diminishes with increasing width of the region. (iv) There is a probability that a particle will be reflected by a barrier even if its energy is sufficient to surmount the barrier.

## 8.4 PARTICLE SCATTERING

An important class of experiments that may require quantum mechanical analysis of particles encountering potentials is particle scattering. The basic idea is to impinge a beam of electrons, atoms, or even molecules on an atomic or molecular target. The particles scatter in a manner that is dictated by the nature of the interaction potential, and the directions in which the particles scatter can be measured. Typically, particle detectors are arranged at regular locations surrounding a scattering target, and these detectors count the number and perhaps the energy of the particles that come in their direction. The reason for doing these experiments is that the data can be analyzed, in principle, to deduce the interaction potential. Such potentials represent fundamental chemical information that can be related to structure and stability questions. For spectroscopic experiments, we analyzed the mechanics of molecular vibration and rotation within the picture of quantum mechanics, and the predictions offered a basis for interpreting the measurements of transition

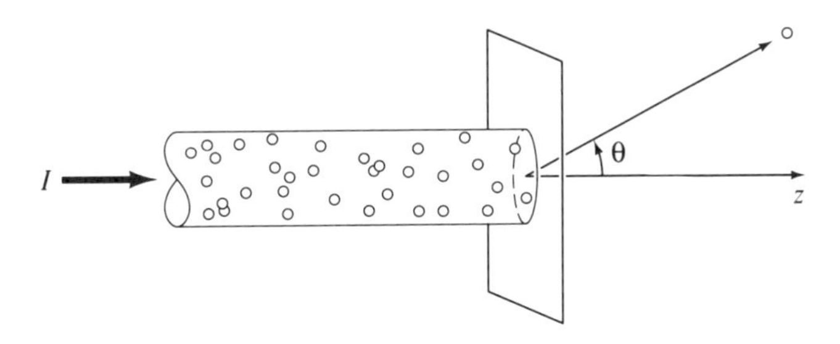


**FIGURE 8.6** Particles in a beam with intensity I impinge on a target and are scattered in different directions. The angle θ is the angle with respect to the direction of the beam, and the angle ϕ (not shown) is the angle about this axis. The intensity of the beam after hitting the target is in general some function of the direction. The direction of the incident particle beam is chosen to be parallel to the z-axis.

---

frequencies and intensities. For scattering experiments, we must analyze the mechanics of the scattering process in order to interpret the data. We shall take a quantum mechanical view, although for atom and molecule scattering, analysis via classical mechanics is often appropriate.

The mechanics of scattering by one particle impinging target must be related to the *scattering cross section*, σ, which gives the intensity of the impinging particle beam and is a function of the orientation angles. As shown in Fig. 8.6, we shall let θ and ϕ be the orientation angles measured relative to the location of the target and the direction of the impinging beam. Thus, we take the phenomenological definition of the *differential scattering cross section* to be

$$\sigma_d \equiv I(\theta, \phi) / I_{initial} \tag{8-36}$$

(This is phenomenological because it is a definition in terms of measured values instead of one arising from a fundamental analysis.) A total cross section is obtained by integrating the differential cross section over all angles. We seek to relate either the total or the differential cross section to the potential of the interaction between the particles in the beam and the target.

There are different types of collision events. *Elastic collisions* are those in which the total kinetic energy of the particle and the target do not change; however, kinetic energy may be transferred from one to the other. *Inelastic collisions* are those in which some of the translational kinetic energy is converted to internal energy of the impinging particle, the target, or both. For example, an atom A colliding with a molecule BC might leave the BC molecule in a higher vibrational state. The energy for the vibrational excitation comes from the kinetic energy of A colliding with BC. If the collision leads to chemical bond breaking or to bond formation, it is a *reactive collision*. We will limit our attention to the quantum mechanics of elastic collisions. We shall also assume that the interaction potential has no explicit time dependence.

We use the time-independent treatment of elastic scattering, and we take the form of the wavefunction for an impinging particle to be that of a plane wave with the particle moving in the direction $+\vec{k}$:

$$\psi_{\text{initial}}(\vec{r}) = c\, e^{i\vec{k}\bullet\vec{r}} \tag{8-37}$$

where c is a constant and $\vec{k}$ is the wavevector. Before the particle encounters the scattering potential, it is a free particle and its Hamiltonian is simply the kinetic energy operator.

$$\hat{H}_{\text{free}} = -\frac{\hbar^2}{2m}\left(\frac{\partial^2}{\partial x^2} + \frac{\partial^2}{\partial y^2} + \frac{\partial^2}{\partial z^2}\right) \tag{8-38}$$

(If the target is fixed, it is appropriate to use the mass of the impinging particle, m, in the Hamiltonian, but otherwise the mass should be the reduced mass, $\mu$, of the two-body system. In that case, the particles are viewed as moving toward each in a coordinate system where the center of mass is fixed.) Since $\vec{k}\bullet\vec{r} = k_x x + k_y y + k_z z$, we may easily apply the free particle Hamiltonian to the wavefunction in Eq. (8-37).

$$\hat{H}_{\text{free}}\,\psi_{\text{initial}} = -\frac{\hbar^2}{2m}\left(-k_x^2 - k_y^2 - k_z^2\right)\psi_{\text{initial}} \tag{8-39}$$

From this, we have the energy of the particle:

$$E = \hbar^2 (k_x^2 + k_y^2 + k_z^2) / 2m = \hbar^2 \vec{k}^2 / 2m \tag{8-40}$$

The wavefunction is an eigenfunction of the momentum component operators, $p_x$, $p_y$, and $p_z$, too. The magnitude of the momentum, $|p|$, is related to the energy via $|p| = \sqrt{2mE} = \hbar\,|\vec{k}|$.

To simplify the analysis, we shall choose the z-axis of the system to be parallel to the incident particle beam, as shown in Fig. 8.6. Thus, the wavefunction in Eq. (8-37) may be rewritten using

$$e^{i\vec{k}\bullet\vec{r}} = e^{ikz} \tag{8-41}$$

In the outgoing region, it is convenient to use the spherical polar coordinates, and then the wavefunction is seen to be a function of only r and θ.

$$e^{i\vec{k}\bullet\vec{r}} = e^{ikr\cos\theta} \tag{8-42}$$

The exponential in r and θ can be written as a superposition of spherical waves. In other words, we can use a series expansion to represent the wavefunction in the outgoing region.

$$\psi_{out} = \sum_{l=0}^{\infty} a_l\, \Gamma_l(r)\; P_l(\cos\theta) \tag{8-43}$$

This is referred to as a *partial wave expansion*. The functions $P_l$ are the Legendre polynomials. The expansion coefficients are $a_l$, and the unknown radial functions are $\Gamma_l$.

At this point, the interaction potential that gives rise to the particle scattering will be assumed to be a function only of a radial coordinate, r. The incident plane wave is scattered by this central potential, V(r), to yield the outgoing wave. We seek the amplitude for the outgoing wave since that will determine the differential cross section, the measurable quantity in a scattering experiment. The Schrödinger equation is

$$\left( -\frac{\hbar^2}{2m}\nabla^2 + V(r) \right) \psi_{out} = E\,\psi_{out} \tag{8-44}$$

For elastic scattering, the energy E must be that of Eq. (8-40), which is $\hbar^2 k^2/2m$. From substituting this for E in Eq. (8-44) and using the relation between $\nabla^2$ and $\hat{L}^2$ of Eq. (5-6b), the following is obtained.

$$-\frac{\hbar^2}{2m}\left(\frac{2}{r}\frac{\partial}{\partial r} + \frac{\partial^2}{\partial r^2} + k^2\right)\psi_{out} + \frac{\hat{L}^2}{2mr^2}\psi_{out} + V(r)\,\psi_{out} = 0 \qquad (8\text{-}45)$$

This equation is separable into a pure radial equation, because each expansion term in $\psi_{out}$ has an angular dependence $P_l(\cos\theta)$ and because of the property of the Legendre polynomials,

$$\hat{L}^2 P_l(\cos\theta) = \hbar^2\, l(l+1)\, P_l(\cos\theta)$$

Upon replacing $\psi_{out}$ by the expansion in Eq. (8-43), we obtain

$$\sum_{l=0}^{\infty} a_l\left[-\frac{\hbar^2}{2m}\left(\frac{2}{r}\frac{\partial}{\partial r} + \frac{\partial^2}{\partial r^2} + k^2 - \frac{l(l+1)}{r^2}\right) + V(r)\right]\Gamma_l(r)\,P_l(\cos\theta) = 0 \qquad (8\text{-}46)$$

This has the form of a sum of radial and angular product functions that we may represent as

$$\sum_{l=0}^{\infty} a_l\,\beta_l(r)\,P_l(\cos\theta) = 0 \qquad (8\text{-}47)$$

A separate equation for the radial functions is obtained by multiplying Eq. (8-47) or (8-46) by $\sin\theta$ and an arbitrary Legendre polynomial and then integrating over $\theta$. For convenience, the left hand and right hand sides of Eq. (8-47) are exchanged as this is carried out.

$$0 = \sum_{l=0}^{\infty} a_l\,\beta_l(r)\int_0^{\pi} P_j(\cos\theta)\,P_l(\cos\theta)\sin\theta\,d\theta$$

$$= \sum_{l=0}^{\infty} a_l\,\beta_l(r)\left(\frac{2\,\delta_{jl}}{2l+1}\right)$$

$$= 2\,a_j\,\beta_j(r)\,/\,(2j+1) \quad\Rightarrow\quad \beta_j(r) = 0 \qquad (8\text{-}48)$$

The result of the integration over θ follows from the properties of the Legendre polynomials.

Recalling what $\beta_l(r)$ represents in Eq. (8-46), we may rewrite the condition of Eq. (8-48) as

$$\left[ -\frac{\hbar^2}{2m}\left( \frac{2}{r}\frac{\partial}{\partial r} + \frac{\partial^2}{\partial r^2} + k^2 - \frac{l(l+1)}{r^2} \right) + V(r) \right] \Gamma_l(r) = 0 \qquad (8\text{-}49)$$

(The subscript j in Eq. (8-48) has been renamed $l$ here.) This differential equation is solved by first defining a new set of radial functions,

$$G_l(r) \equiv r\,\Gamma_l(r) \qquad (8\text{-}50)$$

and then substituting into Eq. (8-49). After simplification (including multiplying the equation by r), the following differential equation is obtained.

$$-\frac{\hbar^2}{2m}\left( \frac{\partial^2 G_l}{\partial r^2} + k^2 G_l - \frac{l(l+1)}{r^2} G_l \right) + V(r)\,G_l = 0 \qquad (8\text{-}51)$$

For the special case of V(r) = 0, the solutions of this equation are spherical Bessel functions.

The asymptotic form (i.e., the limit as $r \to \infty$) of the solutions of Eq. (8-51) are important because the measurement of the scattering cross sections is at large r. For the Bessel functions that satisfy Eq. (8-51) with V(r) = 0, the asymptotic form, which will be designated $\tilde{G}_l^{o}$ is

$$\tilde{G}_l^{o}(r) = g_l \sin\left( kr - \frac{l\pi}{2} \right) \qquad (8\text{-}52)$$

where $g_l$ is a constant. Solving Eq. (8-51) for V(r) being other than zero is more involved than the analysis so far, but the asymptotic solution is quite similar to that of Eq. (8-52).

$$\tilde{G}_l(r) = g_l \sin\left( kr - \frac{l\pi}{2} + \eta_l \right) \qquad (8\text{-}53)$$

$\eta_l$ is called a *phase shift*. It is possible to show that the phase shift is related to the solutions of Eq. (8-51).

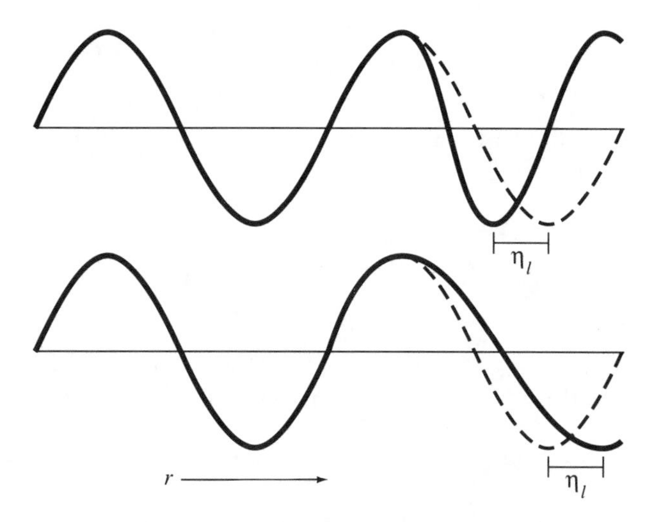


**FIGURE 8.7** Sketch of the form of a wavefunction with a negative and with a positive phase shift.

$$\sin \eta_l = -\frac{2\,m}{k\hbar^2} \int_0^\infty G_l^o(r)\; V(r)\; G_l(r)\; dr \tag{8-54}$$

An approximation of Eq. (8-54) is,

$$\sin \eta_l \cong -\frac{2\,m}{k\hbar^2} \int_0^\infty G_l^o(r)\; V(r)\; G_l^o(r)\; dr \tag{8-55}$$

This is helpful because $G_l^o$ is known even if $G_l$ has not been found. Furthermore, since the integral in Eq. (8-55) involves the square of $G_l^o$, then the integral value will be negative if V(r) is everywhere negative (i.e., if the potential is attractive), and it will be positive if V(r) is positive (repulsive). Thus, if the potential is attractive, the phase shift is positive, whereas if the potential is repulsive, the phase shift is negative. This is illustrated in Fig. 8.7. And of course, if V(r) = 0, the phase shift is zero.

The phase shift of Eq. (8-54) is specific to the $l$ partial wave. However, the complete wavefunction is an expansion in partial waves. So, the scattering at a given angle depends on the square of the total wavefunction (the probability density). The net angular dependence turns out to give

$$\sigma_d = \left| \frac{1}{2ik} \sum_{l=0}^{\infty} (2l+1)(e^{2i\eta_l} - 1) P_l(\cos\theta) \right|^2 \tag{8-56}$$

Notice how the phase shift enters this relation. With this result, we have an outline of the theory needed to deduce a potential energy function, V(r), from experimental differential cross sections. In practice, a trial form of the function V(r) is selected, and then the $G_l(r)$ is obtained that satisfies Eq. (8-51). Then, these are compared with the functions $\overset{o}{G}_l(r)$ [i.e., those for which V(r)=0] and then the phase shift is obtained for each choice of $l$. From that, the sum in Eq. (8-56) may be directly evaluated, and the result is compared with the cross section data. The process is repeated for different choices of V(r) until there is sufficient agreement between the calculated and experimental differential cross sections.

## Exercises

1. The particle-in-a-sloped-box problem is the usual particle-in-a-box problem but with the bottom of the box sloped because of a linear potential of the form V(x) = ax inside the box. Treat this as a perturbation on the usual particle-in-a-box problem and find the energies and wavefunctions correct to first order. Also, find the second-order energy corrections.
2. Find the expectation value $\langle x \rangle$ and the uncertainty $\Delta x$ for the lowest three states of the one-dimensional particle in a box.
3. Express the one-dimensional particle in a box wavefunctions of Eq. (8-5b) as a superposition of complex exponential functions. The general form was given in Eq. (8-2b). With this equivalent form, find the expectation value $\langle p \rangle$ and the uncertainty $\Delta p$ for the lowest three states of the one-dimensional particle in a box.
4. Make a sketch of the energy levels of the one-dimensional particle in a box by drawing a straight horizontal line for each level, placing the

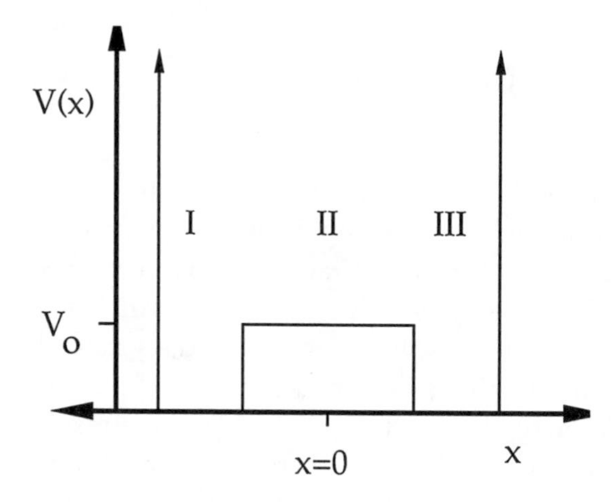


**FIGURE 8.8** A particle-in-a-box problem, but with a barrier step potential. The potential is taken to be infinite, except in regions I, II, and III. In regions I and III, the potential is zero. In region II, the potential is the constant value $V_o$.

lines against a vertical energy scale. Assume the length of the box is *l*. To the left, sketch the same energy levels for the particle in a box if the length is *l*/2, and to the right sketch the same energy levels if the length of the box is 2*l*.

5. Based on the wavefunctions for the harmonic oscillator and the wavefunctions for the particle in a box, construct a sketch to show the qualitative features of the three lowest energy wavefunctions for a particle trapped in a potential, V(x), that is infinite at x = 0, and that for x > 0 is $kx^2/2$ where k is a constant.
6. If the lowest energy state of the particle-in-a-box problem shown in Fig. 8.8 were at an energy slightly greater than $V_o$, in what ways would the wavefunctions of the lowest and first excited states differ from the wavefunctions of the same system but in the absence of the barrier step potential? Use a sketch to show qualitative features.
7. If the particle of the one-dimensional particle-in-a-box problem were an electron, what would the length of the box, *l*, have to be in order for there to be a transition energy of 10,000 $cm^{-1}$ for a transition from the lowest state to the first excited state? What would the value of n

for H-(C≡C)$_n$-H have to be for a linear polyene to have approximately this length?

*8.* Find the first-order energy corrections for the first four states of the particle-in-a-box problem if there is a perturbing potential of the form $V'(x) = 0.1\sqrt{\frac{2}{l}}\,\sin(\pi x / l)$.

*9.* For a particle in a three-dimensional box with equal length sides in the x-, y-, and z-directions, find the degeneracy for each of the first five energy levels.

*10.* Find the degeneracy of the first four energy levels of a particle in a three-dimensional box with the lengths of the sides related as $l_x = 16 l_y = 4 l_z$.

*11.* For the problem represented in Fig. 8.8, set up the connection conditions between regions and determine if $V_0$ is an allowed value for the energy of the particle.

*12.* Verify the derivation from Eq. (8-49) to Eq. (8-51).

★*13.* Repeat Prob. 1 using linear variation theory with a basis set of the first four functions of the usual particle-in-a-box problem.

## Bibliography

(See INTRODUCTORY and INTERMEDIATE texts listed at the end of Chaps. 3 and 4.)

1. W. H. Flygare, *Molecular Structure and Dynamics* (Prentice-Hall, Englewood Cliffs, New Jersey, 1978). This offers an intermediate to advanced level treatment of particle and light scattering. See Chap. 9.

2. L. I. Schiff, *Quantum Mechanics* (McGraw-Hill, New York, 1968). This is a general advanced level text.

3. R. B. Bernstein, *Chemical Dynamics via Molecular Beam and Laser Techniques* (Oxford University Press, New York, 1982). Included in Chap. 5 is an introduction to quantum mechanical scattering that is

beautifully tied to the laboratory techniques for scattering experiments. This is at an advanced or research level.

4. R. D. Levine and R. B. Bernstein, *Molecular Reaction Dynamics and Chemical Reactivity* (Oxford University Press, New York, 1987). This is a research level volume that shows the role of scattering experiments in understanding molecular reactions.

# Chapter 9

# Atomic Structure and Spectroscopy

*An important application of quantum mechanics is in determining the electronic structure of isolated atoms. The atom with the fewest electrons, the hydrogen atom, serves as an important model problem. The quantum mechanical analysis of the hydrogen atom is carried out in detail in this chapter. From that, we explore the qualitative features of the structure of more complicated atoms and their spectroscopy.*

## 9.1 THE HYDROGEN ATOM

The quantum mechanical description of a hydrogen atom is the starting point for understanding the electronic structure of atoms and of molecules. It is a problem that can be solved analytically, and it is useful to work through the details. The hydrogen atom consists of two particles, an electron and a nucleus, which is a proton. Since the proton mass is about 2000 times that of an electron, the proton would be expected to make small excursions about the center of mass of the atom relative to any excursions of the very light electron. That is, we expect the electron to be moving quickly about, and in effect, orbiting the nucleus.

The Schrödinger equation for this two-body problem starts out the same as the general two-body Schrödinger equation [Eq. (6-1)]; however, the potential function, V(r), will be different from that of the vibrating-rotating diatomic molecule. It is an electrostatic attraction of two point charges and its form is

$$V(r) = -\frac{Ze^2}{r} \qquad (9\text{-}1)$$

where Z is the integer nuclear charge, which is +1, and e is the size of the fundamental charge of an electron. This interaction is the product of the charges, in this case +Ze for the nucleus and –e for the electron, divided by the distance between them, r.

The potential function for the hydrogen atom is only dependent on the spherical polar coordinate r. Thus, the separation of variables carried out in going from Eq. (6-1) to Eq. (6-5) remains valid for this problem. That means we immediately know that the wavefunctions for the hydrogen atom will consist of some type of radial function, R(r), times a spherical harmonic function, $Y_{lm}(\theta,\phi)$. To find the radial function, we start with Eq. (6-5) and use the potential function of Eq. (9-1).

$$-\frac{\hbar^2}{2\mu}\left(\frac{2}{r}\frac{\partial R}{\partial r} + \frac{\partial^2 R}{\partial r^2} - \frac{l(l+1)}{r^2}R\right) - \frac{Ze^2}{r}R = ER \qquad (9\text{-}2)$$

$\mu$ is the reduced mass of the two-body system. Notice that with the sizable difference in mass of the two particles, the reduced mass is very nearly equal to the mass of the lighter particle, the electron.

$$\mu = \frac{m_e M_P}{m_e + M_P} = \frac{1836.15\, m_e^2}{m_e + 1836.15\, m_e} = \frac{1836.15\, m_e}{1837.15} \cong m_e$$

$M_P$ is the proton mass and $m_e$ is the electron mass.

In the vibrating-rotating two-body system, it was appropriate to approximate the $l(l+1)$ term in the radial Schrödinger equation; however, that would not be appropriate for the hydrogen atom. The separation distance between the electron and the proton, which is given by the coordinate r, can and does vary widely in the hydrogen atom states, and so there is no basis for using the truncated power series expansion, as was done for the vibrating-rotating diatomic. Fortunately, the differential equation in Eq. (9-2) has known solutions. There are an infinite number of these solutions for each particular value of the quantum number $l$, and we introduce a new quantum number, n, to distinguish these solutions. Both $l$ and n must label the different eigenfunctions, e.g., $R_{nl}(r)$. A condition on this new quantum number, n, that comes about in solving the differential equation is that for some choice of $l$, n may only take on the values $l+1$, $l+2$, $l+3$, and so on. This condition can be inverted so as to relate the value of $l$ to n. Then, we have that for a given choice of n, $l$ can be only 0, 1, 2, ..., or n − 1. Also, n must be a positive integer.

From the separation of variables, we now have that the wavefunctions for the hydrogen atom are

$$\psi_{nlm}(r,\theta,\phi) = R_{nl}(r)\, Y_{lm}(\theta,\phi) \tag{9-3}$$

And the quantum numbers that distinguish the possible states must satisfy these conditions:

$$n = 1, 2, 3, \ldots \tag{9-4a}$$

$$l = 0, 1, 2, \ldots, n-1 \tag{9-4b}$$

$$m = -l, -l+1, \ldots, l-1, l \tag{9-4c}$$

The radial functions, $R_{nl}(r)$, that are the eigenfunctions of Eq. (9-2) may be constructed from a set of polynomials called the *Laguerre polynomials*. The Laguerre polynomial of order k may be generated by

$$L_k(z) = e^z \frac{d^k}{dz^k}(z^k e^{-z}) \tag{9-5}$$

*Associated Laguerre polynomials* may be generated by,

$$L_k^j(z) = \frac{d^j}{dz^j} L_k(z) \tag{9-6}$$

The radial functions, expressed in terms of associated Laguerre polynomials are

$$R_{nl}(r) = \sqrt{\frac{(n-l-1)!}{2n\,[(n+l)!]^3}}\; e^{-\rho/2} \rho^l L_{n+l}^{2l+1}(\rho) \tag{9-7}$$

where ρ is the variable r scaled by the constants Z, μ, e, and ħ that are in the Schrödinger equation and by the quantum number, n.

$$\rho \equiv \frac{2 Z \mu e^2}{n \hbar^2} r \tag{9-8}$$

The square root factor in Eq. (9-7) is for normalization over the range from $r = 0$ to infinity. Recall that in spherical polar coordinates the volume element is $r^2\, dr\, \sin\theta\, d\theta\, d\phi$, and so the normalization of the radial equations obeys the following:

$$\int_0^\infty R_{nl}^2(r)\, r^2\, dr = 1$$

Table 9.1 lists the explicit forms for several of these radial functions.

When the functions of Eq. (9-7) are used in Eq. (9-2), the energy eigenvalue associated with a particular $R_{nl}(r)$ function is found to be

$$E_{nl} = -\frac{\mu Z^2 e^4}{2\hbar^2 n^2} \tag{9-9}$$

Thus, the energies of the states of the hydrogen atom depend only on the quantum number n. The lowest energy state is with n = 1, and for this state, the energy given by Eq. (9-9) is $-109{,}678\ cm^{-1}$. The next energy level occurs with n = 2, and this energy is one fourth ($2^{-2}$) of the lowest energy or $-27{,}420\ cm^{-1}$. Next is the energy level for states with n = 3. This energy is one ninth ($3^{-2}$) of the lowest energy or $-12{,}186\ cm^{-1}$. At n = 100, the energy of the hydrogen atom is $-11\ cm^{-1}$. Clearly, as n approaches infinity, the energy approaches zero. This limiting situation corresponds to the ionization of the atom; the electron is then completely separated from the nucleus, and there is zero energy of interaction.

Since the energy of the hydrogen atom depends only on n, and since, according to Eq. (9-4), there may be several states with the same n, then the states of the hydrogen atom may be degenerate. We can use the rules of Eq. (9-4) to see that the degeneracy of each level is $n^2$, as in Table 9.2.

TABLE 9.1 Hydrogen Atom Radial Functions.

| n, $l$ | | Radial Function $[\rho = 2Z\mu e^2 r/(n\hbar^2)$ and $N = (Z\mu e^2/\hbar^2)^{3/2}]$ |
|---|---|---|
| n = 1 | $l = 0$ | $R_{10}(r) = N\,2e^{-\rho/2}$ |
| n = 2 | $l = 0$ | $R_{20}(r) = \frac{N}{2\sqrt{2}}(2-\rho)\,e^{-\rho/2}$ |
| | $l = 1$ | $R_{21}(r) = \frac{N}{2\sqrt{6}}\rho\,e^{-\rho/2}$ |
| n = 3 | $l = 0$ | $R_{30}(r) = \frac{N}{9\sqrt{3}}(6-6\rho+\rho^2)\,e^{-\rho/2}$ |
| | $l = 1$ | $R_{31}(r) = \frac{N}{9\sqrt{6}}(4\rho-\rho^2)\,e^{-\rho/2}$ |
| | $l = 2$ | $R_{32}(r) = \frac{N}{9\sqrt{30}}\rho^2\,e^{-\rho/2}$ |
| n = 4 | $l = 0$ | $R_{40}(r) = \frac{N}{96}(24-36\rho+12\rho^2-\rho^3)\,e^{-\rho/2}$ |
| | $l = 1$ | $R_{41}(r) = \frac{N}{32\sqrt{15}}(20\rho-10\rho^2+\rho^3)\,e^{-\rho/2}$ |
| | $l = 2$ | $R_{42}(r) = \frac{N}{96\sqrt{5}}(6\rho^2-\rho^3)\,e^{-\rho/2}$ |
| | $l = 3$ | $R_{43}(r) = \frac{N}{96\sqrt{35}}\rho^3\,e^{-\rho/2}$ |
| n = 5 | $l = 0$ | $R_{50}(r) = \frac{N}{300\sqrt{5}}(120-240\rho+120\rho^2-20\rho^3+\rho^4)\,e^{-\rho/2}$ |
| | $l = 1$ | $R_{51}(r) = \frac{N}{150\sqrt{30}}(120\rho-90\rho^2+18\rho^3-\rho^4)\,e^{-\rho/2}$ |
| | $l = 2$ | $R_{52}(r) = \frac{N}{150\sqrt{70}}(42\rho^2-14\rho^3+\rho^4)\,e^{-\rho/2}$ |
| | $l = 3$ | $R_{53}(r) = \frac{N}{300\sqrt{70}}(8\rho^3-\rho^4)\,e^{-\rho/2}$ |
| | $l = 4$ | $R_{54}(r) = \frac{N}{900\sqrt{70}}\rho^4\,e^{-\rho/2}$ |

TABLE 9.2 Energy Levels of the Hydrogen Atom.

| Level | Energy[a] | Allowed $l$'s | Allowed m's | Degeneracy ($n^2$) |
|---|---|---|---|---|
| n = 1 | $E_{g.s.}$ | 0 | 0 | 1 |
| n = 2 | $E_{g.s.}/4$ | 0 | 0 | |
| | | 1 | 1, 0, –1 | 4 |
| n = 3 | $E_{g.s.}/9$ | 0 | 0 | |
| | | 1 | 1, 0, –1 | |
| | | 2 | 2, 1, 0, –1, –2 | 9 |
| n = 4 | $E_{g.s.}/16$ | 0 | 0 | |
| | | 1 | 1, 0, –1 | |
| | | 2 | 2, 1, 0, –1, –2 | |
| | | 3 | 3, 2, 1, 0, –1, –2, –3 | 16 |

[a] The energy is given in terms of the ground state energy, $E_{g.s.}$, which is $-109{,}678\ cm^{-1}$.

## 9.2 PROPERTIES OF THE RADIAL FUNCTIONS

Examination of the radial functions in Table 9.1 reveals that except for the $l = 0$ functions, all are zero-valued at $r = 0$ ($\rho = 0$). For $r > 0$, each $R_{nl}$ function has $n - l - 1$ points where the function is zero-valued. These points are roots of the polynomials in the $R_{nl}$ functions, and they are simply the points where the radial functions change sign. They are nodes in the wavefunctions. Fig. 9.1 is a plot of several of these functions. Since the quantum mechanical postulates tell us that the square of a wavefunction is the probability density, it is also interesting to notice the forms of $R_{nl}^2$, which are shown in Fig. 9.2.

One of the first features of the hydrogen atom where observations must be reconciled with the quantum mechanical picture is the size of the atom. The quantum mechanical description gives a probability distribution for finding the electron located about the nucleus. In analogy with a

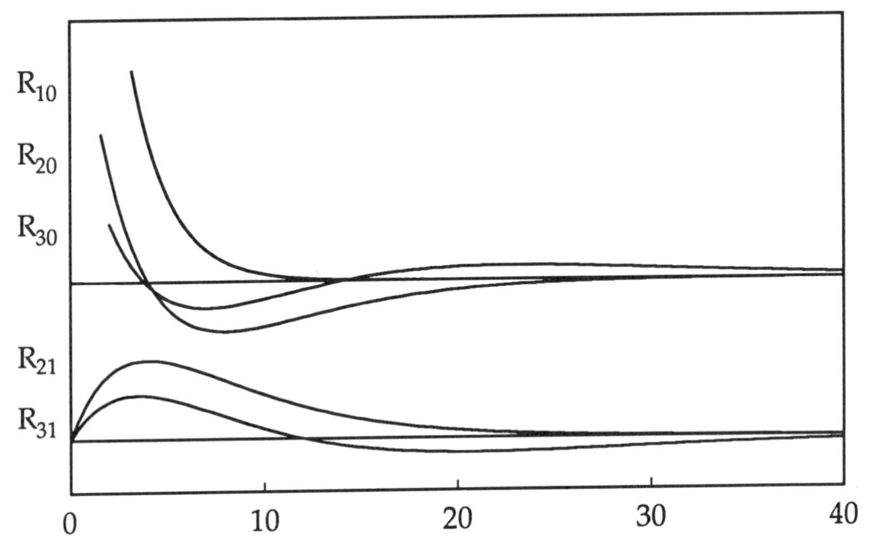


**FIGURE 9.1** Radial functions, $R_{nl}(r)$, of the hydrogen atom. The top three functions are the $l = 0$ functions; the first two $l = 1$ functions are shown below on a different vertical scale. The horizontal scale is in bohr radii which are 0.52918 Å.

classical mechanical system of two particles in an attractive potential, we regard this distribution as corresponding to the electron orbiting the nucleus (or both orbiting the center of mass). From Fig. 9.2, we see that the square of the wavefunction dies away smoothly at large r; it does not end abruptly at some particular value of r. This means that there does not exist a finite sphere that entirely encompasses the hydrogen atom and defines its size. Thus, a different notion of atomic size is required, and perhaps the most reasonable one is the average separation distance between the electron and the nucleus. From the quantum mechanical postulates, such an average may be obtained from the expectation value of r. So, for a hydrogen atom state specified by the quantum numbers n, $l$, m,

$$\langle r \rangle_{nlm} = \int \psi^*_{nlm}\, r\, \psi_{nlm}\, r^2\, dr\, \sin\theta\, d\theta\, d\phi$$

$$= \int r^3 R^2_{nl} dr \int Y^*_{lm} Y_{lm} \sin\theta\, d\theta\, d\phi$$

$$= \beta \frac{n^2}{Z}\left(\frac{3}{2} - \frac{l(l+1)}{2n^2}\right) \tag{9-10}$$

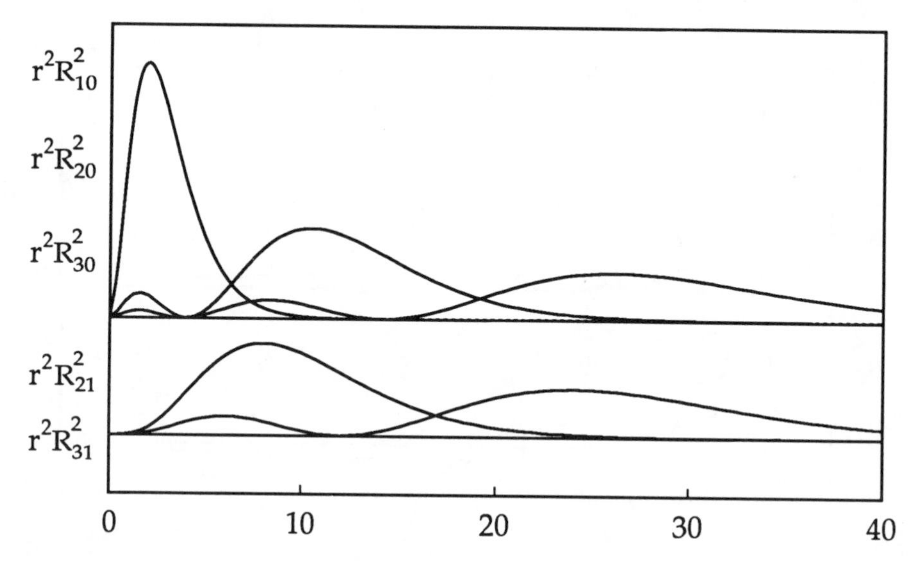


**FIGURE 9.2** Radial probability functions of the hydrogen atom, $r^2R_{nl}{}^2$. (The factor $r^2$ is included because the radial volume element is $r^2dr$.) The number of nodes increases with both n and $l$.

The integration over the radial coordinate has been carried out by employing special properties of the Laguerre polynomials. The integration over the angular coordinates yields one, since the spherical harmonic functions are normalized. The constant β, called the Bohr radius of the hydrogen atom (also designated $a_o$), is obtained from the reduced mass and two fundamental constants and is equal to 0.529177 Å or $0.529177 \times 10^{-8}$ cm.

$$\beta = \frac{\hbar^2}{\mu e^2} \tag{9-11}$$

For the ground state of the hydrogen atom, the average value of the radial coordinate is $3\beta/2$ or about 0.8 Å. This result fits experience showing that atomic dimensions are on the order of Å. On the other hand, a hydrogen atom in a state very near the ionization limit would be significantly larger. With n = 1000 and $l = 0$, the expectation value for r is about 0.01 cm.

The deviation from the average separation distance can be obtained from the square of the expectation value of r and from the expectation value of $r^2$. The formula that is obtained upon integration of the hydrogenic radial equations with $r^2$ is

$$<r^2>_{nlm} = \frac{\beta^2 n^4}{Z^2}\left(\frac{5}{2} - \frac{3l(l+1) - 1}{2n^2}\right) \tag{9-12}$$

For the ground state, this yields $3\beta^2$, and so the uncertainty in a measurement of r would be $\sqrt{3}\,\beta/2$ or about 0.45 Å.

## 9.3 ORBITAL ANGULAR MOMENTUM

The angular parts of the hydrogen atom wavefunctions are the spherical harmonics, which are, of course, eigenfunctions of the angular momentum operators, $L^2$ and $L_z$. The associated eigenvalues, $l(l+1)\hbar^2$ and $m\hbar$, give the magnitude and the z-component of the angular momentum vector arising from the orbital motion. The orbital motion of an electrically charged particle is a circulation of charge, and that must give rise to the type of magnetic field that is associated with a magnetic dipole source. In classical electromagnetic theory, the magnetic dipole moment,* $\vec{\mu}$, from charge flowing through a circular loop is proportional to the current and to the area of the loop, while its direction is perpendicular to the plane of the loop. We may analyze the hydrogen atom's magnetic dipole by considering it to be a charge, –e, flowing around a loop of some radius r, and then generalizing to the true orbital motion. The area of the loop is $\pi r^2$, and the current is the charge times the frequency at which the charge passes through any particular point on the loop (i.e., the angular velocity, $\omega$, divided by $2\pi$).

$$\mu = (\pi r^2)\left(\frac{-e\omega}{2\pi}\right)\frac{1}{c} = -\frac{e}{2c}r^2\omega \tag{9-13}$$

where c is the speed of light. The angular momentum for a particle moving about a circular loop is the particle's mass times the square of the

* Notice that the Greek letter $\mu$ is used for many things, reduced mass, magnetic dipole moment, and electric dipole moment. While this may be confusing, it should be clear from the context which is meant.

radius of the loop times $\omega$, that is, $mr^2\omega$. By collecting $r^2\omega$ in Eq. (9-13), we can introduce the angular momentum, L, and then generalize to orbital motion by allowing it to be the angular momentum vector of the hydrogen atom:

$$\vec{\mu} = -\frac{e}{2mc}\vec{L} \tag{9-14}$$

The magnetic dipole moment vector is proportional to the orbital angular momentum vector. Since angular momentum will be in the units of $\hbar$, it is convenient to collect it with the proportionality factor in Eq. (9-14) and define a new constant, $\mu_B$, called the *Bohr magneton.*

$$\mu_B \equiv \frac{e\hbar}{2mc} \tag{9-15}$$

This is the basic unit or measure for electronic magnetic dipole moments in the same sense that $\hbar$ is the measuring unit for angular momentum.

If an external magnetic field is applied to an isolated hydrogen atom, the effect of the field must be incorporated into the quantum mechanical description. This means that the interaction between the magnetic dipole of the orbital motion and the external field must be added to the Hamiltonian. The classical interaction goes as the dot product of the dipole moment and the field. The quantum mechanical operator that corresponds to this classical interaction is easy to determine because of Eq. (9-14). With $\vec{H}$ as the applied magnetic field,

$$\vec{\mu} \bullet \vec{H} = \mu_x H_x + \mu_y H_y + \mu_z H_z$$

$$= -\frac{e}{2mc}\left(L_x H_x + L_y H_y + L_z H_z\right)$$

Letting the orientation of the field define the z-axis in space means that the field components in the x- and y-directions are zero. Thus, the additional term in the Hamiltonian needed to account for an external field is

$$\hat{H}^{int} = -\vec{\mu} \bullet \vec{H} = \mu_B H_z \hat{L}_z / \hbar \tag{9-16}$$

The wavefunctions for the states of a hydrogen atom in an external uniform magnetic field are eigenfunctions of the original Hamiltonian, which we shall now identify as $H^o$, with $H^{int}$ added to it.

We should realize that since the eigenfunctions, $\psi_{nlm}$, of $H^o$ are eigenfunctions of the operator $L_z$, then they must already be eigenfunctions of $H^o + H^{int}$.

$$(\hat{H}^o + \hat{H}^{int})\,\psi_{nlm} = \left(E_n + m\mu_B H_z\right)\psi_{nlm} \tag{9-17}$$

The eigenenergies now have a dependence on the m quantum number. This means that an applied magnetic field will remove the degeneracy of states with the same n and $l$, but with different m quantum numbers. The separation between the levels will increase with the strength of the applied field, according to Eq. (9-17). We can also see from this result why the m quantum number is often referred to as the magnetic quantum number.

The angular form of the orbital functions, $\psi_{nlm}$, or really of the spherical harmonic functions, is interesting. The $l = 0$ or s orbitals are spherically symmetric, and that means they can be represented as spheres. The form of the higher $l$ orbitals is more complicated. Fig. 9.3 shows the form of the m = 0 spherical harmonics. Notice that the number of nodal planes (planes in space where the function is zero) is equal to $l$. Thus, an $l = 1$ m = 0 or $p_0$ function is zero-valued everywhere in the xy-plane. The sign or phase of the function changes from one side of this plane to the other. The $l = 2$ m = 0 or $d_0$ function has two nodal planes.

Certain linear combinations of the spherical harmonic functions are easy to represent and offer a useful picture to keep in mind. The linear combinations and designations for $l = 1$ and $l = 2$ orbitals are

$$Y_{11} + iY_{1-1} \rightarrow p_x$$

$$Y_{11} - iY_{1-1} \rightarrow p_y$$

$$Y_{10} \rightarrow p_z$$

$$Y_{22} + iY_{2-2} \rightarrow d_{x^2-y^2}$$

$$Y_{22} - iY_{2-2} \rightarrow d_{xy}$$

$$Y_{21} + iY_{2-1} \rightarrow d_{xz}$$

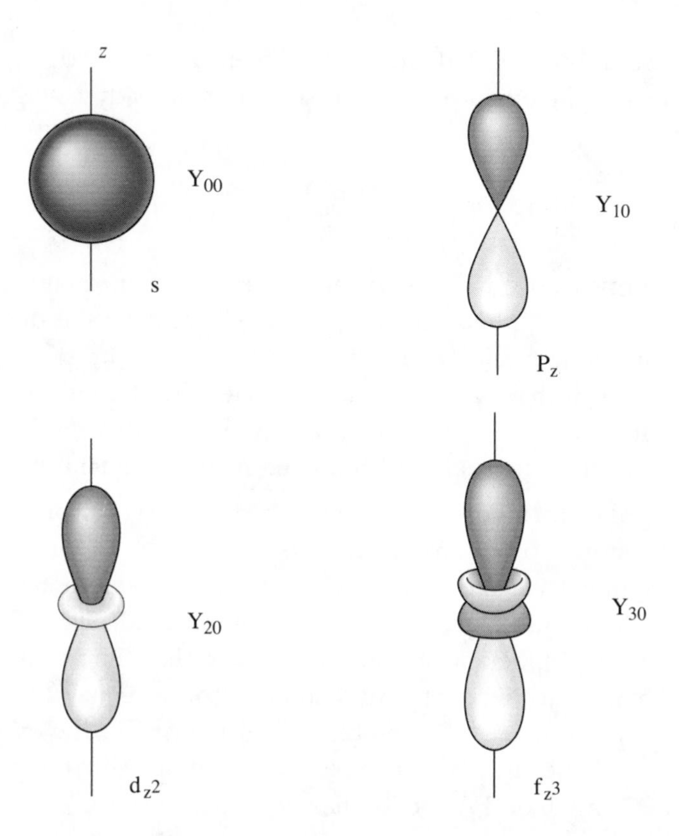


**FIGURE 9.3** Representation of several of the spherical harmonic functions with m = 0.

$$Y_{21} - iY_{2-1} \quad \rightarrow \quad d_{yz}$$

$$Y_{20} \quad \rightarrow \quad d_{z^2}$$

These combinations provide the real parts of the angular functions, and they are represented in Fig. 9.4. Similar combinations can be made for higher *l* angular momentum functions.

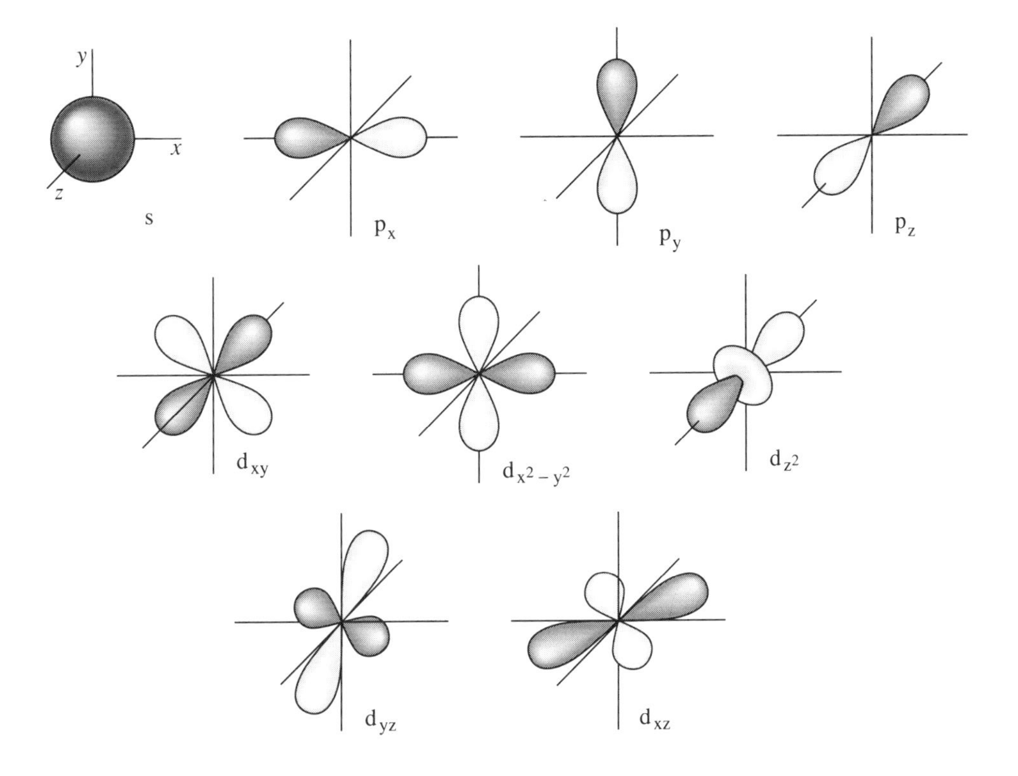


**FIGURE 9.4** Representations of the real parts of the angular functions formed from linear combinations of spherical harmonic functions.

## 9.4 ELECTRON SPIN

Electrons have an intrinsic angular momentum which gives rise to an intrinsic magnetic dipole moment. Very early in the history of quantum mechanics, deflection experiments that measured the magnetic moment of a moving particle by passing it through a magnetic field were able to establish that an electron in an $l = 0$ orbital still possesses a magnetic moment even though $\vec{L} = 0$. Of considerable excitement at the time was the fact that there were two and only two possible z-components of this magnetic moment. Assuming that the magnetic moment was proportional to some angular momentum vector, the questions that arose involved the source of this other type of angular momentum and what was the size of the vector.

The source of the angular momentum is intrinsic to the electron and is referred to as spin, because the spin of a solid body about an axis is a source of angular momentum. However, the term for this intrinsic feature of an electron does not mean we should picture the electron as some mass spinning about its axis. The feature is a more subtle characteristic than that. It can be accounted for with quantum mechanics, but only if the mechanics have been adapted for relativistic effects. So, spin is more a name than a description of this property of an electron.

That there are two and only two possible z-axis projections of the electron's intrinsic magnetic dipole, or equivalently, its spin angular momentum vector, is important. For orbital angular momentum, the magnetic quantum number, m, could take on values ranging from $-l$ to $l$. The number of such values, which is the number of different projections on the z-axis, is $2l+1$, an odd number. For there to be an angular momentum that has an even number of projections, the associated quantum number must be a half-integer. Letting s be the quantum number for spin angular momentum, the relation $2s+1=2$ requires that $s=1/2$. (We say that the spin of an electron is 1/2.) The quantum number that gives the projection of the spin vector on the z-axis is $m_s$, and to avoid confusion at this point, we now designate the orbital angular momentum m quantum number $m_l$.

There are two spin states for an electron. For one, $m_s = 1/2$, and for the other, $m_s = -1/2$. The first state is commonly identified as being the "spin-up" state, and the second is the "spin-down" state. This is because the orientation of the spin vector with respect to the z-axis is in the +z-direction, or up, for the first, and in the –z-direction, or down, for the second. A shorthand designation of these electron spin states is to write $\alpha$ for an electron with $m_s = 1/2$, and $\beta$ for $m_s = -1/2$. $\alpha$ and $\beta$ are meant to be functions in the same way that $\psi$ is a function of r, $\theta$, and $\phi$. However, $\alpha$ and $\beta$ are abstract functions – we will not express them explicitly – of an abstract coordinate, a spin coordinate. $\alpha$ and $\beta$ are orthonormal functions, and that means the following relations hold.

$$<\alpha \mid \alpha> \;=\; <\beta \mid \beta> \;=\; 1 \tag{9-18a}$$

$$<\alpha \mid \beta> \;=\; <\beta \mid \alpha> \;=\; 0 \tag{9-18b}$$

The integration is over the spin coordinate, and again this is not done as explicit integration.

The magnetic moment that is associated with an electron's spin will interact with an externally applied magnetic field. This magnetic moment is proportional to the spin vector. Analogous to Eq. (9-14) is the following,

$$\vec{\mu} = g_e \frac{e}{2mc} \vec{S} \tag{9-19}$$

This relation differs in an important way from Eq. (9-14). It says that the magnetic dipole moment is proportional to the spin vector with not only a factor of e/2mc but also with an additional factor, $g_e$. (Recall that the proportionality factor for orbital angular momentum is e/2mc.) The additional factor is required because the simple picture of circulation of charge that led to Eq. (9-14) does not apply to the intrinsic spin of an electron. The measured value for the dimensionless constant $g_e$ for a free electron is 2.0023.

## 9.5 ATOMIC ORBITALS AND HYDROGEN ATOM STATES

The wavefunctions, $\psi_{nlm}$, for the hydrogen atom are often referred to as orbitals. Generally, orbitals are any functions of the spatial coordinates of one electron, which in this case means the spherical polar coordinates r, θ, ϕ. The orbitals are named with (1) the value of the principal quantum number, n, (2) a letter associated with the *l* quantum number, and (3) a numerical subscript which is the value of $m_l$. The letters* s, p, d, and f were

* The letters s, p, and d originate in the names sharp, principal, and diffuse. These were the terms that were given to absorption and emission lines in the atomic spectra of alkali atoms on the basis of the appearance of those lines, usually on a photographic plate. Lines of similar type formed series, and it was learned that the transition frequencies measured for a series followed a simple mathematical progression. With quantum mechanics the progressions became understandable consequences of the allowed energy levels and wavefunctions. From this association with types of lines come the orbital letters.

associated with the $l$ quantum number as $s(l = 0)$, $p(l = 1)$, $d(l = 2)$, and $f(l = 3)$. The next is $g(l = 4)$, and from then on the series follows the alphabet. Thus, the ground state orbital of the hydrogen atom is named $1s_0$ or just 1s. The n=2 orbitals are 2s, $2p_1$, $2p_0$, and $2p_{-1}$.

The complete wavefunction of a hydrogen atom is a product of the spatial wavefunction and a spin function, either $\alpha$ or $\beta$. So, with the spin of the electron incorporated, the product functions $\psi_{nlm}\alpha$ and $\psi_{nlm}\beta$ (or $1s\alpha$, $1s\beta$, etc.) are referred to as *spin-orbitals*. Spin introduces further degeneracy in the set of eigenfunctions of the original hydrogen atom Hamiltonian, $H^o$. The 1s orbital gives rise to the $1s\alpha$ and $1s\beta$ spin-orbitals, and so these functions are degenerate functions of $H^o$. Basically, spin doubles the number of states.

The magnetic moments arising from electron orbital motion and from electron spin may interact. This feature of atomic structure, and molecular structure, too, is termed *spin-orbit interaction*. Since the magnetic dipoles due to spin and orbital motion are proportional to their respective angular momentum vectors, the interaction is proportional to the dot product of the angular momentum vectors. This interaction is a small perturbation on the $H^o$ description of the hydrogen atom. We treat it phenomenologically at this point by introducing a proportionality constant, $\alpha$, rather than by developing a fundamental expression that would be used in place of $\alpha$. (Notice the context distinguishes this use of $\alpha$ from the spin function $\alpha$.) The perturbing Hamiltonian for spin-orbit interaction, then, is

$$\hat{H}' = \alpha \vec{L} \cdot \vec{S} \qquad (9\text{-}20)$$

We presume that the spin-orbit interaction constant, $\alpha$, is to be determined from some measurement.

Spin-orbit interaction implies a coupling of the two "motions" of spin and orbit. From the discussion in Chap. 5, we would expect that this coupling may mix the hydrogenic spin-orbital states with the resulting wavefunctions no longer assured to be eigenfunctions of the $S_z$ and $L_z$ operators. The interaction Hamiltonian of Eq. (9-20) can be rewritten following Eq. (5-17).

$$\hat{H}' = \frac{\alpha}{2}\left[\hat{J}^2 - \hat{L}^2 - \hat{S}^2\right] \qquad (9\text{-}21)$$

where J refers to the total angular momentum: $\vec{J} = \vec{L} + \vec{S}$. The rules of angular momentum addition give the allowed values of the quantum number J as ranging from $l$ + s downward, in steps of one, to $|l - s|$. For the hydrogen atom's single electron, s = 1/2. Thus, the spin-orbit coupled states of the hydrogen atom will have J quantum numbers equal to $l \pm 1/2$, except if $l$ = 0, then J = 1/2. These spin-orbit coupled states, which we shall designate concisely with the valid quantum numbers in a bra or ket vector, $|n J l s>$ or $<n J l s|$, are eigenfunctions of the operator H' in Eq. (9-21).

$$\hat{H}' |nJls> = \frac{\alpha \hbar^2}{2} \left[ J(J+1) - l(l+1) - s(s+1) \right] |nJls> \tag{9-22}$$

Thus, the energies of the states after accounting for spin-orbit coupling follow from Eqs. (9-9) and (9-22).

$$E_{nJls} = -\frac{\mu Z^2 e^4}{2 \hbar^2 n^2} + \frac{\alpha \hbar^2}{2} \left[ J(J+1) - l(l+1) - s(s+1) \right] \tag{9-23}$$

Notice that the energies are subscripted with n, J, $l$, and s since these are the valid quantum numbers for states with spin and orbital motion coupled.

The *spin-orbit splitting* is the energy difference between states that are otherwise degenerate. The case with n = 2 and $l$ = 1 is an example. In the absence of spin-orbit effects, the six hydrogen atom states, $2p_1\alpha$, $2p_1\beta$, $2p_0\alpha$, $2p_0\beta$, $2p_{-1}\alpha$, $2p_{-1}\beta$, are degenerate. These states may be mixed in some way because of spin-orbit interaction, and the resulting states would be distinguished according to the two possible J values, J = 1 + 1/2 = 3/2 and J = 1 − 1/2 = 1/2. (There are still six states since the J = 3/2 coupling is four-fold degenerate and the J = 1/2 coupling is doubly degenerate.) Using the formula in Eq. (9-22) we can evaluate the spin-orbit energy for the two possible J values.

$$J = 3/2: \quad E(\text{spin-orbit}) = \frac{\alpha \hbar^2}{2} \left[ \frac{3}{2}\left(\frac{3}{2} + 1\right) - 1(1+1) - \frac{1}{2}\left(\frac{1}{2} + 1\right) \right]$$

$$= \alpha \hbar^2 / 2$$

$$J = 1/2: \quad E(\text{spin-orbit}) = \frac{\alpha \hbar^2}{2} \left[ \frac{1}{2}\left(\frac{1}{2} + 1\right) - 1(1+1) - \frac{1}{2}\left(\frac{1}{2} + 1\right) \right]$$

$$= -\alpha \hbar^2$$

$$\text{Energy difference} \quad = -3\alpha \hbar^2 / 2$$

The energy difference in these two spin-orbit energies is the splitting. We can see that if the splitting were to be measured spectroscopically, then the value of the spin-orbit interaction constant, $\alpha$, would be known.

When an external magnetic field is applied to the hydrogen atom in a state for which $l > 0$, the quantum mechanical analysis of the energies becomes more complicated. The complete Hamiltonian will include the spin-orbit interaction, the interaction of the orbital magnetic dipole with the field, and the interaction of the spin dipole with the field. Instead of treating this situation generally, consider a special case: If the external field were so strong that the interaction energies with the field were much greater than the spin-orbit interaction, a good approximate description would be to neglect the spin-orbit interaction entirely. (A better approximation would be to include the spin-orbit interaction via low-order perturbation theory.) Physically, the strong field may be thought of as orienting the individual magnetic dipoles and thereby overwhelming their coupling with each other. A very strong field, then, is said to decouple the spin and orbital magnetic dipoles. On the other hand, a very weak field would not decouple the dipoles, but would interact with the net magnetic dipole that results from the sum of the spin and orbital angular momentum vectors. (Again, perturbation theory could be used for a more accurate energetic analysis.) The spectra of atoms and molecules obtained with an applied magnetic field are called *Zeeman spectra* and are discussed further in Chap. 12.

## 9.6 ORBITAL PICTURE OF THE ELEMENTS

The Hamiltonians for the Schrödinger equations for many-electron atoms are complicated because of electron-electron interaction. (The like electrical charges repel.) This couples the motions of the different electrons, and that precludes separation of variables. However, if the Schrödinger equation were separable into the coordinates of the different electrons, then the form of the resulting wavefunctions would be products of spin-orbitals. Such a product form can serve as a meaningful approximate description, and it is certainly of qualitative use in understanding the electronic structure of the different elements of the Periodic Table. Within this approximate separation, we may build up the electronic structures of atoms by assigning electrons to specific orbitals or spin-orbitals. These assignments are termed *electron occupancies** in that they indicate the spatial orbitals that are filled by the electrons.

The *Pauli principle* for the electrons of atoms and molecules says that no two electrons can occupy the same spin-orbital, and this is obviously important in building up the orbital picture of the elements. As discussed in Chap. 10, this principle goes hand in hand with the indistinguishability of the electrons and with their half-integer intrinsic spin. For now, it is important in saying that each distinct spatial orbital (e.g., 1s, 2s, $2p_1$, $2p_0$, $2p_{-1}$) may be occupied by only two electrons. This is because a given spatial orbital can be combined with either an $\alpha$ or $\beta$ spin function to form two and only two different spin-orbitals.

A set of spin-orbitals with the same n quantum number is referred to as a *shell*, and the set of spin-orbitals with the same n and $l$ quantum numbers is a *subshell*. The Pauli principle leads to the conclusion that an s type ($l = 0$) subshell can have an occupancy of at most 2 electrons. A p subshell ($l = 1$) can have an occupancy of at most 6 electrons, while a d subshell ($l = 2$) can have an occupancy of at most 10 electrons.

In the hydrogen atom, the energetic ordering of the spatial orbitals is clearly according to the principal quantum number, n. To the extent that this holds for many-electron atoms, we should expect that the orbitals are filled in order of the principal quantum number for the ground states of the elements. That is, as one goes through the Periodic Table, the 1s orbital is expected to be filled first, then the 2s and 2p orbitals, and then the 3s, 3p, and 3d orbitals, and so on.

The interaction between electrons affects the energetically preferred ordering and the shapes of the orbitals in several important ways. (It is appropriate to consider these effects even with the assumed separation of

* The term "electron configuration" is also used; however, "electron configuration" can have a more specific meaning, and so the use of "electron occupancy" for the arrangement of electrons in spatial orbitals avoids possible confusion.

variables that underlies the orbital picture.) The first important consequence of electron-electron interaction is *shielding*. We have already seen in Eq. (9-10) that the average electron-nucleus separation distance for a single-electron atom increases as $n^2$. This simply means that an electron in the 1s orbital is closer on average to the nucleus than an electron in the 2s orbital. The 2s orbital function is closer or tighter than the 3s orbital, and so on. In a lithium atom, where we expect two electrons in the 1s spatial orbital, and one in the 2s orbital, the effective potential that the electron in the 2s orbital experiences is somewhat like a nucleus (Z = +3e) with a tight, negative (–2e) charge cloud surrounding it. In other words, we can view the electron-electron repulsion effects on the 2s electron in this system as if it were a one-electron problem with the positive nucleus being of charge less than +3e, and possibly as little as +e. The nucleus is screened or shielded by the inner two 1s electrons. The immediate consequence is that the lithium atom's 2s orbital is more diffuse, or more spread out radially, than is the 2s electron of the unshielded nucleus, i.e., $Li^{2+}$. It is more diffuse because the effective nuclear charge is smaller.

Early in the development of quantum chemistry, J. C. Slater worked out values that give an excellent idea of screening effects for elements in the Periodic Table. In essence, he replaced the nuclear charge, Z, in the hydrogenic wavefunctions with a parameter $\zeta$ (zeta) that was adjustable. $\zeta$'s are less than Z because of screening. Modern computational analysis has led to standardized lists of optimum values for the $\zeta$'s of the elements, and the orbital functions that use these values are termed *Slater-type orbitals* (STO's).

Shielding, and electron-electron interaction generally, distinguish the energies of electrons in different subshells. For instance, the 2s orbital is usually energetically preferred relative to the 2p orbitals. (The degeneracy in the $m_l$ quantum number remains.) So, the electron occupancy of the carbon atom is $1s^2\ 2s^2\ 2p^2$, which means the 1s and 2s orbitals are fully occupied and there are two electrons occupying 2p orbitals. Were the 2p orbitals preferred, the occupancy would be $1s^2\ 2p^4$, but spectroscopic experiments can unambiguously demonstrate that this is not the occupancy for the ground state of carbon. The ordering of filling orbitals generally follows this pattern throughout the Periodic Table:

$$1s \rightarrow 2s \rightarrow 2p \rightarrow 3s \rightarrow 3p \rightarrow 4s \rightarrow 3d \rightarrow 4p \rightarrow 5s \rightarrow 4d \ldots$$

From this, we can build up the likely electron occupancies of the elements, while remembering that the orbital description implies an approximation involving the separation of variables in the Schrödinger equation.

## 9.7 SPIN-ORBIT INTERACTION IN MANY-ELECTRON ATOMS

Spin-orbit interaction is quite observable in the absorption and emission spectra of many-electron atoms, even though the spin-orbit interaction energies are very small relative to the transition energies of the spectral lines. The energetics of the spin-orbit interaction may be understood in the same phenomenological way that was used for the hydrogen atom. That is, we may say that there is an interaction between the magnetic dipoles associated with angular momentum sources and then apply angular momentum coupling rules. The complication, though, is the number of particles and the fact that there may be more than two angular momentum sources.

In light elements, the strongest coupling of magnetic dipoles is between all those associated with orbital motion and between all those associated with spin. In heavy elements, the spin and orbital momenta of individual electrons are the most strongly coupled. So, for light elements, we must apply angular momentum coupling rules to find a total orbital angular momentum vector ($\vec{L}$) and a total electron spin vector ($\vec{S}$). These are then coupled to form a resultant total angular momentum vector ($\vec{J}$). For heavy elements, the orbital and spin vectors of the individual electrons are coupled, just as was done for the single electron of the hydrogen atom, and the resultant vectors from all the electrons are then added to form the total angular momentum vector ($\vec{J}$). We will consider the procedure for the light elements in detail.

For the purpose of working out the angular momentum coupling, electrons in the same subshell are said to be equivalent, while electrons in different subshells are said to be inequivalent. The first situation to consider is that of inequivalent electrons, and the example will be the electron occupancy

$1s^1\, 2p^1\, 3p^1$

While this is not an occupancy encountered for the ground states of any of the elements, it could correspond to an excited state of the lithium atom. The task is to add together the orbital angular momenta and to add together the spin angular momenta. Let us use $l_1$, $l_2$, and $l_3$ for the angular momenta of the three electrons. The important rule to apply is that the quantum number for a resultant angular momentum vector may take on values ranging from the sum of the quantum numbers of two sources being combined down to the absolute value of their difference, in steps of one. Since $l_1 = 0$ and $l_2 = 1$, the quantum number for the vector sum ($\vec{L}_{12}$) of these two momenta can only be one.

$$\vec{l}_1 + \vec{l}_2 = \vec{L}_{12} \quad \Rightarrow \quad L_{12} = 1$$

The angular momentum of the third electron is added to $\vec{L}_{12}$. Applying the same rule, we have

$$\vec{l}_3 + \vec{L}_{12} = \vec{L}_{total} \quad \Rightarrow \quad L_{total} = 1+1, \ldots, |1-1| = 2, 1, 0$$

This says that there are three different possibilities for the coupling of the orbital angular momenta, and the three correspond to resultant angular momentum vectors with L quantum numbers of 2 or 1 or 0.

The coupling of the spins is done in the same way: The spin vectors of two electrons are coupled together to yield a resultant vector, and then this is coupled with the spin vector of the next electron. Clearly, the process could be continued for any number of electrons. Furthermore, since the quantum number for electron spin is always 1/2, the number of possibilities is rather limited. Applying the rule for adding angular momenta to the first two spins gives

$$\vec{s}_1 + \vec{s}_2 = \vec{S}_{12} \quad \Rightarrow \quad S_{12} = \frac{1}{2} + \frac{1}{2}, \ldots, \left|\frac{1}{2} - \frac{1}{2}\right| = 1, 0$$

This indicates that the spins of two inequivalent electrons may be coupled in two ways. Adding the third spin gives

$$\vec{s}_3 + \vec{S}_{12} = \vec{S}_{total} \quad \Rightarrow \quad S_{total} = \frac{1}{2} + 1, \ldots, \left|\frac{1}{2} - 1\right| \text{ and } \frac{1}{2} + 0$$

$$= \frac{3}{2}, \frac{1}{2}, \frac{1}{2}$$

Notice that both possibilities for the value of $S_{12}$ are used to find the possible values of $S_{total}$. Also notice that the resulting values of $S_{total}$ include two that are the same. This simply means that there are two distinct ways in which the spin vectors may be coupled that produce a resultant vector with an associated quantum number of 1/2.

The multiplicity associated with a given angular momentum is the number of different possible projections on the z-axis. This is always equal to one greater than twice the angular momentum quantum number; that is, multiplicity(J) = $2J+1$. Spin multiplicity is equal to $2S+1$, and the names singlet, doublet, triplet, quartet, and so on, are attached to states with spin multiplicities of 1, 2, 3, and 4, respectively. From the spin coupling that was just carried out, we can see that two inequivalent electrons may spin-couple to produce either a singlet state (i.e., S = 0) or a triplet state (i.e., S = 1). Three inequivalent electrons may be coupled to produce a quartet state (S = 3/2) or two *different* doublet states (S = 1/2).

The magnetic moment associated with the net orbital angular momentum, which we shall now call L instead of $L_{total}$, will interact with the magnetic moment arising from the net spin vector, which shall be S instead of $S_{total}$. With the resultant vector designated J, as before, we have

$$\vec{L} + \vec{S} = \vec{J} \qquad \Rightarrow \qquad J = L + S, \ldots, |L - S|$$

From a given value for L and for S, a number of J values may result. These are different couplings. We must also realize that there may be several different possibilities, not just one, for the L value and for the S value. All of these are to be included in finding the J values for the resultant states. Continuing with the example of three inequivalent electrons, we may tabulate the possible J's.

| Value of L | Value of S | Resultant J's |
|---|---|---|
| 2 | 3/2 | 7/2, 5/2, 3/2, 1/2 |
| 2 | 1/2 | 5/2, 3/2 |
| 2 | 1/2 | 5/2, 3/2 |
| 1 | 3/2 | 5/2, 3/2, 1/2 |
| 1 | 1/2 | 3/2, 1/2 |
| 1 | 1/2 | 3/2, 1/2 |

| Value of L | Value of S | Resultant J's |
|---|---|---|
| 0 | 3/2 | 3/2 |
| 0 | 1/2 | 1/2 |
| 0 | 1/2 | 1/2 |

It is clear that quite a number of distinct spin-orbit coupled states may be associated with the electron occupancy of this problem.

*Term symbols* are designations used in atomic spectroscopy to designate different electronic states and energy levels. The term symbols encode the values of J, L, and S. For the value of L, a capital letter is written: S for L=0, P for L=1, D for L=2, F for L=3, and so on, with G, H, I, etc. The spin multiplicity is written as a presuperscript, and the J value is written as a subscript. The form for these symbols, then, is

$$^{(2S+1)}L_J$$

As an example, if L = 1, S = 1/2, and J = 3/2, the term symbol is $^2P_{3/2}$. In the three electron example we have been using, the table of L, S, and J values is perfect for writing down the term symbols for the states. The first line of the table above would translate into the term symbols $^4D_{7/2}$, $^4D_{5/2}$, $^4D_{3/2}$, $^4D_{1/2}$. The next line would give $^2D_{5/2}$ and $^2D_{3/2}$. These are called *Russell-Saunders* term symbols because Russell-Saunders coupling assumes that the individual orbital angular momenta of the electrons are more strongly coupled than the orbital and spin angular momenta. If the spin-orbit interaction is ignored, the terms symbols are written without the J subscript.

Equivalent electrons, those in the same subshell, are more complicated in the analysis of spin and orbit coupling because of the restrictions of the Pauli principle. Essentially, the different ways in which coupling takes place are restricted. As a simple illustration of this, consider the difference between the electron occupancy $1s^2$ and the occupancy $1s^1\ 2s^1$. Because the two electrons in the same 1s orbital must have opposite spins to satsify the Pauli principle, the net spin vector must be zero; that is, S = 0 only. The two inequivalent electrons in the 1s and 2s orbitals, on the other hand, may be spin coupled as a singlet (S = 0) and as a triplet (S = 1) state.

There are a number of schemes available for finding the term symbols for equivalent electron problems. The aim of these schemes is to keep track of the possible electron arrangements among the available spin-orbitals. One easy-to-remember scheme that is also generally applicable starts with making a complete list of all the possible arrangements of electrons in spin-orbitals that correspond to the given electron occupancy. For instance, if there were three electrons in the 2p subshell, they could be arranged among the six spin orbitals in 20 ways. To work out these 20 arrangements, it is nice to use ↑ and ↓ under the column for a particular spatial orbital to indicate that an electron with $\alpha$ and $\beta$ spin, respectively, occupies that orbital. For three electrons in a 2p subshell, the arrangements that are consistent with the Pauli principle are

| $2p_1$ | $2p_0$ | $2p_{-1}$ |
|---|---|---|
| ↑↓ | ↑ | |
| ↑↓ | ↓ | |
| ↑↓ | | ↑ |
| ↑↓ | | ↓ |
| ↑ | ↑↓ | |
| ↓ | ↑↓ | |
| | ↑↓ | ↑ |
| | ↑↓ | ↓ |
| ↑ | | ↑↓ |
| ↓ | | ↑↓ |
| | ↑ | ↑↓ |
| | ↓ | ↑↓ |
| ↑ | ↑ | ↑ |
| ↓ | ↑ | ↑ |
| ↑ | ↓ | ↑ |
| ↑ | ↑ | ↓ |
| ↓ | ↓ | ↑ |
| ↓ | ↑ | ↓ |
| ↑ | ↓ | ↓ |
| ↓ | ↓ | ↓ |

Tables of this sort can be set up systematically for any occupancy, including those with both equivalent and inequivalent electrons. The number of arrangements in these tables is equal to the number of states of the system, although each line of the table does not necessarily correspond to a particular state. The number of states is the sum, for all resultant J values, of the multiplicity in J.

Each row in the table above may be regarded as having a specific z-axis projection of the total orbital and total spin angular momentum vectors. In the first line, there are two spin-up electrons and one spin-down. The net $M_S$ quantum number for this line is 1/2, which is the sum of the three individual $m_s$ quantum numbers. The sum of the individual $m_l$ quantum numbers for the three electrons, which is 1 + 1 + 0 = 2 for the first line, is the net $M_L$ quantum number. In other words, the z-axis projections of the individual electron momenta are added.

The information about the possible z-axis projections obtained from the table of spin-orbital arrangements is sufficient to deduce the possible total L and S quantum numbers. To see how this comes about, let us consider just one type of angular momentum. For a given angular momentum quantum number, J, the largest z-axis projection is with M equal to J. If we had a list of M values, but did not know the J value, we would only need to look through the list for the largest M value to know what J is. For instance, from the following list of M values,

3, 2, 2, 1, 1, 0, 0, 0, –1, –1, –2, –2, –3

we would conclude that the biggest possible J value is 3. Then, we would eliminate from this list the M values 3, 2, 1, 0, –1, –2, –3 because they are the M values that go along with J = 3. The remaining list would be

2, 1, 0, 0, –1, –1, –2

The largest of these values is 2, and so there must also be an arrangement that yields J = 2. If we then eliminate the M values of 2, 1, 0, –1, –2 because they are the values that go along with J = 2, the list will be simply

0

This means J = 0. So, we conclude that the original list of M values is uniquely consistent with J values of 3, 2, and 0.

In our table of electron arrangements, the net $M_L$ and $M_S$ quantum numbers will determine the total L and S quantum numbers. The procedure, shown in Table 9.3 for the $2p^3$ example, is to go through the list to find the biggest $M_L$ and $M_S$ and conclude that there is a possible state with corresponding L and S quantum numbers. Then, entries on the list that go along with this (L, S) pair are eliminated, and the process is repeated to find another (L, S) pair. For each (L, S) pair, we write a Russell-Saunders term symbol to achieve a concise statement of the different electronic states of the system. From Table 9.3, the first (L, S) pair was L = 2 and S = 1/2. The term symbol to be associated with this result is $^2D$, ignoring the J subscripts which may be worked out from the values of L and S. The next pair is L = 1 and S = 1/2, and the term symbol is $^2P$. The last pair is L = 0 and S = 3/2, and the term symbol is $^4S$.

With term symbols as the final result of analyzing the spin and orbital angular momenta couplings in atoms, the values of the quantum numbers L, S, and J are established. The spin-orbit interaction energy, obtained with the Hamiltonian of Eq. (9-21) and given in terms of a phenomenological constant $\alpha$, can be obtained in the same way as in Eq. (9-22).

$$E_{JLS}^{\text{spin-orbit}} = \frac{\alpha \hbar^2}{2} \left[ J(J+1) - L(L+1) - S(S+1) \right] \tag{9-24}$$

The constant $\alpha$, though, is not the same from one problem to another.

There is a very useful set of rules, which are named *Hund's rules*, that predict the energetic ordering of the term symbol states that arise from a given electron occupancy. These were first developed empirically on the basis of atomic spectra. The most important rule is that the states will be ordered energetically according to their spin multiplicity, with the greatest spin multiplicity giving the lowest energy. The second rule is that among states with the same spin multiplicity (and arising from the same electron occupancy), the energetic ordering will be according to the L quantum number, the lowest level being that with the greatest L. In the example of the $2p^3$ occupancy used in Table 9.3, Hund's rules would predict that the $^4S$ energy level will be lower than the $^2P$ and $^2D$ levels. Also, the $^2D$ will be lower than the $^2P$. The term symbol for the ground state of the nitrogen atom, which has an occupancy $1s^2\ 2s^2\ 2p^3$, is in fact $^4S$. Additional rules distinguish among the energies according to the J quantum number, but Eq. (9-24) already gives a way of being quantitative about these energy differences.

TABLE 9.3 Development of Term Symbols From the Electron Occupancy $2p^3$. The numbers $M_L$ and $M_S$ are obtained from the sum of the $m_l$ and $m_s$ quantum numbers of the individual electrons in each row. From the complete list of numbers, the pair with the biggest $M_L$ and then the biggest $M_S$ was selected, and this yielded one value for the pair of quantum numbers (L=2, S=1/2). The dots that follow below represent the table entries that are eliminated to account for all the different states with this L and this S. From the remaining entries the biggest $M_L$ and $M_S$ were selected, and the process repeated. The three sets of (L, S) pairs that were obtained dictate the term symbols.

| $2p_1$ | $2p_0$ | $2p_{-1}$ | $M_L$ | $M_S$ | | | | |
|---|---|---|---|---|---|---|---|---|
| ↑↓ | ↑ | | 2 | 1/2 | ⇒ | (L=2, S=1/2) | | |
| ↑↓ | ↓ | | 2 | −1/2 | | • | | |
| ↑↓ | | ↑ | 1 | 1/2 | | • | | |
| ↑↓ | | ↓ | 1 | −1/2 | | • | | |
| ↑ | ↑↓ | | 1 | 1/2 | ⇒ | | (L=1, S=1/2) | |
| ↓ | ↑↓ | | 1 | −1/2 | | | • | |
| | ↑↓ | ↑ | −1 | 1/2 | | • | | |
| | ↑↓ | ↓ | −1 | −1/2 | | • | | |
| ↑ | | ↑↓ | −1 | 1/2 | | | • | |
| ↓ | | ↑↓ | −1 | −1/2 | | | • | |
| | ↑ | ↑↓ | −2 | 1/2 | | • | | |
| | ↓ | ↑↓ | −2 | −1/2 | | • | | |
| ↑ | ↑ | ↑ | 0 | 3/2 | ⇒ | | | (L=0, S=3/2) |
| ↓ | ↑ | ↑ | 0 | 1/2 | | | | • |
| ↑ | ↓ | ↑ | 0 | 1/2 | | | • | |
| ↑ | ↑ | ↓ | 0 | 1/2 | | • | | |
| ↓ | ↓ | ↑ | 0 | −1/2 | | • | | |
| ↓ | ↑ | ↓ | 0 | −1/2 | | | • | |
| ↑ | ↓ | ↓ | 0 | −1/2 | | | | • |
| ↓ | ↓ | ↓ | 0 | −3/2 | | | | • |

## 9.8 SELECTION RULES AND ATOMIC SPECTRA

The spectra of atoms arise from transitions between the possible electronic states. Generally, transitions from the ground state of an atom to an excited state will require the energy of photons in the visible and ultraviolet region of the electromagnetic spectrum. The selection rule is determined from the matrix element of the dipole moment operator and two atomic state wavefunctions. For the hydrogen atom, this means that transitions are allowed between a state with quantum numbers n $l$ $m_l$ and n' $l'$ $m_l'$ if the following is nonzero.

$$< \psi_{n l m_l} \mid \vec{r} \; \psi_{n' l' m_l'} >$$

This represents three integral values because of the three vector components of $\vec{r}$, and a transition is allowed if any one is nonzero. It turns out that integration over the radial coordinate gives a nonzero result for any choice of n and n', but the angular coordinate integration requires that $l$ and $l'$ differ by 1 for the result to be nonzero. Thus, the selection rule for the hydrogen atom spectrum is stated concisely as

$$\mid \Delta l \mid = 1 \tag{9-25}$$

The ground state of the hydrogen atom is $^2S$, and so this selection rule means that transitions would be allowed only to $^2P$ states. From the excited $^2P$ state, transitions to $^2S$ and $^2D$ states would be allowed. Fig. 9.5 shows the energy levels of the hydrogen atom and the transitions that are allowed and that have been observed and measured.

All the states of the hydrogen atom are doublet states. But in many-electron atoms, there may be different spin states. For light elements, where the spin-orbit coupling is weak, the selection rules for atomic spectra are

$$\Delta S = 0 \tag{9-26a}$$

$$\mid \Delta L \mid = 1 \tag{9-26b}$$

Furthermore, the change in L can come about only if the occupancy change corresponds to a change of 1 in the $l$ quantum number of one and only one electron. That is, $\Delta l_i = 1$ while $\Delta l_{j \neq i} = 0$. For example, a change from the occupancy $1s^2\, 2s^2\, 2p^1$ to the occupancy $1s^2\, 2s^1\, 2p^2$ is a change in one and

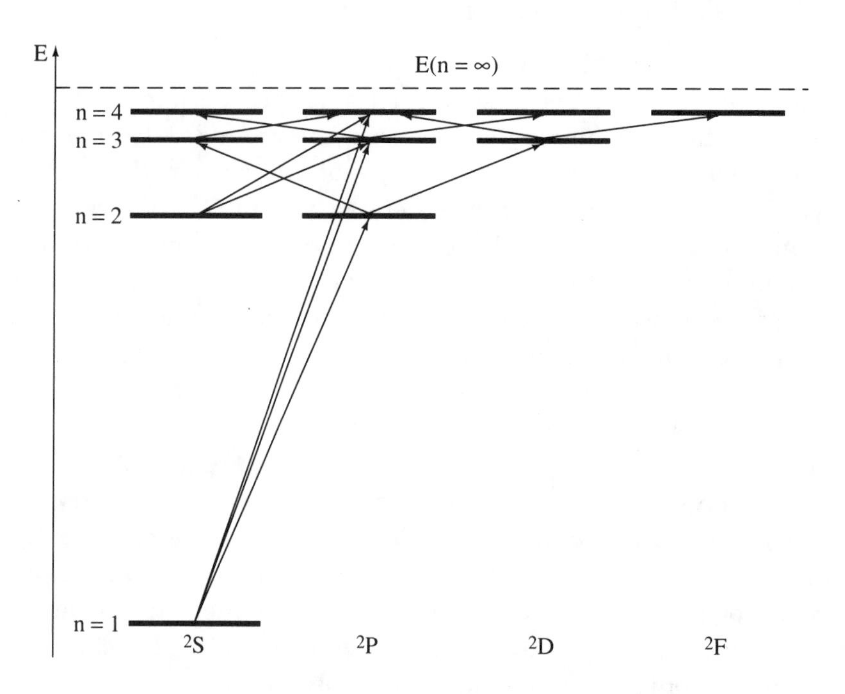


**FIGURE 9.5** The lowest energy levels of the hydrogen atom (horizontal lines) and the allowed transitions.

only one of the electrons' $l$ values of 1. The term symbol for the first occupancy is $^2$P, and so transitions would be allowed to term symbol states arising from the second occupancy, but only if they were $^2$S or $^2$D. If we consider the fine structure of the spectra, which means the small energy differences due to spin-orbit interaction, then a selection on J applies:

$$|\Delta J| = 0, 1 \tag{9-27}$$

This means that J may increase or decrease by one, or it may stay the same.

The selection rules imply that certain patterns of lines will be seen for certain types of transitions. When such patterns are spotted, the assignment of the lines is more clear-cut. For instance, consider an atom with term symbol states, $^3$D and $^3$P, which come about through electron occupancies that differ for one electron for which $\Delta l = 1$. Fig. 9.6 shows a

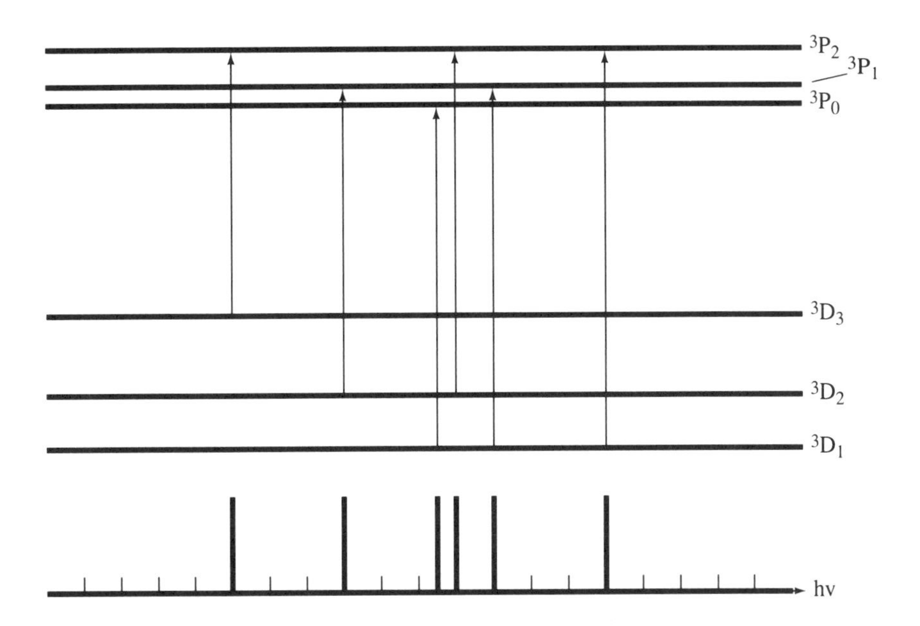


**FIGURE 9.6** Energy levels for a hypothetical atom and the allowed transitions. At the bottom is a stick representation of the resulting spectrum. The horizontal axis of this spectrum is proportional to the transition energies, and so the particular multiplet pattern of lines comes about from all the transitions between a $^3D$ and $^3P$ set of states.

possible energy ordering of the spin-orbit coupled states. With the selection rule of Eq. (9-27), a specific pattern of lines will result, as shown.

The goal of the quantum mechanical analysis of atomic structure is first to be able to account for the spectra of atoms. Atomic spectra had been observed and the wavelengths of absorption lines had been measured before the complete development of quantum theory, but there was no satisfactory way to explain *why* these spectra resulted. Quantum mechanical analysis offers an idea of the electron occupancies, and from that, the L, S, and J values that characterize the allowed states. The selection rules indicate which transitions are likely to be observed as strong transitions. The actual task of using the quantum mechanical information with spectroscopic data is to assign the transition lines to specific initial and final states (initial and final term symbols). This is sometimes like working a puzzle. A tentative assignment for one line might dictate certain other lines that form a series. If those lines are not present in the observed spectra, then the tentative assignment may have to be changed. Once the assignments are made, though, information about the atom, such as the ionization potential, the excitation energies, and the extent of spin-orbit coupling is available via the quantum mechanical analysis.

## Exercises

1. Find the explicit form of the associated Laguerre polynomial $L_3^2(z)$.
2. Show that $R_{30}(r)$ is normalized.
3. Insert $R_{41}(r)$ into Eq. (9-2) and verify that the eigenenergy is that expected from Eq. (9-9).
4. Find the energy of the n = 1 state of the hydrogen atom if the proton mass were infinite. This value is the Rydberg constant, $R_\infty$.
5. Find the energy of a hydrogen atom in a state with n = 100 and $l$ = 0. Also, find the expectation value of r in this state. (Make sure to use the correct reduced mass rather than $m_e$.)
6. The ionization energy of a one-electron atom is the energy required to promote the electron from n = 1 to n = ∞. Find the ionization energy in $cm^{-1}$ for the one-electron atoms, $He^+$, $Li^{2+}$, $C^{5+}$ and $Ne^{9+}$. In what regions of the electromagnetic spectrum (infrared, visible, ultraviolet, etc.) are there photons of energy sufficient to ionize the electron in these species?
7. To get an idea of the contraction in the size of the 1s orbitals with increasing nuclear charge, Z, calculate the expectation value of r for a single electron in a 1s orbital about the nucleus of each of the rare gas elements ($He^+$, $Ne^{9+}$, etc.).
8. Use the table method for electron arrangements to show that S = 0 and L = 0 for an $s^2$, $p^6$ and $d^{10}$ occupancy. Then show that the term symbols

for a $1s^2\ 2p^1$ occupancy are the same as the term symbols for a single electron in a 2p orbital.

9. Find the atomic term symbols for the beryllium states that would be associated with the occupancy $1s^2\ 2s^1\ 2p^1$.
10. Find the atomic term symbols that are associated with an atomic occupancy $1s^2\ 2s^2\ 2p^1\ 3p^1$.
11. Find the Russell-Saunders term symbols for the possible states of the oxygen atom with occupancy $1s^2\ 2s^2\ 2p^4$.
12. Find the atomic term symbol for the ground state of the nitrogen atom.
13. Show that the term symbols for an occupancy $3d^9$ are the same as the term symbols for the occupancy $3d^1$. Then show that the term symbols for the occupancy $3d^8$ and $3d^2$ are the same. What does this suggest?
14. What are the term symbols for the ground state occupancies of Mn and $Mn^+$?
15. What is the expected spin multiplicity for the ground states of halogen atoms?
16. Compute the spin-orbit interaction energy in terms of the parameter $\alpha$ in Eq. (9-24) for the states of an atom with occupancy $1s^2\ 2s^2\ 2p^6\ 3s^2\ 3d^2$.
17. For some given values for the L and S quantum numbers of an atomic state, what is the largest possible spin-orbit interaction energy. Express this in terms of the parameter $\alpha$ in Eq. (9-24). [Hint: Consider how J in Eq. (9-24) is related to the L and S values and attempt to maximize $E^{spin\text{-}orbit}$.]
18. What are the allowed transitions between the ground state of the carbon atom and the excited states associated with the electron occupancy $1s^2\ 2s^1\ 2p^3$?
19. What should be the first allowed transition from ground state neon? (Identify the excited occupancy and the term symbols of the initial and final states.)
20. Find the term symbols for the states of an atom that are assoictaed with an occupancy $1s^2\ 2s^2\ 2p^2\ 3d^1$. (Hint: Though a table procedure may be used for all three p and d electrons, it is more concise to use a table for the two equivalent p electrons and then to use rules for inequivalent electrons to complete the angular momentum coupling.)

## Bibliography

(See INTRODUCTORY and INTERMEDIATE LEVEL texts listed at the ends of Chaps. 3 and 4.)


1. G. Herzberg, *Atomic Spectra and Atomic Structure* (Dover, New York, 1944). This concise and detailed account of atomic structure offers many examples of measured spectra and atomic energy level diagrams for different elements. It covers many more of the details and subtle features of atomic structure than have been discussed here.
2. G. W. King, *Spectroscopy and Molecular Structure* (Holt, Rinehart and Winston, New York, 1964).
3. M. Tinkham, *Group Theory and Quantum Mechanics* (McGraw-Hill, New York, 1964). This is an advanced text that provides a group theoretical basis for understanding features of atomic electronic structure as well as other quantum mechanical problems.

*Chapter 10*

# Molecular Electronic Structure and Spectra

*The electronic structure of molecules is a more complicated problem than the electronic structure of atoms because instead of one central nuclear potential there are several positively charged nuclei distributed in space. Analyzing molecular electronic structure usually begins with an approximate separation of the electronic problem from the problem of nuclear motion. This separation was implicit in the treatment of vibration and rotation in Chaps. 6 and 7, and here we shall see how it comes about. The separated electronic Schrödinger equation is generally approached with a number of approximations, and one of them is the basis for the concept of molecular orbitals. Much of the qualitative understanding of chemical bonding can then be related to the fundamental quantum mechanics of the electronic wavefunctions.*

## 10.1 THE BORN-OPPENHEIMER APPROXIMATION

The general molecular Schrödinger equation, apart from electron spin effects, is

$$(T_n + T_e + V_{nn} + V_{ee} + V_{en})\Psi = E\Psi \qquad (10\text{-}1)$$

where the operators in the Hamiltonian are the kinetic energy operators of the nuclei and the electrons, and then the potential energy operators between the nuclei, between the electrons, and between the nuclei and electrons. The explicit forms of these operators are

$$T_n = -\sum_{\alpha}^{\text{nuclei}} \frac{\hbar^2}{2M_\alpha} \nabla_\alpha^2 \qquad (10\text{-}2)$$

$$T_e = -\sum_{i}^{\text{electrons}} \frac{\hbar^2}{2m_e} \nabla_i^2 \qquad (10\text{-}3)$$

$$V_{nn} = \sum_{\alpha>\beta} \frac{Z_\alpha Z_\beta e^2}{R_{\alpha\beta}} \qquad (10\text{-}4)$$

$$V_{ee} = \sum_{i>j} \frac{e^2}{r_{ij}} \qquad (10\text{-}5)$$

$$V_{en} = -\sum_{\alpha}\sum_{i} \frac{Z_\alpha e^2}{r_{i\alpha}} \qquad (10\text{-}6)$$

There is a repulsive interaction among the nuclear charges and a repulsive charge-charge interaction among the electrons. However, the interaction potential between electrons and nuclei is attractive since the particles are oppositely charged. This particular interaction couples the motions of the electrons and the motions of the nuclei.

The wavefunctions that satisfy Eq. (10-1) must be functions of both the electron position coordinates and the nuclear position coordinates, and this differential equation is not separable. In principle, true solutions could be found, but the task is surely difficult as this is a formidable differential equation to work with. An alternative is an approximate separation of the differential equation based upon the sharp difference between the mass of an electron and the masses of the nuclei. The difference suggests that the nuclei will be sluggish in their motions relative to the electron motions. Over a brief period of time, the electrons will "see" the nuclei as if fixed in space; the nuclear motions will be relatively slight. The nuclei, on the other hand, will "see" the electrons as something of a blur, given their fast motions.

We may express the distinction between sluggish and fast particles in a mathematical form by starting with an expansion of the wavefunction. To do this, let us restrict attention to a diatomic molecule where there is only one coordinate, R, needed to specify the separation of the nuclei. (There are, of course, six nuclear position coordinates, but after separating translation and rotation of the molecule, only one internal coordinate remains.) A power series expansion about some specific point, $R_o$, is

$$\Psi(R, r_1, r_2, \ldots) = \Psi(R_o, r_1, r_2, \ldots) + (R - R_o) \left.\frac{\partial \Psi}{\partial R}\right|_{R=R_o} + \frac{1}{2}(R - R_o)^2 \left.\frac{\partial^2 \Psi}{\partial R^2}\right|_{R=R_o} + \ldots \quad (10\text{-}7)$$

The electron position coordinates are designated $r_1$, $r_2$, etc. We will assume that the first-derivative term in the expansion is the most important and that the higher order terms may be neglected. By definition, the first derivative of the wavefunction is

$$\left.\frac{\partial \Psi}{\partial R}\right|_{R=R_o} = \lim_{\delta \to 0} \frac{\Psi(R_o + \delta, r_1, r_2, \ldots) - \Psi(R_o, r_1, r_2, \ldots)}{\delta} \quad (10\text{-}8)$$

Our physical notion is that the electrons traverse a much greater distance for any interval of time than do the nuclei because of the mass difference. Therefore, for some small increment, $\delta$, the electronic elements of the wavefunction will be much more nearly the same at $R = R_o + \delta$ and at $R = R_o$ than will the nuclear elements. This means that the derivative in Eq. (10-8) is largely determined by the change in the nuclear part of the wavefunction, and that in turn suggests an approximate separation of the electronic and nuclear motion parts of the problem.

To make the approximation that the electronic contributions to the derivative in Eq. (10-8) are zero requires first that the wavefunction be taken to be (approximately) a product of a function of nuclear coordinates only, $\phi$, and a function of electron coordinates, $\psi$, *at a specific nuclear geometry*. At the specific nuclear geometry, $R=R_o$, this is expressed as

$$\Psi(R_o, r_1, r_2, \ldots) \cong \phi(R_o)\, \psi^{\{R_o\}}(r_1, r_2, \ldots) \qquad (10\text{-}9)$$

Notice the superscript on the electronic wavefunction $\psi$. It designates that this function is for $R = R_o$, a specific point. A different R point would imply a different $\psi$ function. From this, we must also have that

$$\Psi(R_o + \delta, r_1, r_2, \ldots) \cong \phi(R_o + \delta)\, \psi^{\{R_o\}}(r_1, r_2, \ldots)$$

$$\left.\frac{\partial \Psi}{\partial R}\right|_{R=R_o} \cong \psi^{\{R_o\}}(r_1, r_2, \ldots) \lim_{\delta \to 0} \frac{\phi(R_o + \delta) - \phi(R_o)}{\delta}$$

$$= \psi^{\{R_o\}}(r_1, r_2, \ldots) \left.\frac{\partial \phi}{\partial R}\right|_{R=R_o} \qquad (10\text{-}10)$$

Considering $\psi$ at all values of R is to view the electronic wavefunction as having a dependence on the nuclear coordinate; the nuclear position coordinates are said to be parameters, and $\psi$ has a parametric dependence.

If the product form of the wavefunction, $\Psi(R, r_1, r_2, \ldots) = \phi(R)\psi^{\{R\}}(r_1, r_2, \ldots)$, is to be used in the Schrödinger equation, then we need to know something about the effect of the kinetic energy operators. Since $T_e$ acts only on electron coordinates, then $T_e\Psi = \phi T_e \psi$. $T_n$ affects both functions since $\psi$, the electronic wavefunction, retains a dependence on the nuclear position coordinates as parameters. However, in Eq. (10-10), the derivative of the electronic wavefunction with respect to the nuclear coordinates was approximated as zero. If the second derivative is also approximated as zero, then

$$T_n \Psi(R, r_1, r_2, \ldots) = T_n \phi(R)\, \psi^{\{R\}}(r_1, r_2, \ldots) \cong \psi^{\{R\}}(r_1, r_2, \ldots)\, T_n \phi(R) \qquad (10\text{-}11)$$

Within this approximation, the Schrödinger equation is

$$\phi T_e \psi + \psi T_n \phi + \phi\psi(V_{nn} + V_{en} + V_{ee} - E) = 0 \qquad (10\text{-}12a)$$

Rearrangement of the terms in this expression leads to

$$\phi\, [\, T_e + V_{ee} + V_{en}\, ]\, \psi + \psi\, [\, T_n + V_{nn} - E\, ]\, \phi = 0 \qquad (10\text{-}12b)$$

Of the operators in brackets acting on $\psi$, only one involves the nuclear position coordinates. It is a term giving the electron-nuclear attraction potential. In $\psi$, the nuclear coordinates are treated as parameters, and the same may be done for this operator term since it acts on $\psi$. This allows us to write a purely electronic Schrödinger equation as

$$[\, T_e + V_{ee} + V_{en}^{\{R\}} \,]\, \psi^{\{R\}} (r_1, r_2, \ldots) = E^{\{R\}} \psi^{\{R\}} (r_1, r_2, \ldots) \quad (10\text{-}13)$$

The energy eigenvalue, E, must also have a parametric dependence on the nuclear positions, as denoted by its superscript. That is, at each different R, there will be a different $V_{en}$, and consequently a different $\psi$ and a different E.

Substitution of Eq. (10-13) into Eq. (10-12b) yields

$$\phi E^{\{R\}} \psi^{\{R\}} + \psi^{\{R\}} [\, T_n + V_{nn} - E \,] \phi = 0 \quad (10\text{-}14)$$

Since $\phi$ is a function only of nuclear position coordinates, then the following separation results.

$$[\, T_n + (V_{nn} + E^{\{R\}}) - E \,] \phi = 0 \quad (10\text{-}15)$$

This is the Schrödinger equation for finding $\phi$. The Hamiltonian consists of a kinetic energy operator for the nuclear position coordinates, the repulsion potential between the nuclei, and an effective potential, in the form of $E^{\{R\}}$, that gives the energy of the electronic wavefunction as it depends on the nuclear position coordinates.

The Born-Oppenheimer approximation leads to a separation of the molecular Schrödinger equation into a part for the electronic wavefunction and a part for the nuclear motions, which is the Schrödinger equation for vibration and rotation. The essential element of the approximation is Eq. (10-11) which is that applying the nuclear kinetic energy operator to the electronic wavefunction yields zero. The physical idea is that the light, fast-moving electrons readjust to nuclear displacements instantaneously. This is the reason the approximation produces an electronic Schrödinger equation for each possible geometrical arrangement of the nuclei in the molecule (e.g., each value of R in the diatomic we considered). We need only know the instantaneous positions of the nuclei, not how they are moving, in order to find an electronic wavefunction, at least within this approximation.

The approximation is put into practice by "clamping" the nuclei of a molecule of interest. That means fixing their position coordinates to correspond to some chosen arrangement or structure. Then, the electronic Schrödinger equation [Eq. (10-13)] is solved to give the electronic energy for this clamped structure. After that, perhaps, another structure is selected, and the electronic energy is found by solving Eq. (10-13) once more. Eventually, enough structures might be treated so that the dependence of the electronic energy on the structural parameters is known fairly well. At such a point, the combination of $E^{\{R\}}$ with $V_{nn}$, which is a simple potential according to Eq. (10-4), yields the effective potential for molecular vibration. It is the potential energy for the nuclei in the field created by the electrons.

The Born-Oppenheimer approximation is one of the best approximations in chemical physics in the sense that it proves to be a valid approximation in most situations. As with any approximation, of course, there is a limitation in its applicability, and in fact, there are special phenomena that are associated with a breakdown in the Born-Oppenheimer approximation. We may understand certain of the formal aspects of dealing with a breakdown by comparing the form of a Born-Oppenheimer wavefunction with a more general form. First, we note that any well-behaved function of two independent variables can always be written as a sum of products of independent functions, as in the following.

$$f(r,s) = \sum_i \sum_j c_{ij} \, g_i(r) \, h_j(s)$$

The sums must be over sets of functions that are complete in the range of the function f. In a power series expansion, for instance, the $g_i$ and $h_i$ functions would be polynomials in r and s, respectively. The $c_{ij}$'s are the expansion coefficients. In a way, the Born-Oppenheimer approximation is attempting to use such an expansion, only to do so with all but one term ignored. The wavefunction is just one product of a function of the nuclear coordinate(s) and a function of the electron coordinate(s).

We realize that Eq. (10-13) may have a number of different solutions. Just as the harmonic oscillator has different energy states, the electrons in an atom or molecule may have different states with different energies available. Also, the nuclear Schrödinger equation, Eq. (10-15), will have a number of different solutions. The set of these solutions is complete in the range of the electronic-nuclear wavefunction, $\Psi$. So in principle, the non-approximate wavefunction could be written as a sum of product functions.

$$\Psi = \sum_i \sum_j c_{ij} \phi_i \psi_j \tag{10-16}$$

The Born-Oppenheimer approximation is to retain only one $\phi\psi$ product, and so the validity of the approximation depends on the coefficients of the other products (in the true $\Psi$) being significantly smaller. When this fails to be the case, the approximation breaks down.

A possible means for dealing with a breakdown in the Born-Oppenheimer approximation is to let the wavefunction include other products. For instance, if the electronic energies of two Born-Oppenheimer electronic states became very close at some point in the vibration of a molecule, then the nonapproximate wavefunction, or the non-Born-Oppenheimer wavefunction, might properly be a superposition (or linear combination) mostly of two products. As suggested by the curves in Fig. 10.1, the two product functions would be the Born-Oppenheimer electronic-vibrational wavefunctions otherwise obtained for the two different electronic states. We might then say that the system has a nonvanishing probability of being in one or the other "Born-Oppenheimer state," or that it could be hopping between. Of course, such a description is entirely within the context of the Born-Oppenheimer approximation since it presumes a separation of the electronic and nuclear motions. So important is the Born-Oppenheimer approximation in chemistry that discussions are very often within the context of the approximation (e.g., separation of electronic and nuclear motion), even if not stated explicitly.

## 10.2 POTENTIAL ENERGY SURFACES

The separation of the electronic and nuclear motion via the Born-Oppenheimer approximation leads to an important concept, that of the *potential energy surface*. A potential energy surface is the dependence of the electronic energy, plus nuclear repulsion, on the geometrical

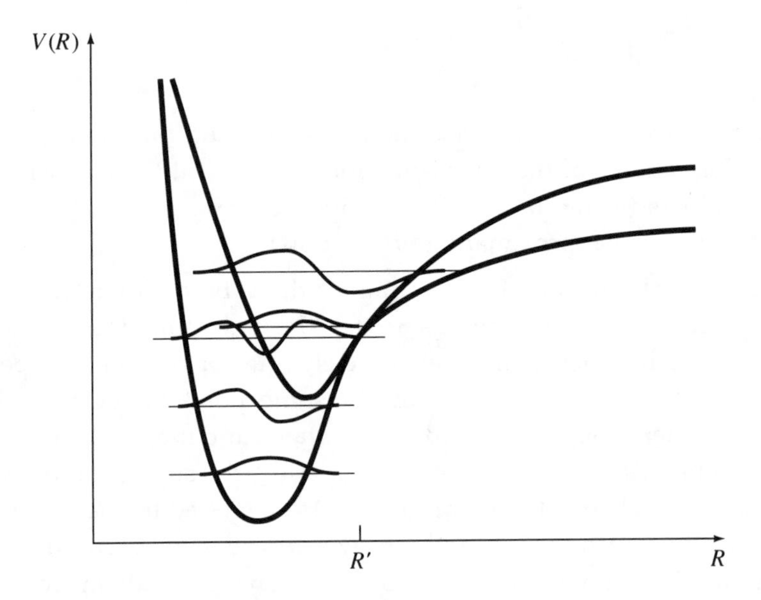


**FIGURE 10.1** A hypothetical set of potential energy curves for a diatomic molecule. The heavy lines represent the potential energy for nuclear motion, the stretching vibration of the diatomic, and the light lines are the vibrational wavefunctions. The potential energy at a given separation distance, R, is the sum of the electronic energy obtained within the Born-Oppenheimer approximation for the nuclei fixed at that separation and the nuclear-nuclear repulsion energy. The closeness of the electronic energies around the distance marked R' and the closeness of the energies of the vibrational states suggest, though they do not ensure, that there may be significant nuclear-electronic effects that make the true wavefunction more of a superposition of the two. This would be a breakdown in the Born-Oppenheimer approximation.

coordinates of the atomic centers of a molecule. It is only within the Born-Oppenheimer approximation that we can follow the dependence of the electronic energy on the atomic position coordinates. That only comes about with the separation of the electronic and nuclear problems.

Potential energy surfaces are the conceptual framework for thinking about and analyzing many problems in chemical physics. The surface, of course, is the effective potential for vibrations of the molecule. But continued to large atom-atom separations, the surface holds the energetic information about reactions and bond breaking. It also provides a picture of the pathways for reactions.

A potential energy surface is a representation of a function of all the internal coordinates of a molecular system. For even a simple molecule such as ammonia, the potential is a function of six coordinates, making it difficult to display or to visualize. We may, however, make a graphical representation of a slice through the multidimensional surface. The slice is the potential function with all but one or two of the coordinates fixed at certain chosen values. If all but one are fixed, the slice is a function of only that one coordinate and its representation is a potential curve. If all but two are fixed, the slice is a function of two coordinates.

Functions of two coordinates are easily represented by contour diagrams, and this is one of the generally used ways of presenting potential energy surface slices. An example is shown in Fig. 10.2. The curves that we see in this figure connect or follow equipotential points on the potential energy surface. Cutting across one of these curves means going "uphill" or "downhill" in energy. Usually, we may identify a lowest energy point on a slice through a potential energy surface (i.e., a two-dimensional contour plot), and in Fig. 10.2, that point lies near the middle of the diagram. The contour plot can only show us that the potential is uphill in two directions away from this point. In the other directions, the ones that were fixed to create the slice, the surface might be sloping upward or downward. Should it be that the surface slopes upward in every direction, then the point is a minimum energy point or *minimum* on the surface. For complicated surfaces, it is possible for there to be several minima, and the one of lowest energy is called the *global minimum*. The global minimum is a point on the surface that corresponds to the *equilibrium structure* of the species.

A slice of a hypothetical potential energy surface of a hypothetical linear triatomic molecule, ABC, is shown in Fig. 10.3. This diagram extends to the regions where the molecule is dissociated into the diatomic AB and atom C and where it is dissociated into the diatomic BC and atom A. The minimum is near the lower left. Stretching either bond, which means increasing either coordinate, is an uphill process. However, there is a unique point in the A-B stretching where the potential will start to be downhill. This point, called a *saddle point*, is where the potential surface

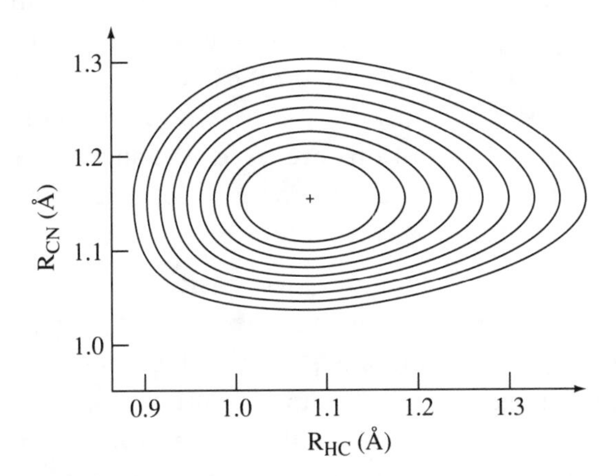


**FIGURE 10.2** A contour diagram of a slice of the potential energy surface of the hydrogen cyanide molecule. The coordinates are the H-C distance and the C-N distance. The bending angle is fixed at 180°, which corresponds to a linear structure. The contours are at energy steps of 500 $cm^{-1}$, and the plot is of the region around the equilibrium structure of the molecule; that is, R(H-C) = 1.07 Å and R(C-N) = 1.15 Å. These contours were determined on theoretical grounds from detailed calculations involving the electronic Schrödinger equation [see C. E. Dykstra and D. Secrest, *J. Chem. Phys.* **75**, 3967 (1981)].

is uphill in each direction except one, and in that one direction, the potential is downhill, either forward or backward.

If we follow the contour plot of Fig. 10.3 to the right of the saddle point, we encounter a trough. This is simply a region where the potential is slowly changing, either upward or downward, along one direction. Eventually, this trough will have a flat bottom and walls that do not change with the A-B separation as the point is reached where the A atom is completely removed from the BC diatomic and there is no interaction. At this limit (far right of Fig. 10.3), a cut through this two-dimensional surface

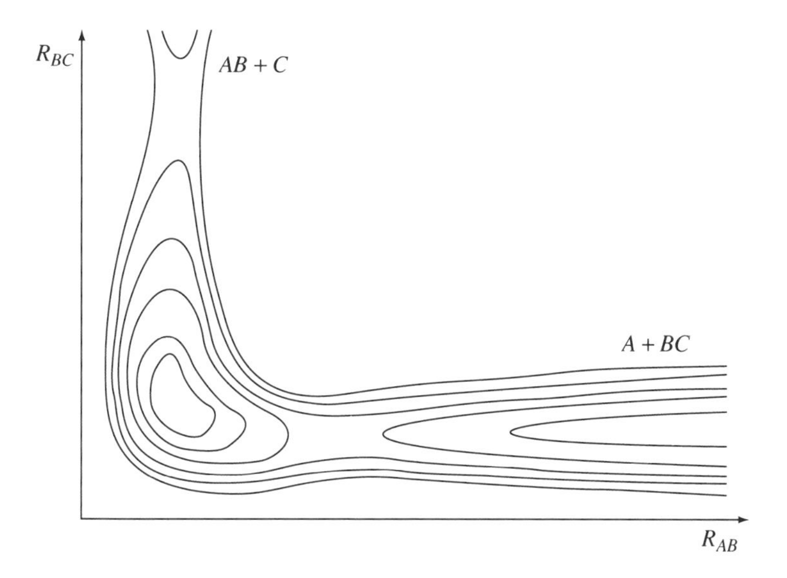


**FIGURE 10.3** A contour diagram of a slice of a hypothetical, but representative, potential energy surface of a triatomic molecule, ABC. The horizontal axis coordinate is the A-B distance, and the vertical axis coordinate is the B-C distance. The bending angle is fixed at 180°. The energy steps between adjacent contour levels are meant to be alike.

for any A-B distance is a potential curve; it is the potential energy as a function of the B-C separation, which is just the stretching potential of the BC diatomic. Likewise, a horizontal cut across the top left of the surface in Fig. 10.3 must be the stretching potential for the AB diatomic.

The bottom of a trough that connects limiting regions such as these with minima or with saddle points is called a *minimum energy path*. That is good term, especially if we consider the potential surface as if it were a physical surface with real hills and valleys. If we were at the bottom of the trough on the far right of the surface in Fig. 10.3 and wished to "hike" to the equilibrium structure on the lower left, the minimum climb would take us right through the saddle point and then down the valley where the potential minimum is found. That is the minimum energy path. Notice that is does not follow any one coordinate direction. At the outset, it follows the direction of the coordinate of the horizontal axis. Then it twists a bit, and if we follow it further, it will follow the direction of the coordinate of the vertical axis.

*Energetic profiles* for reactions, interconversions, and isomerizations are often given as simple curves of potential energy (vertical axis) versus a *reaction coordinate* (horizontal axis). This is not as abstract a notion as it may seem, because the minimum energy path provides a perfectly acceptable reaction coordinate. That is, each step along a reaction coordinate is a step along the minimum energy path. In this sense, energy profiles are simply special slices through potential energy surfaces. Again, the Born-Oppenheimer approximation has provided the basis for an important chemical concept, and it provides the context for discussing details of reaction energetics.

## 10.3 ANTISYMMETRIZATION OF ELECTRONIC WAVEFUNCTIONS

The electrons moving about in an atom or molecule are indistinguishable particles. The mathematical consequence is that the probability density of an electronic wavefunction of an atom or molecule must be invariant with respect to how the electron coordinates are labeled. With $\vec{r}_i$ being the designation of the position vector of the $i^{th}$ electron, interchange of two electrons' position coordinates means swapping $\vec{r}_i$ and $\vec{r}_j$ in $\psi$. For the probability density to be unchanged, the following must hold for any choice of i and j.

$$\psi(\vec{r}_1, \vec{r}_2, \dots, \vec{r}_i, \dots, \vec{r}_j, \dots)^2 = \psi(\vec{r}_1, \vec{r}_2, \dots, \vec{r}_j, \dots, \vec{r}_i, \dots)^2 \qquad (10\text{-}17)$$

For the wavefunction itself, this implies that interchange of a pair of electron coordinates can change the wavefunction only by a factor $e^{i\varphi}$, since $e^{i\varphi}(e^{i\varphi})^* = 1$. This same reasoning applies to a wavefunction of any other kind of indistinguishable particle, and it turns out that the value of $\varphi$ depends on the intrinsic spin of the particles. As will be verified later, for electrons and other half-integer spin particles, $\varphi = \pi$. (For other than half-integer spin particles, such as those with spin = 0, 1, 2, ..., the phase is zero:

$\varphi$ = 0.) Since $e^{i\pi} = -1$, all electronic wavefunctions must satisfy the following condition.

$$\psi(\vec{r}_1, \vec{r}_2, \dots, \vec{r}_i, \dots, \vec{r}_j, \dots) = -\psi(\vec{r}_1, \vec{r}_2, \dots, \vec{r}_j, \dots, \vec{r}_i, \dots) \qquad (10\text{-}18)$$

This is termed an *antisymmetrization* requirement; an electronic wavefunction is antisymmetric (changes sign) with respect to interchange of any pair of electron position coordinates.

To impose Eq. (10-18) on some electronic wavefunction, a special operator, called the antisymmetrizer, can be used. Let us develop the form of this operator, assuming that we have at hand a normalized electronic wavefunction, $\Phi$, for some system, and that $\Phi$ has the form of a product of independent, orthonormal functions, u(r), of the electron coordinates.

$$\Phi(\vec{r}_1, \vec{r}_2, \dots) = u_1(\vec{r}_1)\, u_2(\vec{r}_2) \dots$$

The interchange of a pair of position coordinates is a well-defined mathematical operation, as we have in Eqs. (10-17) and (10-18). Thus, we can define an operator that performs that interchange: $P_{ij}$ will be an operator that interchanges the coordinates of some electron, i, with the coordinates of some electron, j. That is,

$$P_{ij}\Phi(\vec{r}_1, \vec{r}_2, \dots, \vec{r}_i, \dots, \vec{r}_j, \dots) = \Phi(\vec{r}_1, \vec{r}_2, \dots, \vec{r}_j, \dots, \vec{r}_i, \dots)$$

From this we may construct a new function, $\Phi'$, that is assured to be antiymmetric with respect to interchange of particles 1 and 2:

$$\Phi' = \frac{1}{\sqrt{2}}\left(\Phi - P_{12}\Phi\right) \qquad (10\text{-}19)$$

Antisymmetrization is tested by applying $P_{12}$ and finding that the negative of the function results:

$$P_{12}\Phi' = \frac{1}{\sqrt{2}}\left(P_{12}\Phi - P_{12}P_{12}\Phi\right) = -\Phi'$$

Notice that $P_{12}$ applied twice means swapping the coordinates and then swapping them back, and that is an identity operation. The square root of 2 in Eq. (10-19) is introduced so that $\Phi'$ is normalized if $\Phi$ is normalized.

$$< \Phi' \mid \Phi' > \;=\; \frac{1}{2} < \Phi - P_{12}\,\Phi \mid \Phi - P_{12}\,\Phi >$$

$$= \frac{1}{2}\Big[ <\Phi \mid \Phi> \;-\; <\Phi \mid P_{12}\Phi> \;-\; <P_{12}\Phi \mid \Phi> \;+\; <P_{12}\Phi \mid P_{12}\Phi> \Big]$$

$$= \frac{1}{2}[\,1 - 0 - 0 + 1\,] \;=\; 1$$

The cross terms must be zero because of the independence of the two particles' coordinates and the orthogonality of the u(r) functions.

At this point, it is helpful to regard Eq. (10-19) as the application of an operator that antisymmetrizes the function with respect to the interchange of the first two particles.

$$\Phi' \;=\; \left\{ \frac{1}{\sqrt{2}}\left( 1 - P_{12} \right) \right\} \Phi \qquad (10\text{-}20)$$

This operator is one minus the interchange operator with a factor to ensure normalization.

The next step in imposing Eq. (10-18) on $\Phi$ involves electron 3. The wavefunction must be made antisymmetric with respect to interchanging electron 1 with 3 (i.e., applying $P_{13}$), and to interchanging electron 2 with 3 (i.e., applying $P_{23}$). This can be accomplished by a generalization of the operator in Eq. (10-20) to use $P_{13}$ and $P_{23}$, and then by applying it to $\Phi'$.

$$\Phi'' \;=\; \left\{ \frac{1}{\sqrt{3}}\left( 1 - P_{13} - P_{23} \right) \right\} \Phi'$$

$$= \left\{ \frac{1}{\sqrt{3}}\left( 1 - P_{13} - P_{23} \right) \right\} \left\{ \frac{1}{\sqrt{2}}\left( 1 - P_{12} \right) \right\} \Phi$$

We can continue this for all the electrons. For instance, considering the fourth electron will mean using $P_{14}$, $P_{24}$, and $P_{34}$. So, for N electrons, an entirely antisymmetric wavefunction, $\psi$, may be constructed from an arbitrary wavefunction, $\Phi$, by application of a sequence of operators:

$$\psi \;=\; \frac{1}{\sqrt{N!}}\left( 1 - P_{1N} - P_{2N} - \ldots P_{N-1,N} \right) \ldots \left( 1 - P_{13} - P_{23} \right)\left( 1 - P_{12} \right) \Phi \qquad (10\text{-}21a)$$

$$= A_N \Phi \tag{10-21b}$$

The sequence of operators and the collective normalization constant may be considered as one overall antisymmetrizing operator, which we will call the antisymmetrizer for N electrons, $A_N$.

There are two important properties of the antisymmetrizer. First, if it is applied to a wavefunction that is already properly antisymmetric, it will make no change. The implication of this statement is that if the antisymmetrizer is applied twice to an arbitrary function, the same result will be achieved as if it were applied only once; the second application does not do anything. This is the condition of *idempotency*, and $A_N$ is an idempotent operator. The operator equation that expresses this fact is

$$A_N A_N = A_N \tag{10-22}$$

(This is by no means a statement that $A_N = 1$.) The second property is that it commutes with the electronic Hamiltonian. It is in the Hamiltonian, of course, that the indistinguishability of the electrons is apparent. An interchange of electron coordinates in the Hamiltonian leaves the Hamiltonian unchanged. That is, $P_{ij}H = H$ for any choice of i and j electrons. With this, it is easy to demonstrate that the antisymmetrizer for a two-electron system commutes with the Hamiltonian:

$$\left[ H, \frac{1}{\sqrt{2}}(1 - P_{12}) \right] \Phi(r_1, r_2) = \frac{1}{\sqrt{2}} \left\{ H(1 - P_{12})\, \Phi(r_1, r_2) - (1 - P_{12}) H\Phi(r_1, r_2) \right\}$$

$$= \frac{1}{\sqrt{2}} \left\{ H\Phi(r_1, r_2) - H\Phi(r_2, r_1) - H\Phi(r_1, r_2) + H\Phi(r_2, r_1) \right\} = 0$$

It can be proved, as well, that the antisymmetrizer for an N-electron system commutes with the Hamiltonian.

Since the antisymmetrizer commutes with the Hamiltonian, Theorem 4.4 means that wavefunctions can be found that are simultaneously eigenfunctions of the Hamiltonian and of the antisymmetrizer. If the antisymmetrizer is applied to an already-antisymmetric wavefunction, it should give back that wavefunction, and that says merely that the eigenvalue associated with the antisymmetrizer is one. Likewise, if the antisymmetrizer is applied to an arbitrary electronic wavefunction, the resulting function is an eigenfunction of the antisymmetrizer, with eigenvalue of one, because of the idempotency of the antisymmetrizer. Because of this, it is a common approach to use basis functions for variational and perturbative treatments of electronic wavefunctions that are already antisymmetrized.

In Chap. 9, spin-orbital functions were defined for atoms. For molecules, we may construct one-electron spatial wavefunctions by making linear combinations of atomic orbitals (LCAO) and by other means. Spin orbitals for molecules, then, are products of a spatial function of the spatial coordinates of one electron and a spin function, $\alpha$ or $\beta$, for that electron. The orbital picture of the ground state of the ammonia molecule, for example, has 10 electrons in five spatial orbitals. If these are designated $\phi_1$ through $\phi_5$, then the spin-orbitals are

$$\phi_1\alpha,\ \phi_1\beta,\ \phi_2\alpha,\ \phi_2\beta,\ \phi_3\alpha,\ \phi_3\beta,\ \phi_4\alpha,\ \phi_4\beta,\ \phi_5\alpha,\ \phi_5\beta$$

The orbital picture implies that the wavefunction is a product,

$$\Phi(r_1, \ldots r_{10}) \;=\; \phi_1(r_1)\alpha(s_1)\,\phi_1(r_2)\beta(s_2) \ldots\ \phi_5(r_9)\alpha(s_9)\,\phi_5(r_{10})\beta(s_{10})$$

where the abstract spin coordinates are $s_1$, etc. This, of course, is not antisymmetric, but the antisymmetrizer can be applied to it.

It turns out that the function produced by applying the antisymmetrizer to a product of spin-orbitals can be expressed in the concise and useful form of a determinant (see Appendix I). In explaining this, it is convenient to adopt a special shorthand notation. We will use one symbol for a spin-orbital, and here that symbol will be u. A given spin-orbital is a function of the spin and spatial coordinates of some electron, but instead of writing $u(r_i,s_i)$, we will write $u(i)$; that is, just the electron number will be written since that is sufficient to interpret what is meant. In this shorthand notation, the product function given above for ammonia would be written as

$$\Phi(r_1, \ldots r_{10}) = u_1(1)\, u_2(2)\, u_3(3)\, u_4(4)\, u_5(5)\, u_6(6)\, u_7(7)\, u_8(8)\, u_9(9)\, u_{10}(10)$$

In general, the orbital product form of an N-electron wavefunction is

$$\Phi(r_1, \ldots r_N) = u_1(1)\, u_2(2)\ \ldots\ u_N(N) \tag{10-23}$$

Application of the N-electron antisymmetrizer to this product function yields a function that consists of N! products combined together, as there are N! products of permutation operators in Eq. (10-21a). These N! terms can be obtained by expanding the following determinant.

$$\psi = \frac{1}{\sqrt{N!}} \begin{vmatrix} u_1(1) & u_2(1) & u_3(1) & \ldots & u_N(1) \\ u_1(2) & u_2(2) & u_3(2) & \ldots & u_N(2) \\ u_1(3) & u_2(3) & u_3(3) & \ldots & u_N(3) \\ \cdots & \cdots & \cdots & \cdots & \cdots \\ u_1(N) & u_2(N) & u_3(N) & \ldots & u_N(N) \end{vmatrix} \tag{10-24}$$

The diagonal of this determinant follows the product of Eq. (10-23). Each column uses a different spin-orbital function, and each row uses a different electron's coordinates. $\psi$ is called a *Slater determinant* after the inventor of this device, and we have that $\psi = A_N\Phi$.

Slater determinants are functions that are immediately seen to be properly antisymmetric with respect to exchange of the coordinates of any pair of electrons. Such an exchange would correspond to interchanging two rows of the Slater determinant, and it is a property of determinants that their value changes sign if two rows are exchanged. This means that the expanded form of the determinant, which is $\psi$, will have an opposite sign when a pair of electron coordinates are exchanged, and that comes about by the corresponding exchange of rows of the determinant.

There is another important property revealed by the Slater determinant construction. If there are two identical spin-orbitals, then there will be two identical columns in the Slater determinant. It is a property of determinants that their value is zero if two columns are identical. Thus, if we consider two different electrons to be in the same spin-orbital in an original product function [Eq. (10-23)], then antisymmetrization (e.g., the construction of the corresponding Slater determinant) will produce zero: Such a wavefunction is not permitted. Thus, antisymmetrization leads to a requirement that only one electron occupy a particular spin-orbital. This is a way of stating the *Pauli exclusion principle*.

## 10.4 THE MOLECULAR ORBITAL PICTURE

The product form of an electronic wavefunction can only come about through an approximation. The electronic Hamiltonian is not separable into independent terms for different electrons. The electron-electron repulsion operator prevents this separation, and in turn, an exactly determined wavefunction cannot be formed as a product of one-electron functions, e.g., $u_1(\vec{r}_1)u_2(\vec{r}_2)u_3(\vec{r}_3)...$, which is the orbital form. The approximation that leads to the orbital form is an approximation to the electron-electron repulsion. The physical aspect of the approximation is to replace the electron-electron repulsion by an electric field due to all the electrons residing in their particular spin orbitals. Then, each electron looks like its own quantum mechanical system where the potential is that of the electron's charge interacting with the positively charged nuclei and with a fixed field arising from the charge cloud of the electrons in the molecule or atom.

The approximation to the electron-electron repulsion has a mathematical effect of converting an N-electron problem, that of the original electronic Schrödinger equation, to N one-electron problems. The approximation leads to an effective one-electron Hamiltonian, called the Fock operator.

$$\hat{F} = \hat{h} + \hat{g} \tag{10-25}$$

is the operator corresponding to an electron's kinetic energy and its attraction for the nuclei.

$$\hat{h} = -\frac{\hbar^2}{2m_e}\nabla^2 - \sum_{\alpha}^{\text{nuclei}} \frac{Z_\alpha e^2}{r_\alpha} \tag{10-26}$$

$r_\alpha$ is the distance between the a nucleus and the electron. In Eq. (10-25), $\hat{g}$ is the operator associated with the effective field of all the electrons. Of course, an electron cannot be said to interact with itself, and so we really want a field operator for the field of all the *other* electrons. But, it turns out that $\hat{g}$ can be constructed so that upon applying the operator, the self-term, meaning the part for the interaction of the electron with its own share of the field, vanishes. As a result, the field operator $\hat{g}$ is the same for every electron.

The orbital wavefunctions for the atom or molecule are the eigenfunctions of the Fock operator. It is, in effect, the Hamiltonian for the system, though in the sense of a Hamiltonian for one electron at a time.

$$\hat{F} u_i = \varepsilon_i u_i \tag{10-27}$$

The eigenfunctions, $u_i$, are functions of only one coordinate. They are orbitals, and just as in many other Schrödinger equations (e.g., the harmonic oscillator), there may be many different solutions. The eigenvalues associated with the orbitals are labelled by the same index, and they are referred to as *orbital energies.*

Separability of a Schrödinger equation comes about if there are independent additive pieces of the Hamiltonian. The separability implicit in Eq. (10-27) means that the approximation corresponds to a many-electron Hamiltonian that is simply the sum of Fock operators for each electron, and we will designate this Hamiltonian as $\hat{H}_o$.

$$\hat{H}_o = \sum_{v}^{N} \hat{F}_v \tag{10-28}$$

The Fock operators in Eq. (10-28) are subscripted to indicate that each acts on independent position coordinates for the N electrons. Separability also means that the energy is a sum, and in this case, we have

$$E_o = \sum_{i}^{N} \varepsilon_i \tag{10-29}$$

$E_o$ is the eigenenergy of $\hat{H}_o$.

The field that electrons experience in this approximation can only be prescribed when the orbital wavefunctions are known. In other words, Eq. (10-27) is quite different from the other eigenequations we have considered since the operator is dependent on the solutions. We need to know the orbitals that the electrons occupy in order to know $\hat{g}$. The dilemma is that we need to know $\hat{g}$ in order to find the orbitals. Actually, this problem can be solved by a bootstrap procedure. From a set of guess orbitals, a corresponding $\hat{g}$ operator is formed for use in Eq. (10-27). If the orbitals that are then obtained are not the same as the guess, they are used in constructing a new $\hat{g}$ operator. The whole process is repeated again and again until the orbitals used to construct $\hat{g}$ turn out to be the same as the eigenfunctions of Eq. (10-27). This means that $\hat{g}$, which represents the effective field, is not prescribed from the outset but is determined in a *self-consistent* manner. The effective field, then, is usually called a *self-consistent field.*

The orbital energies provide a framework for chemical energetics. For instance, since the net energy needed to form a molecule from separated nuclei and electrons is (approximately) $E_o$, then a reaction energy may be taken to be a difference in electronic energies of reactants and products. An example is the energy for the reaction of methylene and hydrogen to produce methane, $CH_2 + H_2 \rightarrow CH_4$. Were a quantum mechanical calculation of the orbital energies of methane, methylene, and hydrogen to give the following energies,

$$E_o(CH_4) = -27.568 \text{ a.u.}$$
$$E_o(CH_2) = -26.216 \text{ a.u.}$$
$$E_o(H_2) = -1.187 \text{ a.u.}$$

then an appropriate sum and difference could be related to the reaction energy:

$$E_{rxn} = -27.568 - (-26.216 - 1.187) = -0.165 \text{ a.u.} = -104 \text{ kcal/mol}$$

In this way, orbital energies offer a simple way of estimating reaction energies.

Another energetic feature of a molecule, the *ionization potential* (IP), is the energy required to remove an electron. This is simply an orbital energy according to Eq. (10-29). (This is within the approximation of a self-consistent field and the approximation that the orbitals for the original molecule and the ionized molecule are the same.) *Photoelectron spectroscopy* (PES) is an experiment where molecules are irradiated with high-energy photons of fixed energy (i.e., monochromatic radiation). When the photon energies are greater than an ionization potential of the sample, an electron may be ejected, and its kinetic energy will be the difference between the photon energy and the ionization potential. The ejected electrons are energy-analyzed in a photoelectron experiment, and the spectrum of energies represents the body of data that are obtained. Electrons may be ejected from any orbital, and so the photoelectron spectrum has sharp peaks at energies usually associated with ionization from the different orbitals. Far ultraviolet radiation is used for valence orbitals, whereas X-ray radiation is used for core orbitals. The former type of experiment is designated UPS and the latter XPS.

## 10.5 LINEAR COMBINATIONS OF ATOMIC ORBITALS

Molecular orbital pictures have immense conceptual and qualitative value in chemistry. In many cases, this comes about through semi-quantitative use of orbital energies and qualitative thinking about the orbital shapes. The principle of forming *linear combinations of atomic orbitals* (LCAO) to construct molecular orbitals is one important aspect of molecular orbital pictures. The basic idea is to consider orbitals of two interacting fragments of a molecule as independent electron problems. An orbital from fragment A may interact with an orbital from fragment B to produce two new orbitals (one-electron states). The interaction is the element of the Hamiltonian for the system that involves both fragments, and that means it is something that must approach zero as the fragments are pulled further and further apart. The interaction between the parts is taken as a perturbation, and the problem is analyzed with second-order perturbation theory.

The first case for LCAO is the $H_2$ molecule. At infinite separation of the two atoms, the Hamiltonian is a sum of independent hydrogen atom Hamiltonians, as in

$$\hat{H}_o = \hat{H}^{(A)} + \hat{H}^{(B)}$$

where one hydrogen atom is A and the other is B. The interaction that develops when the atoms approach each other is an attraction of each atom's electron for the other's proton plus the repulsion between the two electrons. This interaction is a perturbation, $\hat{H}_1$. There is also a repulsion between the two nuclei, and within the Born-Oppenheimer approximation, this is a constant that is added to the electronic energy at each (fixed) internuclear separation.

In the absence of the perturbation, the energies of the two electrons are the hydrogen atom energies, and it is common to represent this zero-order

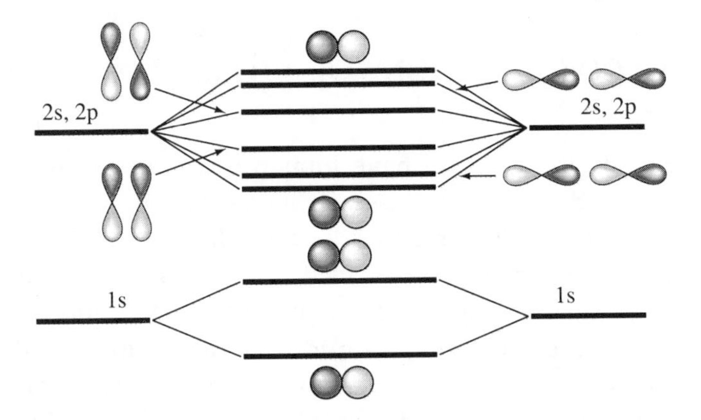


**FIGURE 10.4** An orbital energy correlation diagram for the linear combination of atomic orbitals to form molecular orbitals. The vertical axis is an orbital energy scale, and the leftmost and rightmost horizontal lines represent the orbital energies of two noninteracting hydrogen atoms. In the middle of the figure are two horizontal lines representing the energies of mixed functions of the left and right hydrogen 1s orbitals. Their energies are qualitatively deduced from first-order degenerate perturbation theory which requires that the two mixed states be above and below the energies of the unmixed, degenerate states by equal amounts. The qualitative form of the mixed orbitals is also shown.

situation by two horizontal lines alongside a vertical (orbital) energy scale. Since these two lines place the true energies of the atoms at a large separation, one line is drawn on the far left and the other on the far right, as in Fig. 10.4. In the presence of the perturbation the orbitals of the left atom and the right atom may mix with each other. The qualitative form of this mixing is an additive combination, one plus the other, and the opposite, one minus the other. At the lowest order of perturbation theory that yields a nonzero effect of $\hat{H}_1$, one of the new orbitals is raised in energy, and the other is lowered by a like amount. The one that is raised is an unfavorable LCAO, and it is the one that has a node between the two atoms. It is called an *antibonding orbital*. The one that is lowered is more stable than the original orbitals, and so it is a bonding orbital. Since $H_2$ has two electrons, they will occupy this orbital in the $H_2$ ground electronic state. Of course, there are excited electronic states from the promotion of one or both electrons to the higher lying orbital.

In a more complicated diatomic, the orbitals of atom (fragment) A are paired up with the orbitals of atom (fragment) B on the basis of similar energies, and the same analysis is carried out. Implicit in this picture is that electrons are interacting largely one on one with electrons in the other fragment. This is partly because with low-order perturbation theory, the energetic effects will be small if the orbital energies of the two electrons from the fragments are much different in energy. (A large energy difference would lead to a sizable energy denominator in the perturbation theory expressions for the energy and wavefunction corrections.) Fig. 10.5 puts this into practice for the CO molecule. The orbital energies of carbon are represented by the series of horizontal lines on the left, and the orbital energies of oxygen are on the right. These orbitals correlate with the molecular orbitals whose approximate energies are represented by the horizontal lines in the center. The correlation of atomic orbitals with the molecular orbitals is indicated by dashed lines. Notice that the oxygen 1s orbital is not paired with a carbon orbital. Its orbital energy is much lower than any of the carbon atom's orbitals, and so, from low order perturbation theory, little mixing is expected with the carbon orbitals. This is in keeping with not considering an oxygen 1s orbital to be a valence orbital. It is an orbital that closely surrounds the oxygen nucleus, and we should expect it to be little affected by chemical bonding.

The correlation of atomic orbitals with molecular orbitals, as in Fig. 10.5, ranks the molecular orbitals in terms of their expected energies. The lowest energy electronic state of a molecule is expected to arise when the electrons fill up the molecular orbitals from the bottom up; that is, they occupy the most stable orbitals. A chemical bond is said to exist when two electrons occupy a bonding orbital, which is one that does not have a nodal plane between the two atoms. $H_2$ has one bond. When antibonding orbitals are occupied, the net bonding is taken to be given by a number called the *bond order* (BO):

Bond order = (no. of electrons in bonding orbitals –
no. of electrons in anti-bonding orbitals) / 2

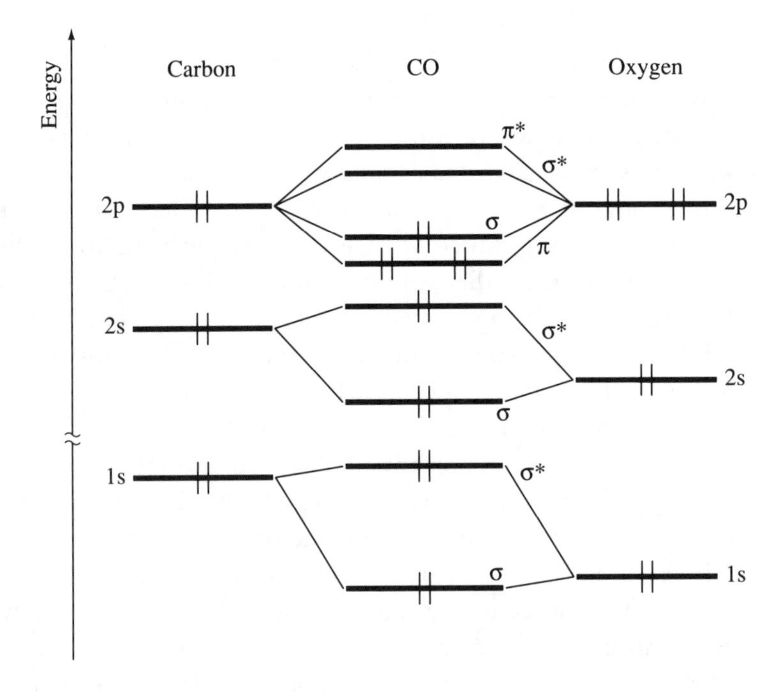


**FIGURE 10.5** An orbital correlation diagram for carbon monoxide. The carbon atomic orbital energies are on the left, and the oxygen atomic orbital energies are on the right. The molecular orbitals that form from mixing of the atomic orbitals are represented by the horizontal lines in the center at their approximate orbital energies in the CO molecule. The vertical lines indicate the orbital occupancy.

Thus, $He_2$ has a bond order of zero because, according to Fig. 10.4, two electrons would be in a bonding orbital just as in $H_2$, but two electrons would be in an antibonding orbital as well. Carbon monoxide has a bond order of 3. The bond order gives a rough idea of the bond strength, or of the relative bond strengths. Table 10.1 is a short list of bond orders that are predicted in this fashion. We may see from this that the weakest bonding tends to be for homonuclear diatomics of elements in the leftmost and rightmost columns of the Periodic Table. The rare gas dimers, of course, have a zero bond order.

TABLE 10.1 Bond Orders of Certain Diatomic Molecules on the Basis of LCAO.

| | Number of Bonding Electrons | Number of Antibonding Electrons | Bond Order |
|---|---|---|---|
| $H_2$ | 2 | 0 | 1 |
| $He_2$ | 2 | 2 | 0 |
| $Li_2$ | 2 | 0 | 1 |
| $Be_2$ | 2 | 2 | 0 |
| $C_2$ | 6 | 2 | 2 |
| $N_2$ and CO | 8 | 2 | 3 |
| $O_2$ | 8 | 4 | 2 |
| $F_2$ | 8 | 6 | 1 |
| $Ne_2$ | 8 | 8 | 0 |
| $Na_2$ | 2 | 0 | 1 |
| $Mg_2$ | 2 | 2 | 0 |
| $S_2$ | 8 | 4 | 2 |
| $Cl_2$ | 8 | 6 | 2 |
| $Ar_2$ | 8 | 8 | 0 |

## 10.6 THE SLATER-CONDON RULES

The orbital energy expression for the total electronic energy, Eq. (10-29), overcounts the electron-electron repulsion. A more rigorous evaluation of the total electronic energy comes from taking the expectation value of the orbital wavefunction with the true Hamiltonian. In relation to the self-consistent field approximation, this amounts to treating the difference between the Fock operator Hamiltonian of Eq. (10-28) and the true Hamiltonian as a perturbation and then including the effect of this perturbation to first order (i.e., an expectation value of the perturbing operator). So, at this stage we must develop the means for finding expectation values with the electronic Hamiltonian.

The Slater-Condon rules are concise expressions that have been derived to express integrals of the electronic Hamiltonian and Slater determinant wavefunctions as integrals over the orbital functions in the Slater determinants. To present the Slater-Condon rules, it is helpful to first understand the integrals over orbital functions. The electronic Hamiltonian consists of two types of operators, and the first type is that of the operator $\hat{h}$ in Eq. (10-26). It involves only one electron's coordinates and is called a one-electron operator. The other type is just the electron-electron repulsion, $1/r_{ij}$, which is a two-electron operator since it involves two electrons' coordinates. With a set of orbitals, $\{\phi_i\}$, which are taken to be real functions for convenience, one- and two-electron integrals are defined for the two operators. The symbol I is often used for the one-electron integral values:

$$I(i \mid j) \equiv \langle \phi_i(\vec{r}_1) \mid \hat{h}(\vec{r}_1)\, \phi_j(\vec{r}_1) \rangle$$

$$= \int \phi_i(\vec{r}_1) \left[ -\frac{\nabla^2}{2} - \sum_\alpha \frac{Z_\alpha}{r_{1\alpha}} \right] \phi_j(\vec{r}_1)\, d\vec{r}_1 \qquad (10\text{-}30)$$

The electron position coordinates corresponding to $\vec{r}_1$ are the variables of integration, and this is regardless of the choice of i and j for the orbital functions. $I(i \mid j)$ should be regarded as a number or value and is not a function. For a set of N spatial $\phi_i$ orbitals, there are $N(N+1)/2$ unique I values, since $I(i \mid j) = I(j \mid i)$.

Since the one-electron integrals require specifying two orbitals (i and j), it is not surprising that a general two-electron integral involves four orbitals. A standard notation and definition is the following.

$$(ij \mid kl) \equiv \langle \phi_i(\vec{r}_1)\, \phi_k(\vec{r}_2) \mid \frac{1}{r_{12}}\ \phi_j(\vec{r}_1)\, \phi_l(\vec{r}_2) \rangle$$

$$= \int d\vec{r}_1\, \phi_i(\vec{r}_1) \left\{ \int d\vec{r}_2 \frac{\phi_k(\vec{r}_2)\, \phi_l(\vec{r}_2)}{r_{12}} \right\} \phi_j(\vec{r}_1) \qquad (10\text{-}31)$$

$\vec{r}_1$ and $\vec{r}_2$ are the variables of integration. Again, with the orbital functions taken to be real, there are integrals that are equivalent:

$$(ij \mid kl) = (ji \mid kl) = (ij \mid lk) = (ji \mid lk)$$

$$= (kl \mid ij) = (kl \mid ji) = (lk \mid ij) = (lk \mid ji) \qquad (10\text{-}32)$$

There are two special types of these two-electron integrals. One type is a *Coulomb integral*, and it has the following form ( i i | j j ). That is, it involves only two orbital functions. It is called a Coulomb integral because it corresponds to the Coulombic repulsion of the charge cloud of one electron in the orbital $\phi_i$ and one electron in the orbital $\phi_j$. This is clear upon using the definition of Eq. (10-31).

$$(ii \mid jj) = \int d\vec{r}_1\, \phi_i(\vec{r}_1) \int d\vec{r}_2\, \phi_j(\vec{r}_2)\, \phi_j(\vec{r}_2) / r_{12}\ \phi_i(\vec{r}_1)$$

$$= \int d\vec{r}_1 \int d\vec{r}_2\ \frac{\phi_i^2(\vec{r}_1)\, \phi_j^2(\vec{r}_2)}{r_{12}}$$

Since $\phi_i^2$ and $\phi_j^2$ may be interpreted as probability densities, then $e\phi_i^2$ is a charge density for one electron and $e\phi_j^2$ is the charge density for the other electron. (Of course, in atomic units, e = 1.) So, the integral ( i i | j j ) is equivalent to an expression for the electrostatic repulsion between two charge distributions or charge clouds. The other special two-electron integral is called an *exchange integral* and is of the form ( i j | i j ). It involves just two orbital functions, though it does not have a simple physical interpretation.

A closed-shell wavefunction is one for which every occupied spatial orbital is fully occupied. Thus, the Slater determinant for any closed-shell wavefunction is $\Phi = \det\{\phi_1\alpha\ \phi_1\beta\ \phi_2\alpha\ \phi_2\beta\ \ldots\ \phi_N\alpha\ \phi_N\beta\}$, and N is the number of occupied spatial orbitals. The first Slater-Condon rule gives the expectation value of the energy of a Slater determinant, and for the case of a closed-shell wavefunction with N occupied orbitals, it is

$$<\Phi \mid H\Phi> = 2\sum_{i}^{N} I(i|i) + \sum_{i}^{N}\sum_{j}^{N} [\,2(ii \mid jj) - (ij \mid ij)\,] \qquad (10\text{-}33)$$

The SCF orbital energy for the $i^{th}$ orbital is

$$\varepsilon_i = I(i|i) + \sum_{j}^{N} [\,2(ii \mid jj) - (ij \mid ij)\,] \qquad (10\text{-}34)$$

and so we can now see that summing the orbital energies for each electron (remembering that there are two electrons in each orbital) will overcount the electron-electron repulsion terms. Relative to Eq. (10-33), this would give an expression that has two times the double-summation term. Therefore, the expectation value expression in Eq. (10-33) is the SCF energy of a closed-shell wavefunction.

The general form of the Slater-Condon rules is in terms of spin-orbitals, the products of spatial functions with either $\alpha$ or $\beta$ spin functions. A way to express the most general Slater determinant is to use a set of spin-orbitals, $\{u_i\}$, that are each products of spatial functions, $\phi$, and spin functions, $\sigma$. Each spin function must be either the $\alpha$ or the $\beta$ function. Thus, $u_i = \phi_i\sigma_i$. For the Slater determinant for a K-electron system, $\Gamma = \det\{u_1\, u_2 \ldots u_K\}$, the first Slater-Condon rule is

$$<\Gamma \mid H\Gamma> \;=\; \sum_i^K I[u_i \mid u_i] \;+\; \frac{1}{2}\sum_i^K \sum_j^K \left\{ [u_i u_i \mid u_j u_j] - [u_i u_j \mid u_i u_j] \right\} \qquad (10\text{-}35)$$

In this expression, the integrations must be over the abstract spin coordinates as well as the spatial coordinates, and the square brackets have been used for the two-electron integrals to emphasize this distinction. If we now insert the product form of each $u_i$ function, the integrals in Eq. (10-35) can be developed in terms of spatial integrals times spin integrals.

$$I[u_i \mid u_i] \;=\; I[\phi_i\sigma_i \mid \phi_i\sigma_i] \;=\; I(\phi_i \mid \phi_i) <\sigma_i \mid \sigma_i> \;=\; I(i \mid i)$$

$$[\, u_i u_i \mid u_j u_j \,] \;=\; (i\,i \mid j\,j) <\sigma_i \mid \sigma_i> <\sigma_j \mid \sigma_j> \;=\; (i\,i \mid j\,j)$$

$$[\, u_i u_j \mid u_i u_j \,] \;=\; (i\,j \mid i\,j) <\sigma_i \mid \sigma_j> <\sigma_i \mid \sigma_j>$$

The $\alpha$ and $\beta$ spin functions are orthonormal functions, and so the spin integrals are either zero or one: $<\alpha \mid \alpha> = 1$, $<\alpha \mid \beta> = 0$, and $<\beta \mid \beta> = 1$. Thus, $<\sigma_i \mid \sigma_i> = 1$ regardless of whether $\sigma_i$ is $\alpha$ or $\beta$, and this has been used for the one-electron integral and the Coulomb integral. For the exchange integral, though, no further simplification can be made until the actual spin functions are specified for each of the spin-orbitals. Saying a wavefunction is that of a closed shell amounts to specifying the spin orbitals, and so Eq. (10-33) is obtained from Eq. (10-35) by properly evaluating all the $<\sigma_i \mid \sigma_j>$ quantities in the exchange integrals and then collecting all the spatial integrals correctly. [Notice, for instance, that N, the number of fully occupied spatial orbitals in Eq. (10-33) would be equal to K/2 according to the way K is used in Eq. (10-35).]

The other Slater-Condon rules give integral expressions between different Slater determinants. For the second Slater-Condon rule, let us introduce a Slater determinant, $\Gamma'$, that is identical to some Slater determinant, $\Gamma$, except that the $u_r$ spin-orbital has been replaced by a spin-orbital, $u_t$, that has not been used in $\Gamma$. We say that $\Gamma'$ is a singly substituted form of $\Gamma$ since there has been a substitution of a single spin-orbital. The Slater-Condon rule is

$$< \Gamma \mid H \Gamma' > \; = \; I[u_r \mid u_t] \; + \; \sum_{i}^{K} \left\{ [u_r u_t \mid u_i u_i] - [u_r u_i \mid u_t u_i] \right\} \qquad (10\text{-}36)$$

Again, the square brackets indicate that both spin and spatial integration is implied.

If there is a Slater determinant, $\Gamma''$, that differs from $\Gamma$ by two spin-orbitals, it is called a doubly substituted determinant. If the $u_r$ spin-orbital in $\Gamma$ has been replaced by $u_t$ in $\Gamma''$, and if the $u_s$ spin-orbital in $\Gamma$ has been replaced by $u_x$ in $\Gamma''$, then the Slater-Condon rule is

$$< \Gamma \mid H \Gamma'' > \; = \; [u_r u_t \mid u_s u_x] - [u_r u_x \mid u_s u_t] \qquad (10\text{-}37)$$

Finally, if there is a Slater determinant, $\Gamma'''$, that differs by three or more spin orbitals from $\Gamma$, then $< \Gamma \mid \Gamma''' > = 0$.

## 10.7 SELF-CONSISTENT FIELD WAVEFUNCTIONS *

The full specification of the self-consistent field (SCF) operator and the Fock operator of Eq. (10-25) has not been given, so far. They are complicated devices that develop from the variational determination of orbital functions. In considering this, we will limit attention to closed-shell wavefunctions, which are those with no partially occupied spatial orbitals. Then, Eq. (10-33) is the expectation value of the energy, as given by the Slater-Condon rules. It is this value that is to be minimized by variation theory. The only possible linear variations of an orbital that will affect the expectation value of the energy are the mixings with orbitals that are unoccupied or empty. It can be shown that any unitary mixing of any two of the occupied orbitals leaves the SCF energy unchanged. If we designate an arbitrary orbital function that is not occupied in $\Gamma$, the closed-shell wavefunction, as $\phi_r$, then one way we may introduce a parameter for the purpose of variational adjustment is the following.

$$\phi_i' = \left(\phi_i + c\,\phi_r\right) / \sqrt{1+c^2} \tag{10-38}$$

That is, the coefficient, c, is a linear expansion coefficient for the mixing of some unoccupied one-electron function $\phi_r$, with the function $\phi_i$. The square root factor ensures normalization. Application of variation theory means that the expectation value of the energy is to be minimized with respect to c.

If the true SCF orbitals were at hand, then the optimum value of the coefficient c would be zero, which is to say that $\phi_i'$ would be the same as $\phi_i$. Thus, to apply variation theory, we need to form the derivative, with respect to c, of the expectation value of the energy for the wavefunction using $\phi_i'$ in place of $\phi_i$. That wavefunction will be designated $\Gamma'$. Then, this derivative, evaluated at c = 0, is set equal to zero to yield the condition for which the energy is a minimum (or maximum) with respect to c.

The differentiation of the energy expression of Eq. (10-33) involves differentiation of one- and two-electron integrals, and it is useful to examine that process first. From Eq. (10-38), we may express integrals involving the orbital $\phi_i'$ in terms of integrals involving $\phi_i$ and $\phi_r$. Using the concise integral notation, we have that

$$I(i'\,|\,i') = \frac{1}{1+c^2}\left[I(i\,|\,i) + 2c\,I(i\,|\,r) + c^2\,I(r\,|\,r)\right]$$

$$(i'i'\,|\,jj) = \frac{1}{1+c^2}\left[(ii\,|\,jj) + 2c\,(ir\,|\,jj) + c^2\,(rr\,|\,jj)\right]$$

$$(i'j\,|\,i'j) = \frac{1}{1+c^2}\left[(ij\,|\,ij) + 2c\,(rj\,|\,ij) + c^2\,(rj\,|\,rj)\right]$$

Each of these integral quantities is a function of the coefficient c, and so they may be differentiated with respect to c. For the one-electron integral, this gives

$$\frac{\partial I(i' \mid i')}{\partial c} \Big|_{c=0} = I(i \mid r)$$

The expectation value of $\Gamma'$, which again is $\Gamma$ with $\phi_i$ having been replaced by $\phi_i'$, can be obtained directly from Eq. (10-33). Then, the first derivative is easily evaluated.

$$\frac{\partial \langle \Gamma' \mid H \Gamma' \rangle}{\partial c} \Big|_{c=0} = 2 \left\{ I(i \mid r) + \sum_j^N [2(ir \mid jj) - (ij \mid rj)] \right\} \quad (10\text{-}39)$$

We conclude on the basis of the variation theorem that this quantity must be zero for the energy to be a minimum at c = 0. That is a condition for finding c.

The bracketed quantity in Eq. (10-39) comes about from the mixing of the $i^{th}$ occupied orbital with the $r^{th}$ virtual or empty orbital [Eq. (10-38)]. Of course, i could be any of the occupied orbitals and r could be any of the available virtual orbitals, and so, Eq. (10-39) represents a set of "k-times-n" conditions where k is the number of occupied orbitals and n is the number of empty orbitals. Thus, a matrix organization of the problem is appropriate, and in particular, we define a matrix **F** with elements from Eq. (10-39).

$$\mathbf{F}_{ir} = I(i \mid r) + \sum_j^N [2(ir \mid jj) - (ij \mid rj)] \quad (10\text{-}40)$$

The condition that must be satisfied for the energy to be variational is that the off-diagonal elements of **F** are zero.* That is the same as saying that we seek a transformation that diagonalizes **F**.

* More precisely, only the off-diagonal elements between occupied orbitals, i, and empty orbitals, r, must be zero. There is no such requirement for the off-diagonal elements where both row and column labels refer to occupied orbitals, or else both refer to empty orbitals. But the invariance of the SCF energy with respect to the mixing of occupied orbitals with other occupied orbitals, and the invariance with respect to mixing of empty orbitals with other empty orbitals, means that we are free to make **F** completely diagonal.

Normally, the matrices in quantum mechanics that one seeks to diagonalize are matrix representations of an operator. The matrix **F** is, in fact, a matrix representation of an operator, and that operator is the Fock operator. We may work backward to extract the operator from the matrix representation, starting with the following statement from the definition of a matrix representation.

$$\mathbf{F}_{ir} = < \phi_i \mid \hat{F}\, \phi_r > \tag{10-41}$$

By inspection, we see that the operator $\hat{F}$ includes the one-electron operator $\hat{h}$, since $< \phi_i \mid \hat{h}\, \phi_r > = I(i \mid r)$. The two-electron integrals in Eq. (10-40) may also be related to operators by using Eq. (10-31).

$$\sum_j^N (i\, r \mid j\, j) = \int d\vec{r}_1\, \phi_i(\vec{r}_1)\, \phi_r(\vec{r}_1) \sum_j^N \int d\vec{r}_2 \frac{\phi_j^2(\vec{r}_2)}{r_{12}} \equiv < \phi_i \mid \hat{J}\, \phi_r >$$

$$\therefore \hat{J} = \sum_j^N \int d\vec{r}_2 \frac{\phi_j^2(\vec{r}_2)}{r_{12}} \tag{10-42}$$

Similarly, we have

$$\sum_j^N (i\, j \mid r\, j) \equiv < \phi_i \mid \hat{K}\, \phi_r >$$

$$\therefore \hat{K} = \sum_j^N \int d\vec{r}_2 \frac{\phi_j(\vec{r}_2)}{r_{12}}\, \hat{P}_{12}\, \phi_j(\vec{r}_2) \tag{10-43}$$

where $\hat{P}_{12}$ is the permutation operator that interchanges the coordinates $\vec{r}_1$ and $\vec{r}_2$ in whatever functions follow. ($\hat{K}$ will be applied to a function of $\vec{r}_1$.) $\hat{J}$ is called a *Coulomb operator* by its association with Coulomb integrals. $\hat{K}$ is called the *exchange operator* given its application involves an exchange (permutation) of electron coordinates. So, the closed-shell Fock operator of Eq. (10-40) must be a simple combination of the one-electron operator and the Coulomb and exchange operators.

$$\hat{F} = \hat{h} + 2\hat{J} - \hat{K} \tag{10-44}$$

Fock operators may also be derived for electronic states that are not closed shells, and they will consist of similar pieces. The operator associated with the electronic self-consistent field is everything except the one-electron operator.

An important feature of the Coulomb and exchange operators is that they depend on the orbital functions for the orbitals that are occupied. They are termed basis-dependent operators, and since the Fock operator is constructed from them, it is a basis-dependent operator, too. This means that the optimum orbitals are eigenfunctions of an operator [Eq. (10-27)] that requires knowing what the optimum orbitals are. While this dilemma may seem unworkable, all that is required is a bootstrap process. From some initial guess of the orbitals, a Fock operator is constructed. The eigenfunctions of this Fock operator become the updated guess set of orbitals. The process is repeated until self-consistency is achieved, which means until the orbitals used to construct the Fock operator are no different than the eigenfunctions.

In most computer implementations of SCF, the orbitals are developed as linear combinations of atomic orbitals of some chosen form. The linear expansion coefficients are the variational parameters. A guess of the optimum linear combinations starts the process. This is a guess of the transformation matrix that takes the basis of atomic orbital functions into the basis of the SCF orbitals. These guess orbitals are used to construct a matrix representation of the Fock operator in the orbital basis, $\mathbf{F}$. The unitary transformation that diagonalizes $\mathbf{F}$ is a transformation among the guess set of SCF orbitals, just as diagonalization normally yields a transformation of the basis functions (see Appendix II). However, because $\mathbf{F}$ is the representation of a basis-dependent operator, it is subject to change with the transformation to the new set of orbitals. So, it must be reconstructed and rediagonalized, and that means that the whole process is iterative. Instead of one diagonalization, there are repeated diagonalization steps. The iterative process continues until the unitary transformation matrix needed to diagonalize the current iteration's $\mathbf{F}$ is within some tolerance limit of being the identity matrix, $\mathbf{1}$. That is the point at which the orbital set is no longer changing and the point at which self-consistency has been achieved. In practice, the number of iterations required for convergence tends to be 5 to 30, though this is very much dependent on the nature of the molecular problem.

## 10.8 SPECTROSCOPIC STATES: SPIN AND SYMMETRY *

The term "spectroscopic states of molecules" refers to the states that are involved in transitions seen by spectroscopic measurement. For molecules containing atoms of other than the very heavy elements, these states have specific net electron spin. They are singlet states, meaning the spin quantum number, S, is zero, or they are doublet states (S = 1/2), or triplet states (S = 1), or states with a still greater spin quantum number. They will also have certain properties that reflect symmetry in the molecule, if any.

From a molecular electron occupancy, we may derive the possible values of S, just as with atoms. Fully occupied orbitals contribute zero to the total spin, since the spins of the electrons in these orbitals are paired. If there were two electrons outside filled orbitals, for instance, and if they were inequivalent, then the possible molecular spin states would be singlet and triplet.

A Slater determinant is not always a valid description of a spectroscopic state because single Slater determinants are not always eigenfunctions of the operator $\hat{S}^2$. That means that there is no S quantum number to assign to a single determinant. However, linear combinations of Slater determinants can be formed in general ways so as to be eigenfunctions of $\hat{S}^2$. (The procedures involve use of the angular momentum coupling coefficients discussed in Chap. 5.) An important example is the singlet and triplet coupling of two inequivalent electrons. Let us use $\phi_1$ and $\phi_2$ as the spatial orbitals occupied by one electron each in some molecule. There are four possible Slater determinants that arise from this occupancy, and expressed in a shorthand notation where only the diagonal elements are listed and the coordinates are suppressed, these are

$$\Gamma_1 = | \phi_1 \alpha \quad \phi_2 \alpha |$$

$$\Gamma_2 = | \phi_1 \alpha \quad \phi_2 \beta |$$

$$\Gamma_3 = | \phi_1 \beta \quad \phi_2 \alpha |$$

$$\Gamma_4 = | \phi_1 \beta \quad \phi_2 \beta |$$

Each of these, as well as every Slater determinant, is an eigenfunction of the operator giving the z-axis projection of the total spin. This operator is the sum of the z-axis projection spin operators for the individual electrons:

$$\hat{S}_z = \sum_{i}^{\text{electrons}} \hat{S}_{i-z} \tag{10-45}$$

The resultant $M_S$ quantum numbers obtained from applying this operator to the four Slater determinants are 1 for $\Gamma_1$, 0 for $\Gamma_2$ and $\Gamma_3$, and −1 for $\Gamma_4$. This reveals that $\Gamma_1$ and $\Gamma_4$ must be eigenfunctions of the operator $\hat{S}^2$ with eigenvalue S = 1. From the two other determinants, we must construct a function for which S = 1 and $M_S$ = 0 and an orthogonal function for which S = 0 and $M_S$ = 0. In this way, the four Slater determinants are spin-adapted into three functions that comprise the members of the triplet and the one function that is the singlet.

Application of the spin-lowering operator (see Chap. 5) to $\Gamma_1$ will produce the state with S = 1 and $M_S$ lowered by one to a value of 0. The lowering operator commutes with the antisymmetrizer, and so we may follow the effect of the lowering operator by applying it to the diagonal of the determinant. With the concise notation for the Slater determinants, we start with the equivalence of the overall lowering operator and the sum of the lowering operators for the individual spins.

$$\hat{S}_- \,|\, \phi_1 \alpha \;\; \phi_2 \alpha \,| = \hat{S}_{1-} \,|\, \phi_1 \alpha \;\; \phi_2 \alpha \,| + \hat{S}_{2-} \,|\, \phi_1 \alpha \;\; \phi_2 \alpha \,| \tag{10-46}$$

Following Eq. (5-26), the left hand side will yield the desired state with S = 1 and $M_S$ = 0, which we shall designate as $\Phi_{10}$. The individual electron-lowering operators will change the α spins to β.

$$\sqrt{2}\; \Phi_{10} = |\, \phi_1 \beta \;\; \phi_2 \alpha \,| + |\, \phi_1 \alpha \;\; \phi_2 \beta \,|$$

Equivalent to this statement is

$$\Phi_{10} = \frac{1}{\sqrt{2}} \left( \Gamma_2 + \Gamma_3 \right) \tag{10-47}$$

The other $M_S$ = 0 function must be a linear combination of the same two Γ functions, and it must be orthogonal to $\Phi_{10}$. This requires that the singlet state, $\Phi_{00}$, be

$$\Phi_{00} = \frac{1}{\sqrt{2}} \left( \Gamma_2 - \Gamma_3 \right) \quad (10\text{-}48)$$

So, we have found that the singlet state is a linear combination of two Slater determinants.

Molecular symmetry refers to the structural equivalence of atoms in a molecule. The two hydrogens in formaldehyde, for instance, are in chemically equivalent positions. A formal consequence is that the molecular wavefunctions must be eigenfunctions of certain abstract operators we define that transform the atoms of the molecule into equivalent sites. A rotation of the formaldehyde molecule by 180° about the C-O axis would interchange the two hydrogens, and so the wavefunction must display this same symmetry by being an eigenfunction of the operator that carries out the rotation. The general analysis uses the mathematics of group theory. However, there are certain key features that can be presented on their own.

Since self-consistent field theory recasts the many-electron problem into many single-electron problems, in effect, we could correctly assume that spatial orbitals as well as the complete wavefunctions will display the molecular symmetry. For linear molecules, the types of orbitals are named using the Greek equivalent of s, p, d, f, etc. They are $\sigma$, $\pi$, $\delta$, $\phi$, etc., and they differ from one another according to the number of nodal planes. These nodal planes include the molecular axis and are uniformly arranged. We can represent the qualitative differences between the orbital types by sketching them from end on, as in Fig. 10.6. Like atomic s orbitals, $\sigma$ orbitals are not degenerate. $\pi$, $\delta$, $\phi$, etc., orbitals, though are doubly degenerate, and so they may be occupied by four electrons. The symmetry properties of the molecular electronic states are designated $\Sigma$, $\Pi$, $\Delta$, $\Phi$, etc.

Symmetry operations in nonlinear molecules include rotations about an axis, reflections through a plane, and inversion through a central point. The orbital designations are a and b for nondegenerate orbitals, e for doubly degenerate orbitals, and t for triply degenerate orbitals. The difference between a and b orbitals is that the a orbital does not change phase upon the associated 180° rotation, but the b orbital does. Various subscripts are attached to indicate other symmetry properties of the orbitals. The molecular electronic states have an overall symmetry, and capital letters (e.g., A, B, E, T) with various subscripts are used. The symmetries of the

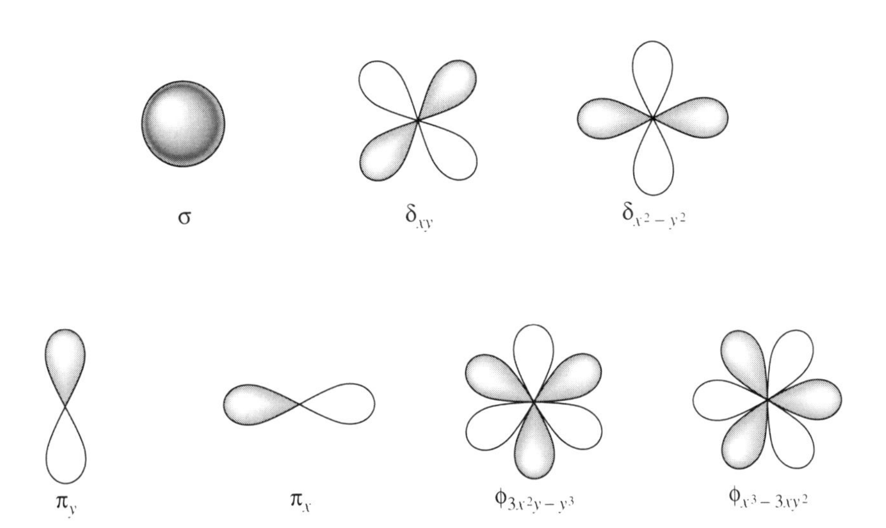


**FIGURE 10.6** End-on view (down the z-axis) of the qualitative forms of the real parts of $\sigma$, $\pi$, $\delta$, and $\phi$ orbitals of a linear molecule.

states are important because detailed selection rules for transitions can be related to the symmetries of the initial and final states.

## 10.9 VISIBLE-ULTRAVIOLET SPECTRA OF MOLECULES

Visible-ultraviolet spectroscopy of molecules is an experimental way to examine the nature of excited electronic states, since typically the absorption of a photon in this energy regime electronically excites the molecule. Probably the strongest selection rule is

$$\Delta S = 0 \tag{10-49}$$

That is, transitions are normally between states of the same spin. Other selection rules relate to the geometrical symmetry of the molecule.

The molecular transitions seen in the visible and ultraviolet regions of the spectrum must be transitions from one rotational-vibrational-electronic state to another. We shall consider this in detail for a hypothetical diatomic molecule for which the ground state and excited state potential curves are those shown in Fig. 10.7. For both electronic state potential energy curves there will be a set of vibrational states and rotational sublevels. Notice that the equilibrium distance is not the same for both curves and that the curvature (i.e., the force constant) is not the same either. Thus, there will be a different vibrational frequency and a different rotational constant for each electronic state. This will have to be taken into account in working out the transition frequencies.

The practice we have followed in understanding molecular spectroscopy has been to work backward, in some sense, so as to predict the frequencies of transition lines that presumably were measured. The goal is always to assign the lines to specific transitions. This enables one to use the measured frequencies to extract the true values for the vibrational frequency, the rotational constant, and other spectroscopic constants. The same procedure holds for electronic spectra.

As seen from Fig. 10.7, a transition energy will depend first on the energy difference between the potential minima of the two states, and this is called the *term value*, $T_e$. This is the contribution to the transition energy that involves only the electronic energies. The second component of the transition energies is that due to the vibration-rotation state energies within each of the two wells. The energies of those states will depend on vibrational and rotational quantum numbers n' and J' for the initial state, and n" and J" for the final state. The simplest prediction of the pattern of transition energies that we might use to assign lines would come from using the lowest order treatment for vibrational-rotational levels of a diatomic (from Chap. 6). At this level, anharmonicity, centrifugal distortion, and vibration-rotation coupling are neglected, and so we may we may develop a concise expression for the transition energies.

$$\Delta E = E_{final} - E_{initial} = E''(n'', J'') - E'(n', J') \qquad (10\text{-}50a)$$

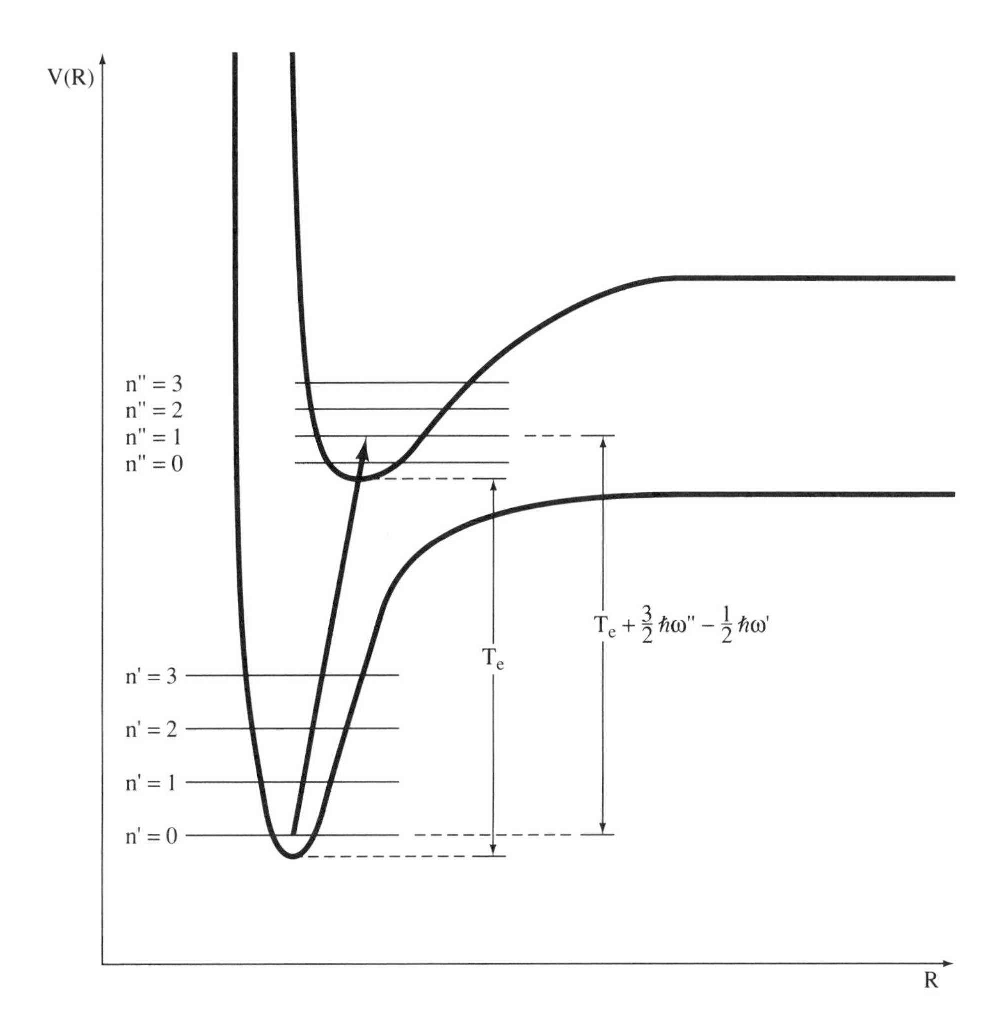


**FIGURE 10.7** Potential energy curves for two electronic states, the ground state and an excited state of the same spin, of a hypothetical diatomic molecule. Each potential may be analyzed independently to yield vibrational-rotational states and energy levels. The lowest four vibrational levels are shown. In an absorption spectrum transitions may originate from the vibrational-rotational levels of the ground electronic state and end in the vibrational-rotational levels of the excited state following the appropriate selection rules. The energies of the transitions will be sums of the energy difference between the bottoms of the potential wells, designated $T_e$, and the difference between the vibrational-rotational state energies within their respective potentials. One such transition is drawn, and the contributions to its transition energy are depicted to the right.

$$\Delta E = T_e + (n'' + \frac{1}{2}) \hbar \omega'' + B'' J''(J'' + 1)$$

$$- (n' + \frac{1}{2}) \hbar \omega' - B' J' (J' + 1) \qquad (8\text{-}50b)$$

The selection rules for J in the case of an electronic transition of a diatomic molecule is that J can change by one or can be unchanged. This gives three cases for the part of the transition energy in Eq. (10-50) associated with the rotational energy, which is $\Delta E^{rot}$.

$$\Delta E = T_e + (n'' + \frac{1}{2}) \hbar \omega'' - (n' + \frac{1}{2}) \hbar \omega' + \Delta E^{rot}$$

$$\Delta J = J'' - J' = 1 \Rightarrow \Delta E^{rot} = B''(J'+1)(J'+2) - B'J'(J'+1) \qquad (10\text{-}51a)$$

$$= (B'' - B') J'^2 + (3B'' - B') J' + 2B''$$

$$\Delta J = J'' - J' = 0 \Rightarrow \Delta E^{rot} = (B'' - B') J'^2 + (B'' - B') \qquad (10\text{-}51b)$$

$$\Delta J = J'' - J' = -1 \Rightarrow \Delta E^{rot} = (B'' - B') J'^2 - (B'' + B') J' \qquad (10\text{-}51c)$$

These transition frequencies correspond to a whole series of lines for each possible n'-to-n" vibrational level changes. From the relative sizes of typical rotational constants, we should expect that $\Delta E^{rot}$ will separate the lines by a small amount in relation to the size of $T_e$. In other words, the spectrum should show a bunch of lines clustered about a transition energy of

$$T_e + (n'' + \frac{1}{2}) \hbar \omega'' - (n' + \frac{1}{2}) \hbar \omega'$$

This feature is called a *vibrational band* of the electronic spectrum. The rotational contributions in Eq. (10-51) allow us to decipher the *rotational fine structure* of the band.

If $B'' \approx B'$, then Eq. (10-51a) will simplify to $2B'(J'+1)$, and this is the usual expression for the R-branch lines in the fine structure of a diatomic IR absorption spectrum. Eq. (10-51b) yields zero for $\Delta E^{rot}$, and so all the lines will be at the same frequency, and they comprise a Q branch.

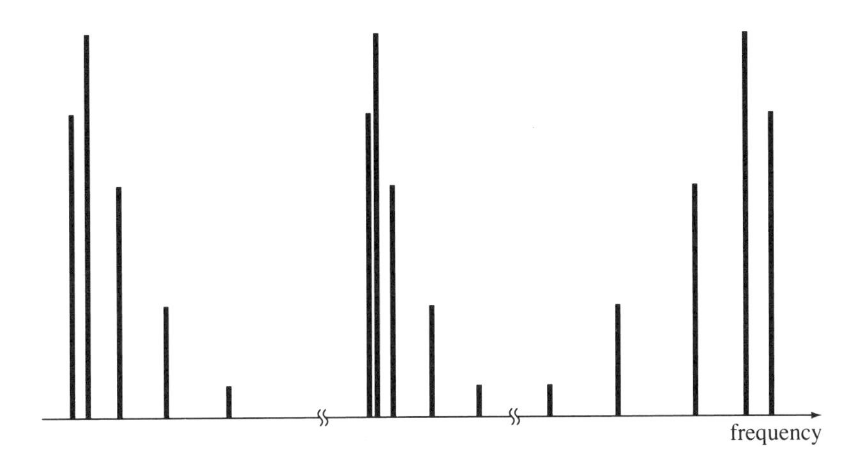


**FIGURE 10.8** The form of the Q branch of an electronic spectrum of a diatomic molecule. The example on the left corresponds to B" > B'. In the middle is the form of the Q branch when B" ≈ B', and on the right is the form when B" < B'. Frequency is increasing to the right in these spectra.

Eq. (10-51c) yields –2B' J', and so these lines will correspond to the P-branch. Generally, B" and B' will be somewhat different, as suggested by the different equilibrium distances for the two potentials in Fig. 10.7. If B" > B', then with increasing J', the P and R branches will extend to higher and to lower frequencies, respectively, and the Q branch lines will spread apart toward higher frequencies. If B" < B', the Q branch lines will be spread apart but will be lower in frequency with increasing J'. Fig. 10.8 is a representation of how differently the Q branch might appear.

The terms that are quadratic in J' in Eq. (10-51) can lead to an interesting feature if the the difference in the ground and excited state rotational constants is sizable. If B" > B', there will in principle be a point at which the quadratic term in Eq. (10-51c) is more important than the linear term. At that point, the lines will be at higher and higher transition energies with increasing J', rather than at lower energies, which is the way the P-branch lines started. This is illustrated in Fig. 10.9. The reversal in the

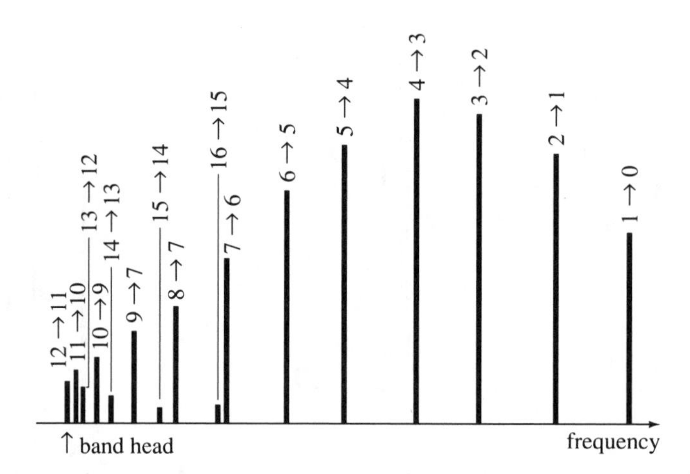


**FIGURE 10.9** Assignment of lines leading to a band head in a P-branch of a hypothetical diatomic molecule. The length of each transition line has been adjusted to reflect the Maxwell-Boltzmann populations of the initial rotational states.

ordering of the lines is called a *band head*. A band head could occur in the R branch if B" < B', so the observation of one of these features could immediately reveal which state had the bigger rotational constant and thus whether the bond length in the excited state was longer or shorter than in the ground state.

Another feature of the rotational fine structure of electronic spectra is that the relative intensities follow the populations of the initial rotational sublevels. These follow the Maxwell-Boltzmann distribution law as mentioned in Chap. 6. This is exactly the same as for infrared spectra, and so we expect an eventual fall-off in the intensities of the lines with increasing J'. Obtaining spectra at several temperatures can help, therefore, in deciphering the rotational fine structure of a band. Cooling the sample will diminish (or eliminate) transitions from the higher rotational levels, and this may reduce the congestion of lines to assign. Heating the sample will extend the P, Q and R branches.

There is no strict general selection rule for the vibrational quantum number in the electronic spectra of a diatomic molecule. Transitions may be seen with n' = n" and with n' different from n" by 1, 2, 3, and so on, and these give the different bands that are observed. There will, however, be differences in the relative intensities of the bands. This is particularly apparent in low-resolution spectra, where the rotational fine structure is not resolved. The intensity differences are understood in terms of the *Franck-Condon overlap* between the initial vibrational state and the final vibrational state. Consider two electronic states with the same force constant and the same equilibrium bond length. Vibrational analysis of each potential would yield essentially the same wavefunctions for the n' = 0 state and the n" = 0 state. The overlap of these two wavefunctions would be nearly unity. On the other hand, the overlap of the n' = 0 wavefunction and the n" = 1 wavefunction would be near zero, just as the overlap of the n' = 0 and n' = 1 wavefunctions would be zero. If the equilibrium length of the excited state were made greater, the overlap between the n' = 0 and n" = 0 wavefunctions would diminish, while the overlap between the n' = 0 and n" = 1 wavefunction might increase. This makes sense from a physical standpoint: With the same equilibrium lengths, the transition is favored because in both states the molecule vibrates about the same equilibrium point. With different equilibrium lengths, the vibrational motion has to make a more drastic change upon the electronic excitation of the molecule, and so the likelihood of the transition is diminished.

Changes in the curvature (force constant and vibrational frequency) of one potential relative to the other can also change the overlap. In a rough way, the intensities of the bands will depend on the overlap between the initial and final vibrational states, and that overlap will depend on the nature of the two potential curves.

*Hot bands* are transitions from excited vibrational levels of the ground electronic state, i. e., n' > 0. They may appear at lower transition frequencies than the transition from n' = 0 to n" = 0.

Gas phase electronic spectra of polyatomic molecules are more complicated than the spectra of diatomics. The number of vibrational modes and the possibility of combination bands will usually lead to numerous vibrational bands, and these may be overlapping. Also, the rotational fine structure tends to be more complicated, as we might expect from the differences between diatomic and polyatomic IR spectra. An

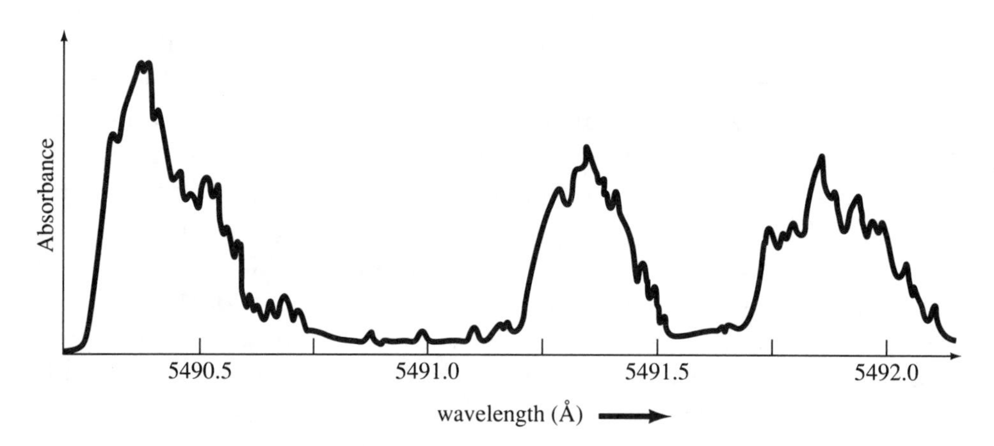


**FIGURE 10.10** Vibrational band structure in the region around 5491 Å of the electronic absorption spectrum of 4-methyl-1,2,4-triazoline-3,5-dione ($C_3N_3H_3O_2$) at room temperature.

example of a typically congested high-resolution absorption spectrum of a small polyatomic molecule is shown in Fig. 10.10. Conventional absorption spectra can prove to be a difficult means of measuring and assigning transitions, and so numerous experimental methods have been devised to select molecules in specific initial states and probe the absorption or the emission spectrum with narrow frequency range lasers. The analysis of the quantum chemistry of the energy levels, though, is the same.

## 10.10 ELECTRON CORRELATION

The self-consistent field approximation provides a very powerful means of working with electronic wavefunctions. Perhaps most important is that it is the basis for the orbital description of molecules, and orbitals have proven quite useful in interpetation of molecular bonding and chemical phenomena. Even so, the SCF approximation is but an approximation, and it does leave errors in the energetics and the charge densities. For instance, it is a typical feature of an SCF wavefunction that electron density is slightly exaggerated in bonding regions, or regions between two nuclei. This is because the self-consistent field is an on-average description of how electrons interact, whereas their instantaneous interactions as real particles – not charge clouds – is significant. This instantaneous interaction is called *electron correlation*, and electron correlation effects are generally any effects from going beyond the SCF approximation.

Electron correlation effects may be inferred from experiment but can not be directly measured. In the laboratory, it is not possible to turn on and turn off the self-consistent field approximation so as to see the differences. Experimental information on correlation effects is usually of the sort where a sharp qualitative difference with an orbital description is detected. This would happen in the case of a serious breakdown of the SCF approximation.

Subtle electron correlation effects are most often understood on the basis of detailed quantum chemical calculations rather than experiment. Wavefunctions and energies are obtained first from SCF calculations and then from treatments that go beyond SCF. The differences are the correlation effects. The exaggeration of electron density in bonding regions, for instance, is corrected as electron correlation is taken into account. As a result, there is often a net effect of electron correlation that amounts to making a bond length somewhat longer and to making the bond less stiff, which means diminishing the force constant. For small covalent molecules, high level treatment of electron correlation shows bond length changes that are mostly 0.01 to 0.03 Å and reduction of harmonic force constants by about 5 to 20%.

Electron correlation will affect the charge distributions in ways that are manifested in the molecular properties. The dipole moment of a molecule is often changed by 5 to 10% because of electron correlation effects. A selection of calculated values for small molecules from the SCF level of treatment and from treatments that carefully include electron correlation effects is shown in Table 10.2. Comparison with measured values provides an assessment of the suitability of the SCF orbital picture in describing molecular systems.

TABLE 10.2 Comparison of Calculated and Measured Values of Certain Molecular Properties. [a]

| | Value From SCF | Value With Correlation | Experiment |
|---|---|---|---|
| LiH equilibrium bond length | 1.6071 Å | 1.5974 Å | 1.5957 Å |
| LiH n = 0 to n = 1 vibrational transition frequency | 1389.5 $cm^{-1}$ | 1358.1 $cm^{-1}$ | 1359.8 $cm^{-1}$ |
| HF n = 0 to n = 1 vibrational transition frequency | 4324 $cm^{-1}$ | 3977 $cm^{-1}$ | 3961 $cm^{-1}$ |
| $H_2O$ harmonic vibrational frequencies | 4132 $cm^{-1}$ | 3830 $cm^{-1}$ | 3832 $cm^{-1}$ |
| | 1771 $cm^{-1}$ | 1677 $cm^{-1}$ | 1649 $cm^{-1}$ |
| | 4236 $cm^{-1}$ | 3940 $cm^{-1}$ | 3943 $cm^{-1}$ |
| $N_2$ quadrupole electrical moment | 0.886 a.u. | 1.085 a.u. | 1.09 a.u. |

[a] These values are collected in Ref. 5 in the bibliography for this chapter.

## 10.11 BOND BREAKING AND THE VALENCE BOND

One of the problems with the SCF approximation is in the description of breaking, or forming, a chemical bond. In the course of breaking a bond, significant changes in the orbital character of the species are not unlikely, and even the form of the wavefunction may not be the same throughout the whole process. For instance, the orbital picture of the nitrogen molecule at equilibrium is a closed shell wavefunction, and the SCF level treatment is reasonable. It dissociates, however, into two identical atoms, and their states cannot be described as closed shells. So, a wavefunction which is appropriate throughout must include several configurations whose importance changes with the geometrical parameters. The use of several configurations means that correlation effects are being accounted for; however, these are nondynamical correlation effects, as opposed to the dynamical correlation* effects discussed in the prior section.

One powerful approach to the problem of bond breaking is the valence bond method. To understand the basic idea of the valence bond picture, let us consider the electronic structure of the ground, singlet state of the hydrogen molecule. The form of the valence bond (or Heitler-London) wavefunction is

$$\psi(1,2) = \frac{N}{\sqrt{2}} \left( a(1)\, b(2) + a(2)\, b(1) \right) \left( \alpha(1)\, \beta(2) - \alpha(2)\, \beta(1) \right) \qquad (10\text{-}52)$$

where we use the shorthand notation of electron numbers 1 and 2 in place of the spatial and spin coordinates of those electrons [e.g., a(1) instead of $a(\vec{r}_1)$ and $\alpha(2)$ instead of $\alpha(\vec{s}_2)$ ]. N is a normalization factor, the spatial functions have been designated a and b, and $\alpha$ and $\beta$ are the usual electron spin functions. This wavefunction is not a Slater determinant, but it is nonetheless properly antisymmetric with respect to particle interchange. In an SCF orbital description, the spatial functions are orthogonal. In the valence bond description, that constraint is relaxed: a and b are allowed to be the same or different spatial functions. This additional flexibility in the wavefunction provides for a better description of the changing electronic structure that occurs during bond breaking and bond formation.

When the two hydrogen atoms in $H_2$ are very far apart, perhaps 1000 Å, there is essentially no interaction. In this limiting case, the electronic structure is the electronic structure of two ground state hydrogen atoms. The wavefunction in Eq. (10-52) can represent this limiting case, which is to say that variational optimization of $\psi$ will yield the correct 1s orbitals. We can see that the wavefunction has this flexibility by replacing a with $1s_A$, meaning a hydrogen 1s orbital on hydrogen A, and by replacing b with $1s_B$, where B is the other hydrogen.

$$\psi(1,2) = \frac{N}{\sqrt{2}} \left( 1s_A(1)\, 1s_B(2) + 1s_A(2)\, 1s_B(1) \right) \left( \alpha(1)\, \beta(2) - \alpha(2)\, \beta(1) \right)$$

* The distinction between dynamical and nondynamical correlation is not sharp. However, as a working definition, nondynamical correlation is manifested as a few configurations entering the wavefunction with sizable importance. For dynamical correlation, there are many configurations with small expansion coefficients and a single dominant configuration.

$$= \frac{N}{\sqrt{2}} \left\{ \left|1s_A \alpha \quad 1s_B \beta\right| - \left|1s_A \beta \quad 1s_B \alpha\right| \right\} \tag{10-53}$$

Since the orbitals $1s_A$ and $1s_B$ are orthogonal, given the large A-B separation distance, the rearrangement of the wavefunction into two Slater determinants in Eq. (10-53) allows us to conclude that this is the form of the wavefunction for two unpaired electrons with singlet coupled spins. In other words, the wavefunction has the flexibility to take on the correct form at the separated atom limit.

The other limiting situation is the equilibrium separation of the protons in the $H_2$ molecule. The molecular orbital SCF picture places the two electrons in one spatial orbital, $\sigma$. We may see that the valence bond wavefunction may take on this form by letting both orbitals a and b in Eq. (10-52) be $\sigma$.

$$\psi(1,2) = \sqrt{2}\, N\, \sigma(1)\, \sigma(2) \left( \alpha(1)\, \beta(2) - \alpha(2)\, \beta(1) \right) \tag{10-54}$$

This is recognized to be the closed-shell SCF determinant written in expanded form.

Showing that the wavefunction has the flexibility to properly describe the separated atoms and the bonded atoms means that a variational determination of the orbitals a and b at each separation distance will provide an appropriate description of the system continuously from one limiting situation to the other. In contrast, the SCF orbital picture is not uniformly appropriate for the breaking of the $H_2$ bond. We may see this by viewing the SCF $\sigma$ orbital to be essentially a linear combination of the two hydrogenic 1s orbitals:

$$\sigma = n\left( 1s_A + 1s_B \right) \tag{10-55}$$

where n is a factor to ensure normalization of the $\sigma$ orbital. If this is substituted into the SCF wavefunction, which is given in Eq. (10-54), the following is obtained.

$$\psi = N' \left( 1s_A(1)\, 1s_A(2) + 1s_A(1)\, 1s_B(2) + 1s_B(1)\, 1s_A(2) + 1s_B(1)\, 1s_B(2) \right)$$

$$\times \left( \alpha(1)\, \beta(2) - \alpha(2)\, \beta(1) \right) \tag{10-56}$$

where all the normalization factors have been collected into N'. This is the form of the SCF wavefunction everywhere, at least while restricting the $\sigma$ orbital to be a linear combination of just the 1s orbitals. So, even at 1000 Å separation, it would have this form. If we compare it with the proper separated limit form of Eq. (10-53), we see a difference. The SCF wavefunction includes two additional terms, $1s_A(1)1s_A(2)$ and $1s_B(1)1s_B(2)$. These terms are interpreted as corresponding to an ionic arrangement of the electron density. The first places both electrons on hydrogen A, leaving bare the proton of hydrogen B. The second term is the opposite ionic form. The nonionic terms, as in Eq. (10-53), correspond to a covalent sharing of the electrons.

This analysis reveals that the SCF description weights the ionic and covalent terms equally for *all* A-B separation distances, whereas at the separated limit, the ionic terms must vanish. So, the flexibility in the valence bond description that comes about from allowing the two electrons of a bonding pair to be in different, nonorthogonal orbitals (i.e., a and b) translates into the flexibility to change the weighting between ionic and covalent parts of the electron distribution. In this way, the valence bond picture is extremely useful for describing bond breaking and bond formation.

With some manipulation, valence bond wavefunctions can be restated in terms of orthogonal orbitals. For the ground singlet state of $H_2$, it can be shown that a description equivalent to Eq. (10-52) is,

$$\psi = c_1 \, | \phi_1 \alpha \;\; \phi_1 \beta | + c_2 \, | \phi_2 \alpha \;\; \phi_2 \beta | \tag{10-57}$$

This wavefunction is a linear combination of two configurations, and their expansion coefficients, $c_1$ and $c_2$, must be obtained variationally. Each configuration places both electrons in the same spatial orbital, and $\phi_1$ and $\phi_2$ are orthogonal. This representation of a valence bond wavefunction proves particularly helpful in computer calculations of the wavefunctions. In the *generalized valence bond* (GVB) method, these ideas have been taken to a powerful computational level that is general with respect to the spin couplings and bonding pairs.

## 10.12 CALCULATION OF CORRELATED WAVEFUNCTIONS *

The most basic scheme for determining correlated electronic wavefunctions employs an expansion of the wavefunction in a basis. The basis must be a set of N-electron functions where N is the number of electrons in the molecule of interest. A standard choice is to use Slater determinants constructed in terms of a set of spatial orbitals arising from an SCF treatment of the system. The flexibility of this determinantal basis will exist for a more concise basis of electron *configuration state functions* (CSF's). These are simply the linear combinations of Slater determinants that are eigenfunctions of the electron spin operator, though they may also have certain properties that reflect the geometrical symmetry of the molecule. To see that the CSF's are a more concise basis than an equivalent set of Slater determinants, consider the Slater determinants formed for a four-electron system where each electron is in a distinct spatial orbital. Since each electron may have an $\alpha$ or a $\beta$ spin, there will be $2^4 = 16$ determinants. However, it can be shown that there are only two linear combinations of these Slater determinants for which the eigenvalue of the spin operator is zero. Thus, if the state of interest were a singlet state ($S = 0$), it would be sufficient to use two CSF's for this occupancy instead of using as many as 16 determinants. So, correlated electronic wavefunctions are very often developed as expansions in a basis of CSF's.

The process of variationally determining a linear expansion of an electronic wavefunction in a basis of configurations is referred to as *configuration interaction* (CI). Of course, this is a diagonalization problem, although it is specialized in needing only the lowest eigenenergy. The matrix to be diagonalized, at least in part, is the matrix representation of the electronic Hamiltonian in the basis of CSF's, or else in a basis of Slater determinants. The Slater-Condon rules, discussed earlier, are employed to determine each Hamiltonian matrix element in terms of the computed one- and two-electron integral values.

The list of CSF's used in a CI expansion is quite important. Typically, the list is so lengthy that it must be truncated to make calculation tractable. While the limits of such truncation have been pushed to over 10 million CSF's (meaning 10,000,000 variational parameters!) in contemporary computational schemes, a judicious basis for truncation remains helpful in CI calculations. The most apparent basis is the level of substitution of orbitals in the configuration. A CI expansion is built around one reference configuration, which is normally the SCF configuration. A consequence of the SCF prescription for the orbitals in the reference configuration is that the Hamiltonian matrix elements with configurations called single substitutions are all zero. This is Brillouin's theorem. To form singly substituted configurations, one spin-orbital in the reference configuration is changed so that the spatial part is an orbital that contained no electrons in the reference. So, truncating the CI expansion at the point of just including the single substitutions is pointless: These configurations will not mix with the reference.

Doubly substituted configurations, those that differ by replacement of two spin-orbitals in the reference configuration, will generally have non-zero Hamiltonian matrix elements with the reference configuration. Triply substituted configurations and still higher order substitutions will have an identically zero Hamiltonian matrix element with the reference because of the Slater-Condon rules, or simply the fact that the Hamiltonian has only two-body operators. As a result, the doubly substituted configurations are the primary correlating configurations; they are the most important. A very common level of CI calculation, then, is CISD, meaning CI with the singly and doubly substituted configurations. The singly substituted configurations are included because they have nonzero Hamiltonian matrix elements with the doubly substituted configurations, and so they mix with the reference configuration indirectly.

If we group the singly (S), doubly (D), and triply (T) substituted configurations together, we may map the regions of the Hamiltonian CI matrix where elements are identically zero. This is shown in Fig. 10.11. We can see that the triply and quadruply substituted configurations will mix only indirectly with the reference configuration. The effects that the inclusion of triply and quadruply substituted configurations yield are termed *higher order correlation effects* because they come about through the higher order substitutions. As more and more configurations are included in the variational CI expansion, the determination of the correlation energy gets closer and closer to the exact value.

Another approach for finding electron correlation effects is perturbation theory, which is usually termed *many-body perturbation theory* (MBPT) to reflect the subtleties associated with treating a system of

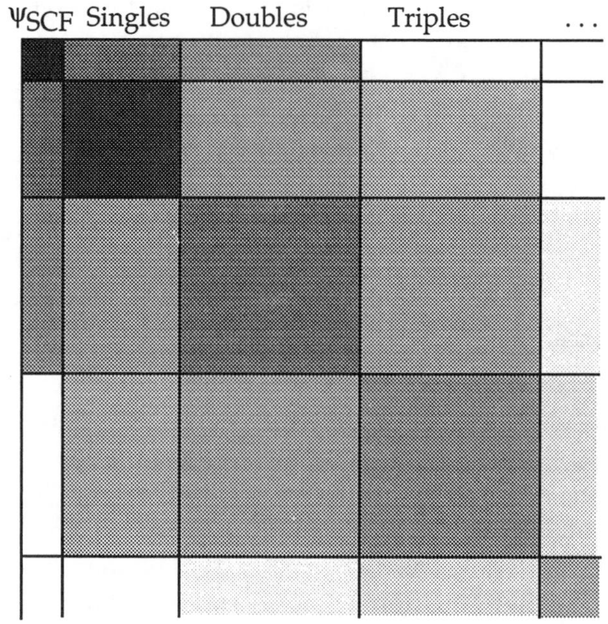

$\Psi_{SCF}$ Singles Doubles Triples ...

**FIGURE 10.11** A schematic drawing of the regions of a Hamiltonian CI matrix organized according to the type of substituted configuration (singly substituted, doubly substituted, and so on) that label the rows and columns. Open regions are those for which all the matrix elements are identically zero. The shaded regions have nonzero elements, but in most of these regions, the nonzero elements tend to be sparse. The shadings give an idea of the relative sparseness of nonzero elements that may be expected.

many interacting electrons. Of course, to apply perturbation theory, we must partition the complete Hamiltonian into a zero-order part and the perturbation. In MBPT, the zero-order part is the sum of Fock operators of Eq. (10-28). The zero-order wavefunctions are the same configurations that are used in a CI expansion, singly substituted, doubly substituted, and so on. Their zero-order energies are simply sums of orbital energies, following Eq. (10-29). Thus, the energy denominators in MBPT, which are differences in zero order energies, reduce to differences in orbital energies.

The perturbation in MBPT is simply the difference between the complete Hamiltonian and the zero-order part. In other words, the perturbation is everything that's left to be included. The perturbation expansion may be continued to any order, and in practice, typical levels are second-, third- and fourth-order corrections to the energy. The convention for designating such calculational treatments is MBPT(n), where n is the order of the correction to the energy. Another name that is widely used for MBPT is Moeller-Plesset theory, and the designation that is used is MP(n).

The map in Fig. 10.11 also serves to show where there are nonzero off-diagonal elements of the MBPT perturbing Hamiltonian, since these must be the same as the off-diagonal CI elements. From this, we may consider the MBPT corrections order by order. An energy of a wavefunction correct through first-order perturbation theory is the expectation value of the wavefunction with the complete Hamiltonian. For electronic wavefunctions, this is simply the SCF energy of the reference configuration. The first contributions to the correlation energy are at second order. At second order, the energy corrections depend on the matrix element between the reference and the other zero-order functions. From the map in Fig. 10.11, we may conclude that only doubly substituted configurations will contribute at second order. Thus, an MBPT(2) treatment includes only double substitutions. At higher orders of treatment, zero-order wavefunctions that do not interact directly with the reference contribute to the energy corrections. So, after second order one designates the configurations used in the MBPT calculational treatment. MBPT(4)-SDQ means the energy was obtained through fourth order and that the zero order functions included the singly, doubly, and quadruply substituted configurations. This is often a very high-level and accurate treatment.

A very sophisticated approach for treating electron correlation evolved from nuclear structure theories, and it is called the *coupled cluster approach* (CCA). This approach focuses attention on how substitutions are made relative to the reference configuration. From a kinematic standpoint, the primary electron correlations are two-electron or pair correlations. That is, it is much more likely for two electrons to be in close proximity at any instant than for three or four or more electrons to be in the same region at the same instant. And of course, it is the electron-electron repulsion, which is strongest for electrons in close proximity, that gives rise to what is meant by correlation. Thus, the main type of four-electron correlations are two simultaneous pair correlations as opposed to four electrons being instantaneously close.

Linearly including double substitutions in a wavefunction can account for the pair correlations, though not the higher order correlations. We may reason from a kinematic standpoint how the main type of higher order correlations may be included. Let us consider the SCF wavefunction for a system of two interacting helium atoms to be $\Phi = \phi_1^2 \phi_2^2$, where $\phi_1$ is essentially localized on one atom and $\phi_2$ is localized on the other. Let us assume that there are two available empty orbitals, $\phi_3$ and $\phi_4$, and that $\phi_3$ is localized on the same atom as $\phi_1$ and $\phi_4$ is localized on the same atom as $\phi_2$. A double substitution that would account, partly, for the correlation of the pair of electrons on the first helium is

$$\Phi_{11}^{33} = \phi_3^2 \phi_2^2$$

A double substitution that would account, partly, for the correlation of the pair of electrons on the other helium is

$$\Phi_{22}^{44} = \phi_1^2 \phi_4^2$$

(Notice that the subscripts and superscripts on the two $\Phi$ functions indicate the substitution.) A correlated wavefunction that includes the main pair correlation effects could consist of the reference plus these two double substitutions.

$$\psi = c_0 \Phi + c_{11}^{33} \Phi_{11}^{33} + c_{22}^{44} \Phi_{22}^{44} \tag{10-58}$$

The expansion coefficients might be obtained variationally or otherwise. Consistent with the view that the primary four-electron correlation comes about as the simultaneous correlation of two separate pairs of electrons, we may incorporate the quadruply substituted configuration of this problem with an expansion coefficient prescribed by the coefficients in Eq. (10-58).

$$\psi = c_0 \Phi + c_{11}^{33} \Phi_{11}^{33} + c_{22}^{44} \Phi_{22}^{44} + c_{11}^{33} c_{22}^{44} \Phi_{1122}^{3344} \tag{10-59}$$

In CI, the coefficient of the quadruply substituted configuration in Eq. (10-59) would be a linear variational parameter independent of the other coefficients. In the coupled cluster approach, there is no need to introduce a new coefficient. The complication is that the wavefunction no longer has a linear dependence on the coefficients.

The generalization of the $He_2$ example just considered is a power series expansion of the substitution operators. Let us define an operator that acts on a configuration to substitute the a and b spin-orbitals for the i and j spin-orbitals:

$$\hat{t}_{ij}^{ab} \Phi = \Phi_{ij}^{ab} \tag{10-60}$$

Let us then define an overall operator that makes all possible double substitutions with each weighted by the appropriate coefficient:

$$\hat{T} \equiv \sum_{i,j} \sum_{a,b} c_{ij}^{ab} \hat{t}_{ij}^{ab} \tag{10-61}$$

Consider the following power series constructed from this operator.

$$\hat{O} = \hat{T} + \frac{1}{2} \hat{T}^2 + \frac{1}{6} \hat{T}^3 + \frac{1}{24} \hat{T}^4 + \ldots \tag{10-62}$$

If this operator were applied to $\Phi$, the SCF reference configuration, the first term would clearly produce all the double substituted configurations weighted by the coefficients in Eq. (10-61). The second term would create double substitutions of double substitutions, which are quadruple substitutions, and each of these would have an expansion coefficient involving products of the coefficients in Eq. (10-61). The third term would create configurations where six spin-orbitals are substituted, and so on. It turns out that the higher order configurations generated with $\hat{O}$ applied to the reference configuration have expansion coefficients that correspond to the simultaneous pair correlations in the same way as the expansion coefficient of the quadruple substitution in Eq. (10-59). So, the idea of the coupled cluster approach is to generate the correlating configurations of the wavefunction with the operator $\hat{O}$.

If we consider the operator $\hat{T}$ as if it were a variable, then the power series expansion in Eq. (10-58) may be written more concisely as the exponential of $\hat{T}$. Thus,

$$e^{\hat{T}} = 1 + \hat{O} \tag{10-63}$$

The coupled cluster wavefunction is concisely written as

$$\Psi_{CC} = e^{\hat{T}} \Phi \tag{10-64}$$

The operator $\hat{T}$ may also be constructed to include single substitutions, triple substitutions, and so on. These will provide for an ever more complete description of electron correlation. The notation in the computational chemistry literature is to use CCD for a coupled cluster wavefunction with doubly substituted configurations in $\hat{T}$, realizing that there are higher order substitutions in $\hat{O}$. CCSD means that $\hat{T}$ included single- and double-substitution operators, and so on.

It is probably fair to claim that coupled cluster treatments comprise the most sophisticated, or at least the most elaborate, calculational approach in quantum chemistry and also that the results attainable with CC can be extraordinarily accurate and reliable. Given the power of modern computers, it is not surprising that many chemical problems are being tackled directly with quantum chemical tools such as GVB, CI, MBPT, and CC.

## Exercises

1. Make a contour plot of the function $V(x,y) = 3x^2 + y^2$ showing the contours for $V = 1.0$, $V = 2.0$, and $V = 3.0$ on an xy-grid. Do the same for the function $V'(x,y) = 3x^2 + y^2 + xy$. In what ways do the contours differ?
2. If the spacing between the contour levels in Fig. 10.2 were 100 $cm^{-1}$, what would be the energy difference between the saddle point and the equilibrium?
3. For a collection of identical spin 3/2 particles, what is the analog of the Pauli exclusion principle? In other words, for the wavefunction to be antisymmetric and to be an orbital product form, how many particles may be found in the same orbital?
4. For a system of four electrons in different spin-orbitals, apply the antisymmetrizer of Eq. (10-21a) to a simple product of the four spin

orbitals and show that this is the same function as the Slater determinant constructed from the orbital product.

5. Verify that a Slater determinant for the lithium atom corresponding to two $1s\alpha$ electrons and one $1s\beta$ electron vanishes.

6. Apply the antisymmetrizing operator to the simple orbital product function $1s\alpha\ 2s\alpha\ 3s\alpha$, or else write out the terms of the Slater determinant with $1s\alpha\ 2s\alpha\ 3s\alpha$ on the diagonal. Then apply the antisymmetrizing operator to each of the terms individually and simplify the result.

7. From qualitative LCAO arguments, predict the bond order of the following diatomics: LiH, HeBe, LiF, $C_2^{2-}$, $CO^+$ and $NeF^+$.

8. Using the equivalence of two-electron integrals over real orbital functions given by Eq. (10-32), determine the number of uniquely valued two-electron integrals that may be written for a set of N orbital functions.

9. Verify that Eq. (10-33) is obtained from Eq. (10-35) for the special case of closed-shell wavefunctions.

10. Carry out the spin integrations implicit in Eq. (10-35) for the special case of a high spin, unpaired occupancy, $\Phi = \det\{\phi_1\alpha\ \phi_2\alpha \ldots \phi_N\alpha\}$, and obtain an expression in terms of integrals over spatial functions.

11. For the closed shell wavefunction, $\Phi = \det\{\phi_1^2\ \phi_2^2\ \phi_3^2\}$, show from the Slater-Condon rules that the expectation value of its energy is indentical to the expectation value of the energy of the wavefunction, $\Phi' = \det\{\phi_1^2\ \phi_2'^2\ \phi_3'^2\}$, where $\Phi'$ is made from the transformed orbitals $\phi_2' = (\phi_2 + \phi_3)/\sqrt{2}$ and $\phi_3' = (\phi_3 - \phi_2)/\sqrt{2}$. (This corresponds to a unitary transformation between the second and third occupied orbitals, and so the equivalence of the energy expressions illustrates that the energy is invariant to the mixing of occupied orbitals with each other.)

12. Find the Fock operator for the special case of a high spin, unpaired electron occupancy where the form of the wavefunction is the Slater determinant, $\Phi = \det\{\phi_1\alpha\ \phi_2\alpha \ldots \phi_N\alpha\}$.

*13. Expand the Slater determinants $\Gamma_2$ and $\Gamma_3$ in Eq. (10-47), and also in Eq. (10-48). Then factor $\Phi_{10}$ and $\Phi_{00}$ so that each is a product of a pure spin function and a pure spatial function. In this form, how do $\Phi_{10}$ and $\Phi_{00}$ differ?

*14. Apply the lowering operator $\hat{S}_-$ to the determinant $| \phi_1\alpha \ \phi_2\alpha \ \phi_3\alpha |$. The resulting linear combination of determinants is a spin eigenfunction with S = 3/2 and $M_S = 1/2$. Apply $\hat{S}_-$ again to produce the function with $M_S = -1/2$. What would the result be if it were applied still again?

15. Repeat the analysis of Eq. (10-51) with a centrifugal distortion term, $D'[J'(J' + 1)]^2$, for the initial state and $D''[J''(J'' + 1)]^2$ for the final state.

16. If the plus and minus signs in the spin and spatial parts of the wavefunction in Eq. (10-56) were interchanged, what spin state would result?

## Bibliography


1. F. L. Pilar, *Elementary Quantum Chemistry*, 2nd ed. (McGraw-Hill, New York, 1990).
2. A. C. Hurley, *Introduction to the Electron Theory of Small Molecules* (Academic Press, New York, 1976). This is an advanced level text on molecular electronic structure.
3. A. Szabo and N. S. Ostlund, *Modern Quantum Chemistry: Introduction to Advanced Electronic Structure Theory* (Macmillan, New York, 1982). This is a thorough, advanced text on the quantum mechanics of electronic structure.
4. R. S. Mulliken and W. C. Ermler, *Diatomic Molecules: Results of Ab Initio Calculations* (Academic Press, New York, 1977). This provides a high-level, detailed examination of the bonding and structure of diatomics.
5. C. E. Dykstra, *Ab Initio Calculation of the Structures and Properties of Molecules* (Elsevier, Amsterdam, 1988).
6. G. Herzberg, *Molecular Spectra and Molecular Structure. III. Electronic Spectra and Electronic Structure of Polyatomic Molecules* (Van Nostrand Reinhold, New York, 1966).

7. M. Tinkham, *Group Theory and Quantum Mechanics* (McGraw-Hill, New York, 1964).
8. R. B. Woodward and R. Hoffmann, *The Conservation of Orbital Symmetry* (Verlag Chemie/Academic Press, New York, 1971).

# Chapter 11

# Magnetic Resonance Spectroscopy

*Electrons and most atomic nuclei possess intrinsic magnetic moments that give rise to an interaction energy with an external magnetic field. The difference between the interaction energies of the different states can be probed spectroscopically, and this is referred to as magnetic resonance spectroscopy. It is an immensely rich and powerful category of molecular spectroscopy that is now used for qualitative and quantitative analysis, determination of molecular structure, and the measurement of reaction rates and molecular dynamics. In the recent history of magnetic resonance, its use has been extended to imaging, and the medical application for diagnostic work is already widespread. This chapter provides the quantum mechanical foundation for understanding magnetic resonance spectroscopy.*

## 11.1 NUCLEAR SPIN

Atomic nuclei consist of neutrons and protons and have a structure that is as rich as the electronic structure of atoms and molecules. In chemistry, though, nuclear structure is often unimportant since it does not change in the course of a chemical reaction. It is normally quite appropriate to consider nuclei to be point-masses with a specific amount of positive charge. However, there are some important manifestations of nuclear structure that have been exploited in developing powerful types of molecular spectroscopies.

Protons and neutrons have an intrinsic spin and an intrinsic magnetic moment. Experiments revealed that there are just two possible orientations of their spin vectors with respect to a z-axis defined by an external field, and this means that the intrinsic spin quantum number is 1/2. In this feature, protons and neutrons are similar to electrons, and all half-integer spin particles are classified as fermions. Bosons are integral spin particles. Whereas the letter S is commonly used for the electron spin quantum number, the letter I is commonly used for nuclear spin quantum numbers. Thus, I = 1/2 for a proton, and the allowed values for the quantum number giving the projection on the z-axis, $M_I$, are +1/2 and −1/2.

The angular momentum coupling of the intrinsic proton and neutron spins in a heavy nucleus is a problem for nuclear structure theory, and it is beyond our area of discussion. However, we may correctly anticipate certain results on the basis of electronic structure. The first of these is that atomic nuclei may exist in different energy states. It turns out that the separation in energy between the ground state of a stable nucleus and its first excited states is usually enormous in relation to the size of chemical reaction energetics. Photons from the gamma ray region of the electromagnetic spectrum may be needed to induce transitions to excited nuclear states, and so these transitions can require hundreds and thousands of times the energy for a transition to an excited electronic state. Consequently, unless something is done to prepare a nucleus in an excited state, we may assume that all the nuclei in a molecule are in their ground state in some chemical experiment.

Another result we may anticipate from electronic structure is the possible values for the total nuclear spin quantum number. For instance, the deuteron (a proton and a neutron) consists of two particles with intrinsic spin of 1/2. Were these two particles electrons, we know that the possible values for the total spin quantum number are one and zero. It turns out for the deuteron that of the two coupling possibilities, $I = 1$ and $I = 0$, the $I = 1$ spin coupling occurs for the ground state. This is a consequence of the interactions that dictate nuclear structure. For chemical applications, the key information is that the deuteron is an $I = 1$ particle.

The intrinsic spins of stable nuclei have been determined experimentally, and the values have been explained with modern nuclear structure theory. So, tables such as that in Appendix IV are available for looking up the spin of a particular nucleus. The rules of angular momentum coupling are an aid in remembering the intrinsic spins of certain common nuclei. For instance, the helium nucleus, with its even number of protons and neutrons, has an integer spin, $I = 0$. In terms of the number of protons and neutrons, the carbon-12 nucleus is simply three helium nuclei. It must have an even-integer spin, and it also turns out to be $I = 0$. The carbon-13 nucleus has an extra neutron and has a half-integer spin, $I = 1/2$.

Nuclei with an intrinsic spin of $I > 0$ have an intrinsic magnetic moment. Just as with electron spin, the magnetic moment vector is proportional to the spin vector. A nuclear magnetic moment can give rise to an energetic interaction with an external magnetic field as well as with other magnetic moments in a molecule, such as that arising from electron spin. The energy differences from interaction of nuclear magnetic moments are the basis for a major class of spectroscopic experiments.

## 11.2 NUCLEAR SPIN STATE ENERGIES

Nuclear magnetic moments are small enough to have an almost ignorable effect on atomic and molecular electronic wavefunctions. On the other hand, the electronic structure has a major influence on the energies of the nuclear spin states. In this situation, it is an extremely good approximation to separate nuclear spin from the rest of a molecular wavefunction. This is because the electronic, rotational, and vibrational wavefunctions can be determined while ignoring nuclear spin, as has been done so far, and then the effect of the electrons can be incorporated as an external influence on the nuclear spin states.

The separated nuclear spin problem is a very special type of problem in quantum chemistry partly because the number of states is strictly limited. For example, a proton is a spin-1/2 particle; this means that $I = 1/2$ and that the spin multiplicity, $2I+1$, is 2. There are only two states, one with $M_I = 1/2$ and one with $M_I = -1/2$. Then, in the hydrogen molecule where there are two protons, the number of nuclear spin states for the molecule as a whole is the product of the spin multiplicities of the two nuclei, i.e., $2 \times 2 = 4$. Clearly, in large, complicated molecules, the number of nuclear spin states may be large, but finite.

Spectroscopy generally is an experiment to observe transitions between quantum mechanical energy levels. Nuclear magnetic resonance (NMR) spectroscopy is concerned with the energies of the nuclear spin states and the transitions that are possible between different states. To work out the energies of the states, we need to understand the interactions that affect nuclear spin state energies and to develop an appropriate Hamiltonian.

Interactions with nuclear spins come about through the intrinsic magnetic moments of nuclei. The nuclear magnetic moment, $\mu$, is proportional to the nuclear spin vector, I, just as the magnetic moment of an electron is proportional to its spin vector. Instead of a fundamental development of the proportionality relationship, we follow a phenomenological approach by simply using an unknown for the proportionality constant, and for now we shall call it $\alpha$.

$$\vec{\mu} = \alpha \vec{I} / \hbar \tag{11-1}$$

As we have already seen with electrons, the interaction energy of a magnetic moment and a uniform external magnetic field, *H*, goes as the dot product of the two vectors. So, the interaction Hamiltonian for a bare nucleus experiencing an applied magnetic field is

$$\hat{H} = -\vec{\mu} \bullet \vec{H} \tag{11-2}$$

If there are several non-interacting nuclei experiencing the field, then the Hamiltonian is a sum of the interactions of each of the nuclei.

$$\hat{H} = -\sum_i \vec{\mu}_i \cdot \vec{H} \tag{11-3}$$

This is the basic form of the Hamiltonian for the nuclear spin state Schrödinger equation.

Nuclei embedded in molecular electronic charge distributions experience an externally applied magnetic field at a slightly altered strength. That is, the electronic motions tend to shield the nucleus from feeling the full strength of the field. There are also situations where the response of the electron distribution to a magnetic field amplifies the strength of the field at the nucleus. Either way, the Hamiltonian in Eq. (11-3) needs to be modified to properly represent nuclei in molecules, as opposed to bare nuclei in space. Again, we may approach this phenomenologically by inserting a correction factor in Eq. (11-3) without yet establishing the fundamental basis for the factor. So, for each different nucleus, there will be a different correction, and this may be expressed as

$$\hat{H} = -\sum_i \vec{\mu}_i \cdot \left(\mathbf{1} - \sigma_i\right) \cdot \vec{H} \tag{11-4}$$

The $\sigma_i$ are the *nuclear magnetic shielding* tensors. Since **1** is the unit or identity matrix, then the multiplication of the matrix quantity in parentheses in Eq. (11-4) by the field vector will lead to a new vector, which is the effective field at the nucleus.

A helpful simplification comes about from breaking up the shielding tensor into isotropic and anisotropic parts:

$$\begin{pmatrix} \sigma_{xx} & \sigma_{xy} & \sigma_{xz} \\ \sigma_{yx} & \sigma_{yy} & \sigma_{yz} \\ \sigma_{zx} & \sigma_{zy} & \sigma_{zz} \end{pmatrix} = \frac{1}{3}\left(\sigma_{xx} + \sigma_{yy} + \sigma_{zz}\right)\begin{pmatrix} 1 & 0 & 0 \\ 0 & 1 & 0 \\ 0 & 0 & 1 \end{pmatrix}$$

$$+ \begin{pmatrix} \frac{2}{3}\sigma_{xx} - \frac{1}{3}\sigma_{yy} - \frac{1}{3}\sigma_{zz} & \sigma_{xy} & \sigma_{xz} \\ \sigma_{yx} & \frac{2}{3}\sigma_{yy} - \frac{1}{3}\sigma_{xx} - \frac{1}{3}\sigma_{zz} & \sigma_{yz} \\ \sigma_{zx} & \sigma_{zy} & \frac{2}{3}\sigma_{zz} - \frac{1}{3}\sigma_{xx} - \frac{1}{3}\sigma_{yy} \end{pmatrix}$$

The isotropic part has equal diagonal elements, and so it has been given as a constant times the unit matrix. The anisotropic part is what remains. At this point, we will ignore the anisotropic part of the shielding tensor, which means that we will take the second matrix on the right hand side of the expression above to be zero. This may be regarded as an approximation, for now, though specific experimental conditions may offer a proper justification. Thus, the isotropic shielding becomes a simple scalar quantity,

$$\sigma^{iso} = \frac{1}{3}\left(\sigma_{xx} + \sigma_{yy} + \sigma_{zz}\right) \qquad (11\text{-}5)$$

Eq. (11-4) simplifies upon neglect of the anisotropic part of the shielding to become

$$\hat{H} = -\sum_i (1 - \sigma_i^{iso})\,\vec{\mu}_i \cdot \vec{H} \qquad (11\text{-}6)$$

The solutions of the Schrödinger equation with this Hamiltonian provide the basic energy level information for NMR spectroscopy.

Let us use Eq. (11-6) to construct an energy level diagram for the nuclear spin states of a somewhat exotic molecule, ethynol, HCCOH. It will have four nuclear spin states because only the protons have non-zero intrinsic spin. Only two quantum numbers are needed to distinguish these states, the $m_I$ numbers for the two particles. We will designate these as $m_{I_1}$ and $m_{I_2}$, and the states will be designated as $|m_{I_1}\ m_{I_2}>$. Using Eq. (11-1) in Eq. (11-6) and letting the applied field define or be applied along the z-axis so that $\vec{I} \cdot \vec{H} = \hat{I}_z H_z$, we have that the spin states are eigenfunctions of the Hamiltonian.

$$\hat{H}\,|m_{I_1}\ m_{I_2}> \;=\; -(1-\sigma_1)\,\alpha_1 H_z\, m_{I_1}\,|m_{I_1}\ m_{I_2}> \qquad (11\text{-}7)$$

$$-(1-\sigma_2)\,\alpha_2 H_z\, m_{I_2}\,|m_{I_1}\ m_{I_2}>$$

The energies of the four states have a linear dependence on the strength of the external magnetic field, and they will separate in energy with increasing field strength. Since both magnetic nuclei are protons in this example, then $\alpha_1$ must be the same value as $\alpha_2$, and we will simply use $\alpha$ for both. A tabulation of the four states' eigenenergies from Eq. (11-7) is

| $m_{I_1}$ | $m_{I_2}$ | Energy |
|---|---|---|
| –1/2 | –1/2 | $\alpha H_z / 2 \left[ (1-\sigma_1) + (1-\sigma_2) \right]$ |
| –1/2 | 1/2 | $\alpha H_z / 2 \left[ (1-\sigma_1) - (1-\sigma_2) \right]$ |
| 1/2 | –1/2 | $\alpha H_z / 2 \left[ -(1-\sigma_1) + (1-\sigma_2) \right]$ |
| 1/2 | 1/2 | $\alpha H_z / 2 \left[ -(1-\sigma_1) - (1-\sigma_2) \right]$ |

Fig. 11.1 is an energy level diagram based on these energies.

In a conventional NMR experiment, transitions between the nuclear spin energy levels are induced by applying electromagnetic radiation perpendicular to the direction of the static magnetic field. The selection rules are that one $m_I$ quantum number can increase or decrease by one while all the other $m_I$ quantum numbers remain unchanged.* In effect, this says that absorption of a photon of energy will "flip" the spin of only one nucleus at a time. From the above tabulation of the energy levels for HCCOH and from Fig. 11.1, we may see that the selection rules correspond to allowed transitions from the lowest energy state to both of the next two energy states and that the transitions from these two states to the highest energy state are allowed. The arrows in Fig. 11.1 indicate these transitions.

The transition energies are obtained by taking differences in the state energies. Using the tabulation above, we may obtain the following transition energies.

| Initial state | Final state | Transition energy |
|---|---|---|
| $\vert\ 1/2 \quad 1/2 \rangle$ | $\vert\ 1/2 \ -1/2 \rangle$ | $\alpha H_z (1-\sigma_2)$ |
| $\vert\ 1/2 \quad 1/2 \rangle$ | $\vert -1/2 \quad 1/2 \rangle$ | $\alpha H_z (1-\sigma_1)$ |
| $\vert\ 1/2 \ -1/2 \rangle$ | $\vert -1/2 \ -1/2 \rangle$ | $\alpha H_z (1-\sigma_1)$ |
| $\vert -1/2 \quad 1/2 \rangle$ | $\vert -1/2 \ -1/2 \rangle$ | $\alpha H_z (1-\sigma_2)$ |

* The interaction that leads to a transition is between the magnetic moment of the nucleus and the magnetic field of the electromagnetic radiation. These are termed "magnetic dipole transitions."

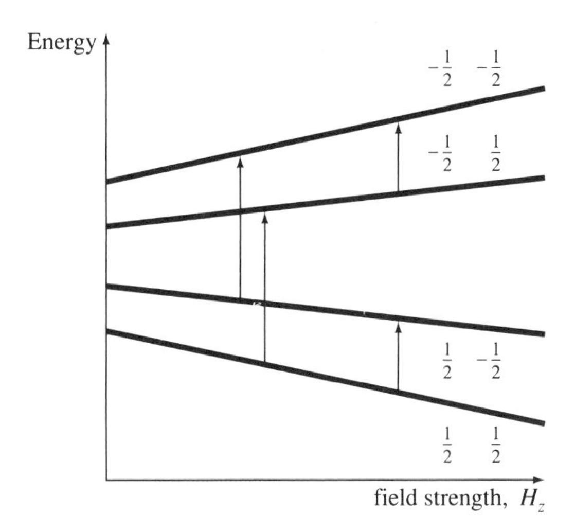


**FIGURE 11.1** Nuclear spin energy levels of a molecule such as HCCOH with two noninteracting protons in different chemical environments. The levels separate in energy with increasing strength of the external field, according to Eq. (11-6). The levels are labelled by the $m_I$ quantum numbers of the two protons. The vertical arrows indicate the allowed NMR transitions. Clearly, the energy of these transitions depends on the strength of the applied field.

This reveals that though there are four possible transitions, there are only two frequencies at which transitions will be detected. Furthermore, measuring the transition frequencies will yield values for $\alpha(1 - \sigma_1)$ and for $\alpha(1 - \sigma_2)$. The typical field strengths used in NMR are such that the transition frequencies of the electromagnetic radiation are in the microwave or radiofrequency regions of the spectrum. These are very low-energy transitions compared to vibrational or electronic excitations of a molecule. The basic experiment can be carried out in two ways. A fixed field strength can be applied and the frequency of the radiation varied until a transition is detected by a change in the power transmission of the radiation. Or, the frequency can be fixed and the power monitored as the field strength is varied. When this sweeping of the field strength brings a transition into resonance with the radiation frequency, a change in the power level of the radiation field is detected. So, transition frequencies may be measured for a given field strength, or field strengths at which transitions occur at a certain frequency may be measured. The latter can be converted to the former.

The information obtained from a low resolution NMR scan is the chemical shielding. It is usually obtained as a shift relative to some standard or reference transition. The chemical shift is designated $\delta$ and is a dimensionless quantity. It is usually a number on the order of $10^{-6}$, and so it is usually stated as being in parts per million (ppm). If the radiation frequency has been varied in the experiment, $\delta$ is given as

$$\delta \cong \frac{\omega_{ref} - \omega}{\omega_{ref}} \tag{11-8}$$

If the field strength has been swept, $\delta$ is given as

$$\delta \cong \frac{H_z - H_{z\text{-ref}}}{H_{z\text{-ref}}} \tag{11-9}$$

The chemical shifts are characteristic of the chemical environment. Thus, NMR spectra may serve as an analytical tool for determining functional groups that are present in a molecule, for instance.

Proton NMR spectra of organic molecules are usually referenced to the proton transitions of tetramethylsilane (TMS). Of course, $\delta$ of the reference, according to Eq. (11-8) or (11-9), is zero. (Another system, with values designated $\tau$, is sometimes used; the values are given as $\tau = 10 - \delta$ with $\tau(\text{TMS}) = 10$ppm.) The NMR signature of protons in different environments in organic molecules is a transition roughly within these ranges ($\delta$ scale):

| | |
|---|---|
| Alkanes | 0 - 1 ppm |
| -C=CH- | 4 - 8 ppm |

| | |
|---|---|
| -C≡CH | 2 - 3 ppm |
| Aromatic | 7 - 8 ppm |
| -OH | 10 - 11 ppm |
| -CHO | 9 - 10 ppm |

More extensive lists of this sort are available.

The proportionality constant introduced in Eq. (11-1) varies from one nucleus to another because of the intrinsic nuclear structure. The proportionality constant, $\alpha$, may be replaced by a dimensionless value, g, for a given nucleus, and $\mu_o$, the nuclear magneton, or basic measure of the size of nuclear magnetic moments.

$$\vec{\mu}_i = \mu_o g_i \vec{I}_i / \hbar \tag{11-10}$$

With the external magnetic field in the z-direction, we may rewrite Eq. (11-6).

$$\hat{H} = -\mu_o H_z \sum_i g_i (1 - \sigma_i) \hat{I}_{i_z} / \hbar \tag{11-11}$$

The variation in the nuclear g values has a much more profound effect on transition frequencies than does the variation in chemical shifts for a given type of nucleus. This is evident from Eq. (11-1): A 10% variation in $\sigma$, given that $\sigma$ is on the order of $10^{-6}$, has much less effect on the energy separation between states than a 10% variation in g. Consequently, NMR spectra are normally nuclei-specific; the instrumentation is set for a narrow range of frequencies to scan (or fields to sweep), and that will be for a certain type of nucleus, such as protons. A different instrumentation setup will be used for carbon-13 nuclei, or for oxygen-17, or for fluorine-19. Of course, even though we may obtain proton NMR spectra independent of obtaining carbon-13 NMR spectra, etc., the complete Hamiltonian of Eq. (11-11) is still at work and still describes the complete problem. Fig. 11.2 illustrates this for the case of the $H^{19}F$ molecule. Both nuclei are spin-1/2 particles. There are four levels, but the two allowed transition frequencies are very different because of the different g values.

As a complicated example of nuclear spin energy levels, let us consider the diimide molecule, HNNH. At this stage, we are assuming no interaction between the magnetic moments of the different nuclei, and so the Hamiltonian of Eq. (11-11) is to be used. In both the trans and cis forms

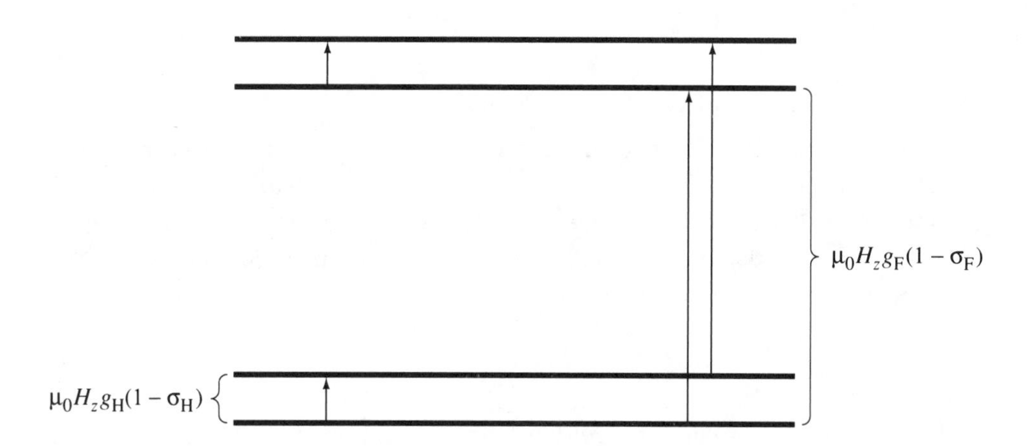


**FIGURE 11.2** An energy level diagram for the four spin states of $H^{19}F$ in an external magnetic field of fixed strength. The allowed transitions are represented by the vertical arrows. Because of the sizable difference in g values, the transition energies for the $^{19}F$ spin flip are much different than the transition energies of the proton. In practice, two different instrumental set-ups are required to observe the two transitions.

of diimide, the nitrogens are in equivalent environments and the protons are in equivalent environments. Nitrogen-14 nuclei have an intrinsic spin of I = 1. The spin multiplicity is 3 since $m_I$ may be –1, 0 or 1, and so with the spin multiplicity of the protons being 2, the number of spin states is $3 \times 3 \times 2 \times 2 = 36$. The energies of these states, obtained via Eq. (11-11) are

$$E_{m_{I_{N-1}} m_{I_{N-2}} m_{I_{H-1}} m_{I_{H-2}}} = -\mu_o H_z \; g_N (1-\sigma_N)\left(m_{I_{N-1}} + m_{I_{N-2}}\right)$$

$$-\mu_o H_z \; g_H (1-\sigma_H)\left(m_{I_{H-1}} + m_{I_{H-2}}\right) \quad (11\text{-}12)$$

where the N and H subscripts indicate the nitrogen and hydrogen atoms, with 1 and 2 used to distinguish the like atoms. The selection rule applied to this problem is that one and only one of the m quantum numbers can change in a transition, and that change may be 1 or –1. This gives 84 different allowed transitions, but if we systematically consider them, we can show that there are only two different transition energies. One will show up in a nitrogen-14 NMR spectrum, and it is at $\mu_o H_z g_N (1-\sigma_N)$. The other will be in a proton NMR spectrum, and it is at $\mu_o H_z g_H (1-\sigma_H)$.

## 11.3 SPIN-SPIN COUPLING

The nuclear spins of different magnetic nuclei may interact and couple much the same as electron spins couple. This will often lead to small energetic effects that tend to be noticeable under high resolution. The interaction that couples nuclear spins depends on their magnetic dipoles and their electronic environments. We may treat this interaction phenomenologically rather than attempting to analyze its fundamental basis. Since the dipole moments are proportional to the intrinsic spin vectors, we may write the spin-spin coupling interaction as proportional to the dot product of two spin vectors. So, for a system of two interacting nuclei,

$$\hat{H}' = \frac{J_{12}}{\hbar^2} \vec{I}_1 \bullet \vec{I}_2 \qquad (11\text{-}13)$$

where J has been introduced as the phenomenological proportionality constant.

The spin-spin coupling interaction of Eq. (11-13) is usually a small perturbation on the energies of the nuclear spin states experiencing the external field of an NMR instrument. So, it is appropriate to treat this interaction with low-order perturbation theory, particularly first order perturbation theory. The expression for the first order correction to an energy is the expectation value of the perturbation. For a system of two nonzero spin nuclei, the spin states are distinguished by the two quantum numbers, $m_{I_1}$ and $m_{I_2}$. The corrections to the energies of these states are given by

$$E^{(1)}_{m_{I_1} m_{I_2}} = \frac{J_{12}}{\hbar^2} < m_{I_1} \; m_{I_2} \mid \vec{I}_1 \cdot \vec{I}_2 \mid m_{I_1} \; m_{I_2} >$$

$$= \frac{J_{12}}{\hbar^2} < m_{I_1} \; m_{I_2} \mid \hat{I}_{x_1} \hat{I}_{x_2} + \hat{I}_{y_1} \hat{I}_{y_2} \mid m_{I_1} \; m_{I_2} >$$

$$+ \frac{J_{12}}{\hbar^2} < m_{I_1} \; m_{I_2} \mid \hat{I}_{z_1} \hat{I}_{z_2} \mid m_{I_1} \; m_{I_2} > \qquad (11\text{-}14)$$

Only the last term in Eq. (11-14) is nonzero, and from evaluating this integral, we obtain a simple result.

$$E^{(1)}_{m_{I_1} m_{I_2}} = J_{12} \, m_{I_1} \, m_{I_2} \qquad (11\text{-}15)$$

This says that the corrections to the energy due to spin-spin coupling between two nuclei will vary with the product of the quantum numbers giving z-component projections of the spin vectors.

As the first example of the effect of spin-spin interaction on NMR spectra we will consider a system with two protons attached to adjacent atoms, such as in HCOOH. The four spin states in the absence of spin-spin interaction would have the following energies, which are the zero-order energies for the perturbative treatment of the spin-spin interaction.

$$E_{-1/2\,-1/2} = \frac{\mu_o H_z g_H}{2} \left[ (1-\sigma_1) + (1-\sigma_2) \right]$$

$$E_{-1/2\;1/2} = \frac{\mu_o H_z g_H}{2} \left[ (1-\sigma_1) - (1-\sigma_2) \right]$$

$$E_{1/2\,-1/2} = \frac{\mu_o H_z g_H}{2} \left[ -(1-\sigma_1) + (1-\sigma_2) \right]$$

$$E_{1/2\;1/2} = \frac{\mu_o H_z g_H}{2} \left[ -(1-\sigma_1) - (1-\sigma_2) \right]$$

The subscripts on the energies are the $m_I$ quantum numbers of the two protons. The first-order corrections, according to Eq. (11-15), are the following.

$$E^{(1)}_{-1/2\ -1/2} = \frac{J_{12}}{4}$$

$$E^{(1)}_{-1/2\ 1/2} = -\frac{J_{12}}{4}$$

$$E^{(1)}_{1/2\ -1/2} = -\frac{J_{12}}{4}$$

$$E^{(1)}_{1/2\ 1/2} = \frac{J_{12}}{4}$$

Notice that the first-order corrections raise the energy of two levels and lower the energy of the other two levels. As seen in Fig. 11.3, these changes in the energy levels "split" the pairs of transitions that were at like energies. The transition energies are

$$\Delta E_a = \mu_o H_z g_H (1 - \sigma_1) + J_{12} / 2$$

$$\Delta E_b = \mu_o H_z g_H (1 - \sigma_1) - J_{12} / 2$$

$$\Delta E_c = \mu_o H_z g_H (1 - \sigma_2) + J_{12} / 2$$

$$\Delta E_d = \mu_o H_z g_H (1 - \sigma_2) - J_{12} / 2$$

Transitions a and b are moved to higher and lower frequencies, respectively, than they would be at in the absence of spin-spin interaction. The same is true for the c and d transitions. The separation between transition a and transition b (i.e., the difference between $\Delta E_a$ and $\Delta E_b$) is the spin-spin coupling constant, $J_{12}$. Likewise, the separation between transition c and transition d is $J_{12}$. Fig. 11.4 is a stick representation of the spectrum that would result.

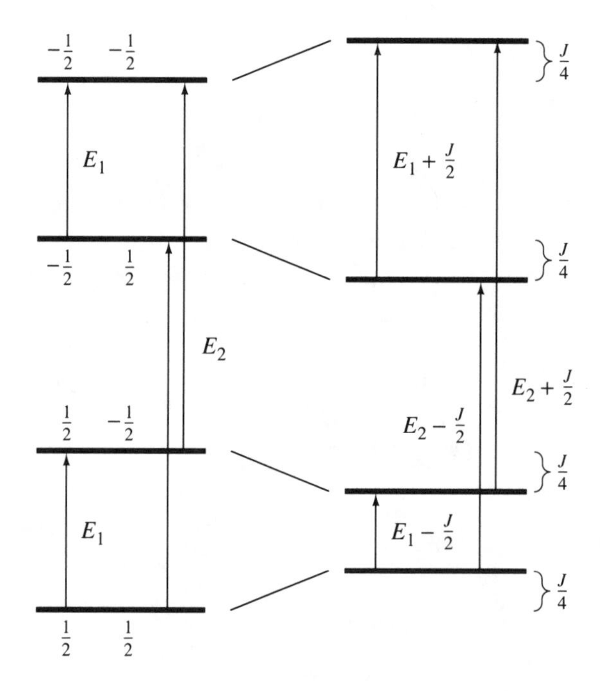


**FIGURE 11.3** Energy levels for a hypothetical molecule with two protons that are nearby but in different chemical environments. On the left are the energy levels that would be found in the absence of spin-spin interaction. On the right are the energy levels with spin-spin interaction treated via first order perturbation theory. The vertical arrows show the allowed transitions, and the lengths of these arrows are proportional to the transition energies. Thus, the effect of the spin-spin interaction is seen to split the transitions, that is, to take each pair of transitions that would occur at the same frequency and shift one to a higher frequency and one to a lower frequency.

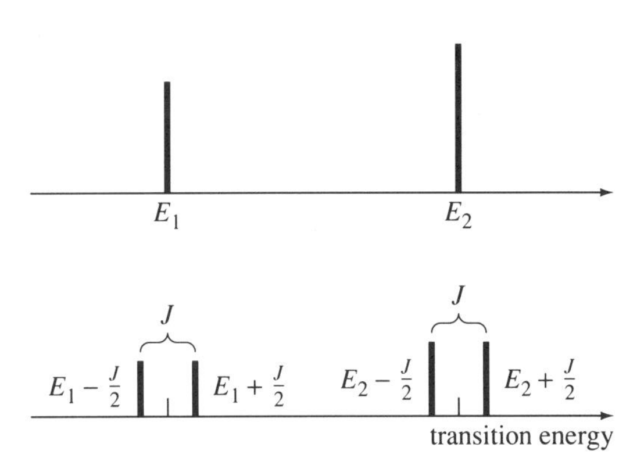


**FIGURE 11.4** Stick representation of the NMR spectra for the hypothetical molecule with the energy levels given in Fig. 11.3. The top spectrum corresponds to the energy levels in the absence of spin-spin interaction, whereas the bottom spectrum includes the effect.

The next example is a hypothetical molecule whose magnetic nuclei are three protons in different chemical environments. The spin-spin interaction Hamiltonian for a system with many interacting nuclei is simply a sum of the pair interactions of Eq. (11-13):

$$\hat{H}' = \sum_i \sum_{j>i} \frac{J_{ij}}{\hbar^2} \vec{I}_i \cdot \vec{I}_j \tag{11-16}$$

So, the first-order corrections to the energy are also sums of pair contributions with the form given in Eq. (11-15).

$$E^{(1)} = \sum_i \sum_{j>i} J_{ij}\, m_{I_i}\, m_{I_j} \tag{11-17}$$

The quantum number subscripts on $E^{(1)}$ have been suppressed for conciseness; they would be the $m_I$ quantum numbers for all the nuclei. Fig. 11.5 shows how the zero-order energy levels for this hypothetical system would be affected by the first-order corrections of Eq. (11-17). The resulting spectrum consists of each original line split into a pair of lines which are then split again. This reveals a simple rule for interpreting proton NMR spectra: A pair of closely spaced lines is likely to have come about because of the spin-spin interaction with a single nearby proton, and if there are other protons nearby, there may be further splitting.

Though we have treated spin-spin interaction phenomenologically (by employing the coupling constant, J, as a parameter instead of deriving it from fundamental interactions), there are certain features that we may anticipate. One of these is that the size of a coupling constant, $J_{12}$, will vary with different pairs of nuclei. It will fall off with increasing separation between the interacting nuclei to the extent that the interaction strength diminishes with increasing separation distance. In practice, proton NMR spectra generally show splitting by protons attached to adjacent atoms. So, we might expect to see effects from proton spin-spin interaction in HCOOH given the proximity of the protons, but perhaps not in HC≡CCOOH. Thus, splittings in spectra are intimately related to molecular structure and may serve to reveal structural features. Also, the lines of both of the interacting nuclei are split by the identical amount (i.e., $J_{12}$), and so finding two pairs of lines with the same splitting may identify two protons that are interacting. It can be a challenging puzzle, but the data from NMR spectra often yield clear-cut structural information.

Another example to consider is one with different kinds of magnetic nuclei, and a simple example is HD (deuterium-substituted $H_2$). The deuterium nucleus has a spin of one ($I = 1$). The $m_I$ values for the deuterium nucleus are 1, 0, and –1. The spin states will be distinguished by the $m_I$ quantum numbers for the deuterium and the proton. From Eq. (11-11) we may write the zero order energies, and from Eq. (11-15), the first-order corrections. We may tabulate these energies and present them in a concise form by listing the possible energy terms at the head of the column, and then with each state as a row in the table, we list the factor that should be applied to the energy term.

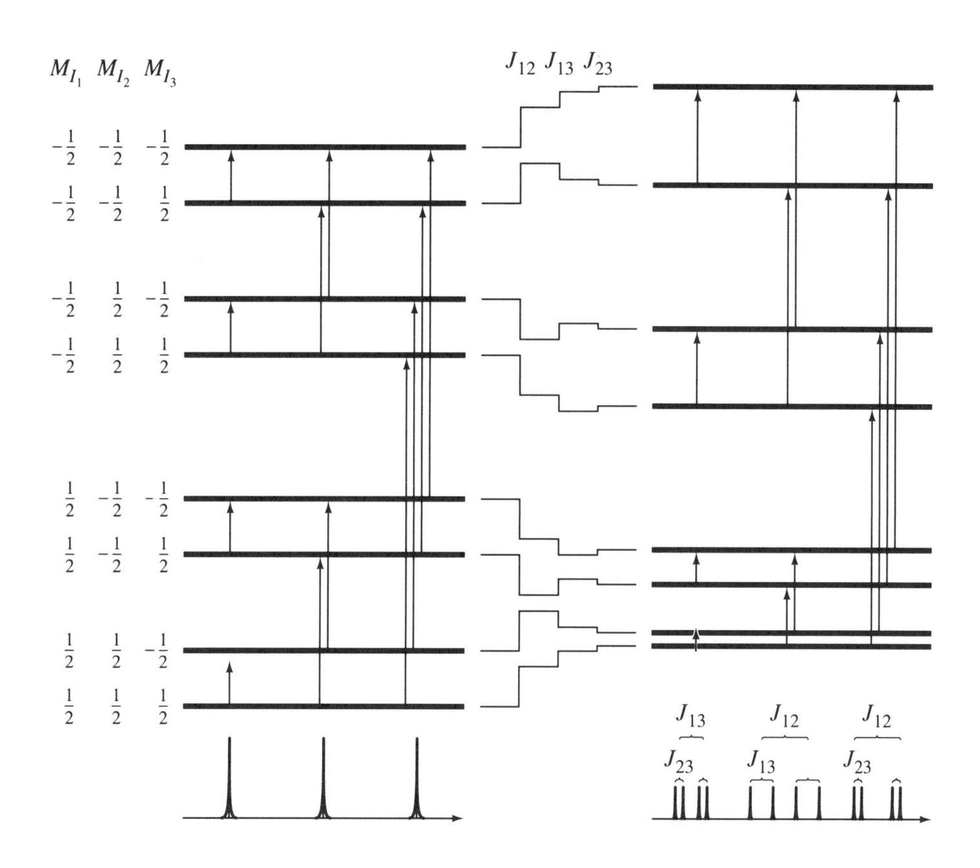


**FIGURE 11.5** Energy level diagram for the nuclear spin states of a hypothetical molecule with three interacting protons in different chemical environments. The energy levels on the left neglect spin-spin interaction. The allowed transitions give rise to a spectrum of three lines, shown as a stick representation at the bottom. These three lines correspond to a spin "flip" of each of the three different nuclei, and their relative transition frequencies give the chemical shift of each. The energy levels on the right have been obtained by first including the 1-2 coupling, then the 1-3 coupling, and finally the 2-3 coupling, assuming $J_{12} > J_{13} > J_{23}$. Transition lines are drawn for the final set of levels, and the resulting stick spectrum is shown at the bottom. The values of the three coupling constants may be obtained directly from the spectrum, as shown.

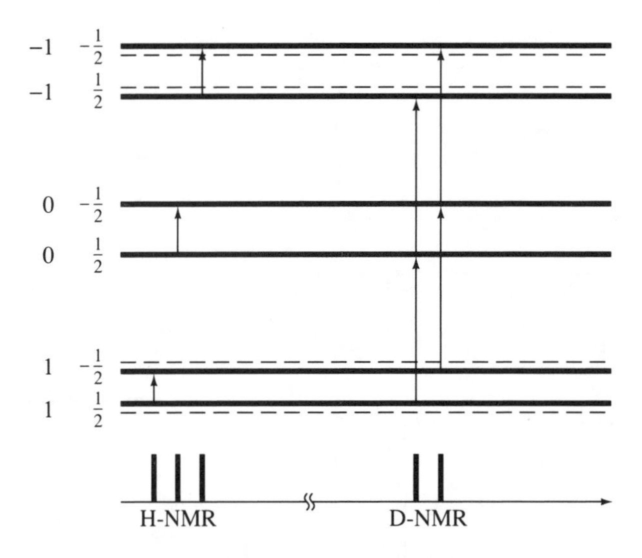


**FIGURE 11.6** Nuclear spin energy levels of the HD molecule in a magnetic field. The allowed transitions, represented by vertical arrows, are those for which the $m_I$ quantum number of the deuterium changes by one, or the $m_I$ quantum number of the proton changes by one. The deuterium and proton stick spectra are shown at the bottom.

| State: $m_{I_D}$ $m_{I_H}$ | Energy terms: $\mu_o H_z g_D (1-\sigma_D)$ | $\mu_o H_z g_H (1-\sigma_H)/2$ | $J_{12}/2$ |
|---|---|---|---|
| –1 –1/2 | 1 | 1 | 1 |
| –1 1/2 | 1 | –1 | –1 |
| 0 –1/2 | 0 | 1 | 0 |
| 0 1/2 | 0 | –1 | 0 |
| 1 –1/2 | –1 | 1 | –1 |
| 1 1/2 | –1 | –1 | 1 |

So, the energy of the state in the second row, for instance, is the sum of the first term plus the second term multiplied by –1 plus the third term multiplied by –1. A sketch of these energy levels is given in Fig. 11.6 along with the stick spectra that are predicted. The spectra show the proton transition line split into three lines, whereas the deuterium transition is

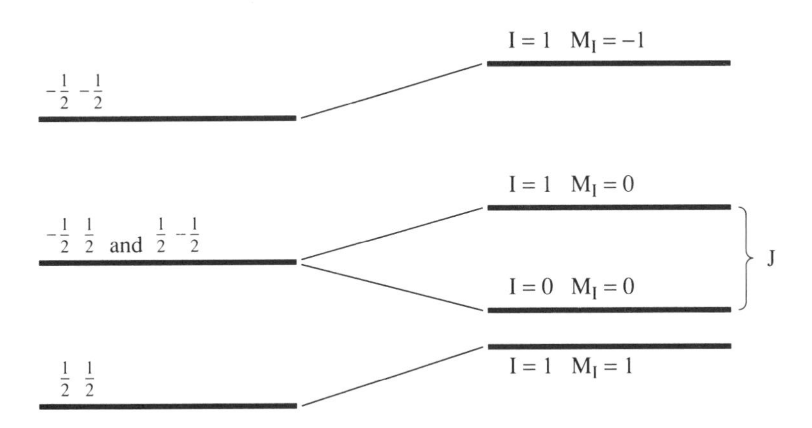


**FIGURE 11.7** Energy levels and transitions for the proton spin states of formaldehyde. On the left are the levels obtained with neglect of spin-spin interaction, and they are labelled by the $m_I$ quantum numbers of the two protons. The middle level is doubly degenerate. If spin-spin interaction is treated with first-order degenerate perturbation theory, the levels on the right result, with an assumed size of the coupling constant, J. A transition from the I = 0 state to an I = 1 state would measure J, but that is a forbidden transition. No splitting of the line in the spectrum occurs.

split into two lines. This reveals a generalization of the rule about nearby protons splitting lines into pairs. It is that a nearby magnetic nuclei will split a transition line into the number of lines equal to the spin multiplicity, 2I + 1, of the nucleus. The deuterium splits the proton transitions into three lines, since $2I_D + 1 = 3$.

Equivalent nuclei are the magnetic nuclei in equivalent chemical environments, such as the two protons in formaldehyde. The chemical shielding for each of these protons must be the same because of the symmetry of the molecule. The quantum mechanics for spin-spin coupling with equivalent nuclei has some complexities that are not encountered with inequivalent nuclei, but a rather simple rule for

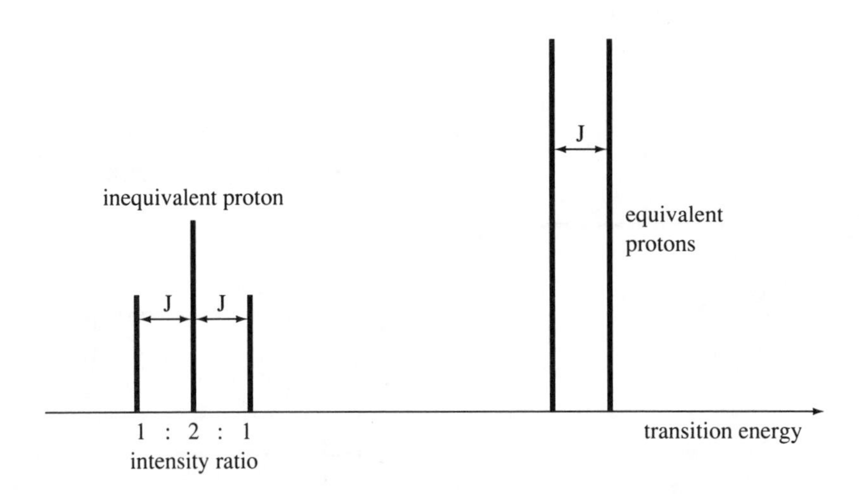


**FIGURE 11.8** The NMR stick spectrum of a hypothetical molecule with two equivalent protons and a third, nearby proton in a different chemical environment. This spectrum may be understood as a combination of the spectrum of a system with an I = 1 particle interacting with the third proton and the spectrum from an I = 0 particle interacting with the third proton. These two pseudoparticles correspond to the possible couplings of the spins of the two equivalent nuclei.

predicting or interpreting spectra emerges just the same. The complexity of equivalent nuclei is spin state degeneracy. The zero-order energies for the proton spin states in formaldehyde with quantum numbers (1/2 −1/2) and (−1/2 1/2) are both zero according to Eq. (11-11). This is the consequence of the $\sigma$ for both nuclei being the same. These two states are degenerate, and degenerate first-order perturbation theory is needed to treat the spin-spin interaction.

The result from degenerate first order perturbation theory for two interacting, equivalent protons amounts to completely coupling their spins. Using the angular momentum coupling rules, two spin-1/2 protons may yield a net coupled spin with I = 1 or with I = 0. The selection rules for NMR transitions of equivalent nuclei turn out to be $\Delta I = 0$ and $\Delta M_I = \pm 1$. Fig. 11.8 shows that with these selection rules, it is not possible to obtain from the NMR spectrum the J coupling constant for the equivalent protons. For formaldehyde, we expect a single proton NMR transition.

The coupling of equivalent nuclei with other nuclei can produce splittings as well as important intensity features in an NMR spectrum. Consider a hypothetical molecular system with two equivalent protons and a third proton in a different environment. The NMR spectrum may be regarded as a superposition of the spectra obtained for the two coupling possibilities (namely, $I = 1$ and $I = 0$) of the equivalent protons. In the case of $I = 1$ coupling, we expect a spectrum that looks like a spin-one particle with a spin-1/2 particle. This is qualitatively the same as the HD molecule. In the case of $I = 0$ coupling, we expect a spectrum that looks like a spin-zero particle with a spin-1/2 particle. The single transition for this case will add to the transition intensity of the middle line of the triplet of lines from the first case. The result is shown in Fig. 11.8. It is that a spin flip by the third proton corresponds to three transition lines in an intensity ratio of 1 : 2 : 1. The equivalent protons' transitions are split into two lines by the third proton's interaction.

## 11.4 ELECTRON SPIN RESONANCE SPECTRA

If a molecule has a nonzero electronic spin (i.e., if $S > 0$), then the associated magnetic moment will interact with an external magnetic field. This is the basis for electron spin resonance (ESR) spectroscopy, which is also called electron paramagnetic resonance (EPR) spectroscopy. The electron spin energy levels are separated by the magnetic field interaction, and the resulting energy differences are probed spectroscopically. The electron spin magnetic moment will also interact with nuclear magnetic moments, and it is this complication that ends up providing much useful information from ESR spectra.

Usually, electronic orbital angular momentum is zero in polyatomic molecules, and so this source of magnetic moments can be ignored. Also, the ESR experiment is usually carried out with a liquid or solid state sample, and so there is no magnetic moment arising from molecular rotation since the molecules are not freely rotating.

The general form for the interaction Hamiltonian between an isotropically* sampled magnetic field $\vec{H}$ and the electron spin is

$$\hat{H} = \frac{g_e \mu_B}{\hbar} \vec{S} \cdot \vec{H} \tag{11-18}$$

where $\mu_B$ is the Bohr magneton [Eq. (9-15)]. If magnetic nuclei are present, the Hamiltonian is a sum of this term and the nuclear Hamiltonian of Eq. (11-11) along with the spin-spin interaction terms. As before, we take the direction of the magnetic field to define the z-axis, and then

$$\hat{H} = \frac{g_e \mu_B H_z}{\hbar} \hat{S}_z - \frac{\mu_o H_z}{\hbar} \sum_i^{\text{nuclei}} g_i (1 - \sigma_i) \hat{I}_{i_z}$$

$$+ \sum_{i>j} \frac{J_{ij}}{\hbar^2} \vec{I}_i \cdot \vec{I}_j + \sum_i \frac{a_i}{\hbar^2} \vec{I}_i \cdot \vec{S} \tag{11-19}$$

This equation introduces the constants $a_i$ phenomenologically; that is, the electron spin–nuclear spin interaction must be proportional to the dot product of the spin vectors, but the fundamental basis for the proportionality constant ($a_i$) is not considered.

The form of the Schrödinger equation that uses the Hamiltonian of Eq. (11-19) is the same as the form of the NMR Schrödinger equation. In both, there are z-component spin operators and spin-spin dot products. The only thing that is different in the ESR Hamiltonian is that one of the spins is electron spin and with it comes a different set of constants (e.g., $g_e$ versus $g_i$). Notice that $\mu_B / \mu_o$ is the ratio of the proton mass to the electronic mass (1836.15).

In NMR, the external magnetic field effect is generally very large with respect to the size of the nuclear spin-spin interaction. This is because of the need to use a strong field in order to make the transition energies great enough to place them in the radiofrequency region of the spectrum. In ESR, the much greater size of $\mu_B$ means that the choice is available to use

* In a crystalline sample, for instance, the interaction cannot be treated as isotropic. Then, $g_e$ must be treated as a tensor, **g**, and the dot product interaction becomes $\vec{S} \cdot \mathbf{g} \cdot \vec{H}$. Notice that the isotropic condition here is analogous to using an isotropic $\sigma$ in place of the NMR shielding tensor, $\boldsymbol{\sigma}$.

weaker fields; at 10,000 G, ESR transitions are typically around 10,000 MHz, whereas NMR transitions are around 10 MHz. Consequently, there are two situations to consider in analyzing ESR spectra, a "high-field" case and a "low-field" case. The high-field case means that the field dominates all spin-spin interactions. The analysis, then, follows that for NMR where the spin-spin terms in the Hamiltonian are treated by first-order perturbation theory. In the low-field case, spin-spin coupling is of much greater relative importance, and first-order perturbation energies are inappropriate.

Let us use the formyl radical HCO• as an example problem for predicting high-field and low-field ESR spectra. In this molecule, the single unpaired electron implies that there is a net electronic spin of 1/2 (i.e., S = 1/2). The proton spin, I, is 1/2, and so there are four spin states for the molecule. With high fields, the Hamiltonian is broken into a zero-order part and the spin-spin perturbation:

$$\hat{H}^{(0)} = g_e \mu_B H_z \hat{S}_z / \hbar - g_H \mu_o H_z (1 - \sigma) \hat{I}_z / \hbar \quad (11\text{-}20a)$$

$$\hat{H}^{(1)} = a \vec{I} \bullet \vec{S} / \hbar^2 \quad (11\text{-}20b)$$

The zero-order wavefunctions, labelled by the quantum numbers $m_S$ and $m_I$, are eigenfunctions of $\hat{I}_z$ and $\hat{S}_z$ (and of $\hat{I}^2$ and $\hat{S}^2$), and the associated energy values are

| State $\mid m_S \ m_I >$ | $E^{(0)}_{m_S m_I}$ |
|---|---|
| 1/2 −1/2 | $g_e \mu_B H_z / 2 + g_H \mu_o H_z (1 - \sigma) / 2$ |
| 1/2 1/2 | $g_e \mu_B H_z / 2 - g_H \mu_o H_z (1 - \sigma) / 2$ |
| −1/2 −1/2 | $- g_e \mu_B H_z / 2 + g_H \mu_o H_z (1 - \sigma) / 2$ |
| −1/2 1/2 | $- g_e \mu_B H_z / 2 - g_H \mu_o H_z (1 - \sigma) / 2$ |

For each of these functions, the first order energy corrections are

$$E^{(1)}_{m_S m_I} = < m_S \ m_I \mid \hat{H}^{(1)} \mid m_S \ m_I >$$

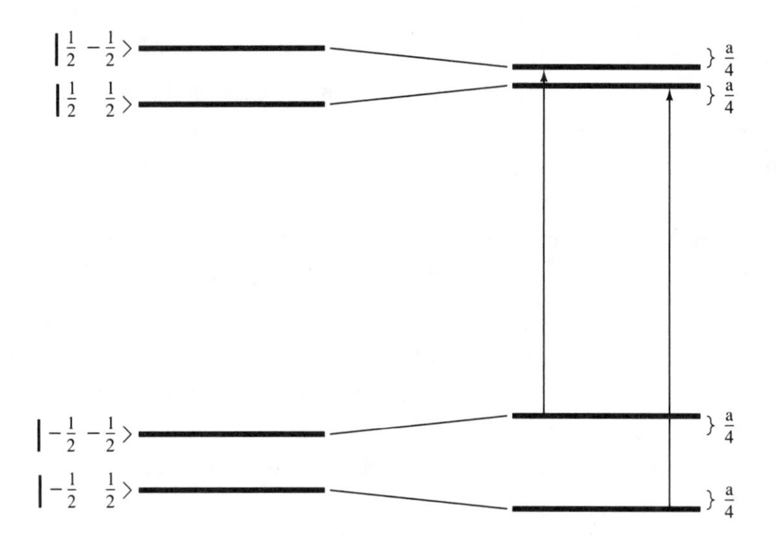


**FIGURE 11.9** High-field energy levels for a system with electron spin S = 1/2 and with a magnetic nucleus with I = 1/2. The levels on the left are the zero-order energies, and the levels on the right include the first-order corrections due to electron spin-nuclear spin interaction. The two vertical arrows show the allowed transitions that would be observed in an ESR experiment. Not shown are the allowed transitions for which $\Delta M_S = 0$ since these are the NMR transitions and occur in a very different energy regime.

$$= \frac{a}{\hbar^2} < m_S\ m_I\ |\ \hat{I}_x\hat{S}_x + \hat{I}_y\hat{S}_Y + \hat{I}_z\hat{S}_z\ |\ m_S\ m_I >$$

$$= a\, m_S\, m_I \qquad (11\text{-}21)$$

A sketch of the zero- and first-order energy levels is given in Fig. 11.9. The high-field selection rules are $|\Delta m_S| = 1$ and $|\Delta m_I| = 0$. Thus, there will be two ESR transitions, and the difference in the two transition energies is simply the coupling constant a.

If there are several magnetic nuclei in a molecule, then the high-field energy level expression that follows from a first-order perturbative treatment of the general Hamiltonian of Eq. (11-19) is

$$E_{m_S m_{I_1} \cdots} = g_e \mu_B H_z m_S - \sum_i^{\text{nuclei}} g_i \mu_o H_z (1 - \sigma_i) m_{I_i} + \sum_{i>j}^{\text{nuclei}} J_{ij} m_{I_i} m_{I_j} + \sum_i^{\text{nuclei}} a_i m_S m_{I_i} \quad (11\text{-}22)$$

Fig. 11.10 gives an example of energy levels that arise from Eq. (11-22) and the resulting form of the spectrum. Notice that there is a different multiplet pattern for the ESR transitions associated with an I = 1 nucleus than with an I = 1/2 nucleus. Indeed, the patterns of lines help identify the type of magnetic nuclei much as they do in NMR spectra, and this holds in the case of equivalent magnetic nuclei, too.

ESR transition energies yield values for each of the $a_i$ coupling constants in Eq. (11-22). The relative size of these constants is significant because the localization of the electron spin density must enter into determining the size of each $a_i$. In fact, ESR provides an ideal means for characterizing organic free radicals. If the measured coupling constants can be properly assigned to atoms in the radical, then one particularly large coupling constant suggests that there is an unpaired electron (the radical electron) that is largely localized at or near that atom. Of course, the coupling constants might indicate that the unpaired electron is delocalized through the molecule, and this will have implications for the electronic structure of the species.

The low-field ESR problem is a more complicated quantum mechanical problem than the high-field case because first-order perturbation theory is not appropriate. A more suitable approach is to use linear variation theory with the Hamiltonian of Eq. (11-19) and with a basis set of independent spin functions, that is, product functions that are eigenfunctions of each of the z-component operators in Eq. (11-19). Then, the application of linear variation theory becomes a problem of matrix diagonalization (see Appendix II). From the energy levels that result, the transition energies are obtained as differences in state energies. In practice, ESR spectra are sometimes analyzed by carrying out this process repeatedly for different choices of the coupling constants until a satisfactory agreement between measured and calculated transition energies is achieved. In this manner, the coupling constants may be extracted from the spectra even without the simplifying aspects of the high-field case.

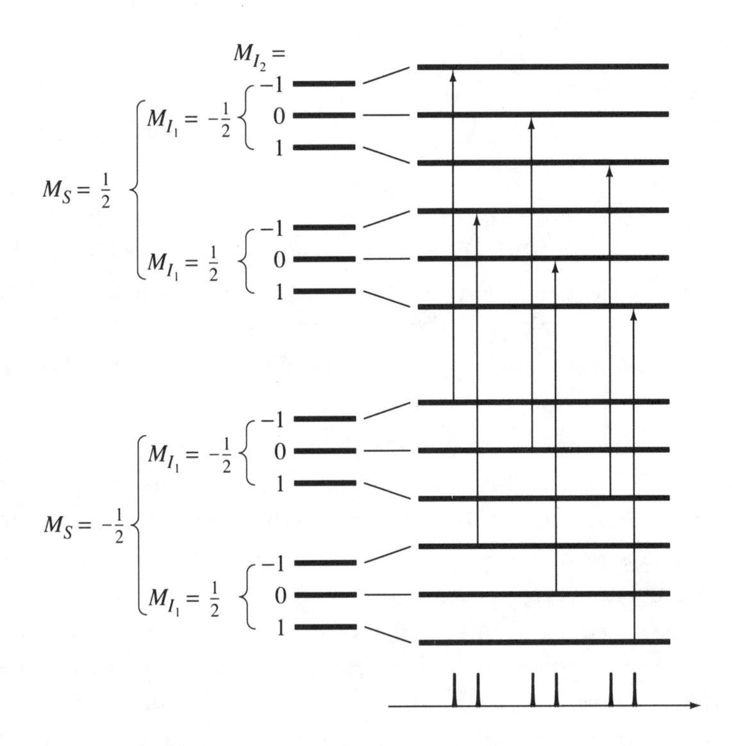


**FIGURE 11.10** High-field energy levels for a system with electron spin S = 1/2 and magnetic nuclei with I = 1/2 and I = 1. On the left are the levels at zero order where only the interaction with the external field is included. On the right are the energy levels according to Eq. (11-22) where the spin-spin interactions have been included by first order perturbation theory. The nuclear spin-spin interaction has been exaggerated relative to typical values in order to show the effect. The vertical lines correspond to the allowed ESR transitions, and below them is a stick representation of the spectrum that corresponds to these transitions.

We may understand some of the features of low-field ESR spectra by considering the low-field energy levels to be intermediate to two limiting cases, the high-field and the zero-field limits. In the absence of an external field, spins are fully coupled. So, for a problem with an electron spin, S, and a single magnetic nucleus with spin I, the rules of angular momentum coupling dictate the energy levels. The total angular momentum is the vector sum of the two spin vectors,

$$\vec{F} = \vec{S} + \vec{I} \tag{11-23}$$

and the associated quantum number F must have the values

$$F = |S - I|, \ldots, S + I$$

Then, as was done for spin-orbit interaction in the hydrogen atom, we may replace the spin-spin operator with the following,

$$\vec{I} \bullet \vec{S} = \frac{1}{2}\left(\hat{F}^2 - \hat{I}^2 - \hat{S}^2\right) \tag{11-24}$$

In the zero-field limit, the Hamiltonian reduces to

$$\hat{H} = \frac{a}{\hbar^2}\, \vec{I} \bullet \vec{S}$$

and so the eigenfunctions of the Hamiltonian are the functions that are simultaneously eigenfunctions of the three angular momentum operators $\hat{F}^2$, $\hat{S}^2$, and $\hat{I}^2$. Therefore, the eigenenergies must be

$$E_{F,S,I} = \frac{a}{2}\left[ F(F+1) - I(I+1) - S(S+1) \right] \tag{11-25}$$

Each of these energy levels has a multiplicity (degeneracy) of 2F + 1.

For the specific situation of an electron spin S = 1/2 and a nuclear spin I = 1/2, Fig. 11.11 shows the energy levels of the limiting cases of no external field and a very strong field. Between these limits, which is the region for low-field ESR, the energy levels must correlate. That is, as the magnetic field is increased from zero to some high value, the energy levels will change smoothly from one limiting case to the other, and this is shown in the middle of Fig. 11.11. As the field is just turned on, the degeneracy in the levels of different F values is removed. For extremely weak fields, the magnetic field may be treated with low-order perturbation theory, and then the separation of levels will depend on the value of $m_F$, the quantum number that gives the net z-component of the total angular momentum. Thus, the F = 1 level will give rise to three levels ($m_F = -1, 0, 1$) that spread apart linearly with the field strength, at least initially. As the field strength increases, the relative size of the nuclear-electron spin coupling begins to diminish and the energy levels start to approach those of the high-field limit. This correlation diagram reveals that with low fields the spectrum has a quite different appearance than at the high-field

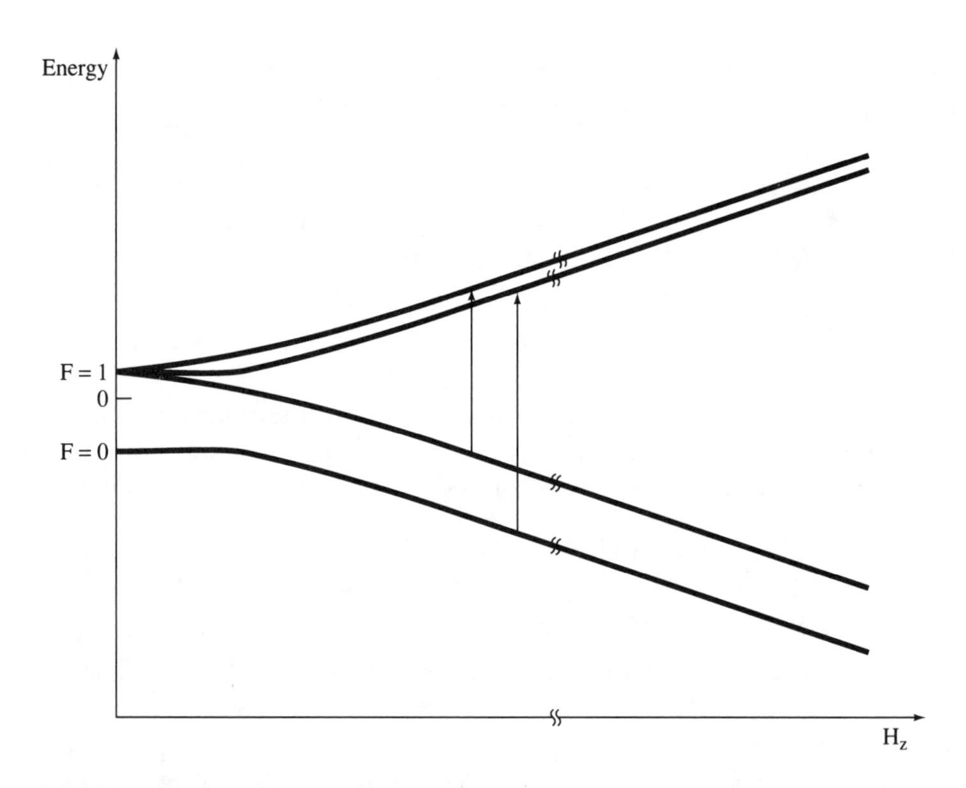


**FIGURE 11.11** The energy levels of a system with S = 1/2 and I = 1/2. On the left are the levels in the absence of an external magnetic field. These are according to Eq. (11-25). The limiting case on the right is that of a high field where spin-spin interaction is well treated by first-order perturbation theory, and so these are the same energy levels as in Fig. 11.9. The intermediate region shows qualitatively how the levels change from one limiting situation to the other as an external magnetic field is applied to the system. The vertical arrows indicate the ESR transitions.

limit. Of course, the coupling constant can be obtained from either a high-field or a low-field spectrum.

## 11.5 RELAXATION AND MAGNETIC RESONANCE IMAGING

As described so far, magnetic resonance is a tool for probing individual molecules. From the spectra, we obtain information that is at the molecular level, with much of it concerning the structures of the molecules. However, there is also important information that develops as a bulk property, and this has become the basis for a powerful application, magnetic resonance imaging. Here, the general idea is to measure a signal that is dependent on the spatial distribution of a species that might otherwise be studied with conventional magnetic resonance spectroscopy. The mathematical task of imaging is to map the signal into a graphical respresentation, an image. We shall only be concerned with how spatial information may be locked into a signal since the imaging step is largely independent. Also, we shall restrict attention to nuclear magnetic resonance imaging, although imaging techniques have been developed for electron spin resonance, too.

As a preliminary illustration of an imaging experiment, consider two long sample tubes filled with ethanol and held parallel in a larger diameter tube that is otherwise filled with benzene. If this is placed in an NMR spectrometer modified so that instead of the static magnetic field being uniform across the sample the field varies in strength from one side to the other, then the ethanol NMR transition signals will vary with the orientation of two small tubes with respect to the field: If they are aligned with the applied field, one sample experiences a stronger field than the other, and so there will be signals at two distinct frequencies. If they are rotated away from this arrangement, the transition peaks will coalesce as the fields experienced by the samples in the two small tubes approach the same strength (i.e., at $90^{\circ}$ from the original orientation). In this way, a spatially varying field strength encodes spatial information into the NMR spectrum.

Classically, a magnetic field, $\vec{H}$, acting on a magnetic dipole moment, $\vec{\mu}$, gives rise to a torque on a magnetic moment. The torque is a force directed toward bringing the magnetic moment into alignment with the field. The torque is proportional to the cross-product of $\vec{\mu}$ and the magnetic field. If the source of the magnetic moment has an angular momentum, which magnetic nuclei do have, then the magnetic field will cause a precession of the angular momentum vector about the field axis, and the time rate of change of the angular momentum vector is equal to the torque. With $\vec{I}$ as the angular momentum, we have that

$$\frac{d\vec{I}}{dt} = \vec{\mu} \times \vec{H} \tag{11-26a}$$

This can be rewritten entirely in terms of $\vec{\mu}$ because $\vec{\mu}$ is proportional to $\vec{I}$.

$$\frac{d\vec{\mu}}{dt} = \gamma\vec{\mu} \times \vec{H} \tag{11-26b}$$

$\gamma$ is the proportionality constant, the magnetogyric (or gyromagnetic) ratio.

There is a bulk property of a sample that corresponds to the magnetic moment of an inidividual molecule called the *nuclear magnetization*, $\vec{M} = (M_x, M_y, M_z)$. It is the sum of the individual magnetic moment vectors per unit volume. So, we may expect that the time evolution of the nuclear magnetization will follow Eq. (11-26b). However, there is a key difference between one isolated molecule and a collection of molecules. In a sample of many molecules, there are interactions between each of the molecules because of the different magnetic moments. Instead of a fundamental approach at this point, we shall accept a phenomenological expression:

$$\frac{d\vec{M}}{dt} = \gamma\, \vec{M} \times \vec{H} - \frac{(M_z - M_o)\hat{k}}{T_1} - \frac{(M_x\hat{i} + M_y\hat{j})}{T_2} \tag{11-27}$$

where $M_o$, $T_1$, and $T_2$ are constants and $\hat{i}$, $\hat{j}$, and $\hat{k}$ are the usual Cartesian unit vectors. The first term on the right hand side comes about from Eq. (11-26), whereas the additional two terms represent the phenomenological description of the effect of interactions between nuclei in the sample.

Writing Eq. (11-27) for each component of $\vec{M}$ yields three equations which are referred to as the *Bloch equations*.

$$\frac{dM_x}{dt} = \gamma(M_y H_z - M_z H_y) - M_x / T_2 \tag{11-28a}$$

$$\frac{dM_y}{dt} = \gamma(M_z H_x - M_x H_z) - M_y / T_2 \tag{11-28b}$$

$$\frac{dM_z}{dt} = \gamma(M_x H_y - M_y H_x) - (M_z - M_o)/T_1 \qquad (11\text{-}28c)$$

These are coupled differential equations, and we approach their solution by considering the specific case where the magnetic field is static and in the z-direction (i.e., $H_x = H_y = 0$) and where $T_1$ and $T_2$ are infinite. In this situation, we have

$$\frac{dM_x}{dt} = \gamma M_y H_z \qquad (11\text{-}29a)$$

$$\frac{dM_y}{dt} = -\gamma M_x H_z \qquad (11\text{-}29b)$$

$$\frac{dM_z}{dt} = 0 \qquad (11\text{-}29c)$$

The solutions are that $M_x$ and $M_y$ are simple oscillating functions, whereas $M_z$ is a constant.

$$M_x(t) = M_x^o \cos(\gamma H_z t) + M_y^o \sin(\gamma H_z t) \qquad (11\text{-}30a)$$

$$M_y(t) = -M_x^o \sin(\gamma H_z t) + M_y^o \cos(\gamma H_z t) \qquad (11\text{-}30b)$$

$M_x^o$ and $M_y^o$ are constants. The solution of Eq. (11-29c) is simply that $M_z(t)$ is a constant, which shall be designated $M_z^o$.

Not requiring $T_1$ and $T_2$ to be infinite, but still assuming the external field is static and in the z-direction, yields a more complicated set of solutions for the nuclear magnetization.

$$M_x(t) = e^{-t/T_2}\left(M_x^o \cos(\gamma H_z t) + M_y^o \sin(\gamma H_z t)\right) \qquad (11\text{-}31a)$$

$$M_y(t) = e^{-t/T_2}\left(-M_x^o \sin(\gamma H_z t) + M_y^o \cos(\gamma H_z t)\right) \qquad (11\text{-}31b)$$

$$M_z(t) = M_o + (M_z^o - M_o)\, e^{-t/T_1} \qquad (11\text{-}31c)$$

The first two of these equations say that the magnetization perpendicular to the axis of the external field decays to zero; that is, for $t >> 0$, $\exp(-t/T_2)$ goes to zero. The component of the magnetization along the field axis, according to Eq. (11-31c), does not go to zero but approaches the constant $M_o$ that was introduced in the Bloch equations. That the time constants for the decay to equilibrium values, $T_1$ and $T_2$, should be introduced as different values now appears consistent with the uniqueness of the axis of the external field.

$M_z(t)$ is unique from another standpoint. The interaction energy with the external magnetic field is proportional to the dot product of $\vec{M}$ and $\vec{H}$. With the applied field taken to be along the z-axis, the interaction energy depends on $M_z(t)$ but not on $M_x(t)$ and $M_y(t)$. This suggests that the time constants $T_1$ and $T_2$ arise from different interaction mechanisms. For the longitudinal direction, the change in magnetization means the energy of interaction changes in time. At the individual molecule level, this implies that some nuclei undergo a spin flip. On the other hand, the transverse changes in magnetization [i.e., $M_x(t)$ and $M_y(t)$] may come about by a mutual change in the precessional motion of two nuclear spins without affecting the energy of the system. This process is referred to as *spin-spin relaxation,* and the longitudinal process is called *spin-lattice relaxation.* The constants $T_1$ and $T_2$ are called *relaxation times* because they are the characteristic lengths of time required for the system to evolve to its equilbrium magnetization.

Relaxation is an important process even in the basic NMR experiment. Let us consider a molecular problem with one spin-1/2 nucleus ($I = 1/2$). There are two spin states, which shall be designated $\alpha$ ($m_I = 1/2$) and $\beta$ ($m_I = -1/2$). In the presence of a static external magnetic field, the energy difference between these states is

$$\Delta E = (m_I^{(\alpha)} - m_I^{(\beta)})\gamma\hbar H_z = \gamma\hbar H_z \tag{11-32}$$

The Maxwell-Boltzmann distribution law tells us that the relative populations of the two states in a sample at equilibrium at temperature, $T$, is,

$$\frac{N_\beta}{N_\alpha} = e^{-\Delta E/kT} = e^{-\gamma\hbar H_z/kT} \tag{11-33}$$

where $N_\alpha$ means the number of molecules in the $\alpha$ state and $N_\beta$ means the number in the $\beta$ state. k is the Boltzmann constant. Because the energy difference is small relative to $kT$ at room temperature, the ratio in Eq. (11-33) is near unity. In other words, there is only a small population difference between nuclear spin states.

The sample magnetization in the z-direction, $M_z(t)$, is related to the population difference. For example, if at some instant $N_\alpha = N_\beta$, then $M_z$ will be zero since the net sum of all the z-components of the individual moments would be zero. Thus, Eq. (11-31c) tells us that the population difference, $n = N_\alpha - N_\beta$, follows the same exponential dependence.

$$n(t) = n_0 (1 - e^{-t/T_1}) \qquad (11\text{-}34)$$

The constant $n_0$ is the long-time value of n(t) [i.e., $n_0 \equiv n(\infty)$]. This means that the rate of change in the population difference is

$$\frac{dn}{dt} = \frac{n_0}{T_1} e^{-t/T_1} = \frac{n_0 - n(t)}{T_1} \qquad (11\text{-}35)$$

The first step follows from differentiation of Eq. (11-34), and the second step results from substituting for the exponential upon rearranging Eq. (11-34).

An oscillating field is used to induce transitions in NMR, and we may assign a transition rate, R, to the probability of inducing a transition from the $\alpha$ state to the $\beta$ state. The quantum mechanical analysis that gives us selection rules and transition probabilities shows that transitions up are as likely as transitions down. So, R is also the rate for transitions from $\beta$ to $\alpha$ that are induced by the applied radiation. This means that under the condition of an oscillating field at resonance (at the right frequency to induce transitions), the rate of change in the population difference requires additional terms.

$$\frac{dn}{dt} = \frac{n_0 - n(t)}{T_1} + \left[ N_\beta R - N_\alpha R \right] - \left[ N_\alpha R - N_\beta R \right] \qquad (11\text{-}36a)$$

The first term in square brackets is the rate of change in the population of the $\alpha$ state. It consists of the number of molecules becoming $\alpha$-state molecules, $N_\beta R$, less the number leaving the $\alpha$ state, $N_\alpha R$. The second term is the rate of change in the $\beta$ state. This simplifies to,

$$\frac{dn}{dt} = \frac{n_0 - n(t)}{T_1} - 2\,n(t)\,R \tag{11-36b}$$

In order to be continuously inducing transitions with the oscillating field, and continuously monitoring it in our NMR experiment, a steady state condition should be achieved where the population difference remains constant: $dn/dt = 0$. Under this condition, we must have that,

$$n(t) = \frac{n_0}{1 + 2RT_1} \tag{11-37}$$

This follows from setting the right hand side of Eq. (11-36b) to zero. Since $n_0$ is the value of n(t) for $t >> 0$, we must conclude that a constant population difference of $n_0$ is achievable only if $2RT_1 << 1$.

The size of R is controlled by the power of the oscillating field (the square of the field strength). So, we may ask what will happen if the power is increased such that $2RT_1$ is no longer very small. If this occurred, the population difference would no longer be constant. The radiation would be pumping molecules to the higher level faster than the spin-lattice relaxation could return them to the lower state. Eventually, the absorption of the radiation by the sample would fall off. This is called *saturation*. Since saturation is related to the power level and the relaxation time, $T_1$, observing saturation is one means of measuring the relaxation time $T_1$. Further analysis, which we shall not carry out, reveals that the lineshape of an NMR peak is related to the relaxation time $T_2$, and lineshape analysis is one means of extracting a value for $T_2$.

More sophisticated means for measuring relaxation times involve a pulse sequence in the oscillating field. That is, the radiation is turned on for a short duration and then turned off for a specific time, and then there is another short pulse. Depending on how the pulse sequence is chosen, the signal from the system will measure the relaxation times.

Magnetic resonance imaging (MRI) uses the magnetization and relaxation of a sample to extract spatial distribution information and thus to construct an image, such as those in Fig. 11.12. As mentioned at the beginning of this section, spatial information may be encoded in the

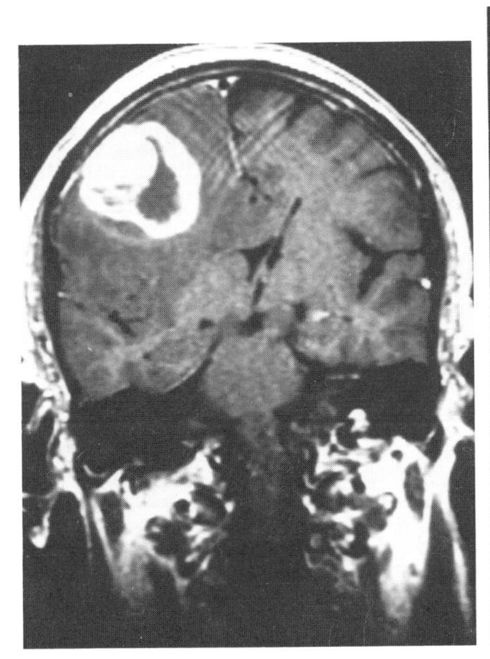

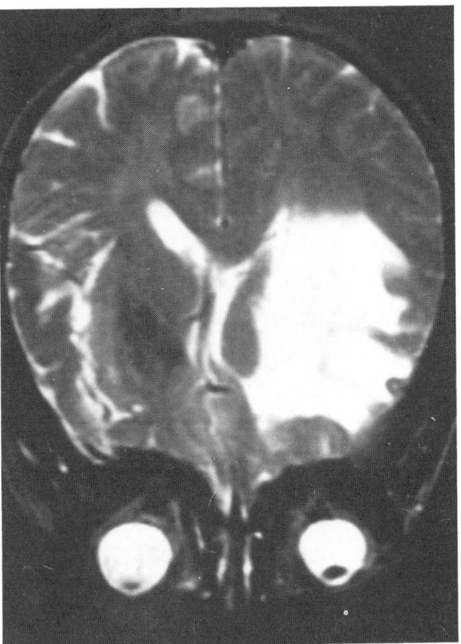

**FIGURE 11.12** On the left is an MRI image of an axial section of a human brain. The image was recorded emphasizing the $T_2$ relaxation time information, and it shows a bright region in the right cerebral hemisphere that is an indication of a tumor. Confirmation of this is obtained from the coronal section image of the same individual shown on the right. Image construction for the coronal section emphasized $T_1$ information. The MRI used to obtain these images operates with a 1.5 Tesla magnet, and the two-dimensional images are based upon signals collected for a region that was 5 mm thick. (Photographs courtesy of the Department of Radiology, University of Iowa.)

measured signals by the use of a field gradient. That is, instead of a uniform static magnetic field, a field which varies linearly with the spatial position coordinates is used. For instance, the field could increase linearly in the x-direction. This encodes spatial information into the sample magnetization, and now $\vec{M} = \vec{M}(x,y,z,t)$. Also, a gradient pulse in the oscillating field may be used to encode spatial information.

The relaxation times, $T_1$ and $T_2$, reflect the molecular environment. Thus, if we are monitoring proton NMR transitions of water molecules in a biological sample, we should expect to find different relaxation times for different structures in the biological sample. If the different times can be associated with specific points in space, then we will have the means for mapping out the biological structures and the boundaries between them. That association can be made because the spatial information has been encoded into the magnetization signal. So, in a rough sense, pulse sequences that give relaxation times combined with gradient fields that serve to identify position make NMR into the very powerful tool for non-destructive imaging called MRI.

## Exercises

1. Assuming all nuclei are in their ground state, how many different nuclear spin states will exist for the following isotopes of water: $H_2O$, $D_2O$, and $D_2{}^{17}O$?
2. Neglecting spin-spin coupling in the example of diimide, HNNH, of the 84 transitions, how many correspond to nitrogen spin flips, and how many correspond to hydrogen spin flips?
3. Neglecting spin-spin coupling, develop an energy level diagram for the nuclear spin states of an isotope of acetylene, $DC^{13}CH$, in a magnetic field. Assume that $g_D > g_C > g_H$. Draw vertical arrows to indicate the allowed NMR transitions.
4. Show that $\langle m_{I_1} m_{I_2} | \hat{I}_{x_1} \hat{I}_{x_2} + \hat{I}_{y_1} \hat{I}_{y_2} | m_{I_1} m_{I_2} \rangle$ equals zero. (First, using the definitions of angular momentum raising and lowering operators, express the operator in the integral in terms of the raising and lowering operators.)
5. Develop a stick representation of the high resolution NMR spectra predicted by first-order treatment of spin-spin interaction of the following molecules.

   *a.* $H_2C^{17}O$ *b.* $H_2C{=}^{13}CH_2$

6. What is the intensity pattern for a proton NMR transition split by interaction with two nearby equivalent $^{14}N$ nuclei?
7. The NMR intensity/splitting pattern for a single magnetic nucleus is 1:1 with one nearby proton; it is 1:2:1 if there are two nearby equivalent protons, and 1:3:3:1 if there are three nearby equivalent protons. What would these patterns become if the protons were replaced by deuterons?
8. If there were four equivalent spin-1/2 nuclei in a molecule and a nearby proton that had a spin-spin interaction with these nuclei, what would be the intensity pattern for the NMR spectrum of that proton?
9. Predict the form of the $^{13}C$ and the proton NMR spectra of $H{-}^{13}C{\equiv}^{13}C{-}^{13}C{\equiv}^{13}C{-}H$ on the basis of a first order treatment of spin-spin coupling.
10. Predict the high-field ESR spectrum of $H^{13}CO\bullet$ and of $H^{13}C^{13}C\bullet$.
11. Predict the high-field ESR spectrum of the radical $H_3C\text{-}CH_2\bullet$ assuming (a) that the $a_i$ coupling constants for $CH_2$ protons are much greater than for the $CH_3$ group protons, and (b) vice versa. Next, assume that the coupling constants are similar in size, although slightly different. Given that this is intermediate between the first two cases, what is the likely form of the spectrum?
12. Generate a version of Fig. 11.11 for which the nuclear spin is $I = 1$ and the electron spin is $S = 1/2$. That is, find the high-field and zero-field limiting cases and connect the energy levels in the intermediate region. What might be the appearance of the low-field ESR spectrum for a system of this sort?
13. Predict the high-field ESR spectrum of $HOO\bullet$, $H^{17}OO\bullet$, and $H^{17}O^{17}O\bullet$.
14. Consider the proton NMR spectra of the acetylenic hydrogens that are embedded in two identical larger molecules (e.g., the spectra of the $HC{\equiv}C{-}$ fragments) that are separated in space in by 1 cm. Make a plot of the separation between the two transition peaks for the two molecules under the condition of a magnetic field gradient along the line connecting the two molecular centers. (Assume a value for the average field strength and plot the separation in the absorption signals as a function of the size of the gradient of the field.)

*15.* Verify that Eqs. (11-30a) and (11-30b) are solutions of the differential equations in Eqs. (11-29a) and (11-29b). Then verify that Eq. (11-31) satisfies the more general differential equations of Eq. (11-28).

## Bibliography


1. M. Karplus and R. N. Porter, *Atoms and Molecules* (Benjamin, Menlo Park, California, 1970). This introductory text provides some of the ideas for understanding the contributing elements of chemical shielding.
2. W. H. Flygare, *Molecular Structure and Dynamics* (Prentice Hall, Englewood Cliffs, New Jersey, 1978). This text offers an intermediate level development of magnetic resonance.
3. F. A. Bovey, L. Jelinski, and P. A. Mirau, *Nuclear Magnetic Resonance Spectroscopy* (Academic, New York, 1988). This provides an up-to-date, thorough coverage of the field including magnetic resonance imaging.

*ADVANCED TOPICS:*

4. C. P. Slichter, *Principles of Magnetic Resonance* (Springer-Verlag, New York, 1978).
5. A. Abragam, *Principles of Nuclear Magnetism* (Clarendon Press, Oxford, 1961).
6. P. Mansfield and P. G. Morris, *NMR Imaging in Biomedicine* (Academic Press, New York, 1982).

# Chapter 12

# Atomic and Molecular Properties

*Atomic and molecular properties may be calculated from the quantum mechanical wavefunctions, and many can be measured or deduced from spectroscopic experiments. These properties characterize the response of the species to various perturbations. This chapter considers the formal quantum mechanical analysis of properties in general, and then it examines the role of properties in spectroscopy involving external electric or magnetic fields. Intrinsic response properties are also important in intermolecular interaction, and an introduction to this topic is presented in the last section.*

## 12.1 PROPERTIES AS ENERGY DERIVATIVES

Molecular properties are values that quantify certain characteristics or features of a molecule. Many properties characterize a molecule's response. For example, the force constant, k, of the bond of a diatomic molecule is the second-order energy response to stretching or compressing the bond. That is, the energy's quadratic dependence on the bond length varies with the value k, where

$$k = \left. \frac{d^2E}{dx^2} \right|_{x=x_{min}}$$

This property is an energy derivative, and in fact, many important properties are also derivatives.

Reponses to applied fields are properties that are various derivatives of the molecular energy with respect to the field strengths. The electrical permanent moments of a molecule's charge distribution (dipole, quadrupole, octupole, etc.) are first derivatives of the energy, while the *polarizability* is a second derivative. The *magnetic susceptibility*, or as we will call it, the *magnetizability*, is a second derivative with respect to the strength of a magnetic field. Responses to oscillating fields will, in general, be dependent on the frequency of oscillation. A polarizability, for instance, may be *frequency-dependent* because it varies with the oscillation frequency of an applied field.

Molecular properties are not the same as bulk properties, such as heat capacities. However, bulk properties must somehow be related to the properties of individual molecules in ways that are the subject of statistical mechanical theories. Our notions of bulk properties and properties of individual isolated molecules are often based on phenomena, since observation usually precedes fundamental theory. Because of this, property definitions are often phenomenological, which is to say that a property is defined in terms of a specific observation or measurement. Even so, a fundamental analysis will usually provide a definition of a property of an individual isolated molecule as some specific derivative of the energy. It is important, then, to take an overview of properties as derivatives.

As discussed in Chap. 4, an energy derivative can be obtained by formal differentiation of the Schrödinger equation. This can be done for electronic, vibrational, and rotational Schrödinger equations; in fact, it can be done for the Schrödinger equation of most any problem.

## 12.2 ELECTRICAL RESPONSE PROPERTIES

Molecular response to electrical perturbations is extremely important in molecular science simply because molecules are distributions of electrical charge. It is their nature to be influenced by electrical potentials or fields. The fields may arise from other molecules, external sources, or electromagnetic radiation. Thus, electrical response properties are important for intermolecular interaction, spectroscopy, aggregation phenomena, and so on.

Electrical response is generally expressed with respect to some *electrical potential* that may arise from a charge distribution or an external source. An electrical potential is a spatially dependent function, V(x,y,z), such that the interaction energy with a fixed point charge, $q_i$, is

$$E_{int} = q_i V(x_i, y_i, z_i) \tag{12-1}$$

where $x_i$, $y_i$, and $z_i$ are the position coordinates of the point charge. The potential may be expressed as a power series expansion about a point in space. Using the coordinate system origin, (0,0,0), as the point of expansion gives

$$\begin{aligned} V(x,y,z) = {} & V(0,0,0) + x\frac{\partial V}{\partial x}\bigg|_0 + y\frac{\partial V}{\partial y}\bigg|_0 \\ & + z\frac{\partial V}{\partial z}\bigg|_0 + \frac{1}{2}x^2\frac{\partial^2 V}{\partial x^2}\bigg|_0 + xy\frac{\partial^2 V}{\partial x \partial y}\bigg|_0 \\ & + \frac{1}{2}y^2\frac{\partial^2 V}{\partial y^2}\bigg|_0 + xz\frac{\partial^2 V}{\partial x \partial z}\bigg|_0 + yz\frac{\partial^2 V}{\partial y \partial z}\bigg|_0 \\ & + \frac{1}{2}z^2\frac{\partial^2 V}{\partial z^2}\bigg|_0 + \frac{1}{6}x^3\frac{\partial^3 V}{\partial x^3}\bigg|_0 + \ldots \end{aligned} \tag{12-2}$$

The first derivatives of V are the components of the electric field vector, and we shall designate them simply as ($V_x$, $V_y$, $V_z$). The second derivatives are the components of the field gradient.

For a classical distribution of N fixed-point charges, the interaction energy of Eq. (12-1) can be summed to give the interaction of the distribution with an external potential. If the power series expansion of Eq. (12-2) is used, the following is obtained.

$$E_{int} = \sum_i^N q_i V(x_i, y_i, z_i)$$

$$= V(0,0,0) \sum_i q_i + V_x \sum_i q_i x_i + V_y \sum_i q_i y_i$$

$$+ V_z \sum_i q_i z_i + \frac{1}{2} V_{xx} \sum_i q_i x_i^2 + \ldots$$

This suggests the definition of the moments of a charge distribution. The zeroth moment is the net charge

$$M_0 = \sum_i^N q_i \qquad (12\text{-}3)$$

The first moment, or the dipole moment, has three elements.

$$M_x = \sum_i^N q_i x_i \qquad (12\text{-}4a)$$

$$M_y = \sum_i^N q_i y_i \qquad (12\text{-}4b)$$

$$M_z = \sum_i^N q_i z_i \qquad (12\text{-}4c)$$

The second moment has nine elements, six of which are unique.

$$M_{xx} = \sum_i^N q_i x_i^2 \qquad (12\text{-}5a)$$

$$M_{yy} = \sum_i^N q_i y_i^2 \qquad (12\text{-}5b)$$

$$M_{zz} = \sum_{i}^{N} q_i z_i^2 \tag{12-5c}$$

$$M_{xy} = M_{yx} = \frac{1}{2}\sum_{i}^{N} q_i x_i y_i \tag{12-5d}$$

$$M_{xz} = M_{zx} = \frac{1}{2}\sum_{i}^{N} q_i x_i z_i \tag{12-5e}$$

$$M_{yz} = M_{zy} = \frac{1}{2}\sum_{i}^{N} q_i y_i z_i \tag{12-5f}$$

The definition of third-moment elements, fourth-moment elements, and so on, follows from the interaction energy expression. Then, the interaction energy can be conveniently expressed as a dot product:

$$\begin{aligned} E_{int} &= M_0 V(0,0,0) + M_x V_x + M_y V_y + M_z V_z \\ &\quad + M_{xx} V_{xx} + M_{xy} V_{xy} + M_{xz} V_{xz} + M_{yx} V_{yx} \\ &\quad + M_{yy} V_{yy} + M_{yz} V_{yz} + M_{zx} V_{zx} + \ldots \\ &= \begin{pmatrix} M_0 & M_x & M_y & M_z & M_{xx} & \ldots \end{pmatrix} \begin{pmatrix} V_0 \\ V_x \\ V_y \\ V_z \\ V_{xx} \\ \ldots \end{pmatrix} \end{aligned} \tag{12-6}$$

(There are different conventions in the definition of the second and higher moments. Often, the factors of 1/2 in Eq. (12-5) and of 1/6 in the third moment are not included. The factors are then inserted when the moment values are used in calculating an interaction.) For a distribution of fixed charges, the first derivative of the energy with respect to an element of the vector $\vec{V}$ in Eq. (12-6), e.g., with respect to $V_x$, $V_y$, and so on, is a particular moment element. For quantum mechanical systems, it remains true that the moments are derivatives with respect to components of the field, or of the field gradient tensor, and so on. We may say that the dipole moment is a property that characterizes the first-order response of the molecular energy to a uniform applied field. The second moment is the first-order response to a field gradient.

In an atom or molecule, the electrical point charges (the electrons and the nuclei) are not fixed, and the wavefunctions may adjust in response to an applied electric potential. Consequently, there will be a change in the energy that is not only first order or linear in the elements of $\vec{V}$, but also quadratic, cubic, and so on. The quantum mechanical analysis of the effect of an applied potential requires adding an interaction to the Hamiltonian. This term uses the electric moment elements as operators. For instance, Eq. (12-5a) serves as an expression for the operator of the x-component of the dipole moment by using –e for the charges of the electrons and letting the $x_i$'s be the x-coordinate position operators of the electrons. The interaction Hamiltonian must be the fundamental interaction of the applied potential with each of the individual particles, and so the operator follows immediately from the classical interaction energy expression.

$$\hat{H}_{int} = V_0 M_0 + V_x \hat{M}_x + V_y \hat{M}_y + V_z \hat{M}_z + V_{xx} \hat{M}_{xx} + \dots \quad (12\text{-}7)$$

The first term is a constant, and so it will add directly to the total energy and can otherwise be ignored.

The Schrödinger equation for some specific choice of the elements of $\vec{V}$ is

$$(\hat{H}_o + \hat{H}_{int}^{\{V\}}) \psi^{\{V\}} = E^{\{V\}} \psi^{\{V\}} \quad (12\text{-}8)$$

where the superscript {V} means that a set of specific values for the elements of $\vec{V}$ were used for the interaction Hamiltonian and that as a result, the eigenenergy and the wavefunction depend parametrically on these values. Now, consider solving Eq. (12-8) for three situations, first where all the elements of $\vec{V}$ are zero, second where the only nonzero element is $V_x = 1$, and third where the only non-zero element is $V_x = 2$. From these three solutions, we can obtain the energy for three values of $V_x$. And we use two of the energy values to find an approximate value of the first derivative.

$$\left.\frac{\partial E}{\partial V_x}\right|_{V_x=0} \approx \frac{E(V_x=1) - E(V_x=0)}{1-0}$$

This is called a finite difference approximation,* and in this case it is an approximation for the x-component of the dipole moment. A finite difference approximation for the second derivative, a property which is called the polarizability, is

$$\left.\frac{\partial^2 E}{\partial V_x^2}\right|_{V_x=0} \approx \frac{\dfrac{E(V_x=2) - E(V_x=1)}{2-1} - \dfrac{E(V_x=1) - E(V_x=0)}{1-0}}{2-0}$$

These values may also be found analytically, without approximation, by using the derivative Schrödinger equation methods outlined in Chapter 4 or by perturbation theory.

Second derivatives of the energy with respect to the elements of a uniform electric field, $V_x$, $V_y$, and $V_z$, make up a tensor (second rank matrix) called the dipole polarizability, $\alpha$.

$$\alpha = \begin{pmatrix} \alpha_{xx} & \alpha_{xy} & \alpha_{xz} \\ \alpha_{yx} & \alpha_{yy} & \alpha_{yz} \\ \alpha_{zx} & \alpha_{zy} & \alpha_{zz} \end{pmatrix} \tag{12-9}$$

$$\text{where} \quad \alpha_{yz} \equiv \left.\frac{\partial^2 E}{\partial V_y \partial V_z}\right|_{\vec{V}=0} \quad \text{etc.}$$

The dipole polarizability may also be defined as a first derivative of the dipole moment because the dipole moment is found by first differentiation of the energy. This means the polarizability tells how the dipole moment vector will change to first order upon application of an external uniform field. We may say that the field induces a dipole, and physically this comes about by a shift in the charge distribution. If we happen to know the dipole polarizability of a charge distribution, then

* For a function f(x), we may approximate the first derivative in the vicinity of two nearby points $x_1$ and $x_2$ as $[f(x_2) - f(x_1)] / (x_2 - x_1)$. This value will approach the true first derivative as $x_2$ approaches $x_1$.

from the definition, we may conclude that the energy of polarization from applying a field is

$$E_{pol} = \frac{1}{2} ( V_x \; V_y \; V_z ) \; \alpha \begin{pmatrix} V_x \\ V_y \\ V_z \end{pmatrix} \tag{12-10}$$

Also, the induced dipole moment is

$$\begin{pmatrix} M_x^{induced} \\ M_y^{induced} \\ M_z^{induced} \end{pmatrix} = \alpha \begin{pmatrix} V_x \\ V_y \\ V_z \end{pmatrix} \tag{12-11}$$

The induced moments are combined with the *permanent moments* (e.g., $M_x$, $M_y$, and $M_z$) to yield the total dipole moment vector in the presence of the field.

An example problem is the dipole polarizability of the hydrogen atom. If a hydrogen atom experienced a uniform field in the z-direction, $V_z$, the interaction Hamiltonian would be

$$\hat{H}' = -e z V_x + e z_p V_x$$

where z is the z-coordinate of the electron and $z_p$ is the z-coordinate of the proton. Taking the proton to be fixed in space makes $z_p$ a constant. So, the interaction with the proton is a constant contribution to the energies of all the states; we may ignore its effect. The coordinate z should be transformed to spherical polar coordinates since they were used to solve for the wavefunctions of the unperturbed hydrogen atom.

$$\hat{H}' = -e V_z \; r \cos\theta \tag{12-12}$$

The first-order perturbation theory correction to the ground state wavefunction would involve atomic orbitals that have a nonzero matrix element with the 1s orbital via the perturbation of Eq. (12-12). The $2p_0$ or $2p_z$ orbital is the lowest energy orbital to have a nonzero matrix element.

$$< \psi_{1s} \mid \hat{H}' \; \psi_{2p_z} > = -e V_z < \psi_{1s} \mid r \cos\theta \; \psi_{2p_z} > \neq 0$$

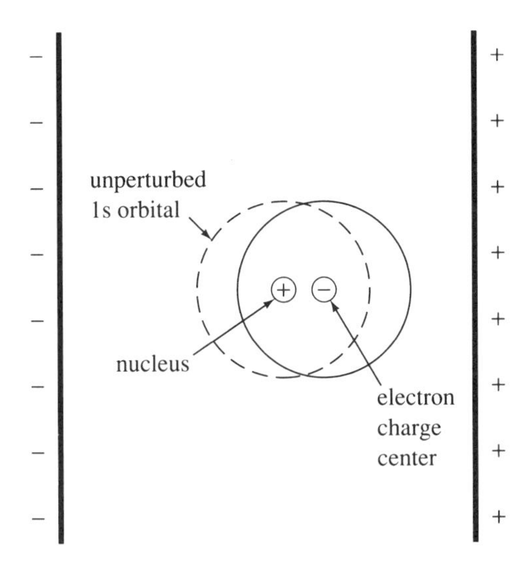


**FIGURE 12.1** The influence of an external electric field in the z-direction on a hydrogen atom. We may think of the field as arising from two oppositely charged parallel plates. We may think of the electron of the hydrogen atom to be attracted toward the positively charged plate, while the proton is attracted toward the negatively charged plate. This represents a separation of the centers of negative and positive charge in the hydrogen atom, as shown schematically. The result is a nonzero dipole moment. The charge distribution of the electron is roughly that of the superposition of a 1s orbital with a 2p orbital along the z-axis. The 2p orbital adds to the wavefunction amplitude on one side of the nucleus and diminishes the amplitude on the other side where it is of opposite phase.

So, at first order the $2p_z$ orbital is mixed with the 1s orbital because of an external field in the z-direction.

Fig. 12.1 illustrates that the mixing of the 1s and $2p_z$ hydrogen atom orbitals will remove the spherical symmetry of the hydrogen atom's wavefunction and charge distribution. The negative and positive charge centers will no longer be at the same point, and so a dipole moment in the z-direction will have been induced. The size of this dipole moment is the polarizability of the atom times $V_z$. We may also see that the polarizability has to do with the ability to mix the 1s and $2p_z$ orbitals. From this, we might compare different atoms, and for instance, we should anticipate that Be is a more polarizable element than He. The He polarizability would be determined largely by the 1s-2p mixing, whereas the Be polarizability would be dictated largely by the energetically more favorable 2s-2p mixing. (The 2s and 2p orbitals are much closer in energy than the 1s and 2p orbitals.)

The isotropic polarizabilities* of the hydrides of the elements carbon through fluorine have a fairly smooth dependence on the atomic number Z of the first-row element. Values obtained from electronic structure calculations are shown in Fig. 12.2. This reflects the declining polarizability as the atomic shell is filled moving to the right in the Periodic Table. To a rough degree, the polarizabilities of small molecules are sums of the polarizabilities of the constituent atoms. For instance, the isotropic polarizabilities of the hydrides, $ABH_n$, where A and B are any of the elements carbon through fluorine and are singly bonded to each other, are very nearly the sums of the polarizabilities of the hydrides of A and of B.

The electrical properties of an atom or molecule are derivatives of the energy with respect to elements of the electric potential. The first derivatives are the permanent moments. If the differentiation is with respect to components of a field, the property is the dipole moment. If it is with respect to elements of the field gradient, it is the second or *quadrupole moment*. The simplest arrangement of point charges possessing a quadrupole moment but with zero dipole is

$$\bullet^{+} \quad O^{2-} \quad \bullet^{+}$$

There are third moments, fourth moments, and so on, and these are defined as first derivatives with respect to the higher order elements of $\vec{V}$.

Second derivatives of the energy are polarizabilities. If the differentiation is with respect to uniform field components, the property is the dipole polarizability. But the differentiation may be with respect to purely field gradient components, and that yields the *quadrupole polarizability*. Just as the dipole polarizability gives the size of the

* The isotropic polarizability is the average of the polarizability tensor elements along the three axis: $\bar{\alpha} = \left(\alpha_{xx} + \alpha_{yy} + \alpha_{zz}\right) / 3$.

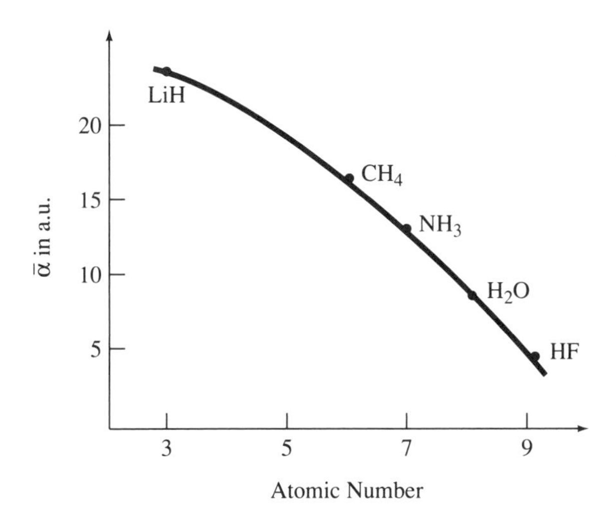


**FIGURE 12.2** Isotropic dipole polarizabilities of first row element hydrides. These values have been obtained from electronic structure calculations of SCF wavefunctions [S.-Y. Liu and C. E. Dykstra, *J. Phys. Chem.* **91**, 1749 (1987)].

dipole induced by an external field, the quadrupole polarizability gives the size of the quadrupole moment induced by an external field gradient. There are also mixed polarizabilities. The dipole-quadrupole polarizability (derivative with respect to field and field gradient components) can give the dipole induced by a field gradient or the quadrupole induced by a uniform field. Then there are third derivatives, which are called *hyperpolarizabilities*, and fourth derivatives, which are called second hyperpolarizabilities, and so on. The list of these properties continues as far as one wishes to differentiate, but in total, they are simply values that characterize the response of an atom or molecule to an electrical potential.

## 12.3 SPECTROSCOPIC STARK EFFECT

In 1913 Johannes Stark found that an electric field causes a splitting in hydrogen atom spectral lines. The effect of an electric field has since proven to be useful in many types of spectroscopy. Because of how gas phase molecular spectra are altered, the Stark effect (the effect of an electric field) can help in assigning spectral lines and in measuring certain molecular properties. Basically, the sample is placed in a cell with two parallel metal plates. Charging the plates, one negatively and one positively, creates an electric field, and spectra may be obtained without the field, or with a field of various, specifically chosen strengths. In molecules, the primary interaction from the external electric field is with the molecular dipole moment. As a result, the Stark effect provides a means for measuring the dipole moment, and sometimes the dipole polarizability.

Stark effect experiments are quite common in microwave spectroscopy and are also carried out in the infrared and visible-UV. In microwave spectroscopy, the Stark effect provides the most direct means for determining the permanent dipole moment because the dipole is little affected by rotational state changes. With electronic excitation, on the other hand, the electric field interaction is quite different in the ground and excited states, and in the end, the Stark effect yields the change in the dipole moment from the initial to the final state. We will consider in detail the quantum mechanics for Stark effect microwave spectra of linear molecules.

The interaction that arises between the molecular dipole moment and an external field is the dot product of the field and the dipole moment vectors. This may be expressed as the product of the magnitudes of the two vectors and the cosine of the angle between them. Taking the direction of the applied field, F, to define the laboratory z-axis, the perturbing Hamiltonian is

$$\hat{H}' = -\mu F \cos\theta \tag{12-13}$$

(The usual convention for fields is such that $F = -V_z$.) $\theta$ is the angle between the molecular axis and the laboratory z-axis, and so it is the same as the spherical polar coordinate $\theta$. $\mu$ is the magnitude of the dipole moment. The effect of this interaction on the rotational energies may be analyzed with perturbation theory.

The first-order perturbation theory corrections to the state energies are the expectation values of the perturbation using the zero order wavefunctions. The rotational wavefunctions of linear polyatomic molecules are the spherical harmonic functions. Thus, the first order energy corrections are

$$E^{(1)}_{J,M} = -\mu F < Y_{JM} \mid \cos\theta \; Y_{JM} > \tag{12-14}$$

Explicit integration will show for any choice of J and M that the integral in Eq. (12-14) is zero; in fact, the integral can be proven to be zero for an arbitrary choice of J and M. This means that the first-order corrections to the rotational state energies of a linear molecule due to the applied field are zero. There is no energy difference at this level, and so a higher order perturbative treatment is needed.*

Second-order perturbation theory gives the energy correction to be

$$E^{(2)}_{J,M} = \mu^2 F^2 \sum_{J' \neq J} \sum_{M'=-J'}^{J'} \frac{< Y_{JM} \mid \cos\theta \; Y_{J'M'} >^2}{E^{(0)}_{J,M} - E^{(0)}_{J'M'}} \tag{12-15}$$

The sum on the right hand side is over an infinite number of J' M' states. However, it can be proven that the integral in the numerator will be zero unless M = M' and $\mid J - J' \mid = 1$. Using $BJ(J+1)$ for the zero order energy of a JM state will simplify the energy denominator, and so, we obtain the general result that

$$E^{(2)}_{j,M} = \mu^2 F^2 \left( \frac{< Y_{JM} \mid \cos\theta \; Y_{J\text{-}1\,M} >^2}{2\,B\,J} + \frac{< Y_{JM} \mid \cos\theta \; Y_{J+1\,M} >^2}{-\,2B\,(J+1)} \right) \tag{12-16}$$

Through fairly sophisticated means, the integrals in Eq. (12-16) can be

* It is only true for linear molecules that the first-order energy correction is zero. For symmetric top molecules and asymmetric top molecules, there is a first-order correction.

evaluated without specifying J and M. Of course, we could also carry out the explicit integration for any particular J M state of interest, but the general result avoids the effort. Without derivation, we state this result as

$$< Y_{J\,M} \mid \cos\theta \; Y_{J+1\,M} > \; = \; \sqrt{\frac{(J+1)^2 - M^2}{(2J+1)(2J+3)}} \tag{12-17}$$

(This expression can be used for both integrals in Eq. (12-16) by utilizing the Hermitian property of the operator cos θ.) With this result, the second-order energy correction is,

$$E^{(2)}_{J,M} = \frac{\mu^2 F^2}{2B(2J+1)} \left( \frac{J^2 - M^2}{J(2J-1)} - \frac{(J+1)^2 - M^2}{(J+1)(2J+3)} \right)$$

$$= \frac{\mu^2 F^2}{2B(2J-1)(2J+3)} \left( 1 - \frac{3M^2}{J(J+1)} \right) \tag{12-18}$$

Eq. (12-18) does not apply to the case where $J = 0$, since there is no lower energy state, or no $J - 1$ state, to mix at second order. However, the correct result with $J = 0$ is obtained by removing the first fraction in parentheses in the first line in Eq. (12-18); that is the term that comes about from mixing with a lower energy state. The effect of the field on spectra can now be determined.

The rotational selection rules were that $\Delta J = 1$ and that $\Delta M = 0$. So, the only allowed transition originating in the ground state (i.e., $J = 0$ and $M = 0$) is to the state $J = 1$ and $M = 0$. To find the effect of the field, at second order, we apply Eq. (12-18) for these two states.

$$E^{(2)}_{10} = \frac{\mu^2 F^2}{10\,B}$$

$$E^{(2)}_{00} = -\frac{\mu^2 F^2}{6\,B}$$

This reveals that the ground state energy will be reduced while the excited rotational state involved in the transition will be at a higher energy due to the field. In total, the transition energy will be increased by $4\mu^2 F^2 / 15\,B$. So, if the transition energy is measured without a field and then remeasured with a field of known strength applied, the difference between the two transition energies will be $4\mu^2 F^2 / 15 B$, at least to the accuracy of this second-order perturbative treatment. In this way, the magnitude, but not the sign, of the molecular electric dipole moment of a linear molecule may be obtained from spectroscopic measurement.

An electric field will interact with a molecule with a second order dependence on the field strength via the dipole polarizability, and with a third order dependence via the dipole hyperpolarizability. This is potentially a complication in using the Stark effect to measure the dipole moment. However, by using several transitions, that is, using $J_{initial} > 0$ transitions, the effects can be distinguished. Then, not only will the dipole moment be obtained but so will the anisotropy in the dipole polarizability tensor.

## 12.4 RAMAN EFFECT AND RAMAN SPECTRA ✯

The interaction of photons with atoms and molecules can lead to absorption of a photon and excitation of an atom or molecule, and this is the process underlying the spectroscopies that have been discussed so far. However, there are other interaction processes, such as the absorption of a photon and the simultaneous emission of another photon at a higher frequency, the same frequency, or a lower frequency. Spectroscopic experiments have been devised for these interactions as well. One of these, Raman spectroscopy, is the observation of low-relative-intensity radiation that is scattered* by an atom or molecule at higher or lower frequency than

* Within the language of the classical description of electromagnetic radiation, a wave may be said to scatter upon interaction with a material, whereas in a complete quantum electrodynamic picture, the interactions involve photons. The classical treatment of the radiation is sometimes sufficient to account for the spectroscopy, which is to say that the wave character of the radiation is the key to the interaction. The classical picture of radiation combined with the quantum picture of atoms and molecules is referred to as a semiclassical picture of electromagnetic interaction, and this picture has been used implicitly in all the earlier discussions of spectroscopy.

the incident radiation.

Raman spectra are typically taken with gas phase and liquid samples, and it is the frequency of the emitted radiation that is monitored. Usually this is done from a direction perpendicular to the beam of incident radiation. The spectra may have features at higher wavelengths than the incident radiation (lower energy photons), and these are called Stokes lines; and it may have features at shorter wavelengths (higher energy photons), and these are called anti-Stokes lines. In the anti-Stokes process, the molecule loses rather than gains energy, and the intensity of these lines is usually less than that of the Stokes lines.

In analyzing Raman spectra, it is not the transition frequency of a line that is important; it is the difference between the line's frequency and that of the incident radiation. This difference corresponds to the energy change in the molecule from the interaction that has occurred, and the energy change is equal to an energy difference between a pair of atomic or molecular energy levels. From a practical standpoint, this is helpful because Raman scattering can be observed with available radiation sources even if their frequency does not coincide with a transition frequency of the sample. So, whereas a laser operating at one fixed frequency could be used in an absorption experiment to detect a transition that occurs only at that frequency, in a Raman experiment it could be used as the incident beam and give a wealth of spectral data. In fact, the information of a full rotational or vibrational spectrum might be obtained from the emitted radiation.

The incident beam in a Raman experiment may have a frequency in the infrared or the visible-UV regions of the spectrum. In a gas phase sample, Raman lines that are very close in frequency to the incident radiation, within 10 wavenumbers for instance, will correspond to pure rotational transitions. Lines that are further removed from the frequency of the incident radiation arise from changes in the vibrational-rotational states. The selection rule for the rotational quantum number is

$$|\Delta J| = 0 \text{ or } 2 \qquad (12\text{-}19)$$

The selection rule for vibrational quantum numbers is the same as for infrared absorption. This means that transitions are primarily those involving no change or else a change of one in a vibrational quantum number, although transitions are not limited to these. (With no change in the vibrational quantum number, the spectrum is the pure rotational Raman spectrum.)

The semiclassical picture of Raman transitions is that the electric field of the incident radiation induces a dipole moment, and the interaction of the induced dipole moment with the field of the radiation leads to a change in the molecular energy state. The interaction operator is

$$\hat{H}' = -\frac{1}{2}\vec{F}\,\alpha\,\vec{F} \tag{12-20a}$$

where $\alpha$ is the dipole polarizability tensor and $\vec{F}$ is the field vector of the radiation. For a diatomic molecule interacting with a field applied along the laboratory z-axis, this interaction simplifies to

$$\hat{H}' = -\frac{1}{2}F_z^2\,\alpha_{zz}\cos^2\theta \tag{12-20b}$$

where the z-axis of $\alpha_{zz}$ is the internal z-axis or molecular axis. (The dot product of a unit vector along the laboratory z-axis and a unit vector along the molecular axis is $\cos\theta$.) The polarizability component, $\alpha_{zz}$, is generally a function of the stretching coordinate, s. Just as a dipole moment may change in the course of a molecular vibration, so too may the other properties including the polarizability.

The selection rules are developed by considering matrix elements with the interaction operator and the vibrational-rotational states of the molecule. So, with nJM as the quantum numbers of the initial state and n'J'M' as the quantum numbers of the final state, the matrix element separates into a product of integrals over the angular functions and the stretching coordinate.

$$<\Psi_{nJ} \mid \hat{H}'\,\Psi_{n'J'}> = -\frac{1}{2}F_{zz} <\psi_n(s) \mid \alpha_{zz}(s)\,\psi_{n'}(s)> <Y_{JM} \mid \cos^2\theta\,Y_{J'M'}>$$

It can be shown that in general the integral over the angular coordinates is zero unless M = M' and unless either J = J' or $|J - J'| = 2$, and this is the selection rule of Eq. (12-19). To understand the integral over the stretching coordinate, we may introduce a power series expansion of the polarizability.

$$\alpha(s) = \alpha(0) + s\left.\frac{\partial\alpha}{\partial s}\right|_{s=0} + \frac{1}{2}s^2\left.\frac{\partial^2\alpha}{\partial s^2}\right|_{s=0} + \ldots \tag{12-21}$$

The subscript on $\alpha$ has been suppressed for conciseness. We now have

$$< \psi_n(s) \mid \alpha(s) \psi_{n'} > \; = \; \alpha(0) < \psi_n(s) \mid \psi_{n'}(s) >$$

$$+ \left. \frac{\partial \alpha}{\partial s} \right|_{s=0} < \psi_n(s) \mid s \, \psi_{n'}(s) >$$

$$+ \frac{1}{2} \left. \frac{\partial^2 \alpha}{\partial s^2} \right|_{s=0} < \psi_n(s) \mid s^2 \, \psi_{n'}(s) > \; + \; \ldots \qquad (12\text{-}22)$$

The first integral on the right hand side will give a nonzero transition probability if $\alpha(0) \neq 0$ and if $n = n'$. This means if the molecule is polarizable at all, which all molecules are, then the selection rule on the vibrational quantum number is $\Delta n = 0$. This is the rule that is associated with the pure rotational Raman spectra. The second term on the right hand side of Eq. (12-22) is nonzero if the first derivative of the polarizability is nonzero and if $|n - n'| = 1$. For a diatomic molecule, the polarizability is always expected to have a non-zero first derivative, and so we arrive at the selection rule that the vibrational quantum number may change by one for allowed Raman transitions. The third term in Eq. (12-22) corresponds to transitions where n changes by 2, but these tend to be much weaker transitions.

In polyatomic molecules, the same selection rules hold for Raman spectra, although the rule for the vibrational quantum number needs to be generalized for the many different modes of a polyatomic. The resulting selection rule may be expressed as

$$\Delta n_i = 0, 1 \quad \text{and} \quad \Delta n_{j \neq i} = 0$$

That is, the vibrational quantum number for one and only one mode may change by one. (Other transitions may be seen but are expected to be less intense.) There is also a requirement that the first derivative of the polarizability along the direction of the $i^{th}$ normal mode of vibration be nonzero. Symmetry arguments are often sufficient to determine whether this is so. For instance, consider the bending vibration of the $CO_2$ molecule. If it were bent $5^o$ from equilibrium, the polarizability would change. However, because the change would be the same amount as if the molecule were bent by $-5^o$, the polarizability function must have a critical point at $0^o$; it must have a zero-valued first derivative at the equilibrium.

In contrast, the dipole moment function, which is zero at $0^o$ and some nonzero value $\delta$ at $5^o$, has a nonzero first derivative because it is $-\delta$ at $-5^o$. So, the bending vibration of $CO_2$ is said to be *infrared-active* but *Raman-inactive*; transitions will be seen in infrared absorption but not in the Raman spectrum.

The symmetric stretching vibration of $CO_2$ is another example. The dipole moment does not change in the course of this vibrational motion, and so this mode is infrared-inactive. However, stretching both bonds by 0.01 Å is different than compressing both bonds by 0.01 Å, and the polarizability changes are different. There must be a nonzero first derivative of the polarizability with respect to this stretching coordinate, and so this mode is Raman-active.

Raman spectra of gas phase samples display vibrational bands analogous to those of infrared absorption spectra. The difference is in the selection rule for J. Whereas absorption spectra will show a P branch ($\Delta J = -1$), possibly a Q branch ($\Delta J = 0$), and an R branch ($\Delta J = 1$), a Raman band should instead have an O branch ($\Delta J = -2$), a Q branch, and an S branch ($\Delta J = 2$). To interpret the branch structure and to assign lines to specific transitions, we may follow the procedure of taking energy differences that led to Eqs. (6-30) and (6-31). From the following energy level expression,

$$\tilde{E}_{n,J} = (n + \frac{1}{2})\,\tilde{\omega} + \tilde{B}_n\, J(J+1) \tag{12-23}$$

we may obtain the energy difference for lines in the O, Q, and S branches:

O branch, $\Delta J = -2$ (Stokes)

$$\Delta\tilde{E}_{n,J \to n+1,J-2} = \tilde{\omega} + \tilde{B}_{n+1}(J^2 - 3J + 2) - \tilde{B}_n(J^2 + J) \tag{12-24a}$$

$$\cong \tilde{\omega} - 4\tilde{B}J + 2\tilde{B} \tag{12-24b}$$

This is the energy difference *relative* to the energy of the incident radiation, and we see that in this branch, lines will be spaced about 4B apart. This is twice the separation of the rotational lines in an infrared absorption experiment.

Q branch, $\Delta J = 0$ (Stokes)

$$\Delta\tilde{E}_{n,J\rightarrow n+1,J} = \tilde{\omega} + (\tilde{B}_{n+1} - \tilde{B}_n) J (J+1) \tag{12-25}$$

$$\cong \tilde{\omega}$$

The lines in the Q branch, as usual, will be very closely spaced and may not even be resolved.

S branch, $\Delta J = 2$ (Stokes)

$$\Delta\tilde{E}_{n,J\rightarrow n+1,J+2} = \tilde{\omega} + \tilde{B}_{n+1} (J^2 + 5J + 6) - \tilde{B}_n (J^2 + J) \tag{12-26}$$

$$\cong \tilde{\omega} + 4\tilde{B}J + 6\tilde{B}$$

The lines in this branch will be separated by about 4B.

From the analysis of the energy differences, we see that it is possible to obtain the same type of information (e.g., rotational constants and vibrational frequencies) from Raman spectra as from absorption spectra. Instrumental capabilities may make one technique more favorable than the other, but the quantum mechanical analysis needed to interpret the spectra follows the same course.

## 12.5 MAGNETIC RESPONSE PROPERTIES

Magnetic response properties characterize the energetics of molecular interaction with magnetic fields. Electrons and nuclei have intrinsic magnetic moments, and their interaction with an external field is the basis for NMR and ESR spectrsocopy. Apart from such intrinsic properties, an external magnetic field may interact with moving charged particles. For molecules, this means both the electrons and the nuclei since the electrons are "orbiting" about the nuclear framework and the nuclei are moving in the course of their vibrations and rotations.

The fundamental interaction Hamiltonian for magnetic fields is developed in terms of the vector potential, $\vec{A}(x,y,z)$. This is the magnetic analog of the electrical potential V(x,y,z) in classical electromagnetic theory. Of course, the vector potential is a three-component vector ($A_x$, $A_y$, $A_z$) where each component is a function of the spatial coordinates x, y, and z. The electrical potential, on the other hand, is a scalar quantity. The vector potentials arising from different magnetic sources, such as external fields and magnetic nuclei, are additive. Each source contributes to the total in a vector sum.

$$\vec{A}(x,y,z) = \sum_{i}^{\text{sources}} \vec{A}_i(x,y,z) \tag{12-27}$$

From classical electromagnetic theory, the vector potential arising from a point magnetic dipole moment $\vec{m}$ is

$$\vec{A}(\vec{r}) = \frac{\vec{m} \times \vec{r}}{r^3} \tag{12-28}$$

where $\vec{r}$ is the position vector originating at the location of the point dipole.

The effect of a magnetic vector potential may be built into a Hamiltonian by a modification of the momentum operator for each charged particle in the following way.

$$\hat{p}_i \rightarrow \hat{p}_i - \frac{q}{c}\vec{A}(\vec{r}_i) \tag{12-29}$$

q is the charge of the particle, and for electrons, it is –e. Since the kinetic energy operator in the Hamiltonian involves the square of the momentum operator of each particle, then the substitution in Eq. (12-29) will generate Hamiltonian terms that are both linear and quadratic in the vector potential.

Magnetic properties are derivatives of the molecular energy with respect to the size of magnetic moments or the strengths of external fields. The relationship between a field, $\vec{B}$, and a vector potential is

$$\vec{B} \equiv \nabla \times \vec{A} = \begin{vmatrix} \hat{i} & \hat{j} & \hat{k} \\ \frac{\partial}{\partial x} & \frac{\partial}{\partial y} & \frac{\partial}{\partial z} \\ A_x & A_y & A_z \end{vmatrix} \tag{12-30}$$

The magnetic field is the curl of the vector potential. This outlines the formal process whereby the Hamiltonian can be written in terms of a vector potential, and given that, differentiation may be carried out with respect to the related elements of $\vec{B}$ or of a magnetic dipole, $\vec{m}$.

The first derivative of the molecular energy with respect to the elements of applied external field, $\vec{B}$, is the magnetic dipole of the molecule, but this can be shown to vanish. The second derivative is the *magnetic susceptibility* tensor, $\chi$, or the *magnetizability* tensor. It is the magnetic analog of the dipole polarizability. The third and higher order derivatives are the magnetic hypersusceptibilities.

NMR parameters are derivatives of the energy, also. The chemical shielding tensor is a mixed second derivative; the differentiation is with respect to the elements of the magnetic field that arises from a nuclear magnetic moment and with respect to the elements of an external magnetic field.

Properties can be identified with still higher derivatives of the energy. For instance, the chemical shielding may be influenced by an electric field. The first-order change in the chemical shielding due to the electric field is a third-derivative property; the differentiation is with respect to the electric field components, components of an external magnetic field, and components of the field arising from the nuclear magnetic moment. This particular property may be called the shielding polarizability. Also, the chemical shielding may be influenced by an electric field gradient. This is important because a typical molecular charge distribution may produce electric fields and/or field gradients that are sizable enough to affect the chemical shifts of nuclei in some nearby molecule by an observable amount.

The formal analysis of magnetic response properties as derivatives of the molecular energy affords the quantum mechanical basis for direct calculation of the properties. This is ultimately connected with the phenomenologically defined properties obtained from spectroscopy.

## 12.6 SPECTROSCOPIC ZEEMAN EFFECT

The Zeeman effect is simply the effect of an external magnetic field on spectra, including rotational, vibrational, Raman, and electronic spectra. The effects include shifts in transition frequencies and splittings of absorption lines. A broadening or splitting of a sodium atomic emission line because of the presence of a magnetic field was the first observation of this effect, and it was made by Zeeman before 1900. The analysis of the Zeeman effect starts with the introduction of a magnetic interaction term into the Hamiltonian.

For an atom with no net electron spin (i.e., S = 0), the magnetic moment due to the electrons results from the orbital angular momentum of the electrons. The interaction Hamiltonian is, as before,

$$\hat{H}' = -\vec{H} \bullet \vec{\mu} \tag{12-31}$$

where $\vec{H}$ is the applied magnetic field and $\vec{\mu}$ is the magnetic moment. Using the proportionality between the magnetic moment and the angular momentum vector, $\vec{L}$, and choosing the field to have only a z-component, this interaction Hamiltonian becomes

$$\hat{H}' = H_z \mu_\beta \hat{L}_z / \hbar \tag{12-32}$$

Since the atomic wavefunctions will be eigenfuctions of $L_z$, the interaction energies will be given by the corresponding quantum number $M_L$.

$$E' = H_z \mu_\beta M_L \tag{12-33}$$

This implies a splitting of energy levels, and this is illustrated for certain term symbol states in Fig. 12.3. The selection rules for unpolarized electromagnetic radiation are $\Delta M_L = -1$, 0, and 1. So, for example, a ${}^1S_0 \leftrightarrow {}^1P_1$ transition will be seen as one line in the absence of a magnetic field, but will be split into three lines if a field is applied. The Zeeman effect, therefore, directly probes the multiplicity of a level.

If an atom possesses a net electron spin, then the magnetic moment is proportional to $\vec{J}$, which is the vector sum of the orbital and spin angular momenta. This means that for the interaction energy of Eq. (12-33), we must use $M_J$ in place of $M_L$. In addition, we must introduce an extra

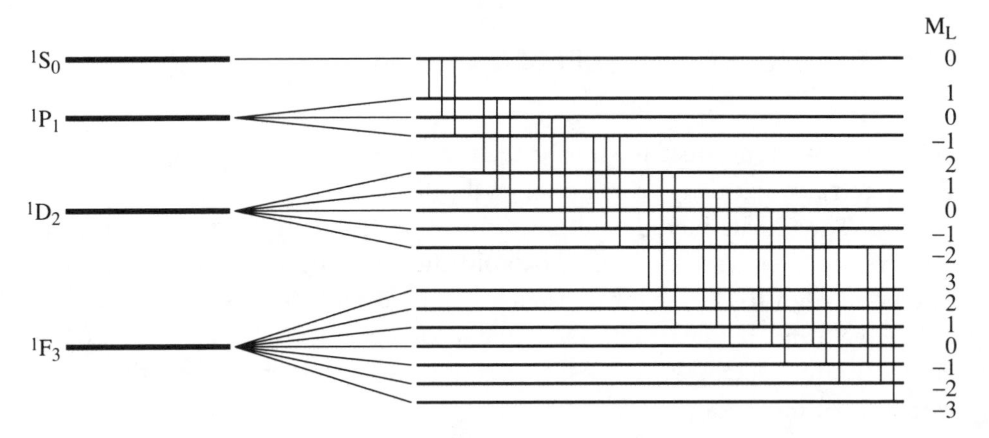


**FIGURE 12.3** Zeeman splitting of $^1S_0$, $^1P_1$, $^1D_2$, and $^1F_3$ atomic term states showing that the degeneracy is removed by a magnetic field. Vertical lines connect levels that differ in $M_L$ by 1, 0 or −1 and correspond to transitions with unpolarized radiation.

constant in the proportionality relation between the magnetic moment and the angular momentum vector. This is because there was a difference in the proportionality relation for spin and orbital angular momentum; for spin, the dimensionless quantity $g_e$ was required. The additional factor will depend on how the spin and angular momentum are coupled together. In other words, it will be dependent on the values of the three quantum numbers L, S, and J. This factor is usually designated $g_J$ and is called the *Lande g factor*. The energy of interaction, in this case from first-order perturbation theory, now becomes

$$E' = H_z g_J \mu_\beta M_J \tag{12-34}$$

where $g_J$ is given by the following expression.

$$g_J = 1 + \frac{J(J+1) + S(S+1) - L(L+1)}{2J(J+1)} (g_e - 1) \tag{12-35}$$

Notice that $g_e - 1 \cong 1$. The interaction energy of Eq. (12-34) implies different splittings for different term symbol states.

The rotation of a molecule gives rise to a magnetic moment, even in the absence of a net electronic angular momentum. Since a molecule is a distribution of negative and positive electrical charges, rotation of the molecule about its mass center amounts to a circulation of charge. That is the origin of the rotational magnetic moment, and it is also points to the fact that the moment is proportional to the rotational angular momentum vector, $\vec{J}$. In this case, the essential proportionality constant is called the *molecular g value*.

$$\vec{\mu} = g \mu_o \vec{J} / \hbar \tag{12-36}$$

$\mu_o$ is the nuclear magneton. We may use this in the general interaction Hamiltonian of Eq. (12-31). Then, for linear molecules and with the field applied in the z-direction, the interaction energy is

$$E' = -g \mu_o H_z M_J \tag{12-37}$$

The energies of a J rotational level are split by a magnetic field according to the quantum number $M_J$. Notice the difference with the Stark effect in rotational spectroscopy where the splitting depended on $|M_J|$ and not $M_J$.

For typical laboratory fields (e.g., 10,000 G), the energy splittings of Eq. (12-37) are small. In effect, molecular g values tend to be relatively small. Often contributing a comparable energy are terms that are quadratic, rather than linear, in the field strength, $H_z$. These terms, which we shall not consider in detail, arise because the magnetic field affects the electronic structure of the molecule. It slightly changes the electron distribution. So, the magnetic moment of the rotating molecule actually has a component that is linear in $H_z$, and that means the interaction energy may have a quadratic (and higher order) dependence. The molecular property identified with this response to a magnetic field is the magnetic susceptibility or magnetizability, $\chi$. Zeeman microwave spectroscopy is often used to obtain information about $\chi$ for small molecules.

## 12.7 WEAK ATTRACTIONS AT LONG RANGE *

Molecules and atoms tend to be attractive at separations that are longer than those of usual chemical bonds. These long-range attractions are not associated with chemical bonds between the interacting species and are instead a consequence of intrinsic response properties of the species. Weak attractions are extremely important in chemistry since they are reponsible for the existence of condensed phases; they are the "weak glue" that is responsible for molecules being held in a liquid or solid state. Chemical phenomena that occur in liquids and solids, including chemical reactions, are often dependent on weak attractive forces, and this seems to be especially true for biomolecular processes where small energetic preferences may control a reaction mechanism.

The form of a weak interaction potential between two molecules is often qualitatively similar to the bond stretching potential of a diatomic molecule. An example is shown in Fig. 12.4. The most important difference between the two types of potentials is in the depth of the potential well. For chemical bonds, the well depth may be on the order of 100 kcal/mol (35,000 $cm^{-1}$ or 4 eV), but for weak "bonds" it may be near zero and up to 10 kcal/mol or so. The weak attraction between two argon atoms, for instance, develops a well depth of about 100 $cm^{-1}$ (0.3 kcal/mol). Furthermore, certain approach orientations may give rise to a potential with no minimum whatsoever, and an example of such a *strictly repulsive* potential appears in Fig. 12.4.

Weak interaction potentials are said to be shallow relative to the potentials for chemical bonds, and this is because of the difference in well depths. Shallowness also implies that the potentials do not change as sharply with the separation coordinate. So, simple functional forms often provide good representations of the true potential. A widely used form for the radial potential between two weakly interacting species is the Lennard-Jones potential.

$$V(r) = D_e \left( \frac{2\, r_{min}^6}{r^6} - \frac{r_{min}^{12}}{r^{12}} \right) \tag{12-38}$$

This is referred to as a "6-12" potential because the dependence on the separation coordinate, r, varies with $r^{-6}$ and $r^{-12}$. $D_e$ is the well depth (a negative value) and $r_{min}$ is the value of r at the potential minimum. The $r^{-6}$ term is attractive because as r becomes smaller, it becomes a larger factor multiplying $D_e$. The $r^{-12}$ term is repulsive because it is of opposite sign. For $r < r_{min}$, the $r^{-12}$ term dominates and the potential function turns upward. That values for $D_e$ and $r_{min}$ can be found such that this potential is a good representation of the true potential function for two argon atoms

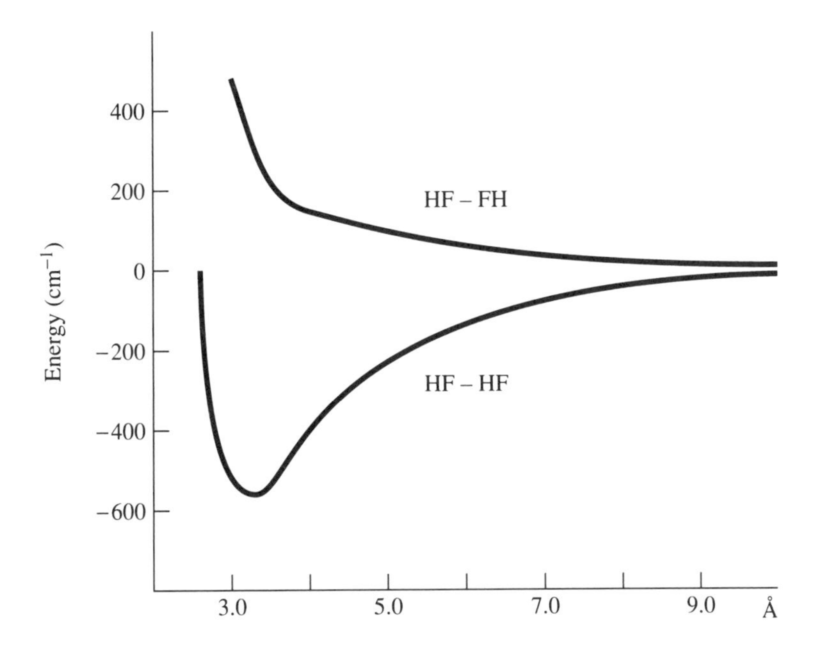


**FIGURE 12.4** Two weak interaction potential curves for the dimer HF-HF. The horizontal axis is the fluorine-fluorine separation distance, and the vertical axis is the energy of the system relative to the energy when the molecules are infinitely far apart. (These potentials are based on a model of weak interaction and are expected to be rather close to the true potentials.) The lower curve corresponds to an approach of the two HF molecules where they are colinear and arranged as HF-HF, whereas the upper curve corresponds to an HF-FH colinear arrangement. The upper curve is strictly repulsive because there is no potential well.

or two neon atoms is important, though it does not tell us the basis for the weak interaction.

Other functional forms are commonly employed to represent weak interaction potentials. One of these amounts to a generalization of Eq. (12-38), with the potential function being a sum of several inverse power terms.

$$V(r) = \sum_{n=3} \frac{C_{2n}}{r^{2n}} = \frac{C_6}{r^6} + \frac{C_8}{r^8} + \frac{C_{10}}{r^{10}} + \frac{C_{12}}{r^{12}} + \ldots \quad (12\text{-}39)$$

In practice, this series is truncated at some point, that point often being after the $C_{12}$ term. A difficulty with inverse power series expansions is that the potentials tend to rise too slowly at close-in separation distances. Thus, another functional form often used consists of a term that is exponential in $-r$ with a term that varies with $r^{-6}$. The exponential term is repulsive and rises sharply, whereas the $r^{-6}$ term provides an attractive contribution. Of course, to use any of these functional forms to represent an interaction potential, we must have a basis for choosing the values of the parameters, such as $C_6$, $C_8$, etc., in Eq. (12-39). In principle, it should be possible to relate the parameters and the functional forms to the fundamental nature of the interaction.

At long-range separations, those greater than several Å, the fundamental nature of the interaction is easier to recognize than at closer separations. At long range, there tends to be only a very weak interaction, and it is largely that of the permanent charge fields of molecules. Other features will have diminished considerably at long range. The charge field interaction may be analyzed via classical electrostatics since it amounts to the electrical interaction energy of two charge distributions (i.e., the molecules) that are well-separated in space. This is where the intrinsic properties of the molecules may be related to the interaction, because the permanent moments of the charge distributions will yield the electrical interaction energy. From classical electrostatics, it may be derived that the interaction of two separated charge distributions, A and B, is

$$E_{int} = \vec{M}^{(A)\dagger} \mathbf{T} \vec{M}^{(B)} \quad (12\text{-}40)$$

where $\vec{M}^{(A)}$ and $\vec{M}^{(B)}$ are the arrays of permanent moment elements of Eq. (12-5) arranged as in Eq. (12-6):

$$\vec{M}^{(A)\dagger} \equiv \left( M_0^{(A)} \; M_x^{(A)} \; M_y^{(A)} \; M_z^{(A)} \; M_{xx}^{(A)} \; M_{xy}^{(A)} \; \ldots \right)$$

**T** is a tensor that is dependent on the vector, $\vec{r}$, that goes from the moment expansion center of A to the moment expansion center of B. Thus, the interaction energy of Eq. (12-40) is evaluated as a row vector times a matrix times a column vector. In practice, the list of moments must be truncated at some point.

The elements of the **T** tensor are derivatives of the position vector $\vec{r}$. Specifically, they come about from applying $(-\nabla)^n$ to $|\vec{r}|^{-1}$ for n=0,1,2,3... . Using a single index designation for the unique elements, the explicit expressions are,

$$T_0 = \frac{1}{|\vec{r}|}$$

$$T_1 = -\frac{\partial}{\partial x}\frac{1}{|\vec{r}|} = \frac{x}{|\vec{r}|^3}$$

$$T_2 = -\frac{\partial}{\partial y}\frac{1}{|\vec{r}|} = \frac{y}{|\vec{r}|^3}$$

$$T_3 = -\frac{\partial}{\partial z}\frac{1}{|\vec{r}|} = \frac{z}{|\vec{r}|^3}$$

$$T_4 = \frac{\partial^2}{\partial x^2}\frac{1}{|\vec{r}|} = -\frac{1}{|\vec{r}|^3} + \frac{3x^2}{|\vec{r}|^5}$$

$$T_5 = \frac{\partial^2}{\partial x\,\partial y}\frac{1}{|\vec{r}|} = \frac{3xy}{|\vec{r}|^5}$$

$$T_6 = \frac{\partial^2}{\partial x\,\partial z}\frac{1}{|\vec{r}|} = \frac{3xz}{|\vec{r}|^5}$$

$$T_7 = \frac{\partial^2}{\partial y\,\partial x}\frac{1}{|\vec{r}|} = \frac{3xy}{|\vec{r}|^5} = T_5$$

where x, y, and z are the components of $\vec{r}$. (The numbering and ordering of the elements may be seen to be following the ordering of the elements in $\vec{M}$.) The **T** tensor is composed of these elements arranged in the following manner.

$$\mathbf{T} = \begin{pmatrix} T_0 & T_1 & T_2 & T_3 & T_4 & \cdots \\ -T_1 & -T_4 & -T_5 & -T_6 & -T_{13} & \cdots \\ -T_2 & -T_7 & -T_8 & -T_9 & -T_{22} & \cdots \\ -T_3 & -T_{10} & -T_{11} & -T_{12} & -T_{31} & \cdots \\ T_4 & T_{13} & T_{14} & T_{15} & T_{40} & \cdots \\ & \cdots & & & & \end{pmatrix} \tag{12-41}$$

From this general form, we may obtain the net distance dependence for various types of interactions. For instance, when molecule A has a nonzero value of $M_0$, the total charge, and molecule B has a nonzero dipole moment element $M_x$, then Eq. (12-40) indicates that the product of these elements and $T_1$ will be a term in the interaction energy. $T_1$ has units of (distance)$^{-2}$, and so we may say that a charge-dipole interaction varies with the distance to the power of –2. Likewise, a dipole-dipole interaction varies as the distance to the power of –3. Clearly, the higher multipole moment interactions fall off more quickly with distance, and so at long range, the truncation to low orders of the multipoles is appropriate. We also see from this analysis that the expansions of interaction potentials [e.g., Eq. (12-39)] as inverse powers of r offers an appropriate form at long range provided that the expansion includes inverse r terms that represent the dominant multipole interactions.

Another electrical interaction is polarization. The charge fields of each molecule will tend to polarize each other's charge distributions. The extent of the polarization will be dependent on the intrinsic polarizabilities of the interacting species. The classical electrostatic equations that give the energetics of polarization happen to be coupled equations because polarization is a mutual interaction. We shall not examine their form or solution; however, we may still appreciate the nature of the polarization process. As molecule A and molecule B polarize each other by their permanent charge fields, the polarization leads to a change in the charge fields. In turn, that means the polarizing effect of each on its neighbor is changed somewhat. In other words, the polarization of A by B ends up having something to do with the polarization of B; in fact, the incremental effect is sometimes called back polarization. A classical electrostatic analysis shows that the polarization element with the smallest inverse-r dependence is via the dipole polarizability, $\alpha$, and it varies with $r^{-6}$. Other polarization effects will enter as $r^{-8}$, $r^{-10}$ and so on. From a classical electrostatic analysis, we may obtain expressions for the coefficients in Eq. (12-39) directly in terms of the intrinsic electrical response properties of the molecules.

The classical analysis of the interaction of permanent charge fields and polarization is a static analysis, which means that the internal charge distributions of the interacting species are presumed to be unchanging in time. However, we know that electrons are circulating about the nuclei in atoms and molecules and that they are not fixed in space. As a result, there may be instantaneous fluctuations in the charge field or in the dipole, quadrupole, and higher moments. As a result, atoms and molecules may have weak, usually attractive interactions from simultaneous fluctations in the charge fields. This is apart from the classical electrical interaction and is usually referred to as *dispersion*. The lowest order long range form of dispersion is as $r^{-6}$. Dispersion is a quantum mechanical feature of the interaction and it can not be measured or deduced directly. However, we can appreciate its role by realizing that for the interaction of two rare gas atoms, such as Ar-Ar, there is no classical electrical interaction because the spherical symmetry of the atomic charge clouds yields no charge field; all the permanent moments of an atom are zero. The attraction between two argon atoms, then, must be almost entirely the result of dispersion. The well depth at the minimum of the Ar-Ar potential is about 100 $cm^{-1}$, and that is characteristic of the size of effects usually attributed to dispersion.

## Exercises

1. Show that the components of the second moment of three-particle charge distribution *a* are invariant to the location of the multipole center, but that those of *b* are not. (Assume that the distances between adjacent particles are all the same.)

*a.* ●$^{+e}$ O$^{-2e}$ ●$^{+e}$

*b.* 

2. Evaluate the first, second, and third moments of the following distributions of fixed point charges:

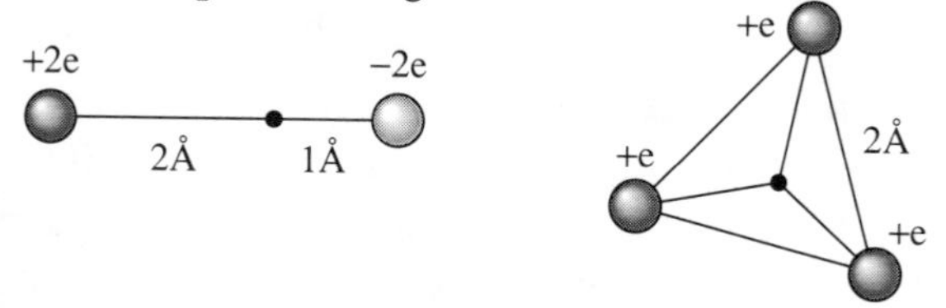


3. For the hydrogen atom, evaluate the integral $< \psi_{1s} \mid r \cos\theta \; \psi_{2p_z} >$

   and then find, as a function of $V_z$, a second-order correction to the energy of the ground state for the perturbation in Eq. (12-12). Consider the zero-order functions to be just the two functions $\psi_{1s}$ and $\psi_{2p_z}$. Find the second derivative of this energy correction with respect to $V_z$ and thereby determine a value for the dipole polarizability of the hydrogen atom. Explain why this is only an approximate value for $\alpha$.
4. Verify by explicit integration that the first-order Stark effect for linear molecules, as given in Eq. (12-14), is zero.
5. Use Eq. (12-18) to predict the effect on the rotational spectrum of a linear molecule experiencing an electric field for the transitions from $J = 1$ to $J = 2$.
6. In terms of the rotational constant, B, what is the separation between the highest frequency line in the O branch of a Raman band and the lowest frequency line in the S branch? How does this compare with the frequency spacing between the P and R branches of an infrared absorption spectrum?
7. Classify the vibrational modes of acetylene as infrared-active or infrared-inactive and as Raman-active or Raman-inactive.
8. In terms of the strength of a magnetic field in the z-direction, $H_z$, determine the splittings of the states of the carbon attom associated with the occupancy $1s^2\,2s^2\,2p^2$.
9. Find the Lande g factors for the term symbols $^3S_1$, $^2D_{3/2}$, and $^4F_{5/2}$.
10. Determine the power of the dependence on separation distance for dipole-quadrupole interactions.

## Bibliography

1. W. H. Flygare, *Molecular Structure and Dynamics* (Prentice-Hall, Englewood Cliffs, New Jersey, 1978).
2. G. W. King, *Spectroscopy and Molecular Structure* (Holt, Rinehart and Winston, New York, 1964).
3. P. W. Atkins, *Molecular Quantum Mechanics,* 2nd ed. (Oxford University Press, New York, 1983).
4. C. E. Dykstra, *Ab Initio Calculation of the Structures and Properties of Molecules* (Elsevier, Amsterdam, 1988).
5. W. Kauzman, *Quantum Chemistry* (Academic Press, New York, 1957).

# Appendix I: Matrix Algebra

Matrices are ordered arrays of constants, variables, functions, or almost anything. A vector in a Cartesian coordinate system is a very simple ordered array; it consists of three elements which are the x-, y-, and z-components of the vector. The order of the elements tells which is which. In the vector (a,b,c), the x-component is a and not b or c. The ordering is established whether one uses a column or a row arrangement of the elements.

$$\text{column: } \begin{pmatrix} a \\ b \\ c \end{pmatrix} \qquad \text{row: } (a,b,c)$$

Of course, there may be more than three elements in certain cases, and so it is convenient to subscript the elements with a number that gives the position in the row or column, e.g., $(a_1, a_2, a_3, \ldots, a_n)$. A single column of elements is termed a *column matrix*. A letter with an arrow is the designation used herein for a column matrix (or vector). Usually the same letter is used for the individual elements, but they are subscripted:

$$\vec{a} = \begin{pmatrix} a_1 \\ a_2 \\ a_3 \\ a_4 \\ \cdots \\ a_n \end{pmatrix} \qquad \text{(I-1)}$$

A row array of elements is a *row matrix*. Such a matrix will be designated in the same way as a column matrix, except that there will be a T (for transpose) as a superscript.

$$\vec{a}^T = \begin{pmatrix} a_1 & a_2 & a_3 & a_4 & \dots & a_n \end{pmatrix} \tag{I-2}$$

Other sets of elements may be ordered by both rows and columns, and these are *square* or *rectangular matrices*. These arrays are designated herein with boldface letters (usually capitals), and the elements are subscripted with a row-column index, that is, with two integers that give the row and column position in the array.

$$\mathbf{A} = \begin{pmatrix} A_{11} & A_{12} & A_{13} & \dots & A_{1n} \\ A_{21} & A_{22} & A_{23} & \dots & A_{2n} \\ A_{31} & A_{32} & A_{33} & \dots & A_{3n} \\ & & \dots & & \\ A_{m1} & A_{m2} & A_{m3} & \dots & A_{mn} \end{pmatrix} \tag{I-3}$$

This is an m-by-n matrix because there are m rows and n columns. Matrices can also be defined with three indices, or more.

Matrix addition is defined as the addition of corresponding elements of two matrices. The following two examples illustrate this definition.

$$\begin{pmatrix} 1 \\ 3 \\ 7 \\ 0 \end{pmatrix} + \begin{pmatrix} -1 \\ 0 \\ 10 \\ 2 \end{pmatrix} = \begin{pmatrix} 0 \\ 3 \\ 17 \\ 2 \end{pmatrix}$$

$$\begin{pmatrix} 0 & 3 & -i \\ 2 & -1 & 5 \end{pmatrix} + \begin{pmatrix} 0.5 & 1 & 1 \\ -2 & 0 & 5 \end{pmatrix} = \begin{pmatrix} 0.5 & 4 & 1-i \\ 0 & -1 & 10 \end{pmatrix}$$

The zero matrix is a matrix whose elements are all zero, and it is designated herein as **0** (bold zero).

Multiplication of a matrix by a scalar (a single value) is defined as the multiplication of every element in the matrix by that scalar. That is,

$$c\,\mathbf{A} = c \begin{pmatrix} A_{11} & A_{12} & A_{13} \\ A_{21} & A_{22} & A_{23} \\ A_{31} & A_{32} & A_{33} \end{pmatrix} = \begin{pmatrix} cA_{11} & cA_{12} & cA_{13} \\ cA_{21} & cA_{22} & cA_{23} \\ cA_{31} & cA_{32} & cA_{33} \end{pmatrix}$$

Multiplication of a matrix by another matrix goes by the "row-into-column" procedure. If **C** = **A B** , then the elements of **C** are given by

$$C_{ij} = \sum_{k=1}^{n} A_{ik} B_{kj} \tag{I-4}$$

where n is the number of columns of **A** *and* the number of rows of **B**. Recall that a vector dot product is,

$$(a_1\ a_2\ a_3\ \ldots\ a_n)\begin{pmatrix} b_1 \\ b_2 \\ b_3 \\ \ldots \\ b_n \end{pmatrix} = a_1b_1 + a_2b_2 + a_3b_3 + \ldots + a_nb_n$$

Thus, Eq. (I-4) says that the i-j element of **C** is obtained by a vector dot product of the $i^{th}$ row of **A** and the $j^{th}$ column of **B**:

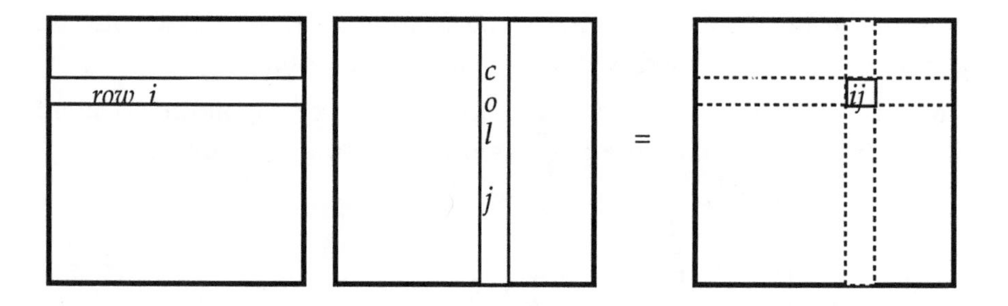


There are several features of matrix multiplication that may not be expected. First, it is not commutative. In general **A B** ≠ **B A**. Second, it is possible for the product of two matrices to be the zero matrix even if both have nonzero elements.

The identity for multiplication of square matrices is a matrix designated here as **1** (bold one). It is a matrix whose elements are all zero except for elements along the diagonal; these are one.

$$\mathbf{1} = \begin{pmatrix} 1\ 0\ 0\ \ldots \\ 0\ 1\ 0\ \ldots \\ 0\ 0\ 1\ \ldots \\ \ldots \end{pmatrix} \tag{I-5}$$

It is easy to show for some arbitrary matrix **A**, that **A 1** = **1 A** = **A**.

The *transpose of a matrix* is formed by interchanging rows with columns. A transpose is designated here with a superscript T.

$$\mathbf{A} = \begin{pmatrix} 1 & 2 & 3 & 4 \\ 5 & 6 & 7 & 8 \\ 9 & 10 & 11 & 12 \\ 13 & 14 & 15 & 16 \end{pmatrix} \quad \Rightarrow \quad \mathbf{A}^{T} = \begin{pmatrix} 1 & 5 & 9 & 13 \\ 2 & 6 & 10 & 14 \\ 3 & 7 & 11 & 15 \\ 4 & 8 & 12 & 16 \end{pmatrix}$$

The complex conjugate of a matrix is the matrix of complex conjugates of the elements of the matrix.

$$\mathbf{B} = \begin{pmatrix} b_{11} & b_{12} & b_{13} \\ b_{21} & b_{22} & b_{23} \\ b_{31} & b_{32} & b_{33} \end{pmatrix} \quad \Rightarrow \quad \mathbf{B}^{*} = \begin{pmatrix} b^{*}_{11} & b^{*}_{12} & b^{*}_{13} \\ b^{*}_{21} & b^{*}_{22} & b^{*}_{23} \\ b^{*}_{31} & b^{*}_{32} & b^{*}_{33} \end{pmatrix}$$

The complex conjugate transpose of a matrix is called the adjoint and is designated with a superscript †. Thus, $\mathbf{A}^{\dagger} = \mathbf{A}^{T*}$.

The inverse of some matrix $\mathbf{C}$ is $\mathbf{C}^{-1}$ if the following is true:

$$\mathbf{C}\mathbf{C}^{-1} = \mathbf{C}^{-1}\mathbf{C} = \mathbf{1}$$

$\mathbf{C}$ and $\mathbf{C}^{-1}$ are inverses of each other.

A *determinant* is a scalar quantity that is expressed in terms of $n^2$ elements where n is the order of the determinant. The elements are arranged in n rows and n columns and placed between vertical bars. Its appearance is similar to a square matrix, but it is something entirely different since it represents a single value. That value is obtained from the $n^2$ elements. Carrying out the evaluation involves using the *minor* of an element of a determinant. To see how this is done, let $d_{ij}$ be an arbitrary element of the determinant $\Phi$:

$$\Phi = \begin{vmatrix} d_{11} & d_{12} & d_{13} & \cdots & d_{1n} \\ d_{21} & d_{22} & d_{23} & \cdots & d_{2n} \\ \cdot & \cdot & \cdot & \cdots & \cdot \\ d_{n1} & d_{n2} & d_{n3} & \cdots & d_{nn} \end{vmatrix}$$

The minor of the element $d_{ij}$ is the determinant that remains after deleting the $i^{th}$ row and $j^{th}$ column from $\Phi$. We shall designate that determinant $D^{ij}$. The value of the determinant $\Phi$ is the sum of the products of elements from any row, or from any column, and their minors with a particular factor of –1:

$$\Phi = \sum_{i=1}^{n} (-1)^{i+j} d_{ji} D^{ji} \quad \text{for } j = 1, 2, \ldots, n$$

$$\Phi = \sum_{i=1}^{n} (-1)^{i+j} d_{ij} D^{ij} \quad \text{for } j = 1, 2, \ldots, n \tag{I-6}$$

Of course, finding values for each of the $D^{ij}$ determinants also requires using one of these equations. The process reaches a conclusion, though, when the minor has one only one element; the value of that determinant (the minor) is that element.

To illustrate the evaluation process, notice that a second order determinant is evaluated from Eq. (I-7) in the following way:

$$\begin{vmatrix} e_{11} & e_{12} \\ e_{21} & e_{22} \end{vmatrix} = (-1)^{1+1} e_{11} E^{11} + (-1)^{1+2} e_{12} E^{12}$$

$$= e_{11} e_{22} - e_{12} e_{21}$$

This is helpful because applying Eq. (I-7) to a third-order determinant (one with three rows and three columns) would give a sum with minors that are second-order determinants; this result can be used in working out the value for a third-order determinant. A fourth-order determinant could be evaluated with minors that are third order, and in such a stepwise fashion, a determinant of any order can be evaluated with Eq. (I-7). When carried out fully, the value of a determinant of order n will be a sum of n! products of its elements.

# Appendix II: Matrix Eigenvalue Methods

There is a linear algebra equivalent of the differential eigenvalue equation. Instead of a differential operator acting on a function, a square matrix multiplies a column vector, and this is equal to a constant, the eigenvalue, times the vector:

$$\mathbf{A}\vec{c} = e\vec{c} \qquad \text{(II-1)}$$

**A** is a square matrix, $\vec{c}$ is the *eigenvector*, and e is the eigenvalue. For an N-by-N matrix, there are N different eigenvalues and N different eigenvectors. That is, there are N different solutions to Eq. (II-1). Thus, the equation is better written with subscripts that distinguish the different solutions:

$$\mathbf{A}\vec{c}_i = e_i\vec{c}_i \qquad \text{(II-2)}$$

Because i = 1, ..., N, Eq. (II-2) does represent the N different equations. However, from the rules for matrix multiplication, it is easy to realize that these equations may be collectively represented by one matrix equation. This is accomplished by collecting the column eigenvectors into one N-by-N matrix.

$$\mathbf{A}\left(\vec{c}_1\ \vec{c}_2\ \vec{c}_3\ \dots\ \vec{c}_N\right) = \left(\vec{c}_1\ \vec{c}_2\ \vec{c}_3\ \dots\ \vec{c}_N\right)\begin{pmatrix} e_1 & 0 & 0 & \dots & 0 \\ 0 & e_2 & 0 & \dots & 0 \\ 0 & 0 & e_3 & \dots & 0 \\ & & & \dots & \\ 0 & 0 & 0 & \dots & e_N \end{pmatrix} \tag{II-3}$$

With the different column vectors set next to each other, we obtain a square array, which will be designated **C**. The rightmost matrix in Eq. (II-3), which will be called **E**, is a *diagonal matrix* since its only nonzero elements are along the diagonal. Because of its diagonal form, matrix multiplication of **E** by **C** just leads to each column of **C** scaled by a corresponding diagonal element of **E**. From this, we have the form of the general matrix eigenvalue equation,

$$\mathbf{A}\,\mathbf{C} = \mathbf{C}\,\mathbf{E} \quad \text{where } \mathbf{E} \text{ is diagonal} \tag{II-4}$$

This equation is solved for some specific matrix **A**, and the solutions are the eigenvectors arranged as columns in **C**, plus the associated eigenvalues arranged on the diagonal of **E**.

It is quite often the case that a matrix whose eigenvalues are sought is real and symmetric, and we will only consider such matrices here. If the matrix **C** in Eq. (II-4) were known, and if its inverse could be found, then multiplication of Eq. (II-4) by the inverse would lead to,

$$\begin{aligned} \mathbf{C}^{-1}\mathbf{A}\,\mathbf{C} &= \mathbf{C}^{-1}\mathbf{C}\,\mathbf{E} \\ &= \mathbf{E} \end{aligned} \tag{II-5}$$

The process of multiplying some matrix on the right by a particular matrix, and multiplying on the left by the inverse of that matrix, is called a *transformation*. A matrix equation remains true upon transformation of each of the matrices in the equation. For example,

$$\text{if} \quad \mathbf{R} + \mathbf{S}\,\mathbf{T} = \mathbf{X}$$

$$\text{then} \quad (\mathbf{C}^{-1}\mathbf{R}\,\mathbf{C}) + (\mathbf{C}^{-1}\mathbf{S}\,\mathbf{C})(\mathbf{C}^{-1}\mathbf{T}\,\mathbf{C}) = (\mathbf{C}^{-1}\mathbf{X}\,\mathbf{C})$$

Notice that if we used a prime to designate a matrix transformed by **C**, e.g.,

$$\mathbf{R'} = \mathbf{C}^{-1}\mathbf{R}\,\mathbf{C}$$

then the example of a transformed matrix equation becomes simply

$$\mathbf{R'} + \mathbf{S'}\,\mathbf{T'} = \mathbf{X'}$$

This has exactly the same form as the original equation, and so "transformation" seems to be appropriate terminology for what has happened. The matrices have been changed by the transformation, but the matrix equation is unchanged in form.

The general matrix eigenequation, expressed in the form of Eq. (II-5), can now be thought of differently: We seek a matrix **C** that transforms the matrix **A** in a very particular way such that the transformed matrix is of diagonal form. The process to accomplish this is referred to simply as *diagonalization* of **A**. If we seek the eigenvalues and eigenvectors of some matrix, then we seek to diagonalize it.

An important category of transformation matrices is that of *unitary matrices*. These are employed for diagonalizing real, symmetric matrices, and we will use **U** to designate a unitary matrix in the following discussion. A unitary matrix is one whose inverse is the same as or equal to its transpose. Or, if the elements of **U** are complex, then unitarity means the inverse is equal to the adjoint:

$$\mathbf{U}^{-1} = \mathbf{U}^{\dagger}$$

The property of unitarity means that a transformation can be written with either the transpose (or adjoint) or with the inverse.

$$\mathbf{U}^{-1}\mathbf{A}\,\mathbf{U} = \mathbf{U}^{T}\mathbf{A}\,\mathbf{U}$$

Another property of a unitary matrix is that the value of the determinant constructed from its elements is equal to one; that is, det(**U**) = 1.

There are a number of ways of finding unitary matrices that diagonalize real, symmetric matrices. A very useful recipe for 2-by-2 matrices uses sines and cosines. Starting with **A** as a matrix we seek to diagonalize

$$\mathbf{A} = \begin{pmatrix} A_{11} & A_{12} \\ A_{21} & A_{22} \end{pmatrix}$$

an angle θ is determined from values in **A**.

$$\tan 2\theta = \frac{2\,A_{12}}{A_{22} - A_{11}} \qquad \text{(II-6)}$$

The transformation matrix is,

$$\mathbf{U} = \begin{pmatrix} \cos\theta & \sin\theta \\ -\sin\theta & \cos\theta \end{pmatrix} \qquad \text{(II-7)}$$

The eigenvalues are

$$\mathbf{U}^T \mathbf{A}\,\mathbf{U} = \begin{pmatrix} e_1 & 0 \\ 0 & e_2 \end{pmatrix} \qquad \text{(II-8)}$$

(The recipe for θ in Eq. (II-6) may be derived by carrying out the matrix multiplication of Eq. (II-8) explicitly in terms of elements of **A** and sin θ and cos θ. This will yield expressions for each of the four elements of the product matrix. Setting the expression for the off-diagonal element to be zero, as on the right hand side of Eq. (II-8), gives the proper condition for θ.)

There is a geometrical analogy to the 2-by-2 diagonalization scheme. The **U** transformation matrix applied to a vector of x- and y-coordinates will rotate the vector about the origin by the angle θ, yielding two new coordinates.

$$\mathbf{U}\begin{pmatrix} x \\ y \end{pmatrix} = \begin{pmatrix} x' \\ y' \end{pmatrix}$$

A consequence of the unitary nature of **U** is that the length of the original vector is unchanged; that is,

$$\sqrt{x^2 + y^2} = \sqrt{x'^2 + y'^2}$$

Therefore, the transformation with **U** is a pure rotation; there is no change in length of any rotated vector. From the geometrical picture of transformations, it is common to refer to unitary transformations as *rotations*. For diagonalization problems, "rotation" is used loosely or abstractly because whatever is being rotated is whatever labels the rows and columns of the matrix being diagonalized. That is not necessarily going to be a set of directions in a geometrical space.

Diagonalization by 2-by-2 rotations is one means of diagonalizing larger problems. The idea is to extract a 2-by-2 from a large matrix and use the recipe above. To see what this involves, we must recall special features of matrix multiplication. Consider an 8-by-8 matrix, **A**. Let us extract the following 2-by-2 from it.

$$\begin{pmatrix} A_{33} & A_{36} \\ A_{63} & A_{66} \end{pmatrix}$$

And let the matrix that diagonalizes this be the following.

$$\mathbf{U} = \begin{pmatrix} a & b \\ c & d \end{pmatrix}$$

In order to apply the transformation given by **U** to the entire **A** matrix, it must be superposed on the unit matrix. This means that we replace the 3-3, 3-6, 6-3 and 6-6 elements of an 8-by-8 unit matrix (i.e., **1**) by a, b, c and d, respectively, from **U**. These replaced elements are in the very same positions as the elements extracted from **A** to make up the 2-by-2 for diagonalization. As illustrated below, this matrix multiplying an 8-by-8 square matrix will alter the values of only the elements in rows 3 and 6.

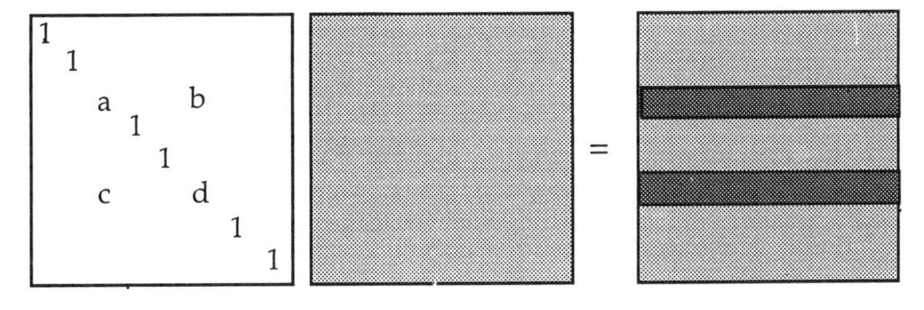

The superposed unit matrix is shown on the left (only nonzero values) multiplying another matrix (shaded). The only elements that are different after the matrix multiplication are those in the rows with a, b, c, and d, and these are indicated by the darkened rows on the rightmost matrix. Were the multiplication done in the reverse order, the elements of the third and sixth columns would be the ones affected:

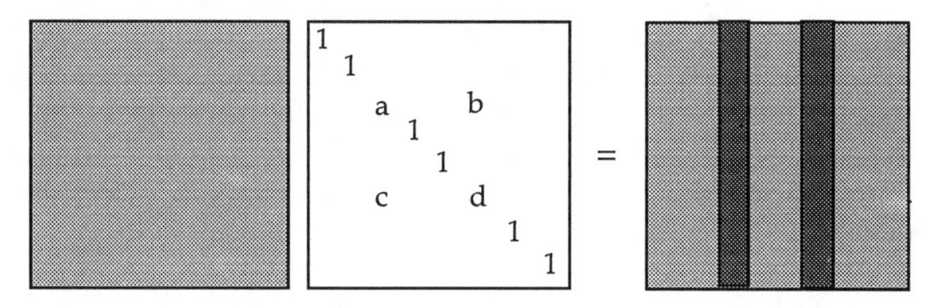


Thus, when the 2-by-2 **U** matrix is properly superposed on the unit matrix and used to transform the original **A** matrix, the result will be the following.

*Transformed* **A** *matrix:*

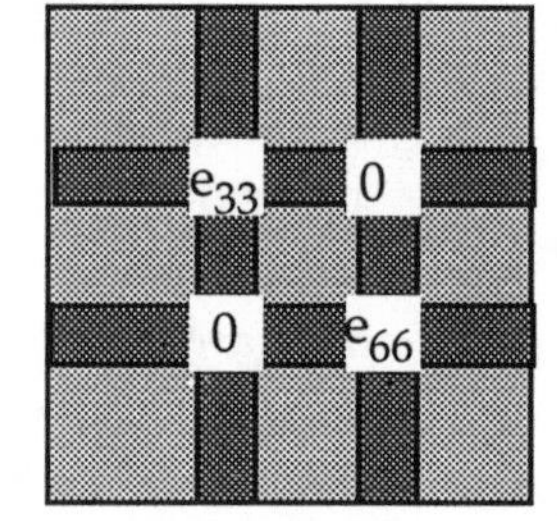


The darkened rows and columns indicate elements whose values are now different from what they were in the original **A** matrix. The eigenvalues of the extracted 2-by-2 part of the **A** matrix are now on the diagonal, and corresponding off-digaonal elements are zero, as indicated. This matrix is closer to being in diagonal form.

In the next step, another 2-by-2 is extracted from **A**, and the same thing is done. A complication is that in this second 2-by-2 rotation, the 3-6 and 6-3 elements that had become zero may change. That is, they may end up being different from the desired value of zero if the second 2-by-2 rotation involves the third or the sixth row and column. However, by continuing the process again and again, it is possible for all the off-diagonal elements to be as close to zero as one wishes. Typically, it is best to select the particular 2-by-2 for each step on the basis of the largest off-diagonal value present. This is a process which is easily coded for computers, but to illustrate how it works and to show all the details, here is an example of the diagonalization of a particular 3-by-3 matrix, **B**.

$$\mathbf{B} = \begin{pmatrix} 5 & -3\sqrt{3} & 2 \\ -3\sqrt{3} & -1 & 2\sqrt{3} \\ 2 & 2\sqrt{3} & 20 \end{pmatrix}$$

Let us extract a 2-by-2 and apply Eq. (II-6) to find the necessary rotation. This 2-by-2 will be from the first two rows and columns of **B**:

$$\begin{pmatrix} 5 & -3\sqrt{3} \\ -3\sqrt{3} & -1 \end{pmatrix}$$

The rotation angle for diagonalizing this 2-by-2 is found to be 30° from the following expression.

$$\tan 2\theta = \frac{2(-3\sqrt{3})}{-1-5} = \sqrt{3}$$

The transformation matrix for the 2-by-2 is placed into a 3-by-3 matrix in the positions corresponding to the 2-by-2 that was extracted from **B**. The rest of this matrix is filled with zeroes off the diagonal and ones along the diagonal. Thus,

$$\mathbf{U} = \begin{pmatrix} \cos\theta & \sin\theta & 0 \\ -\sin\theta & \cos\theta & 0 \\ 0 & 0 & 1 \end{pmatrix} = \begin{pmatrix} \sqrt{3}/2 & 1/2 & 0 \\ -1/2 & \sqrt{3}/2 & 0 \\ 0 & 0 & 1 \end{pmatrix}$$

This is a 3-by-3 unitary transformation matrix.

Transforming **B** with **U** yields a matrix that is closer to being diagonal than the original matrix **B**.

$$\mathbf{U}^{T}\mathbf{B}\,\mathbf{U} = \begin{pmatrix} 8 & 0 & 0 \\ 0 & -4 & 4 \\ 0 & 4 & 20 \end{pmatrix}$$

The next step in the whole process is to extract another 2-by-2 to diagonalize. In this case, the 2-by-2 has to be that in the second and third rows and columns. With this second transformation matrix designated **V**, the rotation angle and the elements of **V** are

$$\tan 2\phi = \frac{2\,(4)}{20-(-4)} = \frac{1}{3} \qquad \therefore\ \phi = 9.22^{o}$$

$$\mathbf{V} = \begin{pmatrix} 1 & 0 & 0 \\ 0 & \cos\phi & \sin\phi \\ 0 & -\sin\phi & \cos\phi \end{pmatrix} = \begin{pmatrix} 1 & 0 & 0 \\ 0 & 0.9817 & 0.1602 \\ 0 & -0.1602 & 0.9817 \end{pmatrix}$$

Applying this transformation to **B**, after transformation by **U**, yields

$$\mathbf{V}^{T}(\mathbf{U}^{T}\mathbf{B}\,\mathbf{U})\,\mathbf{V} = \begin{pmatrix} 8 & 0 & 0 \\ 0 & -4.65 & 0 \\ 0 & 0 & 20.65 \end{pmatrix}$$

Therefore, the eigenvalues of the matrix **B** are –4.65, 8, and 20.65. The eigenvectors are the columns of the matrix that transformed **B** into diagonal form, and by the above expression, the eigenvectors are the columns of the matrix **UV**. The overall transformation is a product of the two rotation matrices.

**B** was chosen for this example in such a way that only two successive rotations were required to completely diagonalize the matrix. In general, the diagonalization of even a 3-by-3 matrix via 2-by-2 rotations may require many 2-by-2 rotations. This happens because a rotation that diagonalizes one 2-by-2 block might have the effect of changing an already-zero off-diagonal element to a nonzero value. This makes the process tedious if not done with a computer but otherwise presents no problem. The whole process can be applied to a matrix of any size; however, there are other algorithms that are computationally advantageous for many diagonalization problems encountered in quantum chemistry.

There is a special recipe, or method, for immediately getting just the eigenvalues of a matrix. The recipe is to subtract an unspecified parameter – call it $\lambda$ – from the diagonal elements and then to set the determinant formed from the elements of the matrix equal to zero. This generates a polynomial in the parameter $\lambda$. The roots of that polynomial are the eigenvalues of the matrix. Here is an example of finding the eigenvalues of a 3-by-3 matrix, **G**, by this scheme.

$$\mathbf{G} = \begin{pmatrix} 5 & 0 & \sqrt{3} \\ 0 & 6 & 0 \\ \sqrt{3} & 0 & 7 \end{pmatrix}$$

The determinant of the elements of this matrix, with λ subtracted along the diagonal, is set to zero.

$$\begin{vmatrix} (5-\lambda) & 0 & \sqrt{3} \\ 0 & (6-\lambda) & 0 \\ \sqrt{3} & 0 & (7-\lambda) \end{vmatrix} = 0$$

Multiplying out or evaluating this determinant yields a polynomial in λ.

$$\lambda^3 - 18\lambda^2 + 104\lambda - 192 = 0$$

The eigenvalues are the roots of the polynomial. In this example, chosen for its simplicity, the polynomial is factorable:

$$(\lambda - 6)(\lambda - 4)(\lambda - 8) = 0$$

Thus, the roots are 4, 6, and 8. These are the eigenvalues of the matrix **G**.

When the eigenvalues are obtained in this manner, the eigenvectors or the transformation matrix that diagonalizes **G** remain unknown. However, the associated eigenvector of any known eigenvalue can always be obtained by going back to the basic eigenvalue equation, Eq. (II-2), and solving it using the known eigenvalues one at a time.

The diagonalization of a complex *Hermitian matrix*, which is one for which off-diagonal elements are related as

$$H_{ij} = H^*_{ji}$$

may require a transformation matrix, **U**, that is complex. Then, the transformation is not that of Eq. (II-8), because the adjoint of **U** must be used instead of just the transpose of **U**. (The *adjoint* of a matrix is the transpose but with the complex conjugate taken of each element; i.e., adjoint(**U**) = $\mathbf{U}^{\dagger} = \mathbf{U}^{T^*}$ .) The eigenvalues of a Hermitian matrix, though, will be real numbers.

*Simultaneous Diagonalization.* A transformation may be found that simultaneously diagonalizes several matrices if they commute with each other. This will not necessarily be a unitary transformation. The problem of simultaneous diagonalization sometimes arises just as a need to transform several matrices to diagonal form. It also arises in a special matrix eigenvalue equation that is really a generalization of Eq. (II-4).

$$\mathbf{A}\,\mathbf{C} = \mathbf{S}\,\mathbf{C}\,\mathbf{E} \quad \text{where } \mathbf{E} \text{ is diagonal} \tag{II-9}$$

Multiplying by the inverse of **C** yields

$$\mathbf{C}^{-1}\mathbf{A}\,\mathbf{C} = (\mathbf{C}^{-1}\mathbf{S}\,\mathbf{C})\,\mathbf{E} \tag{II-10}$$

If the matrix product ($\mathbf{C}^{-1}\,\mathbf{S}\,\mathbf{C}$) were equal to the unit or identity matrix, **1**, then the right hand side would simplify to just **E**, and the problem would be the same as in Eq. (II-5). Notice that if this were true, the matrix **S** would in fact have been diagonalized, its diagonal form being simply that of the unit matrix.

The process of simultaneous diagonalization amounts to finding a transformation where for all but one matrix, the resulting diagonal form is that of the unit matrix. This can be done for any nonsingular matrix by a transformation that is nonunitary. Assuming **S** in Eq. (II-9) is a nonsingular matrix with positive eigenvalues, we can find the unitary transformation that makes it diagonal in the usual way.

$$\mathbf{U}^{-1}\,\mathbf{S}\,\mathbf{U} = \mathbf{D}$$

where **D** is diagonal and **U** is unitary. Using the elements of **D**, a matrix designated **d** is constructed as

$$\mathbf{d} = \begin{pmatrix} 1/\sqrt{D_{11}} & 0 & 0 & \cdots \\ 0 & 1/\sqrt{D_{22}} & 0 & \cdots \\ 0 & 0 & 1/\sqrt{D_{33}} & \cdots \\ & & \cdots & \end{pmatrix} \tag{II-11}$$

This matrix is a diagonal matrix, its elements being the inverse square roots of the corresponding elements of **D**. With this definition, it is easy to see that

$$\mathbf{d D d} = \mathbf{1} = \begin{pmatrix} 1 & 0 & 0 & \ldots \\ 0 & 1 & 0 & \ldots \\ 0 & 0 & 1 & \ldots \\ & \ldots & & \end{pmatrix}$$

This special matrix is equal to its transpose, and so it is not necessary to use a transpose symbol in the above expression. This matrix, though, is not a unitary transformation: If multiplied by its transpose (itself), the result is not the unit matrix. The composite or product transformation of this with **U** is the transformation that takes **S** into the unit matrix, since

$$(\mathbf{dU}^{T})\ \mathbf{S}\ (\mathbf{U\,d}) = \mathbf{1} \qquad \text{(II-12)}$$

If this expression is further transformed with some matrix, **V**, the right hand side will remain as **1** if **V** is unitary. Because $\mathbf{V}^{-1}\,\mathbf{V} = \mathbf{1}$, the transformation will leave the diagonalized **S** matrix as **1**.

The process for solving Eq. (II-9) may now be seen as a three step process:

(1) Find a unitary matrix, **U**, that diagonalizes **S**,

(2) Set up the nonunitary transformation of Eq. (II-11) from the elements of **D**, and

(3) Find another unitary matrix, **V**, such that the composite transformation of **A** takes it into diagonal form.

This process says that the matrix **C** that solves Eq. (II-9) may be constructed as a product of three matrices.

$$\mathbf{C} = \mathbf{U\,d\,V}$$

Susbstituting this into Eq. (II-10) and then using the result of Eq. (II-12),

$$(\mathbf{V}^{T}\,\mathbf{d}\,\mathbf{U}^{T})\,\mathbf{A}\,(\mathbf{U\,d\,V}) = (\mathbf{V}^{T}\,\mathbf{d}\,\mathbf{U}^{T})\,\mathbf{S}\,(\mathbf{U\,d\,V})\ \mathbf{E}$$

$$\mathbf{V}^{T}\,[\mathbf{d}\,\mathbf{U}^{T}\,\mathbf{A\,U\,d}]\,\mathbf{V} = \mathbf{E}$$

The matrix inside the square brackets is the matrix that **V** must diagonalize. Following this three-step process, the eigenvalues of **A** are along the diagonal of **E**, and the matrix **S** is simultaneously diagonal with its eigenvalues all 1.0.

# Appendix III: Table of Integrals

1. General integral relations

$$\int c\, f(x)\, dx = c \int f(x)\, dx$$

$$\int \left( f_1(x) + f_2(x) + \ldots \right) dx = \int f_1(x)\, dx + \int f_2(x)\, dx + \ldots$$

$$\int f(x)\, g(y)\, dx\, dy = \int f(x)\, dx \int g(y)\, dy$$

$$\int \left( f_1(x) + f_2(x) \right)\left( g_1(x) + g_2(x) \right) dx = \int f_1(x)\, g_1(x)\, dx$$

$$+ \int f_1(x)\, g_2(x)\, dx + \int f_2(x)\, g_1(x)\, dx + \int f_2(x)\, g_2(x)\, dx$$

2. Integrals over Gaussian functions ( $c > 0$ )

$$\int_0^\infty e^{-c x^2}\, dx = \frac{1}{2} \sqrt{\frac{\pi}{c}}$$

$$\int_0^\infty x\, e^{-c x^2}\, dx = \frac{1}{2c}$$

$$\int_0^\infty x^2 e^{-cx^2} dx = \frac{1}{4}\sqrt{\frac{\pi}{c^3}}$$

$$\int_0^\infty x^{2n+1} e^{-cx^2} dx = \frac{n!}{2c^{n+1}}$$

$$\int_0^\infty x^{2n} e^{-cx^2} dx = \frac{(2n-1)(2n-3)\dots(3)(1)}{2^{n+1}}\sqrt{\frac{\pi}{c^{2n+1}}}$$

$$\int_{-\infty}^0 x^n e^{-cx^2} dx = (-1)^n \int_0^\infty x^n e^{-cx^2} dx$$

2. Integrals of exp(x) ( $c > 0$ )

$$\int x^n e^{cx} dx = e^{cx} \sum_{j=0}^{n} (-1)^j \frac{n!}{(n-j)!} \frac{x^{n-j}}{c^{j+1}}$$

$$\int_0^\infty x^n e^{-cx} dx = \frac{n!}{c^{n+1}} \qquad (\text{for } n = 1, 2, 3, \dots)$$

3. Integrals over trigonometric functions

$$\int (\sin^2 cx)\, dx = \frac{x}{2} - \frac{\sin 2cx}{4c}$$

$$\int (\sin^3 cx)\, dx = -\frac{(\cos cx)(\sin^2 cx + 2)}{3c}$$

$$\int (\sin^n cx)\, dx = -\frac{(\sin^{n-1} cx)(\cos cx)}{nc} + \frac{n-1}{n}\int (\sin^{n-2} cx)\, dx$$

$$\int c\, x^n \sin(cx)\, dx = -x^n \cos(cx) + n\int x^{n-1} \cos(cx)\, dx$$

$$\int c\, x^n \cos(cx)\, dx = x^n \sin(cx) - n\int x^{n-1} \sin(cx)\, dx$$

$$\int x\,(\sin^2 cx)\, dx = \frac{x^2}{4} - \frac{x\,(\sin 2cx)}{4c} - \frac{\cos 2cx}{8c^2}$$

$$\int x^2\,(\sin^2 cx)\, dx = \frac{x^3}{6} - \left(\frac{x^2}{4c} - \frac{1}{8c^3}\right)(\sin 2cx) - \frac{x\,(\cos 2cx)}{4c^2}$$

# Appendix IV: Table of Atomic Masses and Nuclear Spins

This is a table of the isotopic masses and electron orbital occupancies of most of the non-transition elements of the periodic table. The values of the masses (in amu), the percent natural abundance of each isotope, and the nuclear spins are from the "Table of Nuclides" in *Nuclear and Radiochemistry*, 2nd ed., G. Friedlander, J. W. Kennedy, and J. M. Miller, (John Wiley and Sons, New York, 1964). Unstable isotopes and isotopes with an abundance of less than 0.1% have not all been included. Nuclear g factors [see Eq. (11-10)] are from the table "Nuclear Spins, Moments, and Magnetic Resonance Frequencies" in *Handbook of Chemistry and Physics*, 63rd ed., (CRC Press, Boca Raton, Florida, 1982).

| Element Number | Symbol | Orbital Occupancy | Isotope Mass | Nuclear Spin | $g_N$ | Percent Abund. |
|---|---|---|---|---|---|---|
| 1 | H | $1s^1$ | 1.007825 | 1/2 | 2.79268 | 99.985 |
| | | | 2.014102 | 1 | 0.85739 | 0.015 |
| | | | 3.016049 | 1/2 | 2.97877 | |
| 2 | He | $1s^2$ | 4.002604 | 0 | | 100 |

| Element Number | Symbol | Orbital Occupancy | Isotope Mass | Nuclear Spin | $g_N$ | Percent Abund. |
|---|---|---|---|---|---|---|
| 3 | Li | [He] $2s^1$ | 6.015126 | 1 | 0.8219 | 7.42 |
| | | | 7.016005 | 3/2 | 3.2560 | 92.58 |
| 4 | Be | [He] $2s^2$ | 9.012186 | 3/2 | −1.1774 | 100 |
| 5 | B | [He] $2s^2\,2p^1$ | 10.012939 | 3 | 1.8007 | 19.6 |
| | | | 11.009305 | 3/2 | 2.6880 | 80.4 |
| 6 | C | [He] $2s^2\,2p^2$ | 12.000000 | 0 | | 98.89 |
| | | | 13.003354 | 1/2 | 0.7022 | 1.11 |
| 7 | N | [He] $2s^2\,2p^3$ | 14.003074 | 1 | 0.4035 | 99.63 |
| | | | 15.000108 | 1/2 | −0.2830 | 0.37 |
| 8 | O | [He] $2s^2\,2p^4$ | 15.994915 | 0 | | 99.759 |
| | | | 16.999133 | 5/2 | −1.8930 | 0.037 |
| | | | 17.999160 | 0 | | 0.204 |
| 9 | F | [He] $2s^2\,2p^5$ | 18.998405 | 1/2 | 2.6273 | 100 |
| 10 | Ne | [He] $2s^2\,2p^6$ | 19.992440 | 0 | | 90.92 |
| | | | 20.993849 | 3/2 | −0.6614 | 0.26 |
| | | | 21.991385 | 0 | | 8.82 |
| 11 | Na | [Ne] $3s^1$ | 22.989773 | 3/2 | 2.2161 | 100 |
| 12 | Mg | [Ne] $3s^2$ | 23.985045 | 0 | | 78.70 |
| | | | 24.985840 | 5/2 | −0.8545 | 10.13 |
| | | | 25.982591 | 0 | | 11.17 |
| 13 | Al | [Ne] $3s^2\,3p^1$ | 26.981535 | 5/2 | 3.6385 | 100 |
| 14 | Si | [Ne] $3s^2\,3p^2$ | 27.976927 | 0 | | 92.21 |
| | | | 28.976491 | 1/2 | −0.5548 | 4.70 |
| | | | 29.973761 | 0 | | 3.09 |
| 15 | P | [Ne] $3s^2\,3p^3$ | 30.973763 | 1/2 | 1.1305 | 100 |

| Element Number | Symbol | Orbital Occupancy | Isotope Mass | Nuclear Spin | $g_N$ | Percent Abund. |
|---|---|---|---|---|---|---|
| 16 | S | [Ne] $3s^2\,3p^4$ | 31.972074 | 0 | | 95.00 |
| | | | 32.971460 | 3/2 | 0.6426 | 0.76 |
| | | | 33.967864 | 0 | | 4.22 |
| | | | 35.967090 | 0 | | 0.014 |
| 17 | Cl | [Ne] $3s^2\,3p^5$ | 34.968854 | 3/2 | 0.8209 | 75.53 |
| | | | 36.965896 | 3/2 | 0.6833 | 24.47 |
| 18 | Ar | [Ne] $3s^2\,3p^6$ | 35.967548 | 0 | | 0.337 |
| | | | 37.962725 | 0 | | 0.063 |
| | | | 39.962384 | 0 | | 99.600 |
| 19 | K | [Ar] $4s^1$ | 38.963714 | 3/2 | 0.3910 | 93.10 |
| | | | 40.961835 | 3/2 | 0.2146 | 6.88 |
| 20 | Ca | [Ar] $4s^2$ | 39.962589 | 0 | | 96.97 |
| | | | 41.958628 | | | 0.64 |
| | | | 42.958780 | 7/2 | −1.3153 | 0.15 |
| | | | 43.955490 | | | 2.06 |
| 21 to 30 | Sc to Zn | [Ca] $3d^{1-10}$ | | | | |
| 31 | Ga | [Zn] $4p^1$ | 68.92568 | 3/2 | 2.011 | 60.4 |
| | | | 70.92484 | 3/2 | 2.5549 | 39.6 |
| 32 | Ge | [Zn] $4p^2$ | 69.92428 | 0 | | 20.52 |
| | | | 71.92174 | 0 | | 27.43 |
| | | | 72.9234 | 9/2 | −0.8768 | 7.76 |
| | | | 73.92115 | 0 | | 36.54 |
| | | | 75.9214 | 0 | | 7.76 |
| 33 | As | [Zn] $4p^3$ | 74.92158 | 3/2 | 1.4349 | 100 |

| Element Number | Symbol | Orbital Occupancy | Isotope Mass | Nuclear Spin | $g_N$ | Percent Abund. |
|---|---|---|---|---|---|---|
| 34 | Se | [Zn] $4p^4$ | 75.91923 | 0 | | 9.02 |
| | | | 76.91993 | 1/2 | 0.5325 | 7.58 |
| | | | 77.91735 | 0 | | 23.52 |
| | | | 79.91651 | 0 | | 49.82 |
| | | | 81.9167 | 0 | | 9.19 |
| 35 | Br | [Zn] $4p^5$ | 78.91835 | 3/2 | 2.0990 | 50.54 |
| | | | 80.91634 | 3/2 | 2.2626 | 49.46 |
| 36 | Kr | [Zn] $4p^6$ | 79.91639 | | | 2.27 |
| | | | 81.91348 | 0 | | 11.56 |
| | | | 82.91413 | 9/2 | –0.9671 | 11.55 |
| | | | 83.91150 | 0 | | 56.90 |
| | | | 85.91062 | 0 | | 17.37 |
| 37 | Rb | [Kr] $5s^1$ | 84.91171 | 5/2 | 1.3482 | 72.15 |
| | | | 86.90918 | 3/2 | 2.7414 | 27.85 |
| 38 | Sr | [Kr] $5s^2$ | 83.9134 | | | 0.56 |
| | | | 85.9093 | 0 | | 9.86 |
| | | | 86.9089 | 9/2 | –1.0893 | 7.02 |
| | | | 87.9056 | 0 | | 82.56 |
| 39 to 48 | Y to Cd | [Sr] $4d^{1-10}$ | | | | |
| 49 to 52 | In to Te | [Cd] $5p^{1-4}$ | | | | |
| 53 | I | [Cd] $5p^5$ | 126.90447 | 5/2 | 2.7937 | 100 |
| 54 | Xe | [Cd] $5p^6$ | 127.90353 | | | 1.92 |
| | | | 128.90478 | 1/2 | –0.7725 | 26.44 |
| | | | 129.90350 | | | 4.08 |
| | | | 130.90508 | 3/2 | 0.6870 | 21.18 |
| | | | 131.90416 | | | 26.89 |
| | | | 133.90539 | | | 10.44 |
| | | | 135.90721 | | | 8.87 |

# Appendix V: Fundamental Constants and Conversion of Units

*Systems of Units.* A system of units of measure for mechanical systems may be defined by specifying the unit of measurement for just three physical quantities. For instance, if we decide how mass, length, and time are to be measured, then there will exist a basis for measurement of velocity (i.e., unit length per unit time), momentum, acceleration, energy, and so on. We may specify units of measurement for three other physical quantities to define a system of units, but not more than three.

In the metric system, the three familiar basic units are the gram, the meter, and the second, and the symbols are g, m, and s. Prefixes are used to designate units that are smaller by powers of 10 or larger by powers of 10. The commonly encountered prefixes in chemical physics are the following.

| *A metric prefix of (symbol):* | *Means the prefixed unit is scaled by:* |
|---|---|
| tera (T) | $10^{12}$ |
| giga (G) | $10^{9}$ |
| mega (M) | $10^{6}$ |
| kilo (k) | 1000 |
| centi (c) | 0.01 |
| milli (m) | 0.001 |

| | |
|---|---|
| micro ($\mu$) | $10^{-6}$ |
| nano (n) | $10^{-9}$ |
| pico (p) | $10^{-12}$ |
| femto (f) | $10^{-15}$ |

So, 1 mg is 0.001 of a gram, and 1 kg is 1000 g.

There are two traditional selections of a basic unit for mass and for length that define units of other physical quantitities. The "mks" form uses meters, kilograms, and seconds. The mks energy unit, called the joule, must be the mass unit (kg), the length unit (m) to the second power, and the inverse of the second power of the time unit (s).

$$1\,\text{J} = 1\ \text{kg}\,(\text{m/s})^2 \tag{V-2}$$

The "cgs" form uses centimeters, grams, and seconds. The cgs energy unit is the erg:

$$1\ \text{erg} = 1\ \text{g}\,(\text{cm/s})^2 = 10^{-3}\,\text{kg}\,(10^{-2}\,\text{m/s})^2 = 10^{-7}\,\text{J} \tag{V-3}$$

The conversion between mks and cgs for any other physical quantitites is also obtained by going back to the relative sizes of the mass and length quantities as in Eq. (V-3).

*Fundamental Constants.* Among the "properties of nature," we may include the values of fundamental constants. Three are particularly important in the quantum mechanics of atoms and molecules. These are the speed of light (c), Planck's constant ($\hbar$), and the mass of an electron ($m_e$). The values of these constants in the cgs units are

$$c = 2.997925 \times 10^{10}\ \text{cm/sec}$$

$$\hbar = 1.05450 \times 10^{-27}\ \text{erg-sec} \quad (\text{also used: } h = 2\pi\hbar)$$

$$m_e = 9.1091 \times 10^{-28}\ \text{g}$$

*Frequencies.* Frequencies are associated with a regular oscillation in time. They are expressed in two ways, angular frequencies and linear frequencies. Angular frequencies are in radians per unit time, whereas linear frequencies are in cycles per unit time. Though there is no standard notation convention, it is frequently the case that $\omega$ is used for an angular frequency and $\nu$ is used for a frequency expressed in cycles per unit time.

The relation between the two is simple since the number of radians covered in one cycle must be $2\pi$, the number of radians in a circle.

$$2\pi\nu = \omega \tag{V-1}$$

An oscillation in time (t) that followed a cosine function would be written as either $\cos(\omega t)$ or $\cos(2\pi\nu t)$. That is usually the context that distinguishes angular from linear frequencies when the choice is not stated explicitly. The frequency of electromagnetic radiation is usually specified by the cycles per second, a unit that is called the hertz, and so the values are linear frequencies.

The energy, E, of a photon of radiation is directly proportional to the frequency, and the proportionality constant is Planck's constant:

$$E = \hbar\omega = h\nu \tag{V-2}$$

Thus, the frequency scale of cycles per second may be taken to be equivalent to an energy scale in ergs (via the factor h).

For historical reasons from the development of spectroscopy, another frequency scale has proved to convenient. In this case, the frequencies, $\nu$, are divided by a fundamental constant, the speed of light, c. Then, the frequencies are given in the inverse unit of length, $cm^{-1}$, and 1 $cm^{-1}$ is called a *wavenumber*. This amounts to using the relation between wavelength ($\lambda$) of radiation and frequency, $c = \lambda\nu$, so as to give the inverse wavelength instead of the frequency, the two being proportional:

$$\frac{1}{\lambda} = \frac{\nu}{c}$$

Wavenumbers serve as an energy scale just as well as frequencies do, and it is very common to give atomic and molecular energies in units of $cm^{-1}$.

$$\tilde{E}\,(cm^{-1}) = \frac{E(ergs)}{hc} = \frac{\nu}{c} = \frac{\omega}{2\pi c} \equiv \tilde{\omega}\,(cm^{-1}) \tag{V-3}$$

The symbol $\tilde{E}$ as opposed to just E is sometimes used to indicate that the system of units for the energy is wavenumbers and not ergs. A special symbol, of course, is unnecessary because an energy expression must hold in any system of units.

*Electromagnetic Units and Constants.* From an atomic and molecular view, one of the most basic electromagnetic quantities is the charge of an electron. This fundamental quantity is usually designated e. The size of the charge of an electron in the International System of Units (SI) is $1.6021892 \times 10^{-19}$ coulombs (C). In the cgs-emu (centimeter, gram, second–electromagnetic unit) system, the value is $1.6021892 \times 10^{-20}$ emu, or 10 times the value in coulombs. (This means 1 emu = 10 C; an emu of charge is sometimes referred to as an abcoulomb.) The SI definition of a coulomb is that it is the charge that gives a 1-newton (N) force when placed 1 m from a like charge. Thus, the unit of charge can be expressed in terms of mass, length, and time units. In particular, 1 emu = 1 $cm^{1/2}\, g^{1/2}$.

The SI unit for electrical potential is the volt (V). A particle experiencing an electrical potential will have an interaction energy (in joules) that is equal to the product of the charge in SI units and the electrical potential expressed in volts. Thus, an electron experiencing a one-volt potential will have an interaction energy,

$$E = e\,(1.0\text{ V}) = 1.6021892 \times 10^{-19}\text{ C-V} = 1.6021892 \times 10^{-19}\text{ J}$$

This quantity serves as a measure for a special unit of energy called the electron volt (eV). Simply, 1 eV is defined to be $1.6021892 \times 10^{-19}$ J.

The cgs unit of magnetic induction is the gauss (G). In the SI system, the unit is the tesla (T), and the relation is 1 G = $10^{-4}$ T.

*Atomic Units.* A special system of units called atomic units (a.u.) proves particularly convenient in the quantum mechanics of the electronic structure of atoms and molecules. This is the system of units where the measures of mass, length, and time are chosen such that three fundamental constants, $m_e$, $\hbar$, and e, take on values of exactly 1.0. This may be accomplished because we specify a system of units by three measures, and this choice presents no more than three constraints on those measures. The resulting atomic unit of length is called the bohr (after Niels Bohr) and the atomic unit of energy is called the hartree (after D. R. Hartree).

To illustrate the use of atomic units, let us calculate the ground state energy of the hydrogen atom (under the assumption of an infinitely massive nucleus), which is known to be

$$E = -m_e e^4 / (2\hbar^2) = -2.18 \times 10^{-11}\text{ erg}$$

In atomic units, however, we have that E = –1/2 a.u. Therefore, the conversion between atomic units of energy and ergs is

TABLE V.1 Energy Conversion Factors.

| | | erg | $cm^{-1}$ | eV | kcal/mol |
|---|---|---|---|---|---|
| 1.0 erg | = | 1.0 | $5.0340 \times 10^{15}$ | $6.2415 \times 10^{11}$ | $1.4383 \times 10^{13}$ |
| 1.0 joule | = | $10^7$ | $5.0340 \times 10^{22}$ | $6.2415 \times 10^{18}$ | $1.4383 \times 10^{20}$ |
| 1.0 $cm^{-1}$ | = | $1.9865 \times 10^{-16}$ | 1.0 | $1.2399 \times 10^{-4}$ | $2.8672 \times 10^{-3}$ |
| 1.0 eV | = | $1.6022 \times 10^{-12}$ | 8065.2 | 1.0 | 23.045 |
| 1.0 h | = | $4.3598 \times 10^{-11}$ | $2.1947 \times 10^5$ | 27.212 | 627.1 |

$$1.0 \text{ hartree (h)} = 2\,(2.18 \times 10^{-11} \text{ erg}) = 4.36 \times 10^{-11} \text{ erg}$$

Other energy conversions are given in Table V.1.

*The Electromagnetic Spectrum.* The electromagnetic spectrum has regions that are referred to by names that have arisen, in some cases, because of different instrumental techniques. Table V.2 lists the general designations and gives rough cutoff values for the ranges.

One unit of length that is still widely used for wavelengths is the ångstrom, abbreviated Å. The conversion is 1.0 Å = $10^{-8}$ cm.

TABLE V.2 The Electromagnetic Spectrum.

| Radiation Region | Wavelengths $\lambda$ (cm) | Frequencies $\nu = c/\lambda$ ($sec^{-1}$) | Photon Energy $1/\lambda = \nu/c$ ($cm^{-1}$) |
|---|---|---|---|
| Radio | 10 to $10^6$ | $3 \times 10^4$ to $3 \times 10^9$ | $10^{-6}$ to 0.1 |
| Microwave | 1 to 10 | $3 \times 10^9$ to $3 \times 10^{10}$ | 0.1 to 1 |
| Infrared | $10^{-4}$ to 1 | $3 \times 10^{10}$ to $3 \times 10^{14}$ | 1 to $10^4$ |
| Visible | $4 \times 10^{-5}$ to $10^{-4}$ | $3 \times 10^{14}$ to $7.5 \times 10^{14}$ | $10^4$ to $2.5 \times 10^4$ |
| Ultraviolet | $10^{-7}$ to $4 \times 10^{-5}$ | $7.5 \times 10^{14}$ to $3 \times 10^{17}$ | $2.5 \times 10^4$ to $10^7$ |
| X-ray | $< 10^{-7}$ | $> 3 \times 10^{17}$ | $10^7$ |

TABLE V.3 Values of Constants.

| Constant | Symbol | Value | |
|---|---|---|---|
| Speed of light | $c$ | 2.997925 | $\times 10^{10}$ cm/sec |
| Electron charge | $e$ | 1.602189 | $\times 10^{-20}$ emu |
| | | 1.602189 | $\times 10^{-20}$ cm$^{1/2}$ g$^{1/2}$ |
| | | 1.602189 | $\times 10^{-19}$ C |
| Planck's constant | $\hbar$ | 1.054589 | $\times 10^{-27}$ erg-sec |
| | $h$ | 6.626176 | $\times 10^{-27}$ erg-sec |
| Boltzmann's constant | $k$ | 1.380662 | $\times 10^{-16}$ erg / K |
| | | 0.695031 | cm$^{-1}$ / K |
| Mass of an electron | $m_e$ | 9.109534 | $\times 10^{-28}$ g |
| Mass of a proton | $m_p$ | 1.672649 | $\times 10^{-24}$ g |
| Bohr magneton | $\mu_B$ | 9.274078 | $\times 10^{-21}$ erg / G |
| Nuclear magneton | $\mu_o$ | 5.050824 | $\times 10^{-24}$ erg / G |

## Solutions to Selected Problems in
# *Chapter 2*

1. $H(x, y, z, p_x, p_y, p_z) = T + V$

$$= (p_x^2 + p_y^2 + p_z^2) / 2m + mgz$$

$\frac{\partial H}{\partial q_i} = -\dot{p}_i$ gives

$$\dot{p}_x = 0$$

$$\dot{p}_y = 0$$

$$\dot{p}_z = -mg$$

$\frac{\partial H}{\partial p_i} = \dot{q}_i$ gives

$$p_x / m = \dot{x}$$

$$p_y / m = \dot{y}$$

$$p_z / m = \dot{z}$$

2. $k = 2c$

$x_o = -b / 2c$

$V_o = a - b^2 / 4c$

*4.* From the given relation,

$$Ae^{-i\omega t} + Be^{i\omega t} = A\{\cos(-\omega t) + i\sin(-\omega t)\} + B\{\cos(\omega t) + i\sin(\omega t)\}$$
$$= [A + B]\cos(\omega t) + i[B - A]\sin(\omega t)$$

For Eqs. (2-8a) and (2-8b) to be the same function, it must be that

$A + B = b$ and $B - A = a$.

If $b = 0$, it would mean $A = -B$. And then if $a = 1$, it would mean that $B - (-B) = 1$, or that $B = 1/2$. For the general case with $b = 0$, then $B = a/2$ and $A = -a/2$.

*8.* In Cartesian coordinates, the Hamiltonian for the single particle is

$$H = \frac{1}{2m}(p_x^2 + p_y^2 + p_z^2) + \frac{1}{2}k(x^2 + y^2 + z^2)$$

and if this is transformed to spherical polar coordinates, the potential energy in H simplifies to $kr^2/2$. Using relationships similar to those of Eq. (2-18) to relate x, y, z to r, θ, ϕ (only letting $x = x_2 - x_1$, and so on for y and z) and the method shown in Sec. 2.3 to relate their time derivatives, the translational energy in sperical polar coordinates is

$$T = \frac{1}{2m}(\dot{r}^2 + r^2\dot{\theta}^2 + r^2\sin^2\theta\,\dot{\phi}^2)$$

This expression is also the translational motion due to vibrational and rotational motion for the system of two particles described in Sec. 2.4. So this system of one particle is equivalent to the two particle system where the center of mass translational energy is zero (X, Y, Z are constant).

★*12.* Since **M** is diagonal already, then **D** is the same matrix, i.e., **D** = **M**. Thus,

$$\mathbf{d} = \begin{pmatrix} 1 & 0 & 0 \\ 0 & 1/\sqrt{2} & 0 \\ 0 & 0 & 1 \end{pmatrix}$$

$$\mathbf{X} = \mathbf{d}\,\mathbf{K}\,\mathbf{d} = \begin{pmatrix} 1 & -1/\sqrt{2} & 0 \\ -1/\sqrt{2} & 1 & -1/\sqrt{2} \\ 0 & -1/\sqrt{2} & 1 \end{pmatrix}$$

The squares of the vibrational frequencies are the eigenvalues of **X**. From Appendix II, these may be found as the values of $\lambda$ from the equation where the determinant of $(\mathbf{X} - \lambda\mathbf{1})$ is set to zero:

$$\begin{vmatrix} 1-\lambda & -1/\sqrt{2} & 0 \\ -1/\sqrt{2} & 1-\lambda & -1/\sqrt{2} \\ 0 & -1/\sqrt{2} & 1-\lambda \end{vmatrix} = 0$$

Multiplying out this determinant yields a simple polynomial:

$$(1-\lambda)^3 - (1-\lambda) = 0$$

The roots of this polynomial (the values of $\lambda$ that satisfy the equation) are $\lambda = 0$, 1, and 2. The square roots of these three values are the vibrational frequencies, 0.0, 1.0, and 1.414.

Solutions to Selected Problems in
# Chapter 3

1. Electron's momentum is $p = 0.1 \times 3.00 \times 10^{10}$ cm/sec $\times 9.11 \times 10^{-28}$ g

$= 2.73 \times 10^{-18}$ g-cm/sec

De Broglie wavelength is $\lambda = h/p = 6.63 \times 10^{-27} / 2.73 \times 10^{-18}$ cm

$= 2.43 \times 10^{-9}$ cm

For a marble to have the same wavelength, it must have the same momentum. Thus,

$$(10.0 \text{ g})(v) = 2.73 \times 10^{-18} \text{ g-cm/sec}$$

$$v = 2.73 \times 10^{-19} \text{ cm/sec}$$

2. *b*. The commutator is evaluated by applying it to an arbitrary function.

$$[\hat{B}^2, \hat{A}]\, f(x) = [\frac{d^2}{dx^2}, x]\, f(x) = \frac{d^2}{dx^2}(x\, f(x)) - x\frac{d^2}{dx^2} f(x)$$

$$= \left(2\frac{d}{dx} f(x) + x\frac{d^2}{dx^2} f(x)\right) - x\frac{d^2}{dx^2} f(x) = 2\frac{d}{dx} f(x)$$

Therefore, we have that $[\hat{B}^2, \hat{A}] = 2\, d/dx$ since the net effect of operating with this commutator on any function will yield the same result as operating on the function with d/dx and then multiplying by the number 2.

8. $<p^2>_{n=0} = \int_{-\infty}^{\infty} \psi_0^*(x)\, \hat{p}^2\, \psi_0(x)\, dx$, and likewise for the n = 1 state.

To carry out the evaluation of this integral the momentum-squared operator must be applied to the wavefunction. We can make use of what was already done in the course of applying the Hamiltonian to $\psi_0$.

$$-\frac{\hbar^2}{2m}\frac{d^2}{dx^2}\psi_0(x) = \left[-\frac{\hbar^2}{2m}\beta^4 x^2 + \frac{\hbar^2\beta^2}{2m}\right]\psi_0(x)$$

$$<p^2>_{n=0} = -\hbar^2\beta^4 \int_{-\infty}^{\infty} \psi_0^*(x)\, x^2\, \psi_0(x)\, dx + \hbar^2\beta^2 \int_{-\infty}^{\infty} \psi_0^*(x)\, \psi_0(x)\, dx$$

The second of the two integrals is equal to one because of the normalization condition on the wavefunction. The first integral requires consulting an integrals table or other means. But first, it is helpful to substitute with the variable z.

$$-\hbar^2\beta^4 \int_{-\infty}^{\infty} \left(\frac{\beta^2}{\pi}\right)^{1/4} e^{-\beta^2x^2/2}\, x^2 \left(\frac{\beta^2}{\pi}\right)^{1/4} e^{-\beta^2x^2/2}\, dx$$

$$= -\frac{\hbar^2\beta^2}{\sqrt{\pi}} \int_{-\infty}^{\infty} e^{-z^2/2}\, z^2\, e^{-z^2/2}\, dz = -\frac{\hbar^2\beta^2}{\sqrt{\pi}} \int_{-\infty}^{\infty} z^2\, e^{-z^2}\, dz$$

And from the integral tables in Appendix III and the realization that the integral will be the same value for the integration range 0 to $\infty$ as for $-\infty$ to 0,

$$= -\frac{\hbar^2\beta^2}{\sqrt{\pi}}\frac{\sqrt{\pi}}{2}$$

So, combining the intermediate values,

$$<p^2>_{n=0} = -\hbar^2\beta^2/2 + \hbar^2\beta^2 = \hbar^2\beta^2/2$$

*10.* The uncertainty can be evaluated from the expectation value of the momentum squared and the expectation value of the momentum for a particular state. For the n = 0 state, the expectation value $<p>$ must be identically zero because the integral is over a function that is overall odd with respect to x = 0. From Prob. 8, we have a value for $< p^2 >$, and so

$$\Delta p_{n=0} = \sqrt{< p^2 >_{n=0} - < p >^2_{n=0}} = \sqrt{\hbar^2 \beta^2 / 2} = \hbar \beta / \sqrt{2}$$

Solutions to Selected Problems in

# *Chapter 4*

2. The entire set of wavefunctions would be eigenfunctions of the x operator or of the $p_x$ operator only if they were to commute with the Hamiltonian.

$$[x, H]\, f(x) = [x, \frac{1}{2}kx^2]\, f(x) + [x, -\frac{\hbar^2}{2m}\frac{d^2}{dx^2}]\, f(x)$$

$$= \left(\frac{1}{2}kx^3 - \frac{1}{2}kx^3\right) f(x) - x\frac{\hbar^2}{2m}\frac{d^2}{dx^2} f(x) + \frac{\hbar^2}{2m}\frac{d^2}{dx^2}(x\, f(x))$$

$$= \frac{\hbar^2}{m}\frac{d}{dx} f(x) \neq 0$$

Thus, x does not commute with the Hamiltonian. It can be shown that $p_x$ does not commute with H either. So, the wavefunctions are not eigenfunctions of either of these operators.

5. The energy level expression for an isotropic three-dimensional oscillator is

$$E_{n_x n_y n_z} = (n_x + \frac{1}{2})\hbar\omega + (n_y + \frac{1}{2})\hbar\omega + (n_z + \frac{1}{2})\hbar\omega$$

This results because of separability in the three directions and because the isotropic character of the problem means $k_x=k_y=k_z$, which in turn means the frequencies for vibration in the x-, y-, and z-directions are the same. So, the energy expression may be written simply as

$$E_{n_x n_y n_z} = \left( n_x + n_y + n_z + \frac{3}{2} \right) \hbar \omega$$

To determine the degeneracy of an energy level we must find all combinations of the three quantum numbers that give the same energy, remembering that for a harmonic oscillator the quantum numbers may be zero or any positive integer. From the energy expression, it should be clear that it is just the sum of the three quantum numbers that is important, and we will name this sum N; i.e., $N = n_x + n_y + n_z$. As N increases, the energy increases. The smallest allowed value for N is zero, and then all three of the quantum numbers are zero. This level is not degenerate because there is no other choice of the quantum numbers which gives N = 0. For the first excited state energy level, which means N = 1, there are three different choices of the quantum numbers: $n_x$ could be 1, or $n_y$ could be 1, or $n_z$ could be 1, while the other two are zero. Thus, there are three different states with this energy; the degeneracy of the second level (first excited level) is three. The following list continues this process.

| | $(n_x, n_y, n_z)$ of a state with the given N: | |
|---|---|---|
| N = 0 | (0,0,0) | degeneracy = 1 (nondegenerate) |
| N = 1 | (1,0,0), (0,1,0), (0,0,1) | degeneracy = 3 |
| N = 2 | (2,0,0), (0,2,0), (0,0,2), (1,1,0), (1,0,1), (0,1,1) | degeneracy = 6 |
| N = 3 | (3,0,0), (2,1,0), (2,0,1), (1,2,0), (1,1,1), (1,0,2), (0,3,0), (0,2,1), (0,1,2), (0,0,3) | degeneracy = 10 |

6. It is necessary to examine the energies of the states systematically. With n = 1, the first term in this energy expression is at its lowest value. So, we may start there and find all the energies for the possible values of m. Then, we may repeat this for n = 2, and so on.

n = 1 m = 0 $E/\hbar a = -1$

| | | |
|---|---|---|
| $n = 1$ | $m = 1$ or $-1$ | $E/\hbar a = -3/4$ |
| $n = 1$ | $m = 2$ or $-2$ | $E/\hbar a = 0$ |
| $n = 2$ | $m = 0$ | $E/\hbar a = -1/4$ |
| $n = 2$ | $m = 1$ or $-1$ | $E/\hbar a = 0$ |
| $n = 2$ | $m = 2$ or $-2$ | $E/\hbar a = 3/4$ |
| $n = 2$ | $m = 3$ or $-3$ | $E/\hbar a = 2$ |
| $n = 3$ | $m = 0$ | $E/\hbar a = -1/9$ |
| $n = 3$ | $m = 1$ or $-1$ | $E/\hbar a = 5/36$ |
| $n = 3$ | $m = 2$ or $-2$ | $E/\hbar a = 8/9$ |
| $n = 3$ | $m = 3$ or $-3$ | $E/\hbar a = 77/36$ |
| $n = 3$ | $m = 4$ or $-4$ | $E/\hbar a = 35/9$ |

There are 21 different states because there are 21 different allowed values for the two quantum numbers. From the energy level expression we obtain 11 different energy values, and so there are 11 energy levels. The degeneracies are obtained from counting up the possibilities that yielded each particular energy value. The lowest level or the ground state is not degenerate with any other. The other two $m = 0$ states are nondegenerate, too (degeneracy = 1). There are 4 states with energy of zero, and so this level has a fourfold degeneracy. The remaining seven levels are doubly degnerate (degeneracy = 2). If we add up the degeneracies, $3 \times 1 + 4 + 7 \times 2$, we get 21, the number of states.

8. $$W = \langle \Gamma | \hat{H} \Gamma \rangle = \frac{N^2 \left[ \frac{6\hbar^2}{16m}\sqrt{\frac{\pi}{2\alpha}} + \frac{3k}{32\alpha^2}\sqrt{\frac{\pi}{2\alpha}} \right]}{\frac{N^2 \sqrt{\frac{\pi}{2\alpha}}}{4\alpha}} = \frac{6\hbar^2\alpha}{4m} + \frac{3k}{8\alpha}$$

$$\frac{dW}{d\alpha} = \frac{6\hbar^2}{4m} - \frac{3k}{8\alpha^2} \equiv 0 \text{ for } \alpha_{min},\ \alpha_{min} = \frac{\sqrt{km}}{2\hbar}$$

$$E = W(\alpha_{min}) = \frac{3}{2}\hbar\omega$$

This is the energy of the first excited state of the harmonic oscillator (n = 1). The trial function has the form of the eigenfunction of the first excited state.

*13.* The first-derivative Schrödinger equation with $\alpha = 0$ is

$$H_0^{\alpha}\Psi_0 + H_0\Psi_0^{\alpha} = E_0^{\alpha}\Psi_0 + E_0\Psi_0^{\alpha}.$$

Then, after rearranging, multiplying by $\Psi^*$, and integrating, we get

$< \Psi_0 \mid (H_0 - E_0)\Psi_0^{\alpha} > = - < \Psi_0 \mid (H_0^{\alpha} - E_0^{\alpha})\Psi_0 >$. The left hand side of this expression is zero, and so $E_0^{\alpha} = < \Psi_0 \mid H_0^{\alpha}\, \Psi_0 > = < x^3 >$.

The general second and third derivative equations are

$$(H - E)\,\Psi^{\alpha\alpha} + 2\,(H^{\alpha} - E^{\alpha}) + (H^{\alpha\alpha} - E^{\alpha\alpha})\,\Psi = 0$$

$$(H - E)\Psi^{\alpha\alpha\alpha} + 3\,(H^{\alpha} - E^{\alpha})\Psi^{\alpha\alpha} + 3\,(H^{\alpha\alpha} - E^{\alpha\alpha})\Psi^{\alpha} + (H^{\alpha\alpha\alpha} - E^{\alpha\alpha\alpha})\,\Psi = 0$$

Of course, for this problem, we know that

$$H_0^{\alpha} = x^3$$

$$H_0^{\alpha\alpha} = x^6$$

$$H_0^{\alpha\alpha\alpha} = 0$$

Solutions to Selected Problems in

# *Chapter 5*

5. From Eq. (5-5),

$$\hat{L}^2 \left(\sin^m \theta \cos\theta \, e^{i m\phi}\right)$$

$$= -\hbar^2 \left( \frac{1}{\sin\theta} \frac{\partial}{\partial\theta} \sin\theta \frac{\partial}{\partial\theta} + \frac{1}{\sin^2\theta} \frac{\partial^2}{\partial\phi^2} \right) \left(\sin^m \theta \cos\theta \, e^{i m\phi}\right)$$

$$= -\hbar^2 \left[ m^2 \sin^{m-2}\theta \cos^3\theta - (3m+2) \sin^m\theta \cos\theta \right] e^{i m\phi}$$

$$-\hbar^2 (-m^2) \sin^{m-2}\theta \cos\theta \, e^{i m\phi}$$

with the term in [ ] from the differentiation with respect to $\theta$ and the last term from the differentiation with respect to $\phi$. If the last term is multiplied by $1 = \sin^2\theta + \cos^2\theta$, then it may be combined with the term in [ ], yielding

$$= \hbar^2 (m^2 + 3m + 2) \sin^m\theta \cos\theta \, e^{i m\phi}$$

This shows that the function is indeed an eigenfunction of $\hat{L}^2$. The eigenvalue is $\hbar^2 (m^2 + 3m + 2)$, or $\hbar^2 (m + 1)(m + 2)$.

6. $$[L_x, L_y] = -i\hbar \left[-\sin\phi \frac{\partial}{\partial\theta} - \frac{\cos\phi}{\tan\theta}\frac{\partial}{\partial\phi}\right] -i\hbar \left[-\cos\phi \frac{\partial}{\partial\theta} - \frac{\sin\phi}{\tan\theta}\frac{\partial}{\partial\phi}\right]$$

$$- \left(-i\hbar \left[-\cos\phi \frac{\partial}{\partial\theta} - \frac{\sin\phi}{\tan\theta}\frac{\partial}{\partial\phi}\right] - i\hbar\left[-\sin\phi \frac{\partial}{\partial\theta} - \frac{\cos\phi}{\tan\theta}\frac{\partial}{\partial\phi}\right]\right)$$

which after multiplying out these terms and applying it to an arbitrary function becomes

$$= (-i\hbar)^2 \left[ \sin^2\phi \frac{\partial}{\partial\theta}\frac{1}{\tan\theta}\frac{\partial}{\partial\phi} - \frac{\cos\phi}{\tan\theta}\frac{\partial}{\partial\phi}\cos\phi\frac{\partial}{\partial\theta} + \frac{\cos\phi}{\tan^2\theta}\frac{\partial}{\partial\phi}\sin\phi\frac{\partial}{\partial\phi}\right.$$

$$\left. - \cos^2\phi \frac{\partial}{\partial\theta}\frac{1}{\tan\theta}\frac{\partial}{\partial\phi} + \frac{\sin\phi}{\tan\theta}\frac{\partial}{\partial\phi}\sin\phi\frac{\partial}{\partial\theta} + \frac{\sin\phi}{\tan^2\theta}\frac{\partial}{\partial\phi}\cos\phi\frac{\partial}{\partial\phi}\right]$$

$$= -(i\hbar)\left[-i\hbar\frac{\partial}{\partial\phi}\right] = i\hbar L_z.$$

8. $$Y_{20}(\theta,\phi) = \Theta_{20}(\theta)\,\Phi_0(\phi) = \frac{1}{4}\sqrt{\frac{5}{\pi}}\left(3\cos^2\theta - 1\right)$$

$$\hat{L}_x Y_{20} = -i\hbar\,\frac{3}{2}\sqrt{\frac{5}{\pi}}\cos\theta\sin\theta\sin\phi$$

$$\left\langle Y_{20} \mid \hat{L}_x Y_{20}\right\rangle = \left\langle \text{integral over } \theta\right\rangle \int_0^{2\pi} \sin\phi\, d\phi = 0$$

$$\hat{L}_x(\hat{L}_x Y_{20}) = -\hbar^2\,\frac{3}{2}\sqrt{\frac{5}{\pi}}\left(\sin^2\phi\sin^2\theta - \cos^2\theta\right)$$

$$\left\langle Y_{20} \mid \hat{L}_x^2 Y_{20}\right\rangle = 0$$

10. $$\hat{L}_+ = -i\hbar\left[(-\sin\phi + i\cos\phi)\frac{\partial}{\partial\theta} - \frac{(\cos\phi + i\sin\phi)}{\tan\theta}\frac{\partial}{\partial\phi}\right]$$

$$\hat{L}_+ Y_{21} = \hbar\left[e^{i\phi}\frac{\partial}{\partial\theta} + \frac{i e^{i\phi}}{\tan\theta}\frac{\partial}{\partial\phi}\right] Y_{21} = -\sqrt{\frac{5}{12}}\; 3\cos\theta\sin\theta\,\frac{e^{i\phi}}{\sqrt{2\pi}}$$

$$= 2\hbar \left[ \sqrt{\frac{5}{24}}\ 3 \sin^2\theta\ \frac{e^{i2\phi}}{\sqrt{2\pi}} \right] = 2\hbar Y_{22}$$

$$\hat{L}_+ Y_{22} = \text{constants} \left[ e^{i\phi}\, 2 \sin\theta \cos\theta\, e^{i2\phi} + \frac{i e^{i\phi}}{\tan\theta} \sin^2\theta\, 2i\, e^{i2\phi} \right] = 0$$

*12.* Couple the momenta two at a time. Coupling $j_1$ and $j_2$ results in values of $j_a = 3, 2, 1$. Then, coupling each possible value of $j_a$ with $j_3$ leads to possible values of $J = 4, 3, 2, 3, 2, 1, 2, 1, 0$.

*13.* Two sources may combine such that the total angular momentum quantum number, J, ranges (in steps of one) from the sum to the absolute value of the difference, according to the rule of angular momentum addition. With two hypothetical soucres with quantum numbers of 1/2, the sum is one and the difference is zero. Therefore, 0 and 1 are the possible values for the quantum number of the total angular momentum.

If there is a third source, it adds to the resultant of the first two. Thus, $J^{\text{3-sources}} = \frac{1}{2} + 0 = \frac{1}{2}$ because of the zero sum of the first two sources, and $J^{\text{3-sources}} = \frac{1}{2} + 1, \ldots \left|\frac{1}{2} - 1\right| = \frac{3}{2}$ and $\frac{1}{2}$ because of the case where the sum from the first two sources is one. We interpret this to mean that the three sources may combine to give $J = 3/2$ or they may combine in two *different* ways to give $J = 1/2$.

The fourth source may combine with the 3/2 resultant from three sources to give $J^{\text{4-sources}} = 2$ or 1. It may combine with either of the 1/2 resultants to give $J^{\text{4-sources}} = 1$ or 0. This may be expressed concisely as

$$J^{\text{4-sources}} = 2, 1, 1, 1, 0, 0$$

The allowed possibilities for the resultant J qauntum number with the four sources is this list of integers. Integers are repeated if there are different ways that they may arise.

*15.* The coupled states can be found by applying Eqs. (5-27) through (5-35). Starting with $|3\ 3>_c\ =\ |\ 2\ 1>_u$, application of lowering operators leads to the rest of the $J = 3$ states. Then the $|\ 2\ 2>_c$ state is found using the condition that it is orthogonal to $|\ 3\ 2\ \rangle_c$. The lowering operators are applied once again to generate the rest of the J = 2 coupled states. Finally, since the $|\ 1\ 1\ \rangle_c$ state can result from three uncoupled states ( $|\ 2\ \text{-}1\ \rangle$, $|\ 1\ 0\ \rangle$, and $|\ 0\ 1\ \rangle$), three conditions are required to find the linear combination. Two conditions result since $|\ 1\ 1\ \rangle_c$ must be orthogonal to $|\ 3\ 1\ \rangle_c$ and $|\ 2\ 1\ \rangle_c$, and the third comes from the normalization condition ( $\langle\ 1\ 1\ |\ 1\ 1\ \rangle = 1$).

$$
\begin{vmatrix}
|3\ 3\rangle \\ |3\ 2\rangle \\ |3\ 1\rangle \\ |3\ 0\rangle \\ |3\ \text{-}1\rangle \\ |3\ \text{-}2\rangle \\ |3\ \text{-}3\rangle \\ |2\ 2\rangle \\ |2\ 1\rangle \\ |2\ 0\rangle \\ |2\ \text{-}1\rangle \\ |2\ \text{-}2\rangle \\ |1\ 1\rangle \\ |1\ 0\rangle \\ |1\ \text{-}1\rangle
\end{vmatrix}
=
\begin{vmatrix}
1 & 0 & 0 & 0 & 0 & 0 & 0 & 0 & 0 & 0 & 0 & 0 & 0 & 0 & 0 \\
0 & \sqrt{\frac{1}{3}} & 0 & \sqrt{\frac{2}{3}} & 0 & 0 & 0 & 0 & 0 & 0 & 0 & 0 & 0 & 0 & 0 \\
0 & 0 & \sqrt{\frac{1}{15}} & 0 & \sqrt{\frac{8}{15}} & 0 & \sqrt{\frac{2}{5}} & 0 & 0 & 0 & 0 & 0 & 0 & 0 & 0 \\
0 & 0 & 0 & 0 & 0 & \sqrt{\frac{1}{5}} & 0 & \sqrt{\frac{3}{5}} & 0 & \sqrt{\frac{2}{5}} & 0 & 0 & 0 & 0 & 0 \\
0 & 0 & 0 & 0 & 0 & 0 & 0 & 0 & \sqrt{\frac{2}{5}} & 0 & \sqrt{\frac{8}{15}} & 0 & \sqrt{\frac{1}{15}} & 0 & 0 \\
0 & 0 & 0 & 0 & 0 & 0 & 0 & 0 & 0 & 0 & 0 & \sqrt{\frac{2}{3}} & 0 & \sqrt{\frac{1}{3}} & 0 \\
0 & 0 & 0 & 0 & 0 & 0 & 0 & 0 & 0 & 0 & 0 & 0 & 0 & 0 & 1 \\
0 & -\sqrt{\frac{2}{3}} & 0 & \sqrt{\frac{1}{3}} & 0 & 0 & 0 & 0 & 0 & 0 & 0 & 0 & 0 & 0 & 0 \\
0 & 0 & -\sqrt{\frac{1}{3}} & 0 & -\sqrt{\frac{1}{6}} & 0 & \sqrt{\frac{1}{2}} & 0 & 0 & 0 & 0 & 0 & 0 & 0 & 0 \\
0 & 0 & 0 & 0 & 0 & -\sqrt{\frac{1}{2}} & 0 & 0 & 0 & \sqrt{\frac{1}{2}} & 0 & 0 & 0 & 0 & 0 \\
0 & 0 & 0 & 0 & 0 & 0 & 0 & -\sqrt{\frac{1}{2}} & 0 & \sqrt{\frac{1}{6}} & 0 & 0 & \sqrt{\frac{1}{3}} & 0 & 0 \\
0 & 0 & 0 & 0 & 0 & 0 & 0 & 0 & 0 & 0 & 0 & -\sqrt{\frac{1}{3}} & 0 & \sqrt{\frac{2}{3}} & 0 \\
0 & 0 & \sqrt{\frac{3}{5}} & 0 & -\sqrt{\frac{3}{10}} & 0 & \sqrt{\frac{1}{10}} & 0 & 0 & 0 & 0 & 0 & 0 & 0 & 0 \\
0 & 0 & 0 & 0 & 0 & \sqrt{\frac{3}{10}} & 0 & -\sqrt{\frac{2}{5}} & 0 & \sqrt{\frac{3}{10}} & 0 & 0 & 0 & 0 & 0 \\
0 & 0 & 0 & 0 & 0 & 0 & 0 & 0 & \sqrt{\frac{1}{10}} & 0 & -\sqrt{\frac{3}{10}} & 0 & \sqrt{\frac{3}{5}} & 0 & 0
\end{vmatrix}
\begin{Vmatrix}
|2\ 1\rangle \\ |2\ 0\rangle \\ |2\ \text{-}1\rangle \\ |1\ 1\rangle \\ |1\ 0\rangle \\ |1\ \text{-}1\rangle \\ |0\ 1\rangle \\ |0\ 0\rangle \\ |0\ \text{-}1\rangle \\ |\text{-}1\ 1\rangle \\ |\text{-}1\ 0\rangle \\ |\text{-}1\ \text{-}1\rangle \\ |\text{-}2\ 1\rangle \\ |\text{-}2\ 0\rangle \\ |\text{-}2\ \text{-}1\rangle
\end{Vmatrix}
$$

Solutions to Selected Problems in

# *Chapter 6*

1. Using the power series expansion,

$$\sqrt{a+x} = \sqrt{a} + \frac{1}{2}\frac{x}{\sqrt{a}} - \frac{1}{4}\frac{x^2}{(\sqrt{a})^3} + \frac{3}{8}\frac{x^3}{(\sqrt{a})^5} \cdots$$

to substitute for $\omega_J$ in Eq. (6-17), truncation of the power series after the second term leaves Eq. (6-20).

3. Using the definitions of $Y_{JM}$ given in Eqs. (5-9,10,11) and the recursion relationship for the associated Legendre polynomials,

$$\cos\theta\, Y_{J'M'} = \sqrt{\frac{(2J'+1)(J'-|M'|)!}{2\,(J'+|M'|)!}}\,\Phi_M(\phi)\cos\theta\, P_{J'}^{|M'|}(\cos\theta)$$

$$= \sqrt{\frac{(2J'+1)(J'-|M'|)!}{2\,(J'+|M'|)!}}\,\Phi_M(\phi)\,\frac{1}{2J'+1}\left((J'+1-|M'|)\,P_{J'+1}^{|M'|} + (J'+|M'|)\,P_{J'-1}^{|M'|}\right)$$

$$= \frac{1}{2J'+1}\left(\sqrt{\frac{(2J'+1)\,[(J'+1)^2-|M'|^2]}{2J'+3}}\,Y_{J'+1\,M'} + \sqrt{\frac{(2J'+1)\,(J'^2-|M'|^2}{2J'-1}}\,Y_{J'-1\,M'}\right)$$

With this result, $< Y_{JM} \mid \cos\theta\, Y_{J'M'} >$ reduces to overlap integrals of the spherical harmonics, and from their orthogonality relationships Eq. (6-27) is obtained.

6. $E_{n+1,n} = E_{n+1} - E_n = \omega_e - 2(n + 1)\,\omega_e\chi_e + \{3\,[(n + \frac{1}{2})^2 + (n + \frac{1}{2})\,] + 1\}\omega_e y_e$. Using the given transition frequencies to set up three equations, the three unknowns can be solved for and then $\omega_e$ = 1749.92 $cm^{-1}$, $\omega_e\chi_e$ = 87.50 $cm^{-1}$, and $\omega_e y_e$ = 8.33 $cm^{-1}$. On the other hand, letting $\omega_e y_e = 0$ and setting up two equations in two unknowns leads to $\omega_e$ = 1700 $cm^{-1}$, $\omega_e\chi_e$ = 50 $cm^{-1}$. Alternatively, including all three transitions and fitting them by a least squares analysis with the two term formula leads to $\omega_e$ = 1666.7 $cm^{-1}$ and $\omega_e\chi_e$ = 37.5 $cm^{-1}$.

11. $r_e$ = 1.596 Å.

| J | ΔE |
|---|---|
| 0-1 | 1036.77 ($cm^{-1}$) |
| 1-2 | 1044.81 |
| 2-3 | 1052.72 |
| 3-4 | 1060.55 |
| 4-5 | 1068.36 |
| 5-6 | 1076.22 |
| 6-7 | 1084.21 |

# Solutions to Selected Problems in *Chapter 7*

1. We may solve the simultaneous quadratic equations for the rotational constants using any two isotopic values *only*. With the first two, the solution is $R_{NN}$ = 1.1269 Å and $R_{NO}$ = 1.1911 Å. Very slightly different values may be obtained if the last two values or the first and the third are used. An alternative approach is to treat ($R_{NN}$ + $R_{NO}$) as an independent variable and then to use all three moment-of-inertia equations to solve for the three unknowns. This gives $R_{NN}$ = 1.1331 Å and $R_{NO}$ = 1.1915 Å; however, because ($R_{NN}$ + $R_{NO}$) has been treated independently, its value is not exactly equal to 1.1331 + 1.1915.

3. In the course of the first two stretching modes, the left-right symmetry of the molecule is preserved and so the dipole always remains zero. This means the dipole is unchanging, and so these two modes are infrared-inactive. In the third stretching mode, the symmetry is not preserved. A dipole along the molecular axis will grow in as the molecule vibrates away from its equilibrium structure. The dipole changes in the course of this vibration, and so this mode is infrared-active. The first bending mode is infrared-

active because a dipole will develop that is perpendicular to the molecular axis. The second bending mode preserves the symmetry in such a way that the dipole will remain at zero. It is not active.

5. a. Formaldehyde: There are 3(4) – 6 = 6 modes.

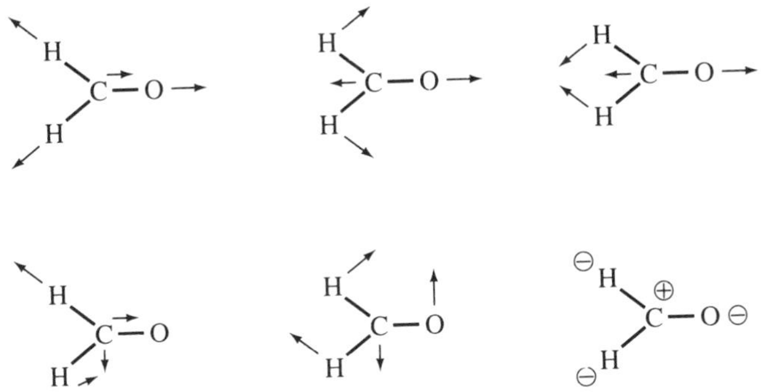

12. The energy levels, which are the eigenvalues of the Hamiltonian matrix, are 0.955, 1.554, 2.717, 3.568, and 4.664. Note that the transition energies are 0.599, 1.163, 0.851, and 1.096, or that pairs of energy levels (0,1 and 2,3) are brought closer together as described in the text.

Solutions to Selected Problems in

# *Chapter 8*

1. $E_n^{(1)} = \langle H^{(1)} \rangle = \langle \psi_n^{(0)} \mid V(x)\, \psi_n^{(0)} \rangle = a \langle \psi_n^{(0)} \mid x\, \psi_n^{(0)} \rangle = \frac{a\,l}{2}$

$$\Psi_n^{(1)} = \sum_j \frac{-16 m l^3 a\, nj}{\pi^4 \hbar^2 (n-j)^3 (n+j)^3}\, \Psi_j^{(0)}, \; j + n = \text{odd}$$

$$E_n^{(2)} = \sum_j \frac{128\, m\, l^4 a^2 n^2 j^2}{\pi^6 \hbar^2 (n-j)^5 (n+j)^5}, \; j + n = \text{odd}$$

2. $\langle x \rangle_n = l/2$ and $\Delta x_n = l\sqrt{\left(\frac{1}{3} - \frac{1}{2(n\pi)^2}\right) - \frac{1}{4}}$ . Thus, for example, if $l = \pi$, then $\Delta x_1 = 0.56786$, $\Delta x_2 = 0.83514$, and $\Delta x_3 = 0.87573$.

4.

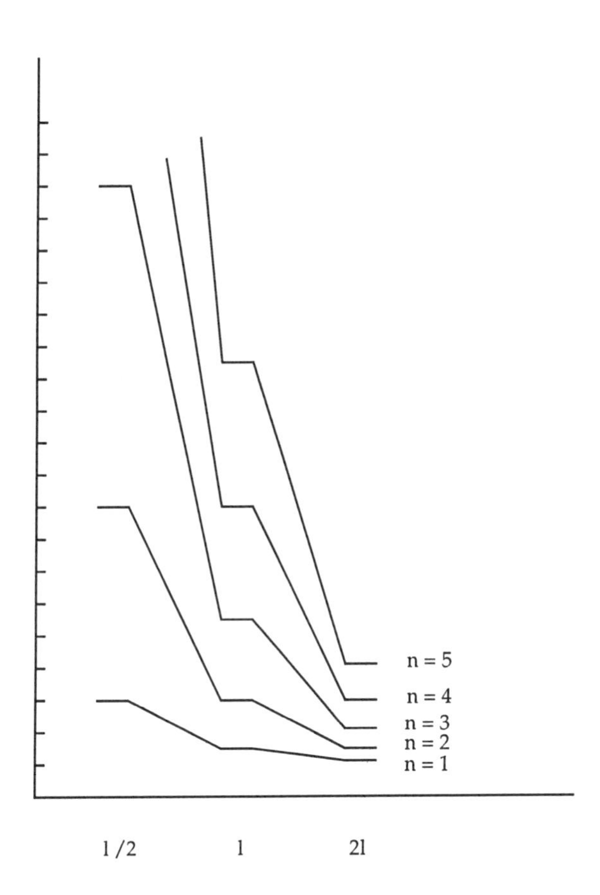


7. $l$ = 9.54Å; n = 4.

11. Let the boundaries of the infinite potential be at ±b, and those of the finite potential $V_0$ at ±a. Then, the connection conditions are

$$\Psi_I(-b) = \Psi_I(b) = 0$$

$$\Psi_I(-a) = \Psi_{II}(-a)$$

$$\Psi_{II}(a) = \Psi_{III}(a)$$

$$\left.\frac{d\Psi_I}{dx}\right|_{-a} = \left.\frac{d\Psi_{II}}{dx}\right|_{-a}$$

$$\left.\frac{d\Psi_{II}}{dx}\right|_{a} = \left.\frac{d\Psi_{III}}{dx}\right|_{a}$$

Solutions to Selected Problems in

# *Chapter 9*

1. $$L_3^0 = e^z \left(\frac{d}{dz}\right)^3 z^3 e^{-z} = 6 - 18z + 9z^2 - z^3$$

$$L_3^2 = \left(\frac{d}{dz}\right)^2 L_3^0 = 18 - 6z$$

6. $E_{ionization} = (Z_{ion} / Z_H)^2 E_H = Z_{ion}^2 E_H$ .

| Ion | $E_{ionization}$ ($cm^{-1}$) |
|---|---|
| $He^+$ | 438,960 |
| $Li^{2+}$ | 987,660 |
| $C^{5+}$ | 3,950,640 |
| $Ne^{9+}$ | 10,974,000 |

8. Since each shell is filled in a $s^2\ p^6\ d^{10}$ occupancy, there is only one entry in the table, with $M_L = 0$ and $M_S = 0$; therefore $L = 0$ and $S = 0$. For an occupancy of $1s^2\ 2p^1$, the table is

| $1s_0$ | $2p_1$ | $2p_0$ | $2p_{-1}$ | $M_L$ | $M_S$ |
|---|---|---|---|---|---|
| ↑↓ | ↑ | | | 1 | 1/2 |
| ↑↓ | ↓ | | | 1 | –1/2 |
| ↑↓ | | ↑ | | 0 | 1/2 |
| ↑↓ | | ↓ | | 0 | –1/2 |
| ↑↓ | | | ↑ | –1 | 1/2 |
| ↑↓ | | | ↓ | –1 | –1/2 |

which leads to the term symbol $^2P$. Since the s shell is full, it has only one possible occupancy, for which $M_L = 0$ and $M_S = 0$. So this table (without the electrons of the s orbital) is correct for a $2p^1$ occupancy, and the term symbol $^2P$ must be correct for a $2p^1$ occupancy as well.

*10.* The resulting terms symbols are $^1D_2$, $^3P_{2,1,0}$, and $^1S_0$.

*12.* The terms symbols for $3d^1$ and $3d^9$ are $^2D$. The term symbols for $3d^2$ and $3d^8$ are $^3F_{4,3,2}$, $^1G$, $^3P_{2,1,0}$, and $^1D_0$. This suggests that an occupancy of n electrons in a shell has the same set of term symbols as an occupancy of n "electron holes" in the same shell.

*18.* The ground state of carbon is $^3P$. The excited states are $^3D$, $^1D$, $^3P$, $^1P$, $^5S$, and $^3S$. Following the selection rules $\Delta S = 0$ and $\Delta L = 1$, the allowed transitions are to $^3D$ and $^3S$.

## Solutions to Selected Problems in
# *Chapter 10*

1.

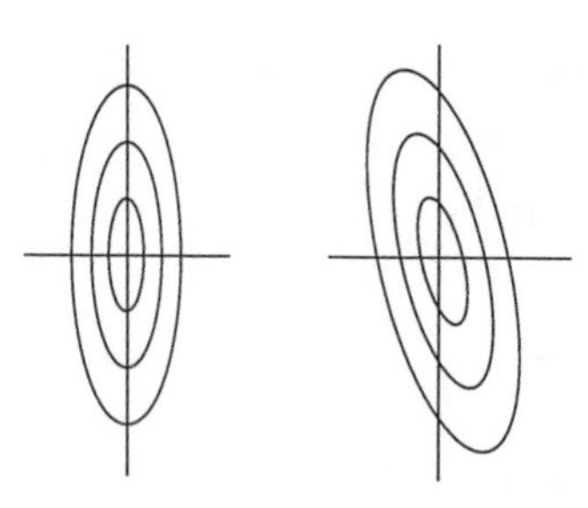

The contours of V' (on the right) are twisted by 13.3° relative to those of V (on the left), and they are larger. The xy term in $3x^2 + y^2 + xy$ can be eliminated by a transformation of the coordinates, just like the cross-term of the coupled oscillator in Sec. 2.3 was eliminated. If this is done, the resultant potential (in the transformed coordinate system) wll be $(4 + \sqrt{5})\, x^2 + (4 - \sqrt{5})\, y^2$. This transformed coordinate system is simply the original system but with the axes rotated by −13.3°.

3. The number of particles allowed per orbital equals the number of different spin states, so for s = 3/2, the exclusion principle would allow four particles [2 (3/2) + 1].

7. LiH, 1; $Be_2$, 0; $N_2$, 3; $F_2$, 1; $NeF^+$, 1.

8. M(M + 1)/2, where M = N(N + 1)/2. This represents the number of unique combinations of the unique pairs of the N functions.

10. $\langle \Gamma | H \Gamma \rangle = \sum_{i}^{K} I(i|i) + \frac{1}{2}\sum_{i}^{K}\sum_{j}^{K}\left\{ (ii|jj) - (ij|ij) \right\}$ . Since all the spin functions are identical, all the integrals over spin functions are 1.0.

12. Using the result from Prob. 10, and following the process of Eqs. (10-40) through (10-44), $\hat{F} = \hat{h} + \hat{J} - \hat{K}$.

Solutions to Selected Problems in

# *Chapter 11*

2. Consider one of the N nuclei. Since it has spin = 1, it has 2 transitions ( $-1 \leftrightarrow 0$ and $0 \leftrightarrow 1$). These transitions can occur for all combinations of the other spin states, so the number of transitions = $2 \times (3 \times 2 \times 2) = 24$. For a hydrogen, which has only one transition, the total number of its transitions is $1 \times (3 \times 3 \times 2) = 18$. The number of nitrogen spin flips = 48, and the number of hydrogen spin flips = 36.

5.

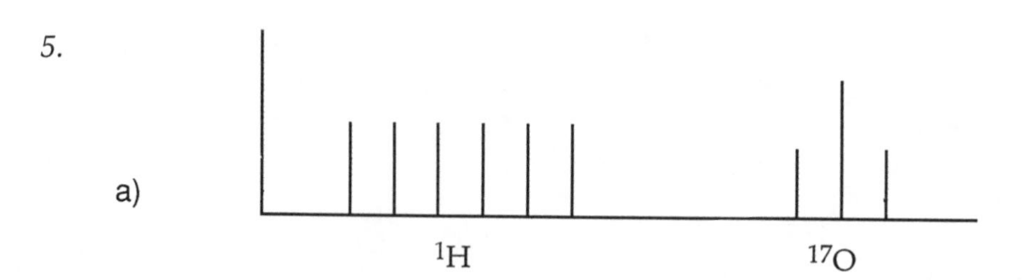

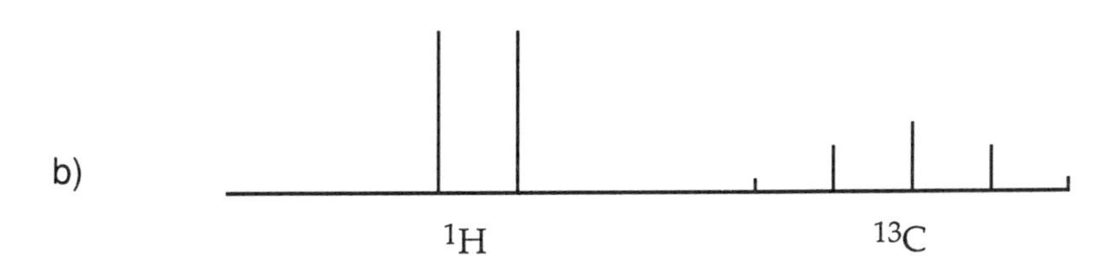


7. One nucleus, 1 : 1 : 1

   Two nuclei, 1 : 2 : 3 : 2 : 1

   Three nuclei, 1 : 3 : 6 : 7 : 6 : 3 : 1

11.

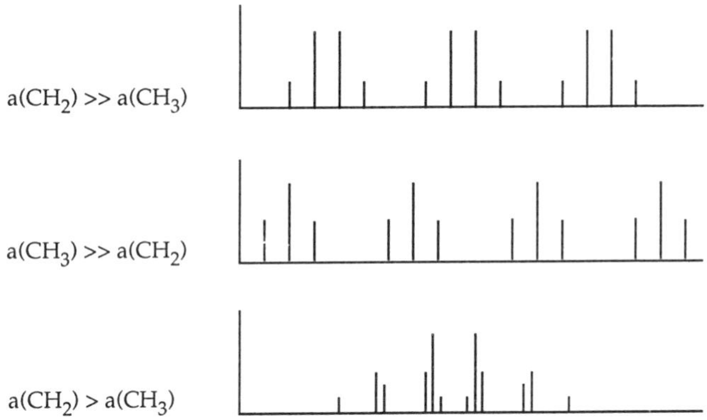

$a(CH_2) >> a(CH_3)$

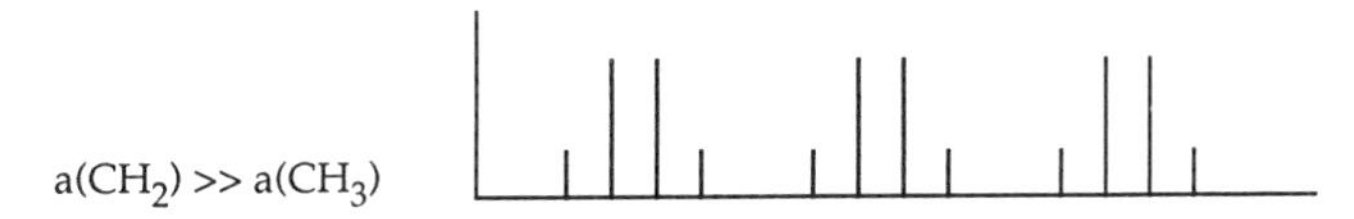

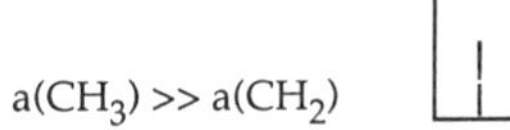


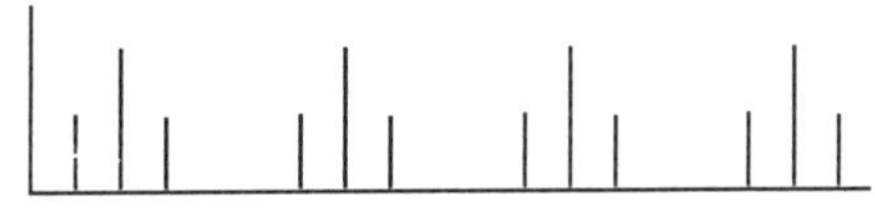

$a(CH_2) > a(CH_3)$

12.

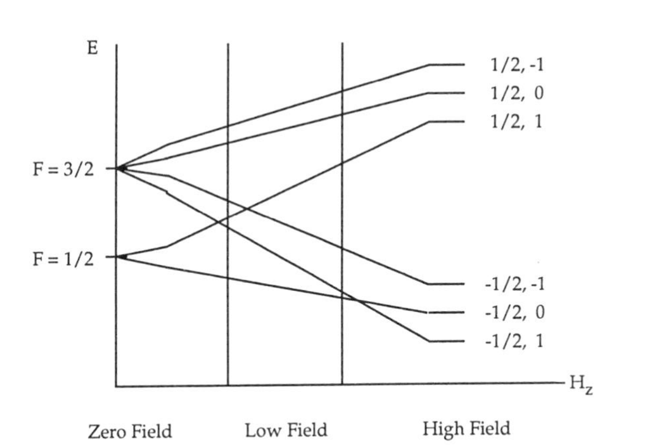

Solutions to Selected Problems in

# *Chapter 12*

3. $$\left\langle 1s \mid r\cos\theta \mid 2p_z \right\rangle = \left\langle R_{00} \mid r \mid R_{10} \right\rangle \left\langle Y_0^0 \mid \cos\theta \mid Y_1^0 \right\rangle$$

$$= \left\{ -4\sqrt{6}\left(\frac{2}{3}\right)^5 \frac{\hbar^2}{m_e\, e^2} \right\} \left\{ \sqrt{\frac{4}{15}} \right\}$$

The first term comes from explicitly integrating the radial functions, and the second from Eq. (12-17). The second order correction to the energy is

$$E^{(2)} = \frac{\left\langle 1s \mid H' \mid 2p_z \right\rangle^2}{E_{1s} - E_{2p_z}} = e^2 V_z^2 \frac{\left\langle 1s \mid r\cos\theta \mid 2p_z \right\rangle}{E_{1s} - E_{2p_z}} =$$

$$-1.1838 \left( \frac{\hbar^2}{m_e e^2} \right)^3 V_z^2 .$$

(Note that the quantity in parentheses is the atomic unit of length, the bohr, $a_o$.) Relating this to Eq. (12-10), $\alpha = 2.3676\ a_o^3$. This is an approximate solution because only one excited state was allowed to enter the perturbed wavefunction. The exact solution, which requires including all excited states in the perturbed wavefunction, is $\alpha = 4.5\ a_o^3$.

*8.* The $1s^2\,2s^2\,2p^2$ occupation generates the states $^1S$, $^3P$, and $^1D$. For $^1D$ and $^1S$, $g_J = 1$ and for $^3P$, $g_J = (g_e + 1)/2 \approx 3/2$. Using Eq. (12-34), the $^1D$ states will be split by $\mu_\beta M_J H_z$, the $^3P$ states will be split by $(3/2)\mu_\beta M_J H_z$, and the $^1S$ state will be unchanged ($M_J = 0$).

*10.* The dipole-quadrupole interaction depends on the inverse of the separation distance to fourth order.

# *Subject Index*

W9-CKM-187
DE LOS SIGLOS
SEÑOR D SEÑORES
a su admirable Luz
COMENZO este gran Imperio
MANCOCAPAC Inga o Monarc
por ficcion de su Madre Mama Huaco fue tenido por hijo del
Sol y dela Cueva Pacarictambo: con que fue aclamado por Rey
en toda la Region del Cuzco, Ciudad y Corte q. e fundo, o acre
cento A los 14 años de su edad se caso con su misma Madre, o ignorante
o sabedor dl parentesco, Puso ley para mayor seguridad de la Real Es
tirpe, q. sus Descendientes se cassasen cõ la hermana mayor Edifico en el
Cuzco el primer Templo al Sol. No tuvo guer. Fue llamado el Sº. Las
Reales Insignias, y Traxe fuerõ al modo q. e repr. la Imagen d cuerpo entero
La muger principal, o Reyna, se llamava COYA, a distincion de otras menõ
princ. llamadas Pallas. La edad d los Ingas segun sus Anales fue tan
larga, como se nota en cada uno.
CAPAC YUPAN QUI INGA V.
IN CA RO CA INGA VI
YAHUARHU ACAC INGA VII
El Arrogante Fue tan fecundo que se di
ze tuvo 600 hijos No tio de Ayo al Here
dero Fundo Escuelas publicas Entro ala
Conquista de los Chunchos Vivio 154 a.
La Coya su Muger Mama Cusi Chim
bo Sucediole su Hijo.
Llamado assi por q. lloro sangre q. nacio
y despues en los Sacrificios Intro duxo los
Ayunos, y Penitencias a su modo Con
quisto a los Parinacochas, Condesuyos y
otros: Vivio 139 años La Coya Mama
Ipa Huaco Sucediole en Vida su Hijo.
MAMA HUA
COYA o REYNA Madr
de Manco Capac Inga
del Peru
HUAYNACA PAC INGA XII
TOPAC YUSIHUAS CAR INGA XIII
ATAHUALL PA INGA XIV
El Desgraciado Hijo de la Coya Mama
Raua Ocllo, La Coya su Muger Mama
Chuquillantu, Mal avenido cõ su herm.
Atahualpa, fue vencido de el y por su
orden muerto, en Jauja por los años de
1533 No tuvo hijo Varon.
El Vencido, Hijo de Mama Chachapo
ya Reyna de Quito Aclamado por R
ey de Cuzco Fue vencido y preso del
Conquistador Dn. Fra. Pizarro y despu
es degollado en Caxamarca hallase que
fue bautizado y se llamo Dn. Juan Año 1533
CARLOS V Rei XV del PERV
El Maximo El Año 1526.
El Marq. Dn. Fra. Pizarro
1542. Renuncio en su Hijo
LUIS I REY XXI del PERU
FELIPE V segunda vez REY XXII del PERU
FERNANDO VI
El Dichoso, Nacio en Madrid a 25 A
g. de 1707. Entro a reynar de 17 añ. por
Enero de 1724 y a los 8 meses mu
rio a 30 de Ag. dexando su sucession
é su Padre y herm. menores cã que
bolvio a empuñar el Cetro su Gran Pa.
Que Dios guarde lleno de Dias Vir
tudes y Triunfos En 24 de Febr. 1725
mando por su Real Cedula executar y
cumplir la del Sr. Carlos II a favor de
los Indios, diziendo: Porconvenira
ssi al Servicio de Dios y mio
El Justo Nacio en Madrid
julio de 1746 y fue procla

There are a number of HORIZON CARAVEL BOOKS published each year. Titles now available are:

BEETHOVEN
THE SEARCH FOR KING ARTHUR
CONSTANTINOPLE, CITY ON THE GOLDEN HORN
LORENZO DE' MEDICI AND THE RENAISSANCE
MASTER BUILDERS OF THE MIDDLE AGES
PIZARRO AND THE CONQUEST OF PERU
FERDINAND AND ISABELLA
CHARLEMAGNE
CHARLES DARWIN AND THE ORIGIN OF SPECIES
RUSSIA IN REVOLUTION
DESERT WAR IN NORTH AFRICA
THE BATTLE OF WATERLOO
THE HOLY LAND IN THE TIME OF JESUS
THE SPANISH ARMADA
BUILDING THE SUEZ CANAL
MOUNTAIN CONQUEST
PHARAOHS OF EGYPT
LEONARDO DA VINCI
THE FRENCH REVOLUTION
CORTES AND THE AZTEC CONQUEST
CAESAR
THE UNIVERSE OF GALILEO AND NEWTON
THE VIKINGS
MARCO POLO'S ADVENTURES IN CHINA
SHAKESPEARE'S ENGLAND
CAPTAIN COOK AND THE SOUTH PACIFIC
THE SEARCH FOR EARLY MAN
JOAN OF ARC
EXPLORATION OF AFRICA
NELSON AND THE AGE OF FIGHTING SAIL
ALEXANDER THE GREAT
RUSSIA UNDER THE CZARS
HEROES OF POLAR EXPLORATION
KNIGHTS OF THE CRUSADES

American Heritage also publishes AMERICAN HERITAGE JUNIOR LIBRARY books, a similar series on American history. Titles now available are:

FRANKLIN DELANO ROOSEVELT
LABOR ON THE MARCH, THE STORY OF AMERICA'S UNIONS
THE BATTLE OF THE BULGE
THE BATTLE OF YORKTOWN
THE HISTORY OF THE ATOMIC BOMB
TO THE PACIFIC WITH LEWIS AND CLARK
THEODORE ROOSEVELT, THE STRENUOUS LIFE
GEORGE WASHINGTON AND THE MAKING OF A NATION
CAPTAINS OF INDUSTRY
CARRIER WAR IN THE PACIFIC
JAMESTOWN: FIRST ENGLISH COLONY
AMERICANS IN SPACE
ABRAHAM LINCOLN IN PEACE AND WAR
AIR WAR AGAINST HITLER'S GERMANY
IRONCLADS OF THE CIVIL WAR
THE ERIE CANAL
THE MANY WORLDS OF BENJAMIN FRANKLIN
COMMODORE PERRY IN JAPAN
THE BATTLE OF GETTYSBURG
ANDREW JACKSON, SOLDIER AND STATESMAN
ADVENTURES IN THE WILDERNESS
LEXINGTON, CONCORD AND BUNKER HILL
CLIPPER SHIPS AND CAPTAINS
D-DAY, THE INVASION OF EUROPE
WESTWARD ON THE OREGON TRAIL
THE FRENCH AND INDIAN WARS
GREAT DAYS OF THE CIRCUS
STEAMBOATS ON THE MISSISSIPPI
COWBOYS AND CATTLE COUNTRY
TEXAS AND THE WAR WITH MEXICO
THE PILGRIMS AND PLYMOUTH COLONY
THE CALIFORNIA GOLD RUSH
PIRATES OF THE SPANISH MAIN
TRAPPERS AND MOUNTAIN MEN
MEN OF SCIENCE AND INVENTION
NAVAL BATTLES AND HEROES
THOMAS JEFFERSON AND HIS WORLD
DISCOVERERS OF THE NEW WORLD
RAILROADS IN THE DAYS OF STEAM
INDIANS OF THE PLAINS
THE STORY OF YANKEE WHALING

A HORIZON CARAVEL BOOK

# PIZARRO

## AND THE CONQUEST OF PERU


*By the Editors of*
HORIZON MAGAZINE

*Author*
CECIL HOWARD

*Consultant*
J. H. PARRY
*Gardiner Professor of Oceanic History and Affairs, Harvard University*

*Published by American Heritage Publishing Co., Inc.*
*Book Trade and Institutional Distribution by*
*Harper & Row*

SECOND PRINTING

Library of Congress Catalog Card Number: 68-9417

AMERICAN MUSEUM OF NATURAL HISTORY, NEW YORK

*Preserved for centuries in arid coastal burial sites, such fabrics as this detail from a Paracas tapestry testify to their makers' astounding artistry and inventiveness. By various techniques, including tie-dying, embroidery, and feather-weaving, Andean artisans produced cloths unequaled to this day. Above, a grotesque monster grasps a struggling human figure.*

# FOREWORD

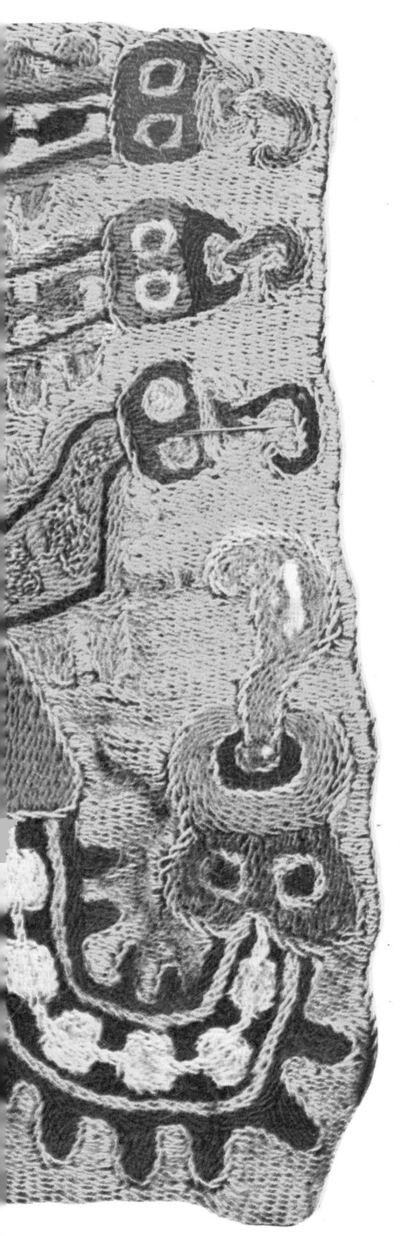

If the conquest of Peru were presented as fiction, few readers would accept the story—so incredible is it. Francisco Pizarro, the grizzled veteran of many Spanish campaigns in the New World, set out in late 1532 to win control of the vast and powerful Inca empire of South America; at his side marched fewer than two hundred men. That Pizarro dared to undertake such a seemingly hopeless task is as amazing as the fact that he succeeded.

Yet Pizarro's band could count on a few critical advantages. Steel armor, gunpowder, and horses were unknown to the natives, and these trappings of sixteenth-century European warfare struck terror in the hearts of the Indians at first encounter. The white man's diseases had preceded him, and thousands of Inca subjects, and at last the native emperor himself, had succumbed. Finally, a contested succession to the Indian throne plunged the empire into civil war just before Pizarro arrived. Although the Spaniards could not know it, the stage was set for history's most dramatic conquest.

The events of the Spanish overthrow of the Inca empire are brutal and violent. Yet the conquistadors were devout and God-fearing Christians. Their country had just emerged triumphant from a centuries-long crusade to expel the Moslems. The colonization of the New World, including suppression of its pagan peoples, was to the Spaniards another crusade upon which God had bestowed His blessing. The conquistadors' religious zeal, military ardor, and lust for gold blinded them to the values of the native civilization they sought to overcome.

Not until historians and archaeologists of later generations studied the vanished Indian cultures of South America did men realize just what it was that the Europeans had so casually destroyed. As an accompaniment to the following narrative, there are many illustrations of surpassingly beautiful Indian artifacts that reveal, almost pathetically, the glories of a vanished civilization. Against the high drama of the Spanish conquest must be placed concern for the tragic destruction of a noble people and their inspiring culture.

THE EDITORS

RIGHT: *Adorned as in life with a mace and necklace, this Chimú silver warrior probably was used as a ceremonial cup.*

MUSEUM OF PRIMITIVE ART

COVER: *The impassive face of an Inca noble, or* Orejon *("big ears"), decorates the side of this colorful* kero, *a painted wooden ceremonial goblet.*

NUESTRA SENORA DE COPACABANA, LIMA: PHOTO, JOSE CASALS

ENDSHEETS: *Manco Capac, founder of the Inca dynasty, and his wife (both in full figure) flank the top row of this eighteenth-century genealogy of Peruvian rulers. The last Inca, Atahualpa, and his Spanish successor, Charles V, are at the end of the second row, while the line ends with Ferdinand VI (1746–1759), called XXIII King of Peru.*

NUESTRA SENORA DE COPACABANA, LIMA: PHOTO, JOSE CASALS

TITLE PAGE: *This silver dish commemorating the conquest shows Francisco Pizarro (right) and Atahualpa (left) surrounded by lush Peruvian foliage.*

LIBRARY OF CONGRESS

BACK COVER: *A detail from the reverse side of Pizarro's banner (see page 69) shows a mounted horseman in full armor.*

MUSEO DEL EJERCITO, MADRID: PHOTO, ORONOZ

# CONTENTS

BIBLIOTHEQUE MINISTERE DE LA GUERRE, PARIS: GIRAUDON

I

# IN SEARCH OF EL DORADO

In 1526 two small caravels flying the Spanish flag lay at anchor off Atacames, a port on the northern coast of modern Ecuador. The men on board were greatly excited by what they saw on land—a town of about two thousand houses, arranged in parallel streets, and bordered by carefully cultivated fields. On the shore, crowds of Indians, dressed in tunics and cloaks of red and purple, gazed with wonder at the strange vessels. Close enough to observe the rich ornaments of the natives, the two middle-aged Spaniards who commanded the expedition, Francisco Pizarro and Diego de Almagro, realized that after years of toil and disappointment they had at last found the land of gold.

DUKE OF ALBA COLLECTION, MADRID: PHOTO, ORONOZ

*A Spanish caravel flying the red cross of Santiago graces another sixteenth-century mariner's map.*

Before the boats of either caravel could be lowered, the natives launched canoes full of fighting men. They wore quilted doublets and were armed with spears, clubs, and swords, all made of wood but tipped or edged with flint or copper. As they circled the ships in their swift craft, they showed no signs of fear. The watching Spaniards could tell by their painted faces and warlike gestures that they came on no friendly mission. Pizarro ordered a boat away to bring some of the Indians aboard, but the canoes shot toward the town, where scores of additional warriors were now being drawn up in battle order.

There were only about one hundred sixty Spaniards on the two small vessels, but they were used to facing odds. Pizarro embarked a landing party, complete with horses for the cavalry, and swinging himself down into the first boat, ordered the sailors to pull for the shore. He hoped that the sight of his formidable following—some at least in armor

*As the conquistadors sailed south from Panama, the land mass of South America was on their left, and the expanse of the Pacific on their right—just as the sixteenth-century French cartographer Le Testu has shown them (opposite), with North at the bottom and the Strait of Magellan at top. Warring natives and land and sea creatures are fanciful embellishments.*

and several of them mounted—would awe the Indians so that he could enter the town without a battle.

The landing of the Spaniards was unopposed, but as soon as the little army began to advance, it was surrounded by warriors. Spears, arrows, and stones filled the air. Although the Spaniards' steel armor protected them, they soon were halted by the sheer mass of Indians ahead. Pizarro, who had no wish for a full-scale battle, ordered his company to retreat. The withdrawal might have turned into a rout had it not been for an almost comic accident. Spanish chroniclers say that Pizarro was finding it difficult to reach his boats when one of his cavaliers fell from his saddle. The Indians, who had never seen a horse before, were so astonished by the sight of one creature breaking into two living parts that they drew back and left open the way to the sea.

Once aboard ship, Pizarro called a council of war with Almagro. The two leaders who confronted their disgruntled soldiers provided a strange contrast. The taller and more impressive of the two, Francisco Pizarro was a lean man with a stern, lined face, graying hair, and a pointed, well-kept beard. Those who knew him said that he was prudent and discerning, even temperate—in almost everything but his appetite for gold. Yet none would question that he was a man who demanded much of himself, and therefore, much of his men.

POMA DE AYALA, *Nueva Coronica* . . .: ROYAL LIBRARY, COPENHAGEN

*Flanking a drummer, Diego de Almagro and Francisco Pizarro carry banners bearing their own names in this drawing from Felipe Poma de Ayala's profusely illustrated codex, or manuscript history, of the conquest. Completed early in the seventeenth century, Ayala's work is a highly important source of information about Inca customs and early history, as well as a record of the Spanish conquest.*

His partner, Diego de Almagro, on the other hand, was more easygoing and lighthearted. He was short and far less noticeable, except for the often gaudy clothes he favored. Unlike Pizarro, who could be quite eloquent at times, Almagro frequently had difficulty in expressing himself. Moreover, his face was disfigured by the loss of an eye, and his whole body carried scars of innumerable heroic campaigns. A little hasty and rather passionate in action, he was at one with the fighting ranks and was more understanding of his soldiers.

Despite such differences, the two men had much in common. Both were about fifty-five years old and had been brought up in poverty, nameless men with no prospects, in their native land of Castile. Both had left home with no regrets, determined to make their fortunes in the New World discovered by Columbus in 1492. They had been soured by years of hard living and perilous adventures among the conquerors of the Spanish Main—years that unfortunately had brought little reward to either man. Although they had proved their mettle in a score of fights and were trusted cap-

DE BRY, *India occidentalis*, 1596: NEW YORK PUBLIC LIBRARY

*Flemish engraver Théodore de Bry's noted work* India occidentalis, *published between 1590 and 1634, reveals, among other things, the ignorance of most Europeans of events in the New World. In the serial actions pictured above, Pizarro's men furl sails (left), set out for shore, and encounter a band of hostile natives (who are inaccurately attired in loincloths rather than in ponchos).*

tains, neither fame nor riches had come their way. When at last their chance had come to outdo Hernando Cortes, who had staggered the world by his amazing conquest of Mexico in 1521, Pizarro and Almagro were aging men. Both realized that they were making their last throw against fate.

The Spaniards who crowded the deck of the caravel off Atacames did not hesitate to take part in the heated discussion between the two commanders. At a time when new ships, methods of warfare, and weapons were being developed in Europe, the Spanish conquistadors had to make do with forces that were more like the bands of freebooters of the Middle Ages. Their vessels were merchant ships with a few small cannon on board; their men either armed themselves as best they could or took what their captain provided. Few had crossbows; even fewer could afford muskets or horses; most of them fought with pikes and swords. No more than a half dozen or so of any given force were gentle-

# A MODEL FOR PIZARRO

When Francisco Pizarro set out from Panama in 1524, the lands he dreamed of securing for the Spanish crown were virtually unknown—but not the means of conquering them. Hernando Cortes' conquest of Mexico a scant three years earlier had shown the world in general, and Spanish adventurers in particular, that it was possible for a handful of determined men to subjugate a vast and populous empire. Pizarro, like Cortes before him, was to ally himself with one faction of the native population, declare war upon the others, and amid such divisiveness, imprison and eventually execute the native king. Opposite, the murder of Montezuma and one of his chiefs by Cortes' henchmen takes place under the baleful gaze of the Aztec god Tezcatilipoca, who viewed the world through a mirror in his foot. Below, native bearers accompany Cortes en route to Mexico City.

BIBLIOTHÈQUE NATIONALE, SERVICE PHOTOGRAPHIQUE, PARIS

BIBLIOTECA LAURENZIANA, FLORENCE

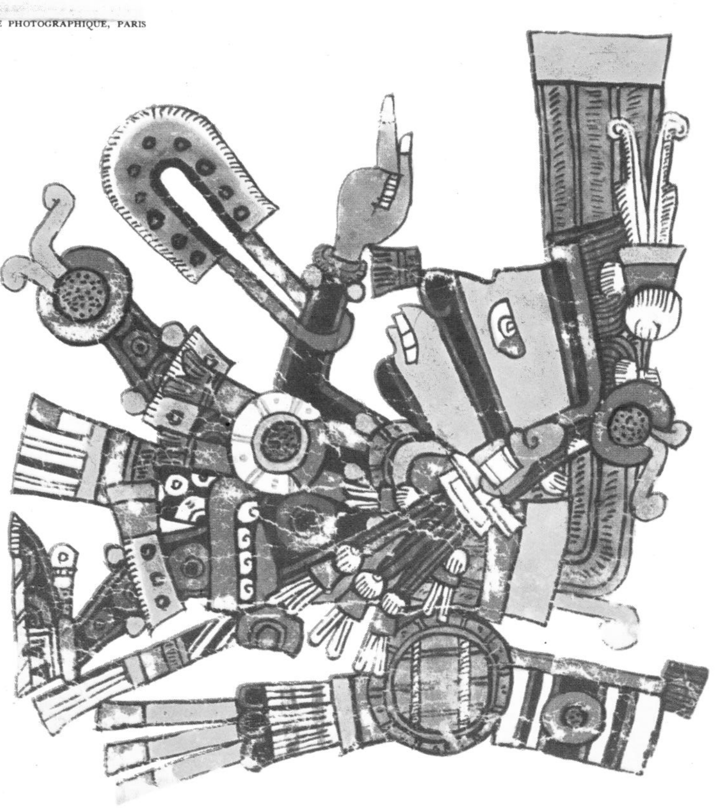

BIBLIOTECA VATICANA, ROME

men, trained in arms, and not many more were professional soldiers. Many were ruffians or misfits, who joined up simply for loot; but some were respectable artisans—cobblers, bakers, carpenters, tailors, blacksmiths—looking indeed for adventure, but hoping also to establish themselves in the business of a young colony once the conquest had been completed. They made splendid fighting material, for they had lived rough lives and knew how to get out of a tight corner. But they were difficult to control when things went wrong—as things did at Atacames—or when an enemy town lay defenseless before them.

All were agreed that they must abandon Atacames. They had no doubt that they could take it by storm, but their losses might be heavy and they would stand little chance of overcoming the armies that probably would be sent against them. The more fainthearted urged Pizarro to give up his attempt to conquer this land and to return to their base of operations at Panama. Pizarro and Almagro firmly opposed this suggestion. For two years they had explored the western coast of South America to find gold, suffering hunger and danger on land and sea. They had fitted out ship after ship, exhausting their resources yet contriving to borrow money when all hope seemed lost. To return now to Panama meant ruin.

*Exploration of the Americas and exploitation of their riches made smithing and shipbuilding—both shown in this early woodcut—two of the chief Spanish industries.*

Almagro proposed that he sail for reinforcements while Pizarro held out on the coast to await his return. A fierce quarrel flared up between the two leaders. Once already on this voyage, Pizarro protested, he and a few dozen men had faced death in the dark forests and the humid ravines of this land. Boa constrictors, alligators, and lurking Indians had killed some of them; the survivors had managed to subsist on the wild potatoes they had rooted up in the forest or the bitter fruit of the mangrove on the shore, where they could escape from the swarms of mosquitoes only by burying themselves up to their necks in sand. Was he to endure these hardships again, he demanded, while his partner sailed away in safety to the comforts of Panama?

Almagro replied fiercely that he would stay and that Pizarro could return to Panama. Pizarro was not so easily appeased, however. Argument turned to bitter accusation, and their hands flew to their swords. At this moment it seemed as if the enmity between the two adventurers would drown all hope of success in blood. Bartholomew Ruiz, the pilot, and Nicolas de Ribera, the treasurer of the expedition, intervened and managed to calm them. But this quarrel was the first open sign of an implacable hatred between

Pizarro and Almagro, a rivalry that was to lead to a civil war between the conquerors and to untold misery for the Indians.

In the end Almagro sailed north in his caravel, while Pizarro landed on the nearby island of Gallo with a handful of unwilling companions. The natives of Gallo fled to the mainland at the first sight of his sails. This was not an unmixed blessing: Although the Spaniards were freed from the danger of attack, they were thrown completely on their own resources. After a few weeks their stores ran out, and they were forced to live on shellfish. Their only shelter from the torrential rain consisted of boughs lopped from the trees. Worst of all, there was no escape from this miserable existence. Pizarro, grim and resolute as ever, had sent his ship back to be refitted at Panama a few days after they had landed—probably because he was afraid that some of the desperadoes in his company might mutiny and seize it.

NEW YORK PUBLIC LIBRARY

*Pizarro, Luque, and Almagro, full partners in the plan to conquer Peru, are pictured in conference by de Bry, who has dressed the destitute adventurers elegantly.*

In the meantime, Almagro had met a cold reception in Panama. The Spanish governor of the colony was shocked at the sight of the scarecrows who returned with him. Matters were made worse when the governor's wife found a message hidden in a ball of cotton that had been presented to her as a specimen of the products of the southern coast. This letter, written by one of the soldiers left behind on Gallo but signed by several of his comrades, described their wretched state, accused Pizarro and Almagro of detaining them against their will, and begged the governor to send a ship to rescue them. The news of this communication soon spread among the colonists, who made great play with the doggerel verse with which it ended:

> Look out, Señor Governor,
> For the drover while he's near;
> Since he goes home to get the sheep
> For the butcher who stays here.

With the jeers of the riffraff of Panama ringing in his ears and the stern refusal of the governor to allow him to raise reinforcements—which would have been almost impossible anyway under the circumstances—Almagro might well have given up hope. Up to that point, only hardship and defeat had greeted his four-year-long search for the gold of Peru.

In 1522 he had joined in a partnership with Pizarro and an influential priest named Hernando de Luque, with the object of launching an expedition south of the Isthmus of Panama. Their lust for riches had been stimulated by

*In his massive nine-volume work, Baltasar Compañon, an eighteenth-century bishop of Trujillo, Peru, assembled pictures of many native Peruvian plants. This illustration includes a lizard about to snare an insect with its tongue.*

rumors of a fabulous kingdom—richer by far than Mexico—that had been reaching the colonists for some time. These tales of a new El Dorado had been confirmed that year by Pascual de Andagoya, who returned from a voyage of discovery that had taken him south along the west coast of South America to a river he called the Birú (a name supposedly later corrupted and applied to Peru). The explorer had traveled inland and met Indian traders who claimed to have come from a land where gold, silver, and precious stones were plentiful. Since Andagoya was a desperately sick man and unable to continue his explorations, the three partners had obtained sanction from the governor to equip an expedition at their own expense to follow up his discovery. Luque, who was the wealthiest of the three, raised the money, while the two soldiers busied themselves finding ships and men.

In November, 1524, one ship and about a hundred fighting men sailed from Panama under Pizarro's command. Almagro was not ready until weeks later. On this first voyage they ran into trouble from the very beginning. Unwittingly Pizarro had embarked at the worst time of the year for a voyage south—the rainy season, when the prevailing winds would be against him; and he was to run into storms and frightening seas throughout the voyage. When at last he reached what he thought was the Birú (the name has been applied to several rivers) and landed with a party to explore, Pizarro found no towns to plunder but only swamps and forests.

A brief march inland proved that the countryside was a pesthole, a swamp of stagnant waters and slimy soils that made travel difficult under the best of circumstances. For men clad in heavy mail, as the Spaniards were, it was nearly impossible. They slipped and slid their way through the muck only to find themselves entangled in thickets, and beyond that they faced the horror of an exhausting hike in unbearable heat over rocky hills.

Soon forced to re-embark, the adventurers left the river and sailed into furious gales, during which the crew pumped for their lives for ten days. On subsequent landings, they were lucky to find a number of small settlements at which the Indians, in their hurry to escape, had left food and crude golden ornaments.

Almagro, following in Pizarro's wake, collected more gold but was struck in the head by an Indian javelin, and after suffering excruciating pain, lost an eye. The grand total of their loot on this first voyage failed either to pay

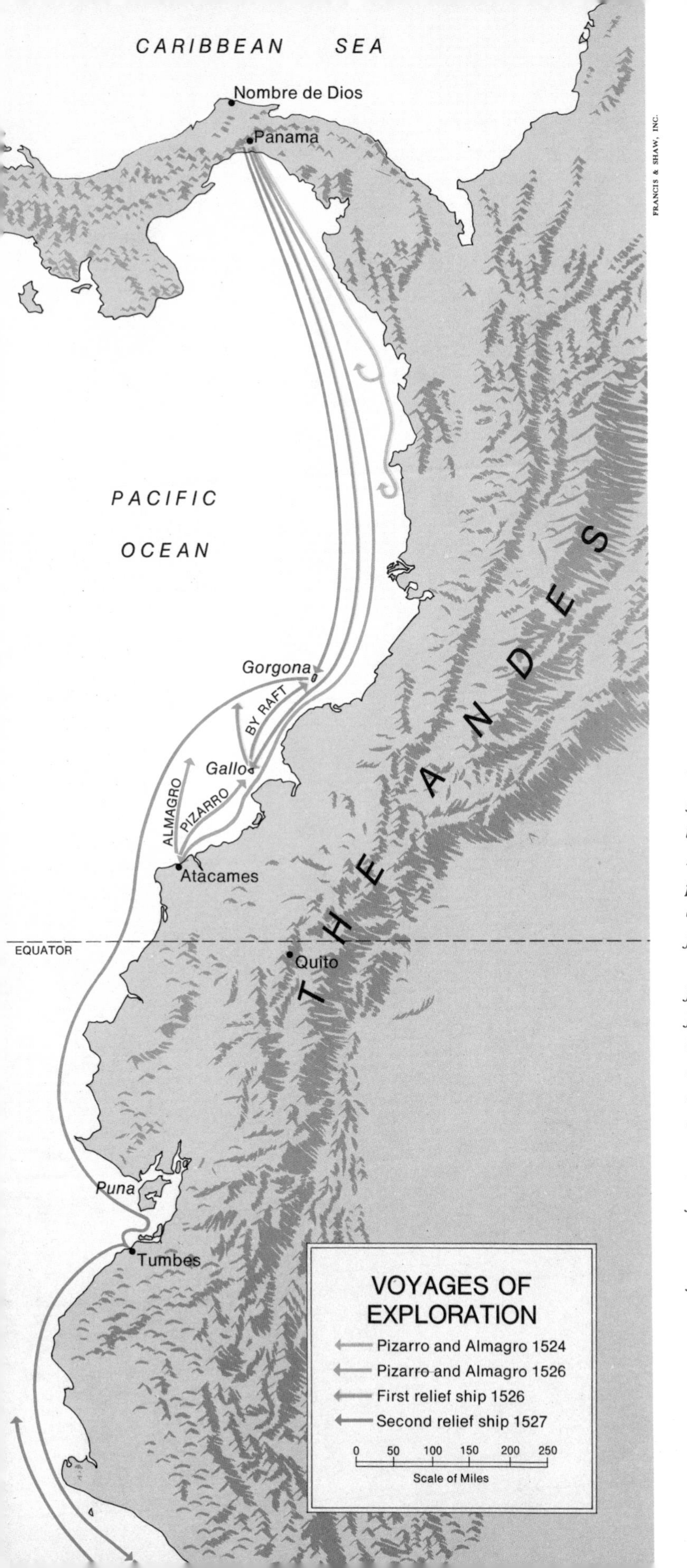




*The highly simplified map at left shows the increasing boldness of the Spanish conquistadors, led by Pizarro and Almagro, in their explorations of the western coast of South America. Two years after a foreshortened initial voyage in 1524 (orange), the two pushed as far south as Atacames (blue) before splitting up—Almagro returning to Panama to seek new backing and Pizarro withdrawing to the island of Gallo. Refusing assistance from the first of two relief expeditions (green), Pizarro—and the twelve who remained faithful to him—pulled back to Gorgona, where, in 1527, the second relief expedition (red) found him. With this ship, Pizarro scouted farther south, returning with sufficient evidence of the wealth of the Inca empire to win the necessary support for his projected conquest of Peru.*

for the expedition or to impress Panama's adventurers.

The two soldiers, loyally supported by Luque, refused to admit defeat. More money was raised, and with great difficulty one hundred sixty men were recruited for a second voyage in 1526, the one that carried them to Atacames.

When Almagro returned to Panama after the adventure at Atacames, few believed his reports of the wealthy town that had given him and Pizarro such a rough reception, and none volunteered to join him in a mission to relieve Pizarro, who was now on the island of Gallo. Moreover, the governor was adamant in his refusal to let more lives be wasted. He dispatched a relief expedition of two caravels to Gallo, giving strict orders to their commander to bring back Pizarro and his followers.

The well-provisioned ships were welcomed rapturously at Gallo. Most of the half-starved survivors never had

PLAUTIUS, *Nova typis* . . ., 1621: NEW YORK PUBLIC LIBRARY

MAGNUS, *Historia de gentibus . . .*, 1555: LIBRARY OF CONGRESS

*Fearsome sea creatures supposedly lurked beneath the surface of all uncharted waters, and fear of an attack, such as the one shown in the 1555 woodcut at right, made voyages all the more terrifying.*

*A century of exploration did little to dispel many European misconceptions about the New World, and docile land and sea monsters populate this 1621 engraving.*

wanted to stay there in the first place, and their dearest wish was to return to civilization at Panama. There was no need to order them aboard. But Pizarro was determined to follow up the discovery at all costs.

The crusty old soldier was a man of few words. Unsheathing his dagger in a dramatic gesture, he drew a line in Gallo's sands, extending from east to west. Then, according to one Spanish chronicle, he raised his voice so that everyone could hear him and know that he meant what he said:

"Friends and comrades," he pointed as he spoke, "on that side are toil, hunger, nakedness, the drenching storm, desertion, and death; on this side, ease and pleasure. There lies Peru with its riches; here, Panama and its poverty. Choose, each man, what best becomes a brave Castilian. For my part, I go to the south." He stepped across the line, and thirteen men followed him. The rest chose to return to Panama.

Rarely in the course of history has such a momentous decision been made with so little justification. Pizarro had seen only the outposts of the land he planned to conquer. He knew nothing of its ruler or of that ruler's armies. More important, he already had been repulsed by the natives of a town on the far north of the empire. Still, against the express orders of his own monarch's governor, and with only thirteen men to follow his lead, he dared to go on with his daring quest.

The captain of the relief ships indignantly protested Pizarro's decision, viewing it as little better than a wish to commit suicide. In an effort to discourage him, the captain refused to leave even one of his ships for Pizarro's use, and the men had difficulty in persuading him to leave them any stores. The captain's arguments had not turned the fourteen from their purpose, although one of them, Ruiz, returned to Panama to help Almagro and Luque outfit an-

THE OLD PRINT SHOP, INC., NEW YORK

other expedition. The hearts of the remaining thirteen must have sunk low, indeed, as they watched the friendly sails drop slowly below the horizon.

Alone again on the barren island, Pizarro knew that activity was required to raise his companions' spirits. He immediately set them to work at building a rough craft in which they could cross to a small, uninhabited island to the north. He was afraid that the natives of Gallo would return when they saw the ships sail away.

The new location, while not inviting, was preferable to the old. Lush forests provided the Spaniards with wood, and there was plenty of game—both hare and pheasant—for the pot. However, the adventurers were still tormented by poisonous insects and drenching rains. Knowing that idleness can lead to despair, Pizarro wisely devised a routine to occupy his companions' time. Each day was started with prayers, the festivals of the Church were always observed, and emphasis was placed on the protection that would be afforded by Heaven to those whose object was to convert the heathen.

Despite the fact that many who sailed to conquer Peru were adventurers of the lowest sort, all were touched by nobler sentiments. Most of them were deeply religious; and intermixed with their greed for gold and an often fiendish cruelty there remained a touch of the old crusading spirit. They believed that in conquering lands assigned to their king by the Pope, they were doing God's will and following the same quest as their fathers, who had driven the Moors from Spain in the preceding century.

The months passed tediously, without a sail breaking the horizon, and in spite of Pizarro's efforts, the men became despondent. They might well have given up all hope had they known what was happening in Panama.

Furious with Pizarro's defiance of his orders, the governor had declared that the thirteen had deliberately thrown away their lives. As far as he was concerned, his responsibility was ended. He would do nothing more to help them, nor would he give Almagro permission to relieve them. For months Luque used all his powers of persuasion, seemingly in vain. He accused the governor of inhumanity in deserting men who were risking their lives in the service of the Spanish monarch, Holy Roman Emperor Charles V, and

*There are no contemporary portraits of Francisco Pizarro, who was in his mid-fifties in 1526—somewhat older than he appears in this 1760 canvas.*

*The wealth of the Inca realm belied the general poverty of its inhabitants—most of whom, like the couple above, squatting in front of their hut, lived simply.*

laid stress on the fact that definite instructions had been given by Charles to his governors to encourage and support exploration. Finally, the governor of Panama yielded. One small ship was permitted to leave with enough sailors to work her, but orders were sent to Pizarro to return to Panama at the end of six months and report to the governor.

Almagro and Ruiz lost no time in equipping their craft, and under the latter's experienced hand, this second relief expedition reached its destination without mishap. It is difficult to imagine the feelings of those forsaken men after seven months in "hell," as the Spaniards called the island, when they saw that ship drop anchor. Their clothes were shredded, and their bodies emaciated; still their eyes burned with determination. This time there was no talk of sailing north. Pizarro realized, however, that with one ship and only a dozen or so fighting men, all thoughts of conquest were out of the question. He had a bare six months to reconnoiter the land and to bring back evidence to convince the governor of the importance of another expedition.

It was to the Indians whom Ruiz had recruited during a side trip south of the equator that Pizarro now turned. These men had by this time learned Spanish and had told the adventurers much about their country. Later they would serve well as interpreters. At their bidding, the little ship set its course far to the south, for a native town called Tumbes, one of the more important ports in the northern part of their emperor's land.

It took three weeks to reach their destination on the Gulf of Guayaquil, where beautiful towns and villages sat snugly in a narrow plain between the Andes and the sea. It was dark when they arrived, and they moored in deep water at the entrance to the bay to wait for morning.

Pizarro, of course, could have no idea of what awaited him in Tumbes. If news of his plundering farther north had reached the town, he could expect the same kind of reception that he had received at Atacames. He therefore planned to keep his forces aboard ship, where he felt certain that they could repel any attack that might be launched. As he sailed across the bay the next morning, his worst fears seemed verified. A fleet of balsas—large rafts made by lashing together trunks of the light balsa tree—had already left port, and as the balsas drew closer, it became more and more obvious that the giant rafts were crowded with warriors. Pizarro did not hesitate. Ordering the helmsman to steer for them, he put his caravel alongside the leading craft. When they were within speaking distance, he had his interpreters question the natives. He learned that he had little to fear. The war party was bound for Puná, a nearby island that was in a state of perpetual warfare with Tumbes.

Somewhat relieved, Pizarro invited some of the chiefs to join him on his ship. The Indians looked with wonder at the strange bearded men and at their shining armor, for in Peru beards were rare, and the uses of iron were unknown. And they were naturally astonished to see their fellow townsmen on board. The interpreters explained to them, however, that the Spaniards were superior beings who would do them no harm, but who had come simply to visit.

Pizarro received the natives courteously and assured them again that his visit was a friendly one. He persuaded them to abandon their raid on Puná and carry back his message of peace and good will to their *curaca*, or chieftain, and to ask him to send much-needed provisions to the ship. As the balsas made for shore, the Spaniards looked with delight at the scene before them. A splendid fort, with a temple nearby, towered above other pleasant buildings, and all stood amid fields of corn, fruit, and vegetables, watered by irrigation canals branching out from the river that ran through the valley.

Soon the balsas returned, laden with animals, fish, and other stores, including a number of llamas, an animal unknown to Europeans. Even more impressive was the appearance of one Indian noble who had come out to visit them. Wearing superbly woven robes, with massive plugs of gold in his ears and precious stones about his neck and arms, he was obviously a man of great importance. The Spanish commander stepped forward to welcome him with suitable ceremony, and for the first time Pizarro stood face to face with an Inca noble.

ALL: ROYAL LIBRARY, COPENHAGEN

*Among numerous agricultural views in Ayala's manuscript are the two at left. One (far left) shows peasants cultivating an irrigated field; the other shows the same field being harvested—by both man and llama—under a fuller moon. In each, an official tends a fire in the background.*

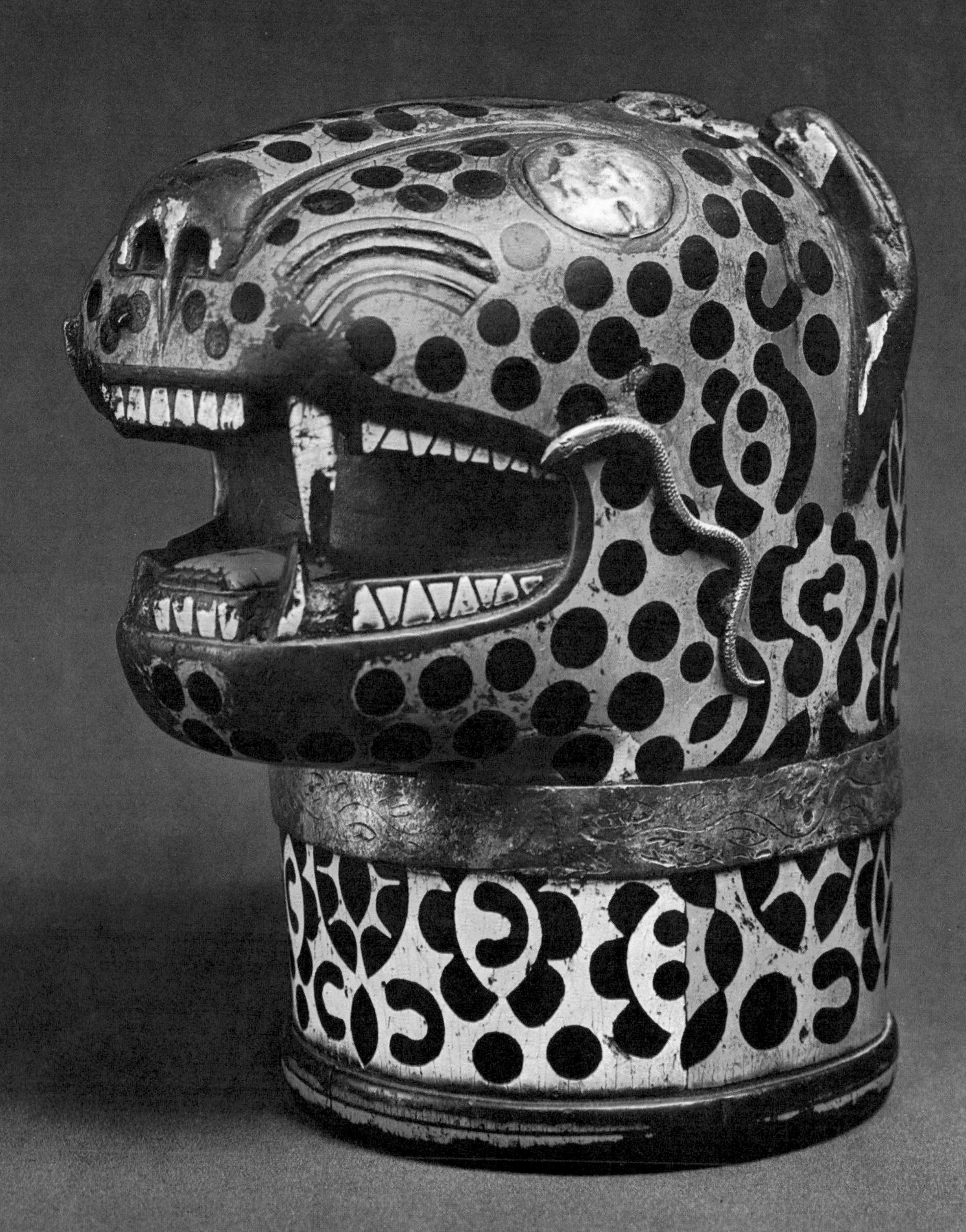

# II

# REALM OF THE INCAS

Pizarro had at last made contact with the Inca empire. In the years that followed he was to learn that its riches far exceeded even his wildest dreams. But if its wealth seemed to belong more to the realm of fairy tale than to hard fact, it was certainly not the only element of fantasy to be found there. Temples with walls encased in gold, splendid palaces with glittering thrones, and massive fortresses provided merely a setting for the empire's ruler, called Lord Inca and believed by his people to be the child of Inti, the Sun God.

The Lord Inca was master of all lands stretching from modern Colombia to Chile, and he was so powerful that not even the greatest chieftain dared to enter his presence without deep obeisance. To show how lowly he was before his lord, an individual would remove his sandals and carry a burden on his back when presenting himself to his monarch, and rarely was he permitted to see his ruler's face. During an audience and while traveling in the royal litter, the Lord Inca generally was hidden from the public by a screen. It was a memorable event when he showed himself to his worshipful subjects.

Even in death he retained this esteemed position in the continuing life of the nation. His mummy was placed in the Temple of the Sun, and his palace was occupied by numerous noble attendants who kept it as if the Lord Inca himself were still alive. On ceremonial occasions his body was carried out into the great square at Cuzco, the Inca capital, to sit in solemn state in the company of fellow Lord Incas who either had preceded him in death or had succeeded him in power. His wealth, his palace, and the reverence paid to his person long outlasted his mortality.

ROYAL LIBRARY, COPENHAGEN

*The mummified remains of Inca rulers were publicly displayed on holy days, as this Ayala shows.*

*Inca artisans used their skills to decorate everyday as well as ceremonial objects. This highly stylized ocelot head, with its silver neckband and a gold serpent in its mouth, served as a* kero, *or wooden drinking cup.*

MUSEUM OF THE AMERICAN INDIAN, HEYE FOUNDATION

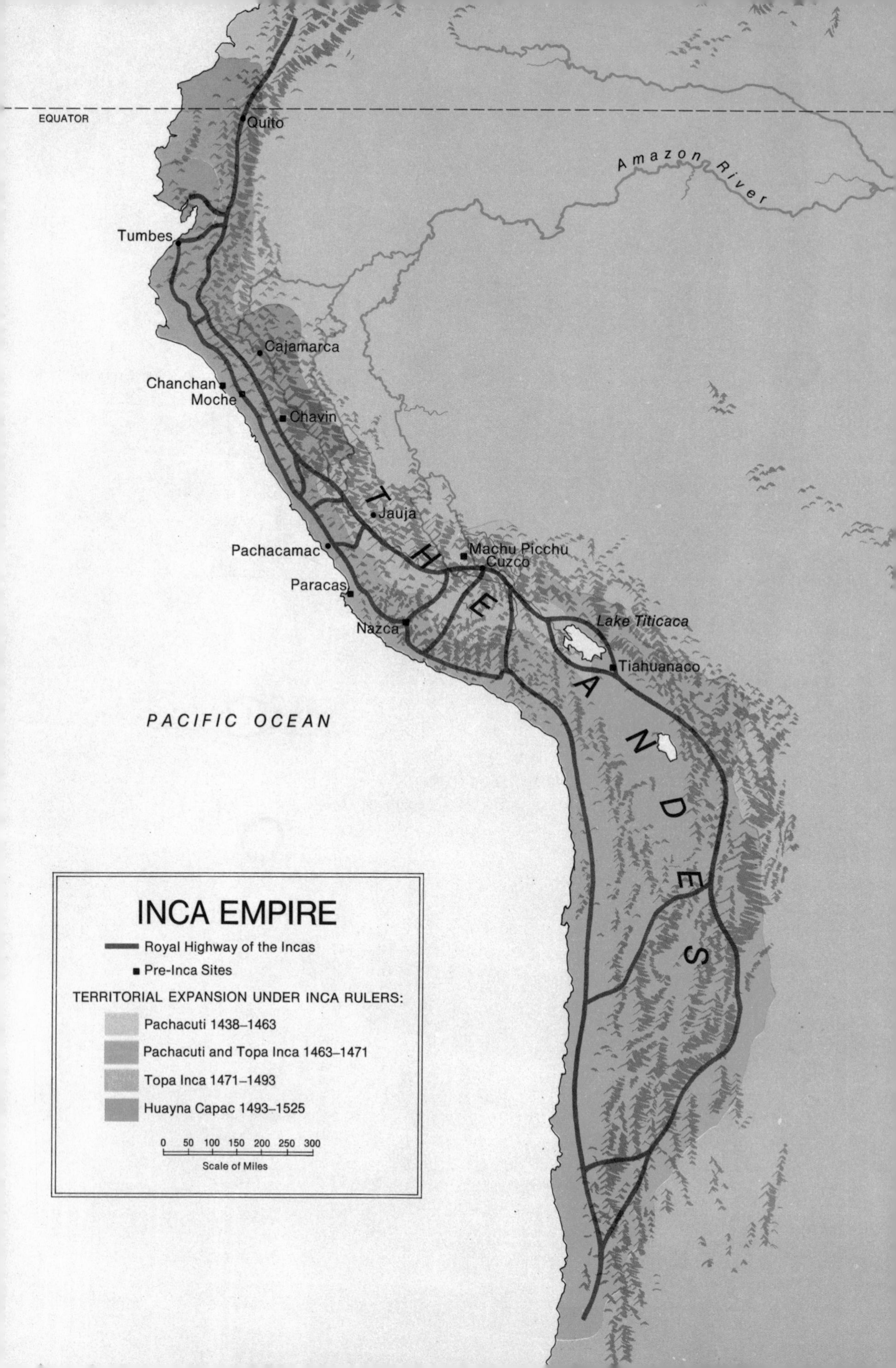

EQUATOR
Quito
Amazon River
Tumbes
Cajamarca
Chanchan
Moche
Chavin
Jauja
Pachacamac
Machu Picchu
Cuzco
Paracas
Nazca
Lake Titicaca
Tiahuanaco
THE ANDES
PACIFIC OCEAN
INCA EMPIRE
Royal Highway of the Incas
Pre-Inca Sites
TERRITORIAL EXPANSION UNDER INCA RULERS:
Pachacuti 1438–1463
Pachacuti and Topa Inca 1463–1471
Topa Inca 1471–1493
Huayna Capac 1493–1525
0 50 100 150 200 250 300
Scale of Miles

The splendor and power of successive generations of these mighty rulers, was not, however, the most remarkable feature of this newly discovered land. The world had known other empires ruled by despots who were feared in life and worshiped in death. But it had no record of such a society as that of the Incas, in which the lives of millions had been totally assimilated into one empire and organized under the strictest of regimens.

From the moment of birth to the day of his death, from sunup to sunset, every loyal subject lived as prescribed by the Lord Inca. Most spent their time working in the fields, men and women together, often singing a hymn to Lord Inca or to Inti, under whose protection they labored. Sometimes a chant simply dramatized their daily routine, as in the following, called "*Jailli,*" which means "triumph" and suggests a victory that comes with work well accomplished.

*QHARICUNA:*
*Ayau jailli, ayau jailli,*
*Kayqa thajilla, kayqa suka!*
*Kayqa maki; kayqa jumpi!*
*WARMICUNA:*
*Ajailli, qhari, ajailli!*

THE MEN:
Ho! Victory, Ho! Victory,
Here digging stick, here the furrow!
Here the sweat, here the toil!
THE WOMEN *(answering)*:
Huzzah, men, huzzah!

TEXTILE MUSEUM, WASHINGTON: PHOTO, LEE BOLTIN

*Heavy gold ear plugs, worn by the Inca and other high officials, were both decoration and a badge of office, and when Atahualpa's ear lobe was torn in battle—an extremely ill omen—he was forced to hide the injury under a cloak. The Chimú plugs at left, 5 1/2 inches in diameter, bear designs on all surfaces, including the whorls.*

Those who did not till the soil worked as craftsmen or provided some other specialized skill or labor. No able-bodied man sat idle, and no one was allowed to be a slacker. Everyone had a definite task to do and did it; punishments were cruel for those who shirked their duties.

Even members of the Inca nobility were not exempt. The Spaniards called them *Orejones*, or "big ears," because the lobes of their ears had been grotesquely stretched out of shape by large ear plugs, which they wore as a mark of their high station. To the nobles fell the more prestigious positions, roles of authority in their religion, army, or state. They alone became the empire's high priests, viceroys, governors, and military commanders. Since the Inca society was based on a rigid, almost static class system, most *Orejones* were princes in their own right, descended from a member of the royal family. But some were the chiefs of conquered tribes, who, for service or loyalty, had been promoted by the Inca. In the course of centuries, many of these men and their ancestors had helped to win for the Inca his great domain, and he, in turn, showered them with honors and wealth. In addition, unlike the common man, they enjoyed certain privileges, a plurality of wives, and exemption from the work tax. Still, each had his assigned duties to perform, and like the common man, an *Orejon* could easily lose his life if he failed to fulfill those duties.

Between the nobleman and the common laborer no middle class of landowners and merchants existed. All land, mines, and herds of llamas and alpacas, the domesticated animals of the Andes, belonged to the emperor, who benevolently divided the giant herds and arable land in three parts to support his empire. The first portion was set aside to provide for his own needs; the second supported the religion, its temples, priests, and attendants; and the third was parceled out annually among the people. The size of

MUSEUM OF PRIMITIVE ART

*For very different reasons, the two highly similar animals opposite were vital to the Inca economy. As these silver statuettes show, the alpaca (left) was prized for its luxuriant fur, while the red-blanketed llama served as a pack animal. Both were considered too valuable to slaughter for meat. The bird motif on the sides of the Chimú jug at right suggests a temple frieze (see also page 44), and the identical scenes, front and back, show a noble sitting on a low throne. The stirrup-shaped spout of this worked silver vessel was also a convenient handle.*

PEDRO ROJAS PONCE

each plot of ground or herd of animals allotted was based on the number of individuals in each family. Where land was involved, the birth of a son earned a father twice as great an increase in his share as did the birth of a daughter.

Both llamas and alpacas were carefully preserved—until the sixteenth century, when the Spaniards came, and unaware of their value, wantonly slaughtered them. The llamas provided the Incas with their only pack animals, and the alpacas produced a fine black or brown wool from which the Indian women made clothes for their families. Seldom was either killed for its meat, and llamas were shorn for their wool only when they died. Their hides then were used to make the sandals that were worn by peasant and noble alike.

Besides these two domesticated animals, vicuñas and guanacos roamed wild through the mountains, grazing the upland regions in vast herds. At regular intervals the royal officials organized great hunts, during which hundreds of these wild beasts were rounded up—the vicuña to be shorn and set free, the guanaco to be killed for its meat. Each subject received his just allotment of the hunt's yield of guanaco, but only the Lord Inca and his nobles were permitted to wear the fine fabric produced from vicuña wool.

Work in the fields was even more highly organized. Farming produced most of the basic foodstuffs on which the Indians depended. When the time came for sowing or harvesting, local officials summoned all peasants in their dis-

BOTH: MUSEUM FUR VOLKERKUNDE, BERLIN

*On terraced mountain slopes, such as those still farmed today at Pisac (opposite), the Incas raised both white and sweet potatoes (represented above on a stirrup-handled water jug). Corn, grown in irrigated fields in the lowlands, was cultivated with great care, as the silver tomb offering below makes evident. It was the vegetables themselves, however—not the precious metals in which they occasionally were represented (left)—that were to be the Americas' most significant contribution to Western Europe. Together they reduced the recurring threat of famine and unreliable harvest yields.*

ROBERT SONIN

ALL: ROYAL LIBRARY, COPENHAGEN

*Corn and potatoes were harvested communally under Inca law, beginning with those fields sacred to the sun, but each worker was the recipient of what his own plot of land produced. Corn (above) was harvested first, and roughly one month later the potatoes were dug and bagged for storage (below).*

tricts to labor, first on the sacred fields, then on those of the Lord Inca. The last fields to be tended were those of the individual or a neighbor away at war or called to work on some government project. To show what high regard he had for such labors, the emperor himself appeared on these occasions, along with his Inca nobility, and inaugurated the work with a golden implement.

In this remarkable empire no one starved. Those unable to farm for themselves—the old, the sick, men whose talents required that they perform other work—were provided for by those strong enough and destined by their low station to cultivate the fields. To see to the proper apportionment of food, two sets of storehouses were erected at many points throughout the empire. Into one went the produce from the state's fields; into the other, the produce from the sacred fields. A portion of the former was parceled out to those in need or not engaged in husbandry. When there was a bad harvest, the royal storehouses were opened also to any Inca subject who had suffered from such crop failure; when the harvest was abundant and the storehouses were overflowing, the excess was divided among the entire population.

In the Andean region the main crop was the white potato, often grown in sheltered mountain valleys as high as fourteen thousand feet above sea level. There terraces were built along the slopes to provide fullest use of even the smallest areas of level land. Many of these terraces with their retaining walls of stone are still in use today. Maize, or corn, was the staple food farther down the mountains, while in the warmer regions below, a great variety of crops were cultivated—several kinds of beans, tomatoes, peanuts, manioc, and sweet potatoes. A highly developed irrigation system brought water to the fields on both mountain terrace and low-lying ground.

All Inca subjects began to work at an early age, and all married by the age of about twenty, under the supervision of an Inca official, who, if necessary, paired off those who had not chosen their own mates. Any able-bodied man was liable to be conscripted to fight, or as part of his *mita*, or work tax, to build roads, storehouses, or fortresses for his emperor. Or he might be required for short periods each year to labor in the mines, a task that was onerous to everyone. Each year, too, Inca officials selected a number of the most promising young boys from each district to serve the Lord Inca as craftsmen or in various menial ways. They were not considered slaves, however, and the ablest of them often rose to high positions in the state. Similarly, the most beautiful or talented girls were chosen to be educated in an *acllahuasi*, or state convent, built in each region. When their course in religion, weaving, and domestic tasks was completed, some were consecrated to the service of the Sun God and took vows of perpetual chastity, while others became the secondary wives of the Inca or his nobles.

Oddly enough, this highly organized and interdependent society was the result of only a century of imperialism, a century during which many different tribes, varied in customs and languages, fell before the Inca sword. Still, all were assimilated into the quickly growing empire with unusual skill and ease. To avoid the possibility of rebellion in newly conquered territories, a practice called *mitima* was carried out. Through it, hundreds of families within the acquired district were transferred to provinces already part of the Inca domain. There they were given new homes and a portion of the fields. Whenever possible, they were placed in provinces similar in climate and natural condition to their old tribal lands. To replace these exiles in the newly conquered territory, a contingent of *mitimaes* then was chosen from among those Inca subjects who had long been faithful to their lord. The *mitimaes* manned the distant garrisons and took up residence with the new tribe to spread the Inca culture, language, and religion, as well as Inca methods of husbandry and craftsmanship.

*Each spring Inca nobles—using crude hoes whose blades were made of gold—formally broke ground for the first planting (above). During the ensuing harvest, the grains gathered from the Inca's own fields were stored in huge government granaries (below) and distributed according to need.*

As a reward for leaving their native land to serve the emperor, the Lord Inca treated the *mitimaes* with special favor. They were allowed more than one wife and given gifts of gold and silver, wool cloth, and feathers. Few empires have organized the mass movement of subjects so successfully or with such desirable results. Disloyal movements among the conquered peoples could be discovered quickly by the *mitimaes* and communicated to the royal officials so that immediate action could be taken.

In addition to the *mitima* program, the sons of the chiefs and nobles of a new tribe were brought to the school at Cuzco, to be educated with the sons of Inca nobles. There a four-year scheme of indoctrination into the Inca way of life was provided. It was the only school that existed in the empire; all other learning among the Inca population came by imitation of one's elders. One of the four years at the Cuzco school was devoted entirely to the study of Quechua, the official language of the Incas and the language still used today by most Peruvian Indians.

In such a vast empire, made of diverse tribal backgrounds, there were bound to be many religious cults. At least two great deities were universally worshiped: Inti, the Sun God, and Viracocha, the Creator God. The Incas were basically sun worshipers and claimed to be descended from Inti, in whose honor Coricancha, the greatest and most splendid temple in the empire, was built. The high priest there was a close relative of the Lord Inca's and presided over all other priests of any rank who served the other districts in the realm. Hundreds of temples existed throughout the land, each attended by priests, whose duties consisted of tending the sacred objects, conducting ceremonies, prophesying, hearing confessions, and making sacrifices.

MUSEO NACIONAL DE ANTROPOLOGIA Y ARQUEOLOGIA, LIMA: PHOTO, PEDRO ROJAS PONCE

Although the Inca religion was by no means as bloodstained as that of the Aztecs of Mexico, human sacrifices occasionally were made to the chief gods. The victims were usually boys and girls of about ten years of age. The girls were selected especially for this purpose, while the boys were offered by their parents. Only the best specimens, physically and mentally, could be offered to the gods, and those chosen considered themselves highly honored.

The whole empire was bound together by a network of roads barely rivaled by those of the Romans. From modern Ecuador to a point not far from the Maule River in central Chile two great highways had been built by order of successive Lord Incas. One ran along the coast from Tumbes, near Ecuador's southern border, to Talca, just north of the Maule, some twenty-five hundred miles. The other wound its way more than three thousand miles through the Andes from Quito in northern Ecuador to Talca. At some points the coastal road was no more than posted sand, but it generally had a width of some twenty-five feet. The variations in the mountain road were just as extreme. Narrow lateral roads, often consisting of steps cut into the mountainsides, connected the two main highways. The wonder with which the Spanish invaders followed these masterpieces of engineering was recounted by one of them, Pedro de Cieza de León, a mounted conquistador, who rode or led his horse along the roads for three thousand miles. He described the road constructed by the eleventh emperor, Huayna Capac, who was ruling when the Spaniards first appeared off the coast of South America:

*The intricately worked handle of the ceremonial knife above represents a scowling, winged deity who wears a headdress and ear plugs of turquoise. At right, a considerably more benign figure sits under a canopy held by four attendants, whose identical tunics, necklaces, and ear plugs, like the chief's own, are decorated with discs probably meant to suggest gold.*

MURUA, *Historia General del Peru . . .*, 1611: BY GRACIOUS PERMISSION OF HIS GRACE THE DUKE OF WELLINGTON, K.G.

UNIVERSITY MUSEUM, PHILADELPHIA

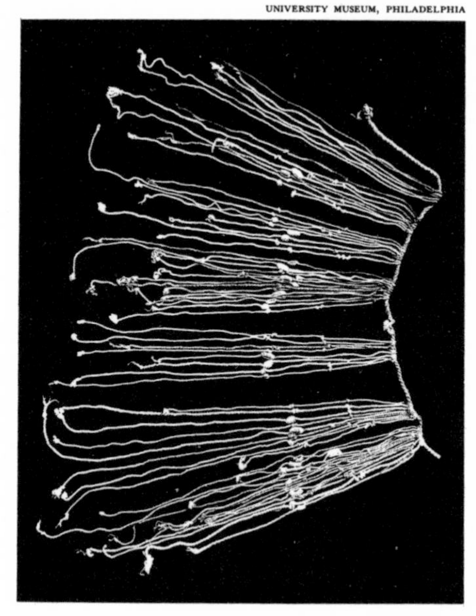

*Having no written language, the Incas relied upon a complex mnemonic system to record historical events and such statistics as harvest yields. Those figures were converted into knots of varying size and shape on a* quipu, *such as the one above. In the water color at left, from the 1611 history of Peru by Fray Martín de Murua, a* quipu *is "read" to Topa Inca Yupanqui, who pushed the boundaries of the Inca empire to their fullest extent.*

Huayna Capac ordered this highway built, larger and wider than that his father had made, as far as Quito, where he planned to go, and that the regular lodgings and storehouses and posts be transferred to it. So that all these lands might know that this was his will, messengers set out to notify them, and then *Orejones* to see that it was fulfilled, and the finest road to be seen in the world was built, and the longest, for it started in Cuzco and went to Quito, and joined that which led to Chile. In the memory of people I doubt there is record of another highway comparable to this, running through deep valleys and over high mountains, through piles of snow, quagmires, living rock, along turbulent rivers; in some places it ran smooth and paved, carefully laid out; in others over sierras, cut through the rock, with walls skirting the rivers, and steps and rests through the snow; everywhere it was clean-swept and kept free of rubbish, with lodgings, storehouses,

*Curiously absent from Inca culture were an alphabet and the wheel, perhaps because the Incas found substitutes for both. Just as the* quipu *eliminated the need for any other means of recording numbers, so the network of Inca footpaths made transportation possible over mountain trails impassable to a wheeled vehicle. In the Compañon illustration at left, three members of a family travel such a road.*

temples to the sun, and posts along the way. . . . The road built by the Romans that runs through Spain and the others we read of were as nothing in comparison to this.*

If the civilization of the Incas was wanting in any major respect, it was in its failure to develop a system of writing. Before the Spanish chroniclers arrived in South America in the sixteenth century, no written testimony of Inca origins and growth existed. Instead, official histories and records were kept almost entirely by memory, not haphazardly, but as was true of so much else in the Inca tradition, in an extremely formal fashion, involving highly trained recorders and a special device known as a *quipu*.

The *quipu* consisted of a series of strings in which knots had been tied to remind its reader of a certain figure, event,

*From *The Incas of Pedro de Cieza de Leon*, translated by Harriet de Onis, edited by Victor W. von Hagen. Copyright 1959 by the University of Oklahoma Press.

or fact. By varying the color and number of the strings and the size, shape, and position of the knots, the "keepers of the *quipus*" were able to commit to memory rather complex and specific information—statistics important to the operation of the empire or concerning the great deeds of an emperor or leader—which they would recite on appropriate occasions.

To a certain extent, the *quipu* system was surprisingly reliable. Yet, inasmuch as the Lord Inca often preferred to glorify the role of his ancestors and his race to the exclusion of all others, the information memorized was sometimes prejudicial, if not downright misleading—a factor that has caused modern historians considerable frustration in their attempts to piece together Inca history. The early tales are almost entirely myth or legend, and none of the dates before the Spanish conquest can be relied upon absolutely.

YALE UNIVERSITY, NEW HAVEN

*In the arid coastal areas, houses frequently had no side walls, and as this stirrup jug suggests, bore designs on the supports and roof.*

One of the richest areas of invention surrounds the origin of the Incas as a people. Perhaps the best known of the several existing myths is that the first chieftain, Manco Capac, with three brothers and four sisters, emerged from a hole in a hill at Paucartambo, about eighteen miles from Cuzco. According to the story, these were the children of Inti, the Sun God, who had created them to conquer the world and to establish his religion throughout it. Inti provided them with a golden staff, made to penetrate and remain fixed in the earth at the site that he had chosen as the capital of their empire. This proved to be at Cuzco, and there, on a spot that later was called Coricancha, Temple of the Sun, Manco Capac and the little band of Indians who had taken him as their leader built their first huts. Of Manco Capac and the two Inca leaders who followed him—Sinchi Roca and Lloque Yupanqui—little can be said. But there is no reason to doubt that they had to fight hard to hold their own against other tribes already settled in the valley.

The Incas were neither the first tribal group to gravitate to the areas we know today as Peru, Ecuador, and Bolivia, nor were they the first great civilization to develop in these regions. Historians place the foundation of the Inca culture at about A.D. 1200 to 1250. As early as 1000 B.C., however, an ancient and impressive civilization, called the Chavín,

*In this water color, Fray Murua has taken several events from the legend of the founding of the Inca dynasty and shown them simultaneously. On the far left, Manco Capac, the first Lord Inca, and his sisters emerge from the ground, while in the background Manco is worshiped by his followers.*

Manco Capac

existed in the northern highlands surrounding the site of Chavín de Huántar. Archaeological excavations in that area have produced excellently carved stone and sculptured heads, as well as fine ceramics and highly developed architectural motifs. A little farther north, in the Chicama Valley, at Cupisnique, the remnants of still another knowledgeable and highly skilled people have been discovered, and their culture is said to have existed side by side with that of the Chavín. Archaeologists, of course, differ among themselves on the dates of these oldest cultures and on the time and extent of those that followed. New excavations seem to change the modern understanding of all these cultures almost annually, and many facts about the fascinating early people of the Andes are hotly disputed by scholars and scientists.

Between 400 B.C. and A.D. 1, the Chavín and Cupisnique cultures seem to have disappeared gradually, and more advanced civilizations, with improved agriculture and sophisticated textile techniques, emerged in their place. The Salinar culture, in the northern coastal region, and the Cavernas, at Paracas, far to the south, near the Pisco Valley, both apparently flourished for several centuries, until about A.D. 400, when the more masterful influences of the Mochica in the north and the Nazca in the south appeared. Textile and ceramic art reached their highest development in these later cultures, and agricultural techniques were refined beyond any degree previously known in these regions. Yet they, too, were destined to be absorbed by more powerful and expanding influences.

About A.D. 400, when the Nazca culture was emerging in the southern coastal area, another culture that would one day achieve even greater technical, if not artistic, advancement was coming into existence at Tiahuanaco, some thirteen miles southeast of Lake Titicaca, high up in the Andes. By the time the Incas first made their appearance in the thirteenth century, the Tiahuanaco influence had spread far across the central Andes, encompassing areas once dominated by the Nazca, and into the northern coastal valleys as well. It carried with it new concepts of organized labor and construction, a broader knowledge of metallurgy and metalworking skills, and a powerfully controlled religious structure. But it provided no really broadly based political organization. That was to come with a far stronger influence in the north, eventually known as the Chimú empire.

By 1200, the boundaries of the Chimú extended from Paramonga to Tumbes and from the coast to the mountain

MUSEO NACIONAL DE ANTROPOLOGIA Y ARQUITECTURA, LIMA: PHOTO, PEDRO ROJAS PONCE

DUMBARTON OAKS, ROBERT WOODS BLISS COLLECTION, WASHINGTON

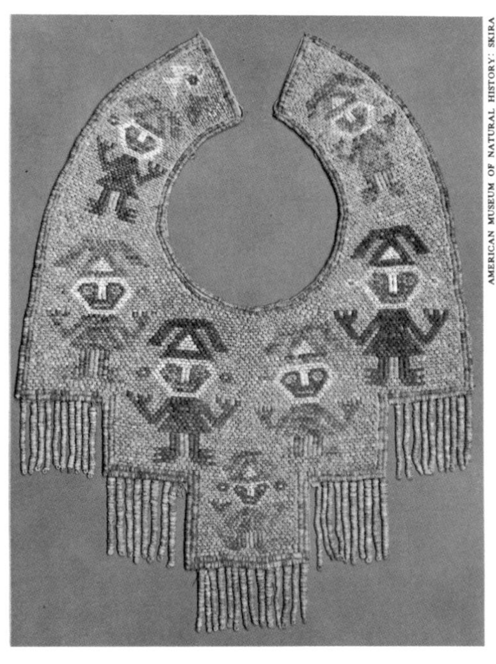

AMERICAN MUSEUM OF NATURAL HISTORY: SKIRA

*Working with an astonishing variety of materials, Peruvian craftsmen created pottery and jewelry of both utility and beauty. The Nazca polychrome pottery fish jug (top), for example, was useful as well as decorative—as was the Tiahuanaco mirror (left), with its mosaic of shell, turquoise, and pyrite, or fool's gold. The ceremonial Chimú collar made of beads of colored shell (above), one of few that have survived intact, features stylized human figures.*

YALE UNIVERSITY

*Houses, temples, and city walls in ancient Peru were banded with delicate friezes. One such relief (above) reconstructed at Chanchan, originally was molded out of clay so damp that it could be worked by hand, without tools.*

interior. The ruins of the Chimú capital, situated at Chanchan, near the site of modern Trujillo and extending over six square miles, even today testify to this civilization's ancient splendor. Excellence in agriculture and architecture, the development of a tight political structure centered about great cities, and the ability to mass-produce items necessary to the survival of its population were characteristic of this society. But in the end, in the late 1400's, the Chimú empire would have to prove its mettle against an expanding kingdom to the south—that of the Incas.

Of Mayta Capac, the fourth Inca chieftain, a few great stories are told, all suggesting, however, that he was powerful enough and aggressive enough to be feared and respected locally. He seems to have forced or persuaded some of the nearby tribes to join the Inca tribal group. He attacked the neighboring Alcahuiza, defeated them, and took their best land.

Like Mayta Capac, succeeding Inca chieftains—Capac Yupanqui, Inca Roca, and Yahuar Huacac—were often at war with their fierce neighbors on the sierra, only occasionally consolidating their slight gains and doing little to spread the Inca domain. It really was not until the eighth ruler, Hatun Tupac Inca, more commonly referred to as Viracocha Inca, came to the throne that the drive for empire began.

Although his territory extended barely twenty-five miles from Cuzco by the time his reign neared its end, it was Viracocha Inca who established the pattern of Inca conquest. Besides taking the customary spoil and captives from conquered adversaries, he imposed a fixed tribute, demanded a labor force to serve him when he needed it, and left a permanent governor in charge of newly gained territories, thereby making them an integral part of his nation. With tribes too strong to conquer he wisely entered into agreements in which each swore to respect the territory of

MUSEUM FUR VOLKERKUNDE, BERLIN

the other. And in and about the Cordilleras war and diplomacy began to go hand in hand.

Pachacuti, son and successor to the Emperor Viracocha, followed his father's lead and quickly became the greatest conqueror and administrator in this long line of able rulers. He and his son Topa Inca Yupanqui expanded the empire until it stretched from northern Ecuador to central Chile, overrunning regions rich in precious metals and exacting a regular tribute, so that Cuzco soon developed into a city of fabulous wealth. It was Pachacuti who built the first great roads and began the system of *mitima*, and it was he who rebuilt Cuzco on a splendid scale to make it worthy of its title as capital of a great empire and the seat of the Lord Inca. By a curious mixture of consideration and generosity toward those who submitted to their rule and ruthless cruelty toward those who rebelled against their establishment, Pachacuti and Topa Inca riveted the Inca yoke to the necks of millions.

In addition, it was during Pachacuti's reign, between 1438 and 1471, that the record of events becomes fairly reliable. Although on occasion a choice still must be made between conflicting accounts, it is generally accepted that Pachacuti himself led the Inca armies against the nearby provinces surrounding Cuzco, pressing as far to the north as the lower Urubamba Valley and as far to the south as the

*Unlike the deliberately stylized and highly impersonal art of many primitive societies, the Mochica effigy pots below almost certainly represent real subjects—as the distinctive features of these vendors holding their wares suggest.*

STAATLICHES MUSEUM FUR VOLKERKUNDE, MUNICH

*Colorful variations of a single figure fill this embroidered shroud from the Paracas peninsula. Each bizarrely costumed figure holds a trophy head.*

shores of Lake Titicaca. His brother, General Capac Yupanqui, then subdued the tribes in the more distant north, to the beaches of Lake Junín, and his son Topa Inca Yupanqui carried the conquest to its farthest bounds.

Sent by his father to prove himself, Topa Inca marched north, first subduing the fierce Cañari. Then turning to the powerful kingdom of Quito, he overthrew the proud forces there. His final campaign came against the strong and independent empire of Chimú, which proved no match for the indomitable forces from the south. Its fall marked the end of the last great independent nation in either Peru or Ecuador. All was now under the Lord Inca's sway.

MUSEUM OF THE AMERICAN INDIAN, HEYE FOUNDATION

*Peruvian women seldom were idle and often spun as they walked. Such industriousness, still seen today, is captured in the vessel above, which shows a woman with her cone of wool and her spindle.*

When he finally became emperor in 1471, Topa Inca consolidated his conquests in the north and subdued the states on the coast, south of Chimú. Yet it was a vast and varied area to rule, and in 1493, when his son Huayna Capac inherited the throne, the empire was already seething with revolt. While the Spaniards were listening to the first stories of the fabulous kingdom of gold, the Lord Inca was straining his resources to the breaking point in crippling battles against his many rebellious subjects. In addition, the years of warring were exacting a dreadful toll among the Inca's best troops, and discontent spread quickly even to his personal regiments of Inca nobles.

Toward the end of his reign, Huayna Capac—the eleventh Lord Inca—was further troubled by rumors of pale-skinned, bearded men who had appeared on floating objects out of the northern sea. Legend said that the God Viracocha had passed out of sight over this same sea when his work of creation was completed. If these beings were the servants of Viracocha, their coming might mean disaster for his house. To add credence to such fears, there was news of an overwhelming calamity, an invisible enemy—probably smallpox, measles, or scarlet fever—that was stalking the land and bringing terror and death to his people. Finally, it struck the emperor himself, about the year 1525.

In the kingdom of Quito, now modern Ecuador, far from his capital of Cuzco, with thousands dying around him, Huayna Capac made a last disastrous decision. He reversed the established order of succession, decreeing that a younger son, Ninan Cuyoche, should succeed him, rather than the heir apparent, Huáscar. Both were legitimate heirs, whereas Huayna Capac's favorite son, Atahualpa, had been born to a secondary wife.

Unfortunately, before he could be crowned, Ninan Cuyoche died of the same pestilence that had taken his father, leaving Huáscar to rule without a rival. Still, there was Atahualpa, whom Huayna Capac had placed in command of the veteran Inca army, then in Quito.

Fearing the newly gained power of his jealous half-brother, Atahualpa refused to return to Cuzco for either his father's funeral or Huáscar's coronation. As a result, Huáscar suspected the brave soldier of disloyalty. Still smarting at the affront of his father's last-minute decision to leave the throne to another, Huáscar now feared that he had a second rival in Atahualpa, the darling of the northern army and a man too powerful to be destroyed without war. Moreover, at Atahualpa's side stood his father's two best generals, Quizquiz and Challcochima.

Some years passed in uneasy peace, and it is uncertain that Atahualpa ever intended to rebel and establish a separate kingdom in Ecuador. At first he seems to have given his allegiance to Huáscar. But the brutal and unjust treatment of envoys sent by Atahualpa to Cuzco set off the spark. Both brothers began to marshal their forces, and soon war was inevitable.

The generals of the northern army pledged themselves and the support of their troops to Atahualpa, now proclaimed ruler of the empire of Quito, and by the time Pizarro and the forces of his third expedition were threading their way once more along the coast of South America,

*Grace and utility are combined in this Mochica pottery trumpet. The clay horn, with its mouth shaped into a growling feline head and its neck decorated with leopardlike spots, was used by the Peruvians in both military processions and religious fetes.*

ART INSTITUTE OF CHICAGO: GIRAUDON

Atahualpa's generals already had won a series of brilliant victories. After defeating the Cañari, who had sided with Huáscar, and taking dreadful vengeance on young and old, they overwhelmed an imperial army farther south and occupied Cajamarca. Then, shattering Huáscar's armies at Mantaro, they headed for the Apurímac River, the last natural barrier on the way to Cuzco.

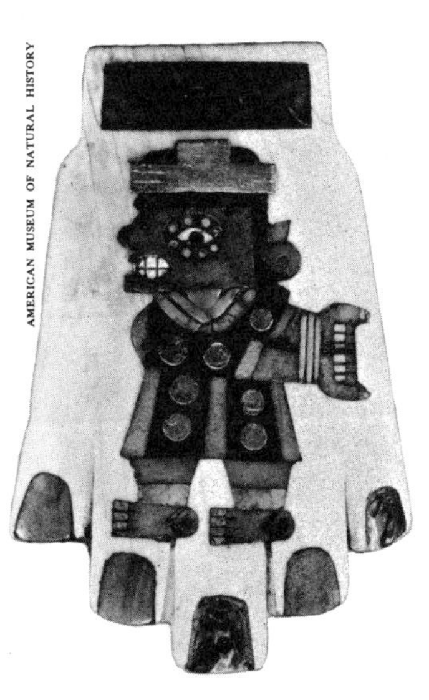
AMERICAN MUSEUM OF NATURAL HISTORY

Huáscar himself commanded the *Orejones* and his southern allies at Cotabamba, in the final battle in defense of his seat of government, and on the first day he succeeded in smashing Atahualpa's advance guard. On the second day, however, Huáscar led his troops into ambush. Eight hundred *Orejones* and Inca warriors fell around him. It was the end: he was dragged from his litter by Challcochima and led away a prisoner.

With their emperor held captive and their main armies destroyed, the Inca nobles who had supported Huáscar fell back to Cuzco to listen to the deceitful offers of peaceful settlement made by Atahualpa's generals. Unarmed and trusting, these nobles came to the victors' camp outside the city to hear the terms. Not until then was the generals' perfidy laid bare. Files of warriors closed in around the defenseless *Orejones* and forced their leaders to suffer the punishment of common criminals—by beating them with clubs. A more cruel punishment was reserved for the others: heavy stones were dropped on their backs, and the crippled survivors were held for further torture. Afterward, the vanquished warriors were allowed to return to their city, but a strong garrison marched with them.

*In another of Ayala's drawings, the fettered Huáscar is led into the capital by supporters of Atahualpa, as one of them signals their arrival with a conch shell. The inlaid shell hand (above) shows a similarly trussed man.*

News of the defeat and capture of Huáscar traveled swiftly to Atahualpa in his headquarters at Cajamarca, and orders were dispatched almost immediately to his commanders at Cuzco to kill Huáscar's family. Brothers, sisters, wives, and children were slaughtered before the eyes of the captive emperor, and their contorted bodies were left in the great square at Cuzco until they could be impaled on stakes in and around the city as a dreadful warning to any who thought of revolt. A reign of terror followed: all the known loyalists were dragged from their houses and butchered.

By 1532, the power and the glory of the lords of Cuzco were destroyed. Tens of thousands of the best fighting men of the empire had died on the battlefields or in the massacres that followed Atahualpa's victories. Thus, when Pizarro and his men prepared to begin the conquest of the empire of gold, a devastating civil war had already robbed it of much of its strength and purpose.

III

# DESIGN FOR CONQUEST

As his little ship rode at anchor off Tumbes in 1527, Pizarro knew nothing of the trouble brewing between the rival brothers in Peru. He treated the Inca noble who had come aboard with great courtesy, showing him about the ship and answering his questions. After the Inca had taken dinner with the Spaniards—finding their wine greatly to his taste—he invited them to visit the city.

Rather than go himself, Pizarro sent one of his most trusted cavaliers, Pedro de Candia. Curiously, Pedro was not Spanish; he was a Greek from the island of Crete. An enormous man with a swarthy complexion and fine, handsome features, Pedro put on full armor and took his harquebus with him when he was rowed ashore. The Indians were astonished to see this splendid figure wading onto the beach, and they were terrified, too, when at their request he fired his gun and shattered a piece of wood.

ROYAL LIBRARY, COPENHAGEN

*Their fears overcome by curiosity, a throng of Indians examine a Spaniard's strange foreign garb in another of Ayala's drawings.*

Once Pedro had allayed their fear of his weapon, however, they led him to a fine mansion, built of stone and adobe, painted in bright colors, and thatched with straw—the residence of the local Inca governor, or *curaca.* With porters at the door and liveried attendants standing by, the *curaca* himself greeted the cavalier, and his servants offered Pedro food and drink from vessels of gold and silver. Pedro was staggered by the signs of wealth around him. Through his interpreter he learned of a mighty emperor who lived in even greater splendor in a city to the east, high up in the vast range of mountains that stood like a protective wall against any invader.

Escorted by Indians dressed in gaily colored mantles and shirts, Pedro was shown the city's fortress, with its triple lines of defense and its numerous and well-disciplined gar-

*The conquistadors, used to lower altitudes and warmer climates, found the awesome barrier of the Andes, mountains second only to the Himalayas in height, almost as difficult to conquer as the mountains' Inca inhabitants.*

rison; the Temple of the Sun, decorated inside with plates of gold and silver and other handsomely wrought ornaments; and the House of the Virgins of the Sun, where the most beautiful girls from the surrounding countryside lived to serve their god. Well satisfied with the tour, the cavalier returned to his ship.

On hearing Pedro's report, the eager soldiers forgot their hard months at sea and long waits on barren islands. They had found the land of gold and were eager to share in its riches. But Pizarro—ever able to embody both a religious fervor and a greed for gold—on the one hand gave thanks to God for bringing him and his men to a land where heathen souls could be saved, while on the other hand, he complained bitterly because he had too few soldiers to sack the Indian town. Cool scheming soon took the place of emotion, however, and he issued strict orders that every Indian met during the rest of the cruise be treated kindly and that any soldier who robbed or mistreated a native would be subjected to harsh punishment. In this way Pizarro planned to gain the confidence of the Indians and make his task easier when he later returned with fire and sword.

Once more, Pizarro put to sea and headed south, skirting the sandy wastes of the Sechura Desert for a hundred miles or so and sometimes being driven out to sea by heavy gales. But always within sight were the towering Andes, eliminating the helmsman's need of star or compass to keep to his course. From time to time the adventurers dropped anchor off isolated valley settlements, and everywhere they found kindness and hospitality. Balsas laden with presents of fruit, vegetables, and fish were brought alongside the little ship, and friendly Indians eagerly boarded the vessel to meet these strange, courteous men from the sea. Word had reached them along the Incas' splendid highway that the *Viracochas*, or "Children of the Sun"—as the natives called the Spaniards—were friendly visitors whom they could safely trust.

Wherever he landed, Pizarro found further proof that he was on the frontier of a well-organized and highly developed society. Elaborate canal and aqueduct systems, suggesting an impressive knowledge of engineering, had turned semideserts into fruitful fields. The great highway of

BRITISH MUSEUM

*Navigation by observation of the stars was an exact science, as shown in this 1542 chart, used to plot a bearing on the North Pole.*

*The same simple but highly serviceable balsa-log rafts that carried the Incas to coastal islands—and possibly as far west as Tahiti—also were used for fishing. In Bishop Compañon's vivid water color, two Indians haul in the seine (center) while their companions steady the small craft.*

the Incas, bordered here and there by low walls and shade trees and always in good repair, promised a convenient invasion route. Yet the awesome, white-topped crests of the Andes, standing between the coast and the heart of the empire, must have sent a chill through even the boldest of the Spaniards.

Most of the Indians lived in single-roomed adobe huts—not unlike those used by Europe's peasantry—yet the travelers were reassured by the adobe brick temples, obviously built by master masons and imaginatively decorated with geometric designs and animals in bas-relief. Inside were plates of gold and silver and other delicately worked ornaments.

Not until he had sailed to a point about nine degrees south of the equator did Pizarro yield to the arguments of his men and turn his ship for home. He had seen enough: This was the land he meant to conquer. Calling at various points on their return voyage, the explorers again were entertained hospitably. Often they exchanged small presents of steelware and European cloth for vases of gold and samples of the superb woven fabrics made by the Indians. Pi-

DR. JUNIUS BIRD

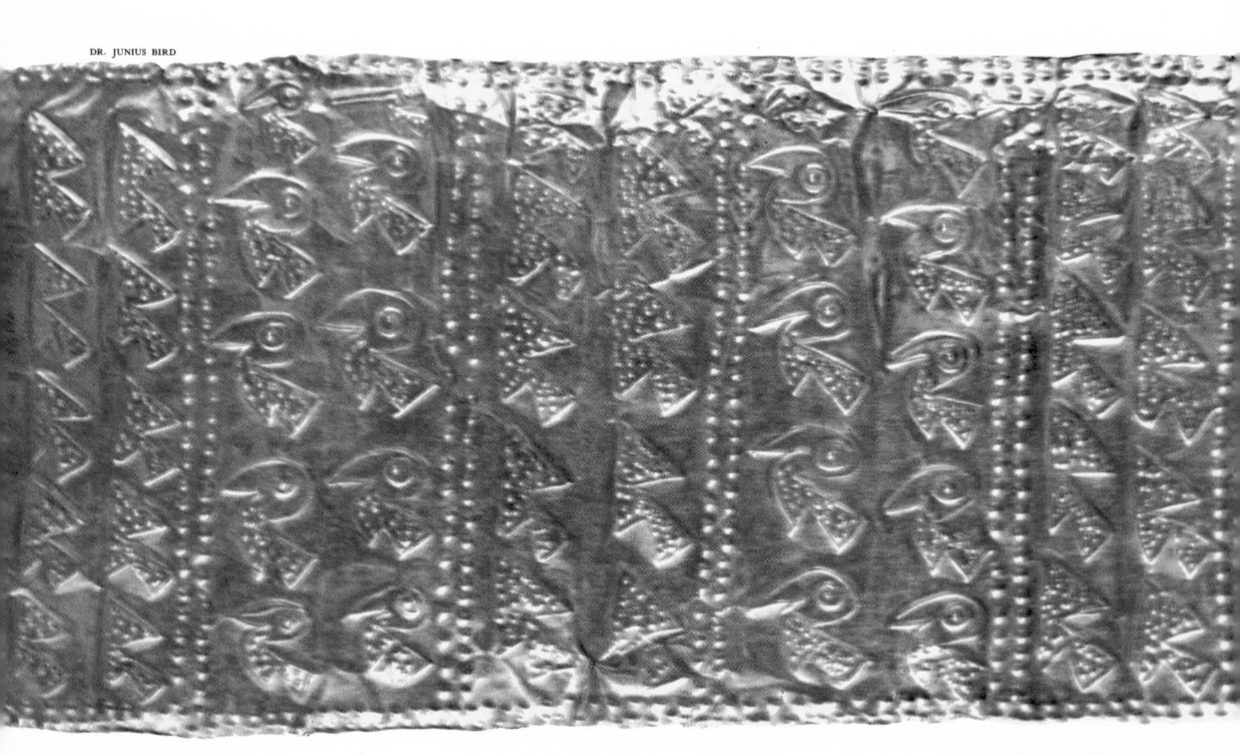

*One of the few Inca gold objects to escape the Spaniards' smelting pots was this nearly twenty-inch-long fragment. It repeats the familiar stylized bird motif seen on the silver jug on page 31 and the frieze on page 44.*

zarro felt certain that such souvenirs would eliminate any doubt that he and his men had indeed found their *El Dorado*.

Pizarro allowed two of his followers to stay at Tumbes until he returned, and the *curaca* sent a few of his young men to sail with the Spaniards. One of these youths, nicknamed Felipillo by Pizarro and his men and later trained as an interpreter, was to play a sinister part in the tragic end of his people's empire.

The wind was fair for the Isthmus, and after more than eighteen months' absence, Pizarro and his devoted band once more anchored in the harbor at Panama. Eager crowds met the explorers, and even those who once had mocked them now gathered to marvel at the llamas, rich cloths, and exquisite gold and silver ornaments that they had brought back.

No longer the laughingstock of Panama, Pizarro and Almagro were on the way to becoming its heroes, and Hernando de Luque rejoiced in their triumph. The cleric had staked not only most of his own money, but his reputation as well, on their risky enterprise. Now he felt certain that he could raise even more funds to equip the necessary force to conquer Peru.

Only the governor remained aloof. Receiving the three partners coldly, he refused to sanction another voyage and declared that he "had no desire to build up other states at the expense of his own." Nor would he throw away any more lives than had been sacrificed already for such a "cheap display of gold and silver toys and a few Indian sheep." Without the support of the emperor's chosen representative in Panama, they were stalemated. They could raise neither the men nor the money for another expedition.

Luque once more saved the day. If the governor would not listen to their arguments, he told his two cohorts, they must appeal to Emperor Charles V himself. It was a suggestion eagerly seized upon by Pizarro and Almagro. But the problems were to find someone to represent them at court and to raise enough money to pay for such an embassy.

Luque, whose education, character, and social standing suited him for the task, was tied to Panama by Church duties; Almagro, a nameless man, with neither education nor prudence, was bound to make a poor impression at the formal Spanish court. There remained only Pizarro. Certainly his appearance was in his favor, and when deeply moved, he could speak amazingly well. Yet he, too, was a man of low birth, unable to write, a man who had spent his whole life in rough company and lacked the manners of a

ESCORIAL: PHOTO, ORONOZ

BIBLIOTHEQUE NATIONALE, PARIS

# MASTER OF EUROPE

Only six when he inherited the Netherlands from his father, Charles V controlled most of Europe by the age of twenty. He and his deranged mother, third child of Ferdinand and Isabella, were appointed co-regents of Spain in 1516, and three years later Charles succeeded his paternal grandfather, Maximilian, as Holy Roman Emperor. To consolidate his far-flung empire, he married his sisters and offspring into every major European royal house—so that blood ties, rather than common principles of administration, united the empire. (The allegorical figure above, whose head is Spain and body, Europe, suggests Charles' immense power.) Despite religious wars at home, Charles (seen on the dappled stallion in the sixteenth-century tapestry at left) maintained an unflagging interest in New World colonization.

BOTH: HERRERA, *Decades*. . . , 1728. INSTITUTO DE CULTURA HISPANICA, MADRID: PHOTO, ORONOZ

*These two silhouetted details from Antonio de Herrera's history of the conquest show Francisco Pizarro (opposite) receiving the Capitulation, or grant to conquer all Peru, from Charles V (above).*

gentleman. The thought of rubbing shoulders with Spain's grandees or addressing its king was more terrifying to him than warfare in a savage land. Still, he was a determined man, and determination alone could carry him through.

Forgetting past quarrels, and believing that no one else could be trusted with the mission, Almagro urged Pizarro to go. Luque disagreed. He distrusted Pizarro. Even after it had been decided, and Pizarro had given his word that he would protect the interests of all, Luque's parting words suggested that he was hardly convinced. "God grant, my children, that one of you may not defraud the other of his blessing," he prayed.

Still, Luque borrowed the money for the journey—1,500 ducats, or about $3,000—and in the spring of 1528 Pizarro left Panama, accompanied by Pedro de Candia. Taking along the Indians from Tumbes and the llamas and artifacts that he had brought from Peru, he crossed the Isthmus and embarked for Spain at Nombre de Dios. After a rather uneventful voyage, he reached Spain in early summer.

More than twenty years had passed since Pizarro had last set foot in his native land. But he was soon welcomed at court, and his simple, straightforward account of the sufferings of himself and his men was said to have brought tears to his sovereign's eyes. Certainly, Pizarro's determination must have impressed the young monarch, not to mention the thought that the riches of the Indies, of which he had heard so much, might at last begin to flow into his treasury. Charles V needed all the gold he could lay his hands on in order to pursue his ambitions in Europe. Already, as Holy Roman Emperor, his power and influence extended throughout Germany and Italy, as well as Spain and the Low Countries. Now he was turning an envious eye toward what remained of the Continent.

Without delay, Charles V instructed his Council of the Indies to examine Pizarro's claims and to give its advice on what action he should take. In matters concerning the Spanish colonies in the New World, the council was omnipotent, a department of government directly responsible to the emperor. Despite such power, it was a conscientious body, and as such, often slow in reaching its conclusions, as in the case of Pizarro.

For months the anxious soldier waited for a decision, his funds running low. Finally, he was forced to petition the queen, left as regent, to spur the council to action. The result was the "Capitulation," which she signed on July 26, 1529, almost a year after Pizarro's arrival. In it Pizarro was

granted the right not only to conquer Peru, but also, with that accomplished, to assume the title of governor of the new settlements and of captain-general for life. He was allowed full power over the administration of such colonies and full responsibility for maintaining law and order in them. He also was authorized to build fortresses and to reward his men with land grants. For all this, he was to receive a salary of 725,000 maravedís (about $3,500) a year.

His fellow explorers—Almagro, Luque, Ruiz, and Pedro de Candia—were mentioned also. Almagro was to be made commander of the fortress of Tumbes with a salary of 300,000 maravedís—a poor recompense for all his efforts—and Luque was to be named Bishop of Tumbes and Protector of the Indians of Peru, with a salary of 1,000 ducats (about $2,000). Ruiz's honors included both the title of Grand Pilot of the Southern Ocean and a generous salary, and Pedro de Candia was cited to become commander of the artillery. None of these salaries, however, was to be paid out of the royal purse. All were to come from revenues gained in Peru.

The Capitulation included, in addition, long sections on how the land was to be governed, cautioning Pizarro to protect the Indians from extortion and ill-treatment and ordering him to take with him a number of priests and monks to convert the natives and to look after their well-being. Finally, it was stated that within six months of the signing of the agreement, Pizarro was to have raised in Spain a force of at least one hundred fifty well-equipped men, to be joined by an additional one hundred men from the New World, once he had returned to Panama. More immediately, in recognition of his achievements, Pizarro was honored by being made a knight of Santiago and by a grant of elaborate armorial bearings.

It was a moment of triumph for a man of such poor beginnings, and not surprisingly, he decided to return to the province where he had grown up—disowned by his father, a colonel of infantry in the Spanish army, and neglected by his mother, a poor girl of Trujillo. This time friends were much easier to find, and some who previously had avoided him now eagerly claimed him as kinsman. His four half-brothers also came to see him, and in the end, to join him. For a man of such cold nature, Pizarro greeted them warmly, eventually made them officers in his service, and afterward helped them to make their fortunes in Peru.

Martín de Alcántara, related to Pizarro on his mother's side, became the adventurer's closest companion and would one day die fighting to defend him from an assassin's sword. Gonzalo and Juan Pizarro, young, fearless, and high-spirited, were related to Pizarro on his father's side. Both later made their marks as popular cavalry commanders. All three were far more likable and generous than their famous half-brother, but they lacked Francisco Pizarro's judgment and firmness and were not really suited to independent commands.

The last half-brother, Hernando Pizarro, was a man of far different quality. Older than the others and the only one of legitimate birth, Hernando was proud, cold-blooded, and grasping. He was a tall man and well built, but of so unattractive and scornful a countenance that other men instinctively looked away from him. Neither honor nor pity could dissuade him from taking revenge on his enemies, and his presence among the conquistadors led to bitter feuds and bloodshed. Yet Hernando was to become Francisco's right-hand man, skillful in military matters and capable of acting with firmness and good judgment in positions of the highest trust.

One more kinsman, named Pedro Pizarro, joined Francisco as his page. Then a boy of fifteen, Pedro was destined not only to survive all the campaigns, but also to write a detailed account of them years later.

A few arms and military stores were all that Charles V contributed toward the great enterprise, which was to bring such vast wealth to the Crown. In fact, it is said that without the generous help of Hernando Cortes, who had recently returned home in triumph from his conquest of Mexico, the expedition to Peru might have foundered even at this point for lack of sufficient funds. Certainly, Pizarro had not seen the end of his troubles. Failing to recruit the one hundred fifty men demanded by the Capitulation, he was forced to sail hurriedly from Seville in January, 1530, to prevent the inspectors sent by the Council of the Indies from discovering the truth.

Meanwhile, Luque and Almagro had waited through the long months with growing impatience, fearing always that Pizarro had failed. In the nearly two years since Pizarro's departure for Spain, the ferment caused by the discovery of Peru had long since died down, and they knew that men in Panama would risk neither their money nor their lives on this account, unless the powers of Church and State were behind it. Then, one day, news reached Panama that Pizarro had anchored at Nombre de Dios with three small ships and a crew of armed men. Immediately, the two

*By the late eighteenth century, when this Spanish map was drawn, Panama's heavily settled waterways had become the crossroads for sea traffic to and from the wealthy Viceroyalty of Peru.*

partners left to cross the Isthmus, eager to know of Pizarro's success—or failure.

Delighted at the sight of the little armada, Luque and Almagro asked Pizarro to tell them on what terms the emperor had sanctioned their expedition. Painfully aware that he had not kept faith with his old comrades, Pizarro explained the Capitulation with downcast eyes. Almagro was furious. His words and tone were bitter as he accused Pizarro of breaking his trust, and he refused to listen to any of Pizarro's attempts to appease him. Neither the excuse that these were the emperor's terms nor the promise that, once Peru was conquered, all would share equally in the honors

ARCHIVO DE INDIAS, SEVILLE: MAS

could cool the fiery Almagro's anger. Later, at Panama, the quarrel became more bitter when the arrogant Hernando Pizarro, thrusting himself forward, made it quite clear that he had nothing but contempt for the ugly little adventurer who stood in the way of his brother's ambitions.

Once more, it was the prudent Luque who interceded, just in time, to quench the hostile fires and save the entire undertaking. To patch up the quarrel, Luque persuaded Pizarro to promise to petition the emperor for a separate territory for Almagro to rule and to grant no office to his own brothers or any others until his partner had been satisfied. It also was agreed, and most solemnly, that all booty would be divided equally among the three shareholders, once the emperor's share of one fifth had been deducted. Making the best of a bad job, Almagro allowed himself to be placated.

The joint efforts of all three men failed to raise the additional hundred men, however. Soldiers in Panama had seen too many broken and disappointed adventurers come ashore from Pizarro's voyages to enlist lightheartedly under his banner. But some of his old company signed on again, determined to see the adventure through, and a number of poor settlers from Nicaragua joined the expedition. These, plus the recruits from Spain, brought the total to about one hundred eighty—a pitiful company to attempt the conquest of an empire.

Yet this was the most powerful force Pizarro had ever commanded, and all were rather well equipped by colonial standards. Although the majority of the men carried only swords and pikes, there were at least twenty or so with crossbows and a few with harquebuses. Twenty-seven war horses and two falconets—a light, mobile cannon—were included. The latter were to be secret weapons for demoralizing the natives.

The assembly of this force took the rest of that year, and not until January, 1531, was everything ready. The little army, its armor flashing and banners waving, marched proudly to Panama's cathedral. There, their flags were consecrated, and at Mass, every man among them partook of the sacrament to prepare himself properly for this "crusade against the infidel." The Spanish adventurer was no hypocrite: He firmly believed that God was on his side.

The anchors soon were weighed, and the three ships, under the command of Francisco Pizarro, sailed out of the harbor, taking the adventurer on his third and last voyage to conquer Peru. Almagro stayed behind to recruit.

*Favoring pageant over precision, de Bry has shown Luque blessing Almagro and Pizarro jointly as they depart for Peru—although their departures, amid dockside bustle much like that at left, were actually some time apart.*

It originally was Pizarro's plan to strike his first treacherous blow at Tumbes, but contrary winds and tides saved the friendly city for a time. Instead, at the end of a thirteen-day journey, the three vessels had gone no farther than the Bay of St. Matthew, some sixty-five miles north of the equator. There Pizarro unwisely decided to continue his advance overland, while his ships kept pace offshore. Knowing this inhospitable, trackless coast from previous voyages, he must have made the decision reluctantly, probably hoping to find gold along the way. If so, he was bitterly disappointed.

For days the heavily armored Spaniards forded or swam streams swollen by winter rains, only to find themselves struggling painfully through muddy swamps. Yet always Pizarro cheered the men on, displaying heroic determination to succeed at all costs. His great powers of leadership won for him the trust, if not the affection, of all who followed him.

MUSEO DE AMERICA, MADRID: PHOTO, ORONOZ

DUMBARTON OAKS, ROBERT WOODS BLISS COLLECTION, WASHINGTON

LEE BOLTIN

*The Indians' strong sense of design extended to weapons as well as household objects (see page 43). Even the least decorative of the maceheads above, a blunted star, is attractive as well as lethal. Oxidation has given the bird-shaped bronze macehead a patina unlike that of the stone star and cat's head, also of bronze. The ubiquitous feline head also appears on the macehead at far right.*

Unexpectedly, the heartbreaking march proved worthwhile. Nearing exhaustion, the troops stumbled on a large settlement in the province of Coaque, where, according to Pedro Pizarro, they "fell on them [the natives], sword in hand," and won for themselves a rich "store of gold and precious stones."

Happily they piled their plunder in a clearing, and after the emperor's fifth had been withdrawn, Pizarro distributed the remainder among his followers. Naturally, he awarded more to the officers and cavaliers than to the rank and file. It was a method of dividing the spoils that was to be rigorously enforced throughout the conquest; any man who concealed treasure did so at the risk of his life.

As soon as the awards had been made, Pizarro sent gold valued at 20,000 *pesos de oro* (about $120,000) back to Panama, hoping to entice other men to follow him to Peru. All three ships were ordered to return with the riches, to provide transport for these hoped-for reinforcements. As it turned out later, this particular scheme worked.

Yet all this new-found treasure failed to alleviate in the smallest way the hardships and exertions of their unenviable journey. The path to conquest that the Spanish forces determined to follow all too often was lost in moving sands that swirled in the wind and blinded and choked both man and beast. When the sand was not a problem, it was the tropical sun, beating down mercilessly on their steel mail and quilted doublets, until those who wore them were almost suffocated by the heat. Even worse things lay ahead. When the men seemed near the limit of endurance, a terrible epidemic of ulcers broke out, covering the sufferers' bodies with masses of great sores. Lanced to relieve the pain, the sores bled so profusely that many of the soldiers died, while others were left so weakened by the experience that they could scarcely stagger on. Even the sight of a ship bringing supplies failed to cheer the men. Most already were cursing the day they had joined the march and the gold they had come to seek.

A treasurer, an inspector, a comptroller, and other important officials appointed by the Council of the Indies to protect the emperor's rights, and a few supplies, were all that had come with the first ship. A second vessel, arriving a few days later, brought reinforcements—some thirty men and an experienced captain named Sebastián de Belalcázar.

The march along the coast ended at the Gulf of Guayaquil, opposite the island of Puná, whose inhabitants, the Spaniards knew, were perpetually at odds with the people of Tumbes. Since Puná provided a good base from which to mount his planned attack on his old friends at Tumbes, however, Pizarro showed no reluctance to accept the invitation of Puná's leaders to visit their island. There he and his men settled down in comfortable quarters, with plenty of food, to await more reinforcements and the end of the rainy season. There, too, Pizarro probably first learned of the war between Huáscar and Atahualpa.

Unfortunately, his stay was not to be uneventful. Hearing of his arrival and confident of his protection, some men from Tumbes marched brazenly into the enemy camp one day to visit Pizarro and stayed on. Naturally, Pizarro's warm treatment of them bitterly offended the men of Puná. It was a strained situation at best, made only worse by Pizarro's interpreters, also from Tumbes, who constantly warned him that the islanders were renowned for their treachery. When a rumor circulated that the Puná chiefs were planning an attack, disaster struck. Seizing the leaders and holding a speedy inquisition, in which all of the conversation passed through the interpreters from Tumbes, Pizarro decided that the Puná men were guilty and foolishly handed them over to the men from Tumbes for punishment. The latter, of course, slaughtered their ancient enemies on the spot—a cruel betrayal that enraged the people of Puná beyond repair. The men of the island seized their weapons, and the Spaniards soon found themselves embattled.

Pizarro's forces stood firm, and when the day was over, the conquistadors had won their first engagement, at little cost. Yet Pizarro's judgment was badly at fault. He had lost the chance of enlisting good warriors to help him attack the mainland by treacherously condemning their chiefs, and he had left himself open to guerrillalike attacks from those who

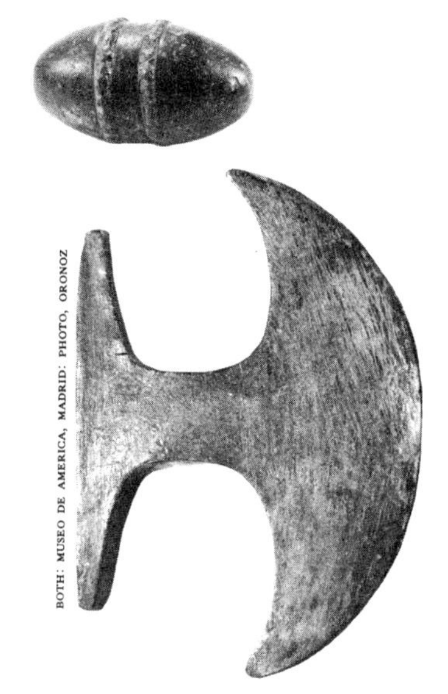

BOTH: MUSEO DE AMERICA, MADRID: PHOTO, ORONOZ

*The graceful curves of the axe head above are ample evidence of Peruvian weaponmakers' superior craftsmanship. By contrast, the tomahawk of the North American Indian is inferior in both shape and sharpness. The projectile at top has been carefully slotted to fit into the strap of a sling.*

MUSEO NACIONAL DE ANTROPOLOGIA Y ARQUITECTURA, LIMA: PHOTO, PEDRO ROJAS PONCE

*The two distinct levels apparent in the dwellings modeled on the sides of this Andean pottery jug have led experts to speculate that Peruvian villages, if not the houses themselves, were built in steps up the mountains. Such dwellings had windows (from which spectators peer), balconies, and colorful exteriors.*

had survived the pitched battle. In his conquest of Mexico, Cortes had shown far greater wisdom in managing to persuade such warlike tribes to march under his banner against those he sought to conquer.

Ultimately, Pizarro was rescued from this unpleasant position by the arrival of two ships with a hundred reinforcements and a number of horses, commanded by Captain Hernando de Soto, "a handsome man, dark in complexion . . . of cheerful countenance, an endurer of hardships and very valiant," according to one chronicler. Certainly, in 1531, De Soto was in the prime of life and noted for his superb horsemanship and daring. In the end, he would prove himself more chivalrous and merciful than most of his hardhearted companions.

ROYAL LIBRARY, COPENHAGEN

*Weeping Indians huddle inside as Francisco Pizarro puts their hut and temporary refuge to the torch in this Poma de Ayala drawing.*

Now powerful enough to make an attempt on the mainland, Pizarro ordered his little army ferried across the bay to Tumbes. There, unaccountably, his landing was resisted. Perhaps events at Puná or tales of the looting of Coaque had put the Indians on their guard. At any rate, three Spaniards were killed before the natives were finally driven off. Their *curaca*, when later captured, protested that the attack had been made by lawless men, contrary to his orders; Pizarro, who had learned his lesson at Puná, accepted the explanation without further question and allowed the inhabitants to return to their town.

The conquistadors were bitterly disappointed by what they found at Tumbes, however. The "war of two brothers" had left the once-splendid city ransacked of its wealth. Only the temple, the fortress, and several of the bigger houses still stood. The rest was rubble. The gold, the silver, the precious stones had vanished, as had the two men Pizarro left behind on the earlier voyage. Only a scroll of paper remained as evidence of their stay. "Know, whoever you may be that may chance to set foot in this country," it read, "that it contains more gold and silver than there is iron in Biscay." But Pizarro's soldiers, discontented after facing two battles and finding no reward, refused to believe that the scroll was authentic and jeered that it was a forgery, produced by their captain to encourage them to carry on a hopeless search.

Not until May, 1532, did Pizarro lead his main force out of Tumbes, leaving a few troops to keep communications to the coast open. For some weeks, he surveyed the country, enforcing the strictest discipline on his men. Everywhere, he proclaimed that he had come as the representative of "the Holy Vicar of God and of the sovereign of

Spain" and demanded that the natives submit to these authorities. The Indians, not knowing in the least what he meant, willingly complied.

About ninety miles from Tumbes he chose a spot for the first Spanish colony in Peru, named it San Miguel, and with the help of the Indians, built a fort, a church, a warehouse, and a hall there. Then, to emphasize the fact that they had come to stay, he distributed land among his followers, giving them legal rights over the Indians.

Having done what he could to secure his base, and having waited in San Miguel for some weeks without receiving the necessary reinforcements, the conquistador made the momentous decision to push on. By this time, it was September, 1532, and his force boasted less than two hundred men—sixty-seven of them cavalry. With these, he set out to meet the mighty Inca ruler. About fifty men were left behind to hold San Miguel. Unquestionably, Pizarro must have known that there could be no retreat now: Either victory or death awaited these Spaniards in the sierra.

By the time Pizarro left San Miguel, Atahualpa already had won the civil war and had decided to stay on at Cajamarca until the disturbing strangers he had been warned about either withdrew from his country voluntarily or were driven out. He had learned of men and horses being killed on Puná and of the happenings at Coaque. Prudent general that he was, he called his warriors of the western seaboard to his banner—and wasted no time in doing so, since many of them were untrustworthy and might just as easily join the Spaniards. That done, he simply waited.

Pizarro's plans were equally vague. He felt that in order to conquer Peru, he had to do little more than strike at its master. The staggering success enjoyed by Cortes when he seized the Aztec emperor had provided Pizarro with a model, and now he hoped to bring off a similar coup, but nothing more. He had plotted no definite moves.

Resolutely, he led his army toward Cajamarca, passing through numerous lovely valleys and into a land gay with flowers, orchards, and fields well watered by elaborate irrigation systems. Everywhere, friendly Indians provided them with food, and when night came, with shelter. After marching for five days, Pizarro rested his troops in one of the valleys and offered to allow any soldier whose heart was

CIEZA DE LEON, *La Chronica del Peru*, 1554: NEW YORK PUBLIC LIBRARY

*Made chronicler of the Indies by La Gasca in 1548, Cieza de León six years later began issuing his* Chronica del Peru, *from which this woodcut of masons building an Andean settlement is taken.*

*Francisco Pizarro's once-colorful banner, now more than four centuries old, is preserved in Spain. Its central escutcheon, like that of Pizarro's sovereign (page 133), includes the lions of León and castles of Castile.*

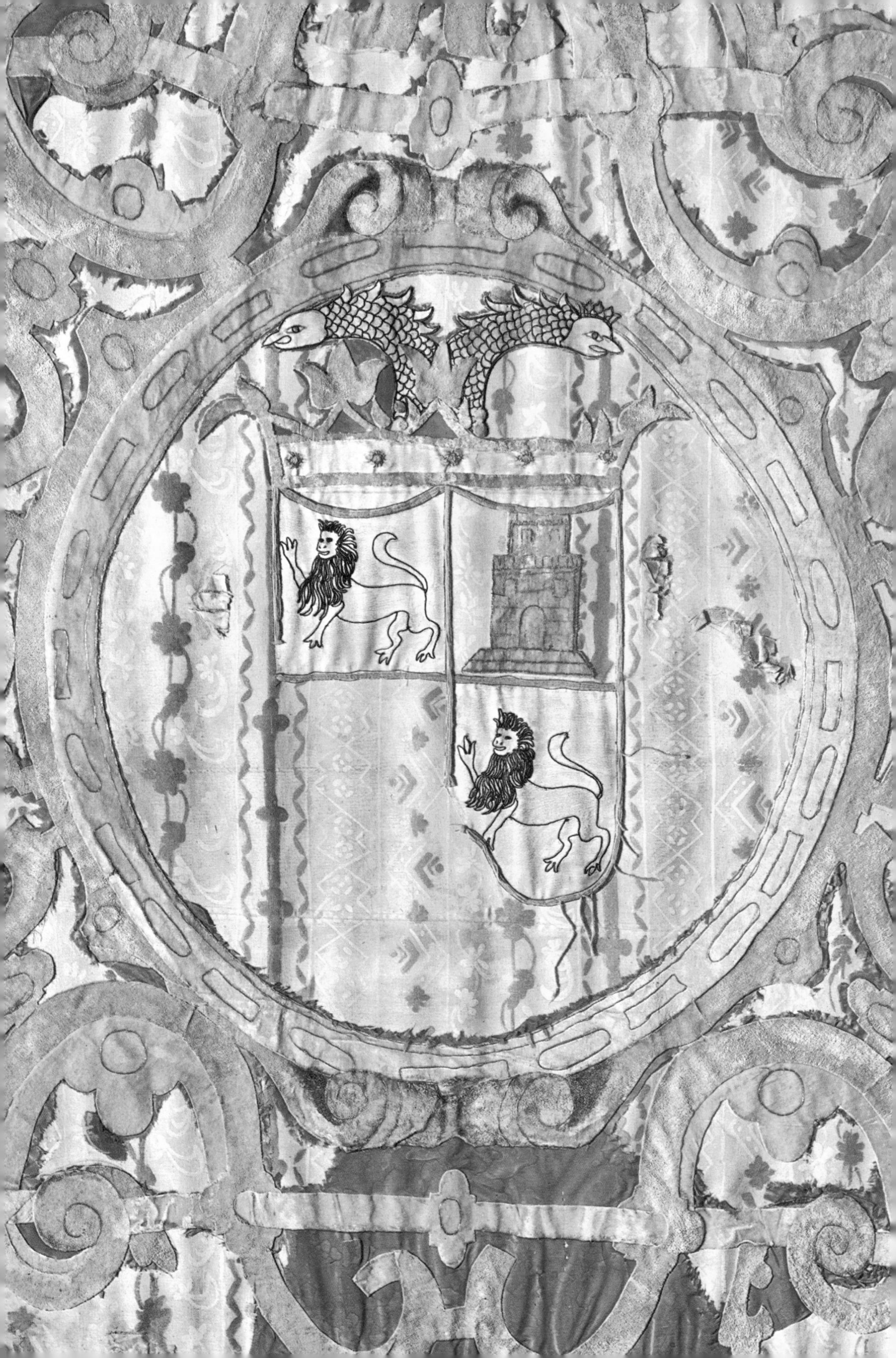

not truly in the adventure to return to San Miguel. Though short of men, Pizarro preferred a small band of loyal and dedicated followers to a larger group divided by uncertainty. Nine took advantage of his offer.

Resuming their march, the adventurers now approached the Cordilleras. There they saw houses, so expertly constructed of solid stone that it was difficult to see where they were joined, and rest houses, well stocked with food to provide for their needs. While they met with no hostility, Pizarro periodically heard conflicting reports from the Indians concerning the Inca leader's intentions and the location and size of his army, and wisely he sent off a trusted native as a messenger to the Lord Inca's camp. Known only to a chosen few, the messenger's real task was to see whether the fortresses along the mountain highway were manned and to send back word by swift runners. Having thus taken all the precautions he could, Pizarro advanced to the base of the Andes.

ROYAL LIBRARY, COPENHAGEN

*War news, official edicts, and other vital information all were relayed along the Inca highways by teams of runners, one of whom signals his arrival with a shell trumpet in Ayala's drawing above.*

Even the boldest of the conquistadors must have looked with fear and awe on that stupendous mountain barrier now to be scaled. Before them lay the foothills of dark forests, interspersed with cultivated terraces, and higher up a chaos of rock. Towering above all were the great and frightening crests, glittering white with snow. Some of Pizarro's men were for taking the broad road to Cuzco rather than meet the challenge of the mountains. But their leader knew that to turn aside now was to fail. "God . . . will humble the pride of the heathen, and bring him to the knowledge of the true faith, the great end and object of the Conquest," he reassured his men, urging them all the while to go forward. Inspired by his words, the soldiers swore to follow him—wherever he might lead.

At dawn the next day, the commander led an advance party into the mountains, while Hernando Pizarro followed with the remainder of their force. The path was even more perilous than they had feared. To avoid great jutting buttresses of rock, the track twisted and turned and climbed so abruptly in places that the troopers, leading their horses, managed to move on only with the greatest difficulty. Yet, somehow, the company progressed, sometimes passing under great crags, sometimes edging along narrow ledges, where one false step meant death in the ravines below.

Always they looked anxiously ahead and every moment feared an attack, which they were powerless to resist. Once they came on a stone fortress guarding a steep and narrow gorge, and as they climbed slowly upward, they saw that it commanded an angle of the pass where a few men could hold an entire army at bay. When no defenders appeared on its ramparts, the whole company went forward with lighter hearts, at last convinced that the Inca had decided not to lay an ambush for them on the mountain.

The following day they climbed through the icy, barren regions, and both men and horses suffered intensely from the cold and from lack of oxygen at such altitudes. Not until they had reached the crest of the Cordilleras did they halt to rest, huddling around campfires in an effort to warm themselves. It was at this point that one of the Indians sent forward earlier rejoined them and reported that the road was clear and that an envoy was on his way from Atahualpa with handsome presents and a desire to know when the Spaniards would reach Cajamarca. When the envoy arrived, Pizarro assured him that the Spaniards would lose no time on the way.

It took the company two days to thread its way across the Cordilleras. On the third they began their descent on the eastern side. Now Pizarro was uneasy. Another Indian he had sent forward some days earlier had returned with the news that the Inca had refused to see him; the messenger was certain that a trap had been set. Had Pizarro known that Atahualpa had given orders to his general, Ruminavi, to place himself and a thousand men between the advancing Spaniards and the coast, he would have been more than alarmed.

Seven days after they had first begun their climb, Pizarro and his followers stood staring in wonder at the beautiful valley of Cajamarca that lay below them. The broad river flowing through cultivated fields of bright and varied hues presented a striking contrast to the dark mountains rising on every side, and in the midst of all a little city stood, with white houses shining in the sun. But a few miles across the valley a less pleasing sight met their eyes. Thousands of tents and pavilions covered the ground where the Inca's army was impressively encamped.

"It filled us all with amazement," wrote one of the conquistadors, "to behold the Indians occupying so proud a position. . . . The spectacle caused something like confusion and even fear in the stoutest bosom. But it was too late to turn back. . . ."

Undaunted, Pizarro called his little army to battle formation and advanced down the slope toward the city, banners flying and Spanish armor flashing boldly in the sunlight.

Ante suyo
calla vaya
collas
por. alos lados tocan
dole dos. flautas que llaman

IV

# FALL OF AN EMPIRE

There was no sign of life in Cajamarca as the Spaniards approached the town. Rather than resist the invaders, the people of the countryside had withdrawn at Atahualpa's order. Now none remained to witness the pomp with which Pizarro and his tiny army entered the city. Only the legions of the Inca, camped in the distance, were there to view the procession. Weary and frightened as the Spaniards were, they made a brave show; Pizarro knew that to do less would mean their doom.

For the first time, horses' hoofs clattered in the deserted streets, and empty houses echoed back the heavy tread of armed men. Two fortresses protected the city: one on high ground, with a triple wall around it; the other dominating the open space of the great square, where the Spanish troops came to a halt. A temple, rising above a grove of trees, with a convent nearby, could be seen on the outskirts of the town. But everywhere an eerie silence pervaded.

As soon as they reached the square, Pizarro sent his brother Hernando and Hernando de Soto with a cavalry guard into Atahualpa's camp. Francisco was anxious to discover the Lord Inca's intentions and probably hoped to stave off any attack by a show of strength. Certainly the cavalcade caused some astonishment when, with arms clashing and trumpets blaring, it swept through the Inca outposts. It was the Spaniards, however, who were the more disturbed by what they saw. Atahualpa's was a large, well-equipped, thoroughly disciplined army.

As the Spanish captains drew rein before the Inca legions, the *Orejones* came forward to lead them into the Lord Inca's presence. Even among the splendidly dressed nobles who made up his entourage, Atahualpa was easy to identify.

MUSEUM OF THE AMERICAN INDIAN, HEYE FOUNDATION

*Members of the Inca's retinue—the litter bearers opposite, as well as the silver Panpiper shown above—wore intricate headdresses.*

*There is an Oriental delicacy to Fray Murua's water color of the Lord Inca (left), wearing his red* borla, *or royal fringe, and being borne on a litter.*

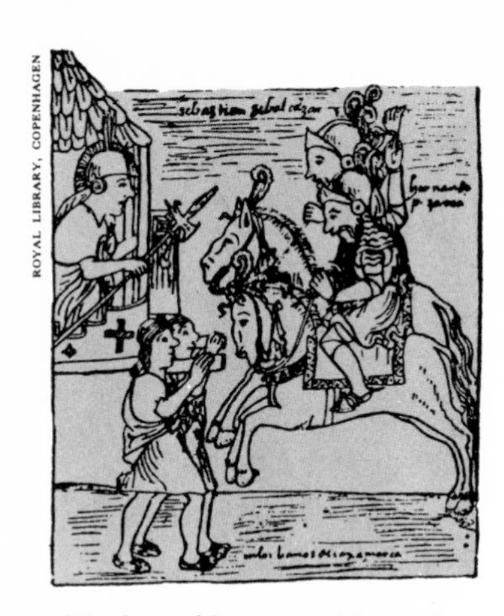


*Displays of horsemanship, such as the one above, put on by Hernando Pizarro and Sebastián de Belalcázar for Atahualpa, made shrewd use of the Indians' surprise and awe at first seeing men on horses.*

He had the assurance and dignity of one accustomed to giving orders, and on his forehead he wore the *borla*, or crimson fringe, as a sign of his sovereignty. He was about thirty years old, well built, and might have been considered handsome except for his bloodshot eyes, which gave him a rather fierce aspect.

Atahualpa received the strangers without apparent emotion, and after listening to what they had to say through their interpreter, dismissed them, briefly stating that he would visit them on the following day. Several times during the audience, however, De Soto noticed the ruler glancing with interest at his fiery war horse, and eager to impress Atahualpa and his troops, the cavalier suddenly darted forth to display his horsemanship. Setting spurs to his charger, he galloped across the plain, put the animal through certain paces, urged his horse into a gallop once more, and charged straight toward the emperor. Only at the last moment, and when he was within a few feet of Atahualpa, did De Soto rein his steed, forcing the animal back onto its haunches. It was an exciting performance, throughout which the Inca sat motionless. No flicker of feeling appeared on his saturnine face. Only after the Spaniards had gone did he react, and those of his courtiers who had drawn back in fear at the sight of the oncoming beast paid for their cowardice with their lives.

Riding back through the long lanes of Indian tents and endless columns of warriors, the Spanish adventurers must have felt dejected. They were sorely outnumbered. Few thought that they could escape from the valley alive. It was a despondency that infected their comrades once they had returned to Cajamarca. Yet Pizarro felt neither fear nor dejection. Instead, he was deeply satisfied that at last he was about to come face to face with Atahualpa. For he was, above all, a man of action, and he immediately began planning his capture of the Inca ruler.

The stratagem was a desperate one, full of danger and reminiscent of Cortes' capture of Montezuma. The plot was to ambush Atahualpa when he visited Cajamarca on the following day. There seemed to be no alternative. Certainly, retreat would eventually have brought about a clash with a foe experienced in mountain warfare—a situation in which Spanish arms would have proved least effective. Open attack on the great army encamped nearby seemed equally foolhardy. A third option, to remain in Cajamarca and trust to Atahualpa's friendship, was the least attractive. Pizarro was convinced that the Lord Inca had lured them to Cajamarca simply to seize their weapons and kill them. Since none of his officers disputed these arguments, Pizarro gave each of them detailed instructions for the action he planned for the following day.

Next, he tried to restore confidence among the men. Moving from group to group, he begged them to trust in him and in their own strong arms—and more than that, in God. He reminded them that they were fighting to convert souls to Christianity. The Church had blessed their crusade. With God on their side, they had no need to fear even swarms of pagans. Convinced that their cause was just, the Spaniards took heart and told Pizarro that he could depend on them to the death.

As dawn broke on November 16, 1532, a trumpet called the Spaniards to arms, and their commander bluntly made his plan known to the men in the ranks. Many turned pale with fear as he spoke.

The halls around the square, with doorways high enough and wide enough for a horseman to pass through easily, were ideal for ambush. In two of them he ordered the cavalry to station itself—one detachment under Hernando Pizarro and the other under Hernando de Soto. A third hallway was to be occupied by most of the infantry. In addition, Pizarro selected twenty of his best swordsmen to follow him and sent a few more to establish the falconets on a tower overlooking the square. Finally, he instructed the cavalry to hang bells on their horses' breastplates to add a frightening noise to their already terrifying appearance,

ART INSTITUTE OF CHICAGO: PHOTO, DR. JUNIUS BIRD

*Tales of strange beasts brought to Peru by the Spaniards spread rapidly—and often inaccurately—as a bronze relief of a horse with llama's hoofs, by an artist with little idea of how to depict the man astride it, shows.*

THE CLEVELAND MUSEUM OF ART

and he cautioned every man to remain hidden from view until he heard a musket fired. This would signal all to rush out from their posts and strike down the Indians.

The troops ate a hearty meal to give them strength for the coming encounter, and their spirits were raised by the celebration of a solemn Mass, during which God's help was requested in this fight against the heathen. At last, they took up their positions in the hallways to await Atahualpa's entry into the city.

For hours, men and horses stood poised for instant action. But it was afternoon before the watchmen on the tower shouted that the Indian army had begun its march. By then the morning's elation had long since evaporated; the report carried to Atahualpa by spies that the strangers were cowering in the hallways around the square was, according to Pedro Pizarro, "not far from the truth."

During the night the Inca monarch had added four thousand men to Ruminavi's forces in the north, positioned there to cut off the Spaniards' retreat, and he had deliberately delayed his march to the city in order to arrive there at nightfall. He, too, had a plan. Even now his progress across

NEW-YORK HISTORICAL SOCIETY

*La Conquista del Peru*, 1534: NEW YORK PUBLIC LIBRARY

La conquista del Peru.
llamada la nueva Castilla. La q̃l tierra por diuina voluntad fue marauillosamente conquistada en la felicissima ventura del Emperador y Rey nuestro señor: y por la prudencia y esfuerço del muy magnifico y valeroso cauallero el Capitan Francisco piçarro Gouernador y adelantado de la nueua castilla, y de su hermano Hernando piçarro, y de sus animosos capitanes z fieles y esforçados compañeros q̃ cõ el se hallaron.

*The 1534 woodcut at right, title page of an anonymous chronicle attributed to a member of the expedition, shows Atahualpa (seated on a litter) being offered a Bible by the priest Valverde. The portrait of Atahualpa (above) gives the Lord Inca a Madonna's features and an exotic costume. He may have ridden in a litter like the one seen at left, which is decorated with figures of painted metal and inlaid wood, and which has slots for bearers' poles.*

the plain was slow. Spanish nerves were being stretched to the breaking point.

As the great procession drew nearer, those on the watch-tower could see Atahualpa seated on a golden throne in an open litter carried by his nobles, the golden ornaments of the *Orejones* blazing like the sun. Ahead of the Inca, a corps of servants swept every hint of rubbish from the road.

Suddenly, to everyone's surprise, the magnificent array ground to a halt, and a messenger came forward to inform Pizarro that Atahualpa would wait until morning to enter the city. Dismayed at the thought of another night of waiting, Pizarro sent back word that a banquet had been prepared and that the Spaniards would be disappointed if the Lord Inca were not there to share it. Strangely, the message caused Atahualpa to continue his march, accompanied by only a few thousand unarmed or lightly armed *Orejones*.

Just before sunset the great cavalcade began to pass through the city gates, its members singing songs of triumph and preparing the way for their emperor. Indian ranks lined the sides of the square, allowing room for the Inca and his retinue in the center. About five or six thousand of the

Indians had filed into the area before the chanting ceased, and all stood in silence while their emperor impatiently demanded to meet the strangers. At his bidding, two figures emerged from the dark buildings and moved slowly toward him. One was a Dominican friar, Vicente de Valverde; the other, his interpreter.

Valverde told Atahualpa that the Spaniards had come a great distance to teach him and his people the true faith and then proceeded to explain Christian philosophy. He ended by stating that the Pope, as God's vice-regent on earth, had entrusted the conquest and conversion of the Inca lands to the mightiest monarch in the world, Charles V, King of Spain and Holy Roman Emperor. He begged Atahualpa to forsake his false gods for the true God of Charles V.

Atahualpa probably understood only the last part of the discourse, for he replied furiously that he would be no man's subject. "Your Emperor may be a great prince," he acknowledged, ". . . and I am willing to hold him as a brother. As for the Pope of whom you speak, he must be mad to talk of giving away countries that do not belong to him. . . ."

Moreover, Atahualpa refused to abandon his own gods in deference to new ones and asked on what authority the friar made such demands. When in reply Valverde pointed to a Bible in his hand, Atahualpa took it, flipped through its pages, and flung it into the dust.

"Tell your comrades," he exclaimed, "that they shall give me an account of their doings in my land. I will not go from here till they have made me full satisfaction for all the wrongs they have committed." The friar picked up the book from the ground and hurried back, urging Pizarro to act at once.

This was the moment that Pizarro had long awaited. He waved a white scarf, and the signal shot rang out. It was quickly answered. From the tower above the crowd spurts of flame blazed forth, accompanied by a thunderlike roar. Then other flashes came from the doorways. Around the Lord Inca, men fell as if struck down by magic, victims of the shot or bolts from the silent crossbows.

Suddenly, figures were rushing out from the doorways, shouting the war cry, "Santiago and at them!" and seconds later the cavalrymen followed, charging out on horseback and cutting down or trampling all who stood in their path. There was no escape. Piles of Indian dead and wounded soon blocked every exit. Wherever the terrified natives turned, they faced death, until at last, in frenzy, they burst through a wall enclosing one side of the square; the survivors streamed out, pursued by the Spanish horsemen.

Hundreds of the Indians fled, but the Inca nobles remained, still forming a living wall around their master. Some threw themselves at the horsemen, trying to tear the merciless troopers from their saddles. Others simply died where they stood, protecting their lord to the end. As one fell, another took his place. So great were their numbers that the Spaniards began to fear that, with night falling fast, the Inca might yet escape. Some of the cavaliers panicked and decided that the only solution was to kill Atahualpa himself. Pizarro's band, meanwhile, was cutting its way through the throng and had almost reached the royal litter when the commander realized the Inca's danger.

"Let no one, who values his life, strike at the Inca," Pi-

*Cavalry, firearms, and a small cannon all figure in the version of Atahualpa's capture below, although some details are inaccurate. In reality, the Inca was captured inside Cajamarca after he had rejected the priest's exhortation.*

COURTESY OF HIS EXCELLENCY, CELSO PASTOR, PERUVIAN AMBASSADOR TO THE UNITED STATES, WASHINGTON

zarro shouted, and stretching out his arm to check the furious assault, he took a blow on the hand from a Spanish sword. It was the only wound suffered by a Spaniard that day. Soon after, several of the nobles who supported the imperial litter were struck down, and only Pizarro and his companions saved Atahualpa from falling heavily to the ground. The captive emperor was led off to a nearby building, and the conquerors went on killing until darkness fell and they could no longer see.

In this terrible half hour the Inca empire fell. More than two thousand Indians, many of the highest rank, were massacred. As news of the disaster reached the army outside the city, panic spread through the ranks, and the sierra was soon dotted with groups of fleeing soldiers. Ruminavi fled too, moving northward and taking his army with him.

That night the Spaniards offered thanksgiving to Providence for their miraculous victory, and Pizarro turned to his duties as "host" and entertained the Inca at a banquet in a hall overlooking the place where the Indian dead still lay.

The next morning, the dead were removed from the square and given a decent burial, lingering enemy forces were dispersed from the valley, and Atahualpa's camp was looted. No resistance was met. Disregarding a cold-blooded suggestion that Indian prisoners be put to death, or at least rendered harmless by having their hands cut off, Pizarro released them, after each of his men had chosen what servants he wanted from among them. Then he ordered that herds of llamas be brought to the city for food and gave permission to break open the imperial storehouses and take what was needed.

Realizing that he was too weak to move on to Cuzco immediately, Pizarro set to work to make Cajamarca a suitable headquarters for his army during its wait for reinforcements. A church was established, the city walls were rebuilt, and Atahualpa was placed in a spacious residence and treated with great respect. Though closely guarded, the Inca monarch retained all the outward show of his sovereignty. His privacy was strictly preserved, and he was allowed to wear his imperial regalia, including the *borla*, or royal fringe. In addition, he was waited on by his numerous wives, who served his food on gold and silver plate, and his subjects entered his room only when summoned.

As was customary, the Lord Inca, set apart from his people, continued to treat his subjects with haughty reserve. Yet in the company of the Spaniards he relaxed. He learned

NEW YORK PUBLIC LIBRARY

ROYAL LIBRARY, COPENHAGEN

*While Atahualpa remained in Pizarro's indulgent custody (above), attended by his wives and staff, a ransom of unparalleled size gradually was accumulated (left). As temples and palaces were stripped of their ornamentation, much of the finest Inca metalwork was smelted down and shipped off to Spain. De Bry has misrepresented the shape, but probably not the size or quantity, of those objects.*

to throw dice and to play chess. He even seems to have enjoyed a joke with his captors. He listened attentively when Friar Valverde or Pizarro tried to convert him to Christianity, and he was strongly moved by the latter's argument that the god whom he worshiped could not be the true God if he had allowed Atahualpa to be captured by his enemies. The Inca did not, however, accept the Christian faith, but he did become quite friendly with the two Hernandos—Pizarro and De Soto.

If Atahualpa feared anything, it seemed to be that his half-brother Huáscar might learn that he had been seized, escape his own captivity, and take control of Peru. Such danger made it imperative that Atahualpa recover his freedom as quickly as possible.

MUSEO DE AMERICA, MADRID: PHOTO, ORONOZ

*A sense of humor, common to Inca artisans, is seen best in their decoration of domestic objects, such as this* kero, *or wooden drinking cup. On its side, an Inca noble sits under a parasol—and drinks from a* kero.

Since the looting of Coaque, reports had reached the Lord Inca of the Spaniards' love for gold. Now, living in their company, he realized just how dominant that passion was and decided to play on this weakness. In return for his liberty, he offered to fill a room, to as high as he could reach, with the precious metal. The room he chose was, according to some accounts, twenty-two feet by seventeen feet in size.

Pizarro thought at first that Atahualpa was joking, but he was tempted to believe that the wealth of this empire had not been exaggerated. Almost in disbelief he drew a line around the room at the spot touched by the Inca and had a legal document drawn up agreeing to Atahualpa's proposition. That settled, Atahualpa also promised to fill another, smaller room twice over with silver objects. All this was to occur within two months—a short period, indeed, to collect what was to become the most stupendous ransom in history.

For centuries the will of the Lord Inca had been law, and even now that Atahualpa was held prisoner, the obedience of his people was absolute. With astonishment the Spaniards had watched the greatest of his nobles humbling themselves before him, taking off their sandals and fastening burdens to their backs in a sign of subservience. At times, so many nobles visited Atahualpa that the conquerors feared that he might be plotting against them. Now, however, his power was to be used to enrich them, and they eagerly awaited the results.

*Although noblewomen occasionally rode in litters, they most often were seen traveling exactly as Murua has shown here: on foot, accompanied by offspring and servants (one of whom is carrying a parasol). A large shawl pin holds the woman's heavy cloak in place.*

Soon after Atahualpa sent forth his command, groups of men and llamas set out for Cajamarca, laden with golden vessels and plates stripped from their local temples and palaces. But the distances were great, and weeks passed before the golden stream began to trickle into the city.

Carefully registered and placed in the room under guard of the conquerors were massive plates of gold, some weighing as much as seventy-five pounds; fine ornaments superbly worked by master craftsmen into the shapes of birds and beasts and flowers; and dishes and cooking utensils. The treasure heap was growing gradually.

Unhappily for the royal captive, however, the avarice of the Spaniards knew no bounds. The greater the loot, the more they demanded. Some even suggested that Atahualpa was deliberately slowing down the collection of gold to gain time to plot his own escape, and rumors of Inca armies massing began to haunt the Spaniards' days and nights.

Eventually, even Pizarro was affected by the growing tensions. He accused Atahualpa of plotting against him, and the Inca, surprised by the charge, quickly pointed out that his fate was in the Spanish commander's hands, not vice versa. Since he would forfeit his life at the first sign of hostilities, it made no sense for him to take part in such a conspiracy. Moreover, he declared that his subjects would not dare to move without his orders and that the delay was only natural. While a message might reach Cuzco from Cajamarca in five days, he explained, a porter laden with gold would require weeks to make the return journey. He urged Pizarro to send his own men out to satisfy himself that no hostile bands were moving on Cajamarca. Atahualpa especially begged him to send other Spaniards to Cuzco to make sure that the priests of the great temples there did not hide their treasures. He offered safe conduct for the Spaniards sent on such missions.

Pizarro acted quickly on the advice. He sent his brother Hernando with twenty horsemen and a few supporting infantry to Huamachuco, a nearby town rumored to be a center of revolt. After receiving word that all was well there, he sent Hernando on to Pachácamac to seize the treasures of its temples.

Pachácamac was situated on the coast, about four hundred miles away, and it was January, 1533, before Hernando reached the city. He made straight for the temple. Brushing aside the priests who tried to bar his way, he ascended through covered passages to the sanctuary of the dread deity, violating the sanctity of the place by his very presence. Almost immediately an earth tremor was felt, and the superstitious Indians, believing that Pachácamac was about to destroy its invaders, fled in terror. The Spaniards had no such fears, however. Eagerly they followed their leader into the holy of holies, expecting to find great spoil. Instead, they found only a place of worship, a foul-smelling den, reeking of the blood of animals and men who had been sacrificed to the gods there. In a dark corner stood a clumsy wooden idol; it was the god of the temple—a profane god to the angry soldiers.

Without hesitation the Spaniards dragged the hideous figure from its shrine and broke it into pieces in front of the awe-struck Indians. On top of the hill the conquerors then placed a large cross, and when the confused Indians came to submit to these white beings, more powerful than their gods, Hernando tried to convert them to Christianity, teaching them the sign of the cross to scare away devils.

But Hernando had come too late for the gold. Only about 80,000 *pesos de oro* ($480,000)—a pittance compared to the vast sums the Spaniards had envisioned—was gleaned from the town of Pachácamac. The priests of the temple had carried off most of the treasure and had either buried or hidden it once they had learned of the invaders' mission.

Still, the trip brought some good. Before he left the city, Hernando discovered the whereabouts of Challcochima. A kinsman of the Lord Inca, the brilliant general who had led Atahualpa's troops to victory over Huáscar, Challcochima was stationed at Jauja, about a hundred miles to the east, across the mountains, with an army of 35,000. Here was a prize worth more than gold. Without thought to the odds, Hernando immediately turned his troops toward the east to confront the Inca general and lead him back to Cajamarca.

Again the Spaniards were forced to scale the high Andes, passing over points nearly sixteen thousand feet above sea level. Many suffered from a painful altitude sickness, as well as a consuming fatigue. Still, they toiled on, and when the rough mountain rocks wore the horses' shoes down to their hoofs, Hernando had the shoes renewed with silver. In the end, Hernando's audacity and his army's resolve paid off. Reaching Jauja, they caught Challcochima in a state of indecision as to what action to take and half persuaded and half bullied him to join them on their return march to Cajamarca.

The journey to Cuzco—more than six hundred miles—was not nearly so arduous for the three soldiers who were sent on the mission to the capital. Carried swiftly along the great highway on litters, they were treated everywhere as superior beings and were provided with food and lodging fit for princes. When the soldiers arrived, the people of Cuzco made haste to fete their visitors and to bring them the riches that they requested.

Up to this time the city had been in the hands of Quizquiz, another of Atahualpa's generals, who ruled with an iron grip, and the people of Cuzco believed that the Spaniards would restore Huáscar to the throne. Quizquiz was, naturally, openly hostile, but he, too, obeyed his master's orders and allowed the soldiers to loot the capital.

Unfortunately for the pious inhabitants of Cuzco, it was not long before they learned their mistake. For the soldiers stripped not only palaces and public buildings of their riches, but the houses of dead kings as well. They even forced the Indians to help them rob the most sacred sanctuary, the Temple of the Sun. Every precious ornament was seized, and crowbars were used to wrench gold friezes from walls. Then, insolent and contemptuous of the Inca nobles, the

*Like the Incas, the Chimú tribesmen were idolaters who worshiped crude wooden shrine figures, such as the one above—a 29 1/4-inch-high figure holding what appears to be a tumi, or ceremonial knife.*

INTERNATIONAL FILM FOUNDATION, INC.: PHOTO, JULIEN BRYAN

*Broad and well paved in the foothills, the Inca highways became mere cobblestone paths near the mountaintops. Many, like the one above, are still used today by Peruvian shepherds moving their flocks from pasture to pasture.*

Spaniards misused the *Orejones'* daughters and burst into state convents to violate the women there. After a week had passed, only fear of Atahualpa's vengeance prevented the men of Cuzco from killing the three emissaries. But by then, the Spaniards' task was completed, and they returned to Cajamarca with their plunder.

Both missions, therefore, had been successful. Hernando's long reconnaissance had proved that no general attack was being prepared, and both had shown that Atahualpa had kept his word. Yet the young man's life still hung by a thread. Shortly after the ransom compact was made between him and Pizarro, the captive learned that the Captain-General—as Pizarro was now often called—intended to have Huáscar brought to Cajamarca. Afraid of what problems Huáscar's arrival in Cajamarca might cause him, Atahualpa gave secret orders to have his brother killed.

The arrival of Almagro in Cajamarca in February of 1533, with one hundred fifty infantry and fifty cavalry, provided an additional threat to Atahualpa's life. Until then, Pizarro could not advance on Cuzco. He needed Almagro's reinforcements and help. Now he was ready, and Almagro, astonished to see such treasure, was eager to press on to get his share of the spoils. Only two things prevented such a move: one was the necessity of deciding what to do with Atahualpa, and the other was to complete the collection of his ransom. Once the latter had been accomplished, some hinted, Atahualpa's death would be the only realistic solution to the former problem.

In addition, the tense and quarrelsome relationship that had grown up between Hernando Pizarro and Almagro made it imperative that Francisco do something soon. But all Francisco's persuasive powers could not repair the unpleasantness that existed between the two men, and he feared that the success of the whole enterprise was endangered by his brother's difficult nature. Once the treasure was disposed of, he decided, Hernando could be charged with delivering Charles V's share to Spain. Then peace could be kept with Almagro. Still, another darker and less honorable motive also may have led Pizarro to this decision. Hernando was known to be the friend of Atahualpa, and he might have interfered with the Captain-General's plans for dealing with the Inca.

Although the line drawn some months earlier had not yet been reached, the gold came almost to the mark. Pizarro decided to wait no longer before determining each man's share. All sorts of objects were involved—plates, goblets, ewers, salvers, many of such delicate craftsmanship that Pizarro planned to send them intact to Charles V.

After a selection had been made, the rest was melted down into standard-size gold bars, and each bar was weighed in the presence of the royal inspectors. The silver was treated in the same way. In both cases, Indian smiths were forced to destroy their own handiwork.

The distribution of this vast amount of plunder, estimated by modern authorities to be worth \$8,165,674, was solely in Pizarro's hands. Having assembled the Spaniards and solemnly prayed to Heaven for guidance and wisdom, he made known his awards, refusing, as he did so, to recognize the claims of Almagro's men to share equally in the prize. In the end, they reluctantly agreed to accept a much smaller sum of 20,000 *pesos de oro* (or about \$120,000). Although disappointed, they contented themselves with the thought of future spoils in Cuzco.

When Pizarro announced the rewards for those who had marched with him to Cajamarca, there was a breathless hush. For himself, he set aside 57,222 *pesos de oro* (about \$343,332); the Lord Inca's golden throne, worth 25,000 pesos (\$150,000), and 2,350 marks of silver (or about

ROYAL LIBRARY, COPENHAGEN

*As Poma de Ayala traveled along the Inca highway to Cajamarca, he noted the number of way stations and royal residences, recording each as a symbol in his logbook. Buildings that the Spaniards had converted into monasteries or churches are marked with a cross.*

$9,400). Hernando received 31,080 pesos ($186,480) and 2,350 marks of silver ($9,400); and De Soto, 17,740 pesos ($106,440) and 724 marks of silver ($2,896). Most of the sixty horsemen received 8,880 pesos ($53,280) and 362 marks of silver ($1,448) each, and about twenty of the infantry were allotted half that amount, while the remainder received shares three-eighths the size of the cavalry's share. Some 15,000 pesos ($90,000) was set aside for the garrison at San Miguel, and 2,220 ($13,320) for "the first Christian temple in Peru." No mention is made of the sums Pizarro paid to Almagro or set aside for Luque's estate. Unhappily, Luque had died before Almagro left Panama and never knew of the success of this great venture.

DR. JUNIUS BIRD, AMERICAN MUSEUM OF NATURAL HISTORY

*A hook-nosed Inca scowls on this small gold ornament that perhaps capped a wooden ceremonial wand.*

Only one obstacle remained—Atahualpa. The treasure divided, he demanded his freedom and was supported in his demands by De Soto and other cavaliers. But Pizarro refused to honor his pledge. While formally acknowledging that the Inca had paid his ransom, he declared that reasons of state necessitated his being kept under guard.

The Captain-General and Almagro were in a quandary. They were afraid that Atahualpa might be rescued if they took him on the difficult march to Cuzco and felt that if they freed him, he might rally his armies and seek revenge.

Again rumors of an impending attack by hostile Indian armies spread through the camp. Most contemporary chroniclers blame Felipillo for the malicious tales. From the moment that Atahualpa had demanded the young interpreter's death for violating one of the royal concubines—a just punishment under Inca law—Felipillo had evidenced a vicious hatred for the emperor. Since Felipillo was indispensable to the Spaniards, he was spared, and after that he missed no opportunity to undermine the Lord Inca.

By this time, Atahualpa's most influential friend, Hernando Pizarro, had been sent to Spain, and De Soto, with a few cavalrymen, was selected to reconnoiter the area where the armies supposedly were massing. With both men out of the way, Pizarro suddenly yielded to the insistent demands of Almagro, the royal officials, and most of the troops. The Inca was brought to trial.

That expediency, rather than justice, was the reason for this decision is made clear by the charges. Only one—that Atahualpa had attempted to raise a revolt against the Spaniards—made any sense at all, and even that charge turned out to be unfounded. The rest of the charges accused the Inca of such crimes as murdering his brother Huáscar, having more than one wife, and squandering public funds. All of these, with the exception of Huáscar's murder, were perfectly acceptable under Inca law. The Spaniards had no right either to try or to punish him for such practices.

The result was a mock trial with a foregone conclusion. On August 29, 1533, Pizarro and Almagro, sitting as judges, found Atahualpa guilty—of how many of the charges is unknown—and sentenced him to be burned at the stake. Execution was to take place that very night in the public square—conspicuously, before De Soto's return.

The trumpet sounded in the square at Cajamarca. Night had fallen, and torches shone on the grim faces of the Spanish soldiers who stood at arms as Atahualpa was led from his quarters, chained hand and foot. By his side walked Friar Valverde, trying to the last to convert him. But it was not until the Inca was bound to the stake and the executioners prepared to light the faggots that the priest succeeded. He promised Atahualpa that if he became a Christian and accepted baptism, he would be garroted rather than burned—a somewhat easier and less humiliating execution. In the midst of his sorrowing people, the last Lord Inca went to his death with calm fortitude.

Two days later De Soto returned from his reconnaissance. He had found no signs of hostile forces. Shocked and indignant to learn of Atahualpa's execution, he went to see Pizarro, whom he found dressed in deep mourning. Boldly, De Soto accused the Captain-General of injustice and blamed him for not sending his prisoner to Spain to be tried by Charles V. Pizarro admitted his haste in punishing the Inca and claimed that the fault lay with his officers. They, in turn, laid all the blame at their commander's feet.

*Out of ignorance or revulsion, de Bry has shown Atahualpa (right) being garroted by Indians rather than by the Spaniards, who actually performed the execution. Held on his throne and heavily fettered, the impassive Lord Inca suffers slow strangulation (a means of execution that he had chosen over being burned alive at the stake).*

The execution of Atahualpa not only caused quarrels among the conquerors, but also failed in its object. Without an emperor, the structure of the Inca state threatened to collapse. No longer could a Spaniard travel through the land in comfort and safety. Porters heading for Cajamarca turned back when they heard of the Lord Inca's death and often buried what gold they carried in such remote spots that presumably it still lies hidden.

Pizarro was quick to realize what had happened, and he selected Toparca, one of Atahualpa's brothers, as the new Lord Inca. Toparca was crowned with all the customary ceremony, and the Indian nobles paid homage to him.

Hostile war bands still roamed the countryside when Pizarro set out for Cuzco in September, 1533. Burned villages and wrecked bridges hindered his march. At Jauja he was confronted by a vast array of warriors waiting to give battle, and once again the iron nerve and dauntless courage of the Spaniards—now about five hundred strong—won the day. Wading or swimming across the swollen river outside the town, the invaders made straight for the enemy, advancing apparently unharmed through a storm of spears, stones, and arrows. The Indians fled in panic, and the Spanish cavalry took a dreadful vengeance on the fugitives.

Pizarro stayed at Jauja for some time after that to rest his troops, but De Soto was sent ahead with sixty of his cavalry as a scouting party. Thrusting on toward Cuzco, the little band crossed the Apurímac River before meeting serious resistance. There, among the rocky defiles of the Andes, they were assailed by thousands of Indians. Only darkness brought an end to the furious encounter.

During the night, Almagro arrived with a relief party, and by morning the Indians were gone. Discovering that their enemies had miraculously increased overnight, they had melted away into the mountain mists, and the way to Cuzco was left open.

At Jauja, Pizarro was deeply perturbed by the growing resistance of the Indians. He was convinced that Challcochima was behind it. But the Inca general swore that he knew nothing of such plots. Despite Challcochima's protests, Pizarro believed that the general had sent agents to Quizquiz, urging him to attack the Spaniards, and he ordered that the general be put in irons and strictly guarded. Shortly afterward, the puppet emperor Toparca died, and Challcochima, accused of his murder, was executed.

The remainder of the journey to Cuzco was militarily uneventful. Joining forces with Almagro and De Soto, the

*Where the facts were sketchy or insufficient, European artists often supplied them from their active imaginations. Thus, in this illustration from an account of the conquest printed in 1621, Manco is shown entering Cuzco in a wheeled carriage drawn by goats and followed by balding courtiers.*

PLAUTIUS, *Nova typis. . . .*, 1621: NEW YORK PUBLIC LIBRARY

conquistadors advanced to a beautiful valley called Jaquijaguana, within fifteen miles of the capital. A halt was called so that the men might rest, and it was there that a man named Manco, a younger brother of Huáscar, appeared in the Spanish camp to claim the throne.

Pizarro was delighted. He had learned by experience that the Indians were easier to control through a powerful Lord Inca. Thus, the thousands of Indians who gathered in the streets of Cuzco on November 15, 1533, to watch the triumphant entrance of the Spaniards into the city also witnessed the arrival of a new monarch. Many trembled with dread as the Spanish trumpets blared, the earth shook under the tread of the war horses, and men with pale faces and dazzling armor invaded the Incas' sacred city.

V

# REVOLT OF THE INCAS

A child of the Sun was fasting in one of Cuzco's temples. Manco, the son of the great Huayna Capac, had been accepted by the nobles of the imperial city as their Lord Inca and was undergoing the customary purification ceremony before being invested with the royal fringe. Beyond the temple walls, his long-suffering subjects prayed, too, that with this coronation their time of trouble would be ended. The Indians had been cruelly persecuted under Atahualpa and robbed by the brutal conquerors. Now the Spanish commander appeared to have relented and given them a prince of their own to rule them.

In the great square, bordered by temples and palaces, a vast throng assembled one day early in 1534. The Spanish troops, with shining armor and waving banners, were drawn up in the center, and around them were ranked the Inca nobles and people. Accompanied by the royal mummies of his ancestors, carried on litters by their retainers, Manco appeared before his subjects. Then Mass was celebrated, and the royal fringe was placed on his brow, not by the high priest of the Temple of the Sun, as was customary, but by Francisco Pizarro. The *Orejones* bowed before their new Lord Inca and promised to serve him faithfully, and a Spanish lawyer read a proclamation declaring that Manco, Lord Inca of Peru, held his high office under the supremacy of Emperor Charles V. When this speech had been translated, the Indians promised loyalty to Charles V and to his governor, Pizarro.

Joyous festivities followed the coronation. For days the exhilarated people of Cuzco celebrated their release from oppression with heavy drinking and ceremonial dancing.

BIBLIOTECA DEL PALACIO, PATRIMONIO NACIONAL, MADRID: PHOTO, ORONOZ

*During the revels that followed the coronation of a new Lord Inca, ceremonial dancers, such as the one above dressed as a lion, may have entertained the celebrants.*

*Artworks were the Incas' archives—the visual record of military as well as court activities (see page 82). Therefore, the frantic battle scene opposite, depicted in vibrant color on the side of a* kero, *or ceremonial drinking cup, may possibly represent an actual event in the long civil wars.*

The Spaniards celebrated, too. Pizarro had his puppet emperor, and the Indians had their Lord Inca again. Both were to be bitterly disappointed.

On their first entry into the city several weeks earlier, the Spaniards had been astonished and delighted by what they had seen. Gateways of colored marble and massive public buildings of stone, with walls often painted or stained in bright colors, made a welcome change from the bleak landscape of the sierra. They had been especially impressed by the elaborate plan on which the city had been rebuilt by the emperor Pachacuti. The paved streets followed a regular pattern, always crossing at right angles, and from the central square, four roads led to the main highways of the empire. A pure stream running under many bridges provided the city with water.

MUSEUM OF PRIMITIVE ART: PHOTO, LEE BOLTIN

*Incas never ceased to be farmers—even in death. Graveboards shaped like agricultural implements often are found at Andean burial sites, mute testimony to the dead men's chief activity while alive. Such boards (above) were handsomely decorated with carved designs.*

Above the city, on a spur of the Cordillera, the massive fortress of Sacsahuamán stood guard over the capital. Built of great blocks of hewn stone, so skillfully put together that it was difficult to insert the blade of a knife between them, Sacsahuamán had never been needed since the time it was completed a few years before the coming of the Spaniards.

As soon as the conquerors had camped in the square, Pizarro forbade looting. Despite such edicts, however, the eager soldiers made haste to plunder the religious houses and the palaces of the Incas. They even ripped the jewels and golden ornaments from the mummies in the temple of Coricancha and forced open graves to despoil the dead. Indians who were suspected of concealing treasure were tortured in the hope of making them reveal their hiding places. Some of the robbers were lucky, being led by their victims to remote caves where they found vases of pure gold, life-size llamas and human figures of the same metal, as well as dresses made from beads of gold. Once more the Indian goldsmiths were put to work to destroy the artistic creations of generations, and another mass of ingots was prepared for the conquerors. The treasure was probably of about the same value as that distributed at Cajamarca, although some say that it was only half as large. In any event, it proved far less enriching to its shareholders, since it was to be divided among nearly five hundred men, rather than about two hundred men, as had been the case at Cajamarca.

It was an enormous reward, but ironically, few of the adventurers benefited from their new wealth. Some, it is true, returned to Spain to live as gentlemen for the rest of their lives, but more lost their shares in feverish gambling, and the rest found that gold was of little value in a land where European goods were in short supply and Indian products were taken without payment. Indeed, there was no coinage in Peru; all work was done for the community, and what little trade there was took the form of barter. The men who stayed had cheap food and slaves to wait on them, but they had to pay a fantastic price for a good horse, a sword, or a pair of shoes.

Pizarro did not long enjoy his conquest in peace. He had just set up a Spanish municipality in Cuzco—in March, 1534—with its own magistrates and councilors, when he received news of a threat in the north. Pedro de Alvarado, governor of Guatemala and a conquistador who had fought with distinction under Cortes in Mexico, had heard of the vast riches pouring out of Peru and had quickly decided to win a share for himself by taking Quito in the northern part of the Inca empire, To counter this invasion, Almagro took a small body of cavalry and marched north, determined to engage the support of Sebastián de Belalcázar and his troops at San Miguel before moving on to Quito.

But Belalcázar had already acted. With about one hundred forty Spaniards and a strong force of Indian allies, he had captured Quito for himself. Later, when Almagro arrived in Quito, the cavalier denied that any treachery was involved in his act, and together he and Almagro waited for Alvarado to reach them in the high sierra. Some historians suggest that personal greed had driven Belalcázar on this foray to the north, but as there was no treasure to be found there, his reasons for going to Quito must remain speculative, if suspect.

Unfortunately for Alvarado and his five hundred followers, the route they chose was a disastrous one. They had set out for Quito from the Bay of Caráques by the most direct path—through the ice and snow of the high Andes. So slow was the going and so fierce was the mountain cold that many of the men froze to death in their saddles. A quarter of Alvarado's Spaniards and most of his Indian allies died from cold or starvation on the way. Alvarado's forces still outnumbered those of the two captains who barred his entrance into Quito, however, and in the end neither side wished to put the issue to the sword. An agreement soon was reached. In return for treasure worth about a half million dollars, Alvarado promised to return to his province of Guatemala, leaving his ships, soldiers, horses, and ammunition. His men, who had made friends with Almagro's followers and had heard of their new riches, were eager to stay, and most of them took service under Almagro. Pi-

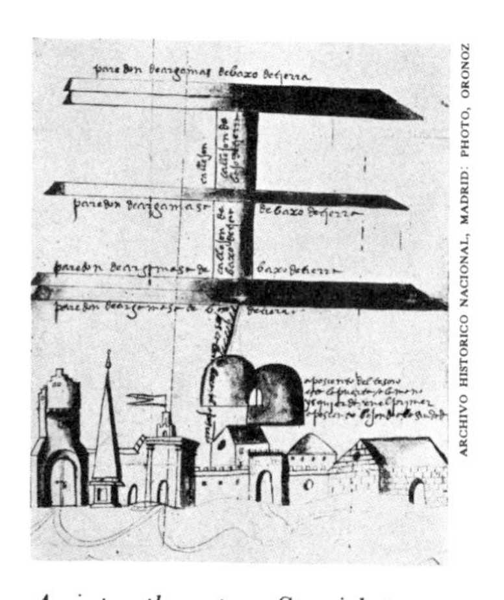

ARCHIVO HISTORICO NACIONAL, MADRID: PHOTO, ORONOZ

*A sixteenth-century Spanish treasure seeker thought he had hit upon a method of finding Inca gold. Beneath this fanciful Peruvian town, he speculated, lay a tunnel complex (top) in which the conquered natives had craftily concealed riches from the white man.*

zarro, of course, ratified the favorable pact. The danger was past. Now, indeed, Peru seemed safely in his hands.

Hernando Pizarro's journey to Spain seemed to bolster Francisco's position. When Hernando's galleon, the *Santa María del Campo*, docked in Seville in January of 1534, hundreds came to stare in wonder at the ingots of gold and silver that were lifted from its hold, and he was received with great honor by Charles V. Moreover, the men whom Pizarro had allowed to sail for home with his brother flashed their gold in every corner of Spain. Thus, before departing once again for Peru, Hernando was able to recruit one of the finest armies to sail for the Indies in many years.

Hernando also brought back with him honors and rewards for the conquerors: Francisco's territory was extended two hundred miles south, and he was made a marquis; Almagro was given an independent command over land stretching six hundred miles to the south of the boundary of Peru; and Hernando himself was made a knight of Santiago, one of the most coveted military orders in Spain.

Not trusting Hernando, Almagro had sent a confidential agent to the Spanish court to make sure that the greedy Pizarros did not once more grab all the rewards. Almagro's representative returned from Spain before Hernando did and brought news of the bequest to his chief at Cuzco. Almagro lost no time in claiming dominion over Cuzco on the ground that it lay within his awarded territory. Since South America had not been accurately surveyed by that time, no one really knew where boundary lines ran, and only with some difficulty did the confused Captain-General persuade his old partner to relinquish Cuzco, at least until the king's decision should be known. Meanwhile, he encouraged Almagro to lead an expedition to conquer Chile, which all agreed fell within his sector.

Leaving Cuzco under the command of his brother Juan, Francisco then set out for the coast, where on January 6, 1535, he had founded his capital, *El Ciudad de los Reyes*, The City of the Kings. (It later became known as Lima, from a corruption of "Rimac," the name of the river on which it was situated.) At Lima, Pizarro turned his great energies to the task of building a splendid city. For the capital of his domain, he had decided against Cuzco. High up in the mountains, it was too far from the sea for him to keep in touch with the home government or with the merchants of Seville, who he knew would soon be looking for trade in Peru. The spot he chose was only about six miles from the sea. As soon as the city was laid out, with straight wide

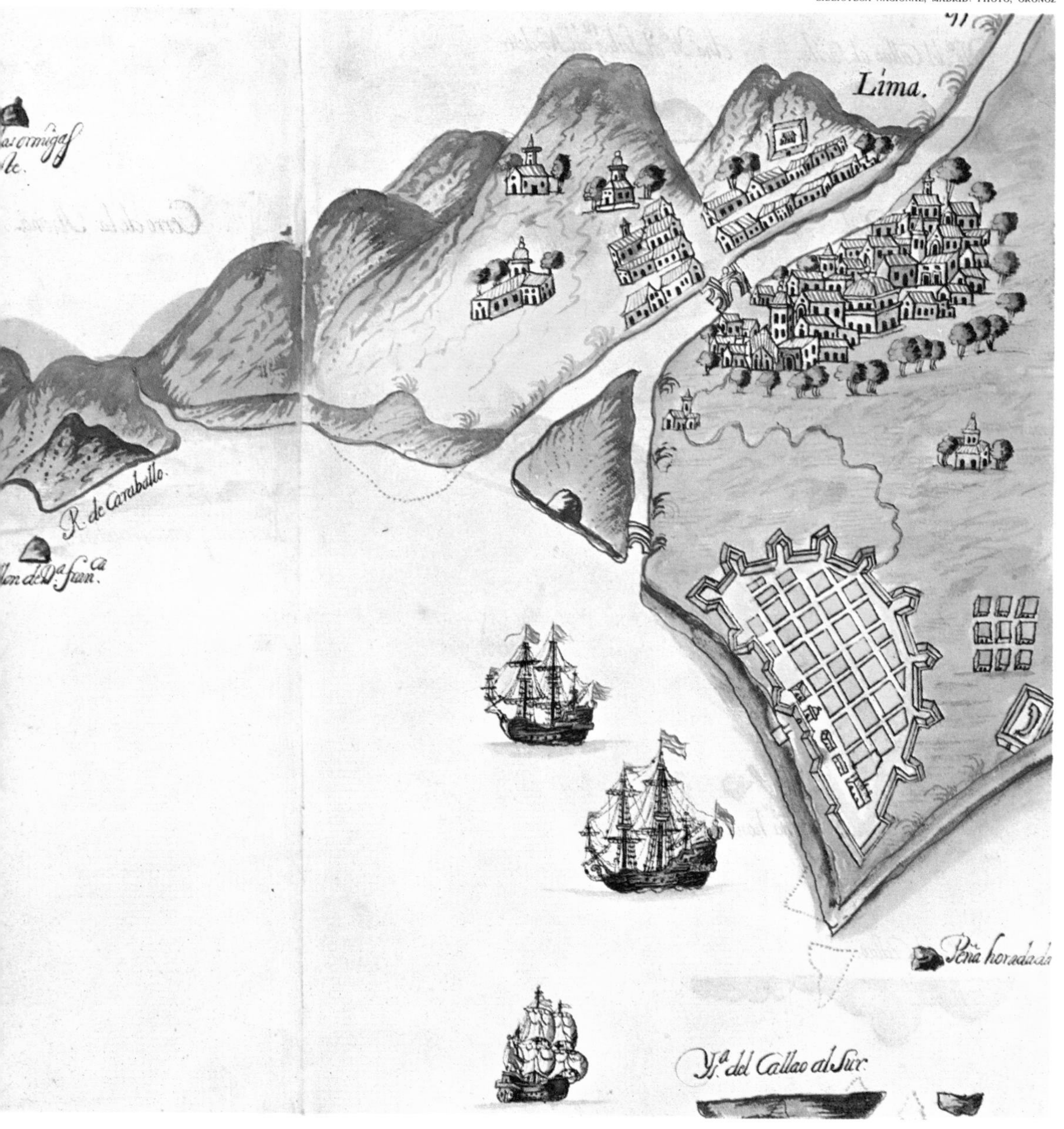


streets far enough apart to allow for gardens and public squares, Pizarro put both Indians and Spaniards to work. A cathedral, a viceroy's palace, municipal halls, and other imposing public buildings were soon being constructed.

Almagro's expedition to Chile had left the garrison of Cuzco dangerously weak. Brave and popular as Juan Pizarro was with his own troops and many of the Indians, he was also overconfident and unwise in matters of diplomacy. He allowed the Lord Inca to be treated with "contemp-

*To the Spaniards, control of the coastal areas meant control over the interior, and they mapped the coastline of Peru with far greater accuracy and care than they did the inland regions. Thus, in the navigator's map above, Lima and the fortified port of Callao are shown in detail, while the lands beyond the first mountain peaks are only sketchily represented.*

tuous indifference," and Manco, who had petitioned Pizarro in vain for real power, soon realized that he was no more than a tool to be used as the conquerors wished. The lowliest Spanish soldier could insult him or his nobles without fear of punishment, and his people were treated like slaves.

Goaded to fury by the way in which their houses, their gold, and even their wives were seized at will by these interlopers, the Indians only awaited the word of their new emperor to take up arms. But Manco was cunning as well as brave, and he wisely bided his time until the conquerors were divided, with Almagro en route to Chile and Francisco Pizarro occupied at Lima. To Juan Pizarro and his arrogant cavaliers, who had won Peru so easily, the natives seemed no more dangerous than a flock of sheep, and no watch was kept on the emperor and his nobles.

Meeting in secret, the Inca chieftains made plans for a general uprising. Messengers were sent to prepare the Indians throughout the land. Manco had only to escape from Cuzco to command armies already mustering.

Certainly the young ruler should have had no difficulty in escaping from Cuzco, except for one apparently disregarded fact. In the city were one thousand Cañari warriors, who hated the Incas and were the allies of the Spaniards. Learning of Manco's flight, these allies immediately reported the news to Juan. The commander led a troop of horsemen in pursuit and was lucky enough to capture Manco before he had gone very far. The Lord Inca was led back a prisoner and placed in the fortress under a strong guard. It looked as if the rebellion had been squelched.

*Though often highly decorated, Inca furniture was generally very simple in design, and this polychrome wooden stool is certainly no exception. Two snarling feline figures, each with a spotted coat and curled tail, support a plain but gracefully curved wooden seat.*

FIELD MUSEUM OF NATURAL HISTORY, CHICAGO

Meanwhile, Hernando Pizarro had reached Lima, where he formally placed in his brother's hands the documents setting out the royal awards and instructions. Still afraid that Almagro might make another attempt to claim Cuzco, Francisco sent the more resolute and dependable Hernando to take command of the city from Juan. Although he was overbearing and openly contemptuous of low-born adventurers like Almagro, Hernando was more considerate toward the Indians than were most of the conquerors. He at once set Manco free and began treating him respectfully. Taking advantage of Hernando's friendship, the Lord Inca further ingratiated himself with the Spaniard by leading him to several hidden piles of treasure. When he

BIBLIOTECA DEL PALACIO, PATRIMONIO NACIONAL, MADRID: PHOTO, ORONOZ

*Peruvian children often re-created events of the conquistadors' era through informal dramatizations. In Bishop Compañon's water color, children re-enact a court scene as part of a ceremony celebrating the execution of the last Inca ruler, Atahualpa. The lizards and toads, both common in Andean countries, were spectators, not participants.*

had won the complete confidence of Hernando, Manco offered to bring to him a golden statue of his father, Huayna Capac, that he claimed was hidden in a cave high up in the Andes. The unwary commander allowed Manco to go on this errand, accompanied by only two soldiers.

When the Inca failed to return by the end of a week, Hernando realized that he had been tricked, and he sent his brother Juan with sixty horsemen to bring Manco back. They galloped through a countryside that was strangely still and deserted. About eighteen miles from the city, they met the two men who had accompanied Manco and learned that the whole country was up in arms. The Inca, who had sent the soldiers back unharmed, was preparing to attack Cuzco. Determined to carry out his mission, Juan led his men on until they reached Yucay, almost due north of the capital. There, on the opposite bank of a deep but narrow stream, Manco waited for them.

AMERICAN MUSEUM OF NATURAL HISTORY

*Feathers, often incorporated in the design of ceremonial garments, are crudely suggested by the knobs on this silver statuette's poncho and headdress. The head of the mace held by the warrior bears a resemblance to the one on page 64.*

Never before had the conquerors faced an Inca army with its emperor at its head. The forces that they had overcome after the murder of Atahualpa had been isolated war bands led by chieftains whose sole object was to stop the invaders. Since the Spaniards' horses and weapons, seen for the first time, were objects of superstitious awe, it is not surprising that the primitive warriors took to their heels when the strangers' armor appeared to be invulnerable to Indian spears and arrows. Now all this had changed. Horse and steel were still deadly, but the brave Indian warrior knew that he was fighting an enemy who could be killed.

Juan's horsemen plunged into the stream and swam their horses across through the usual hailstorm of stones and arrows that at least wounded, although they rarely killed. As soon as they reached the bank, the Spaniards were furiously attacked by the Inca troops, who surrounded the little band and struck at them with copper-tipped lances, battle axes, and maces. Thick doublets of quilted cotton, shields covered with skins, and metal helmets gave some protection from Spanish steel, and for a few minutes the horsemen were in danger of being overwhelmed. Then Spanish character and the crusading zeal that led them to think they were fighting God's battle against the infidel enabled them to rally.

To the battle cry of "Santiago," Juan's force formed a closely packed column of steel and charged deep into the enemy ranks. The Indians, unable to withstand the shock, gave ground, but they fought on furiously even as they retreated, bringing down men and horses. Slowly, still a disciplined army, they withdrew into the narrow defiles, where they were safe from the Spanish cavalry.

That night the Spaniards slept at the foot of the mountain, ready to leap to arms at the first alarm. Never before had they had to fight so hard for victory. But Juan was determined to lay hands on the Lord Inca, hoping that the Indians had learned their lesson from the slaughter of the previous day. He soon found that he was mistaken. On the next morning, when he advanced to the mountainside, thousands of warriors barred the way. Huge rocks crashed down on the Spaniards, and showers of missiles bruised and wounded the weary soldiers. For two or three days Juan stuck grimly to his task, charging and driving back any enemy force that came down to the plain, but he dared not venture into the ravines to be trapped.

Juan Pizarro was saved from having to admit defeat by orders to return to Cuzco at once. Pursued by Indians yelling in triumph, he again crossed the river and swiftly led his troops back to the capital, which he found besieged by tens of thousands of the enemy. Without pause, the Spanish horsemen closed ranks and prepared to cut a passage through the sea of warriors. The Indians, however, drew back and let them pass; the hostile ring closed ominously the moment they were through.

The lion-hearted resistance of the Spaniards in Cuzco during the following months—February to August, 1536—is one of the epic stories of history. Although he had barely two hundred Spanish soldiers and a thousand Cañari allies to oppose the vast Inca army, Hernando refused to be hemmed in by Manco's forces. Often he led his horsemen in surprise attacks, killing great numbers of the enemy, though failing to break the siege.

On the day following Juan's return, the Incas launched a full-scale assault on the invaders. The Spaniards faced a new terror. The enemy archers and slingers shot thousands of flaming missiles into the thatched roofs of the buildings. Soon the whole city was a raging inferno, and dense clouds of smoke choked and blinded the besieged. Temples, palaces, and huts crashed down as flames destroyed the woodwork of their interiors. Only the solid stonework of the palace of Viracocha, where some of the Spaniards were billeted, and the great open space of the main square, where the rest were quartered, saved the conquerors from being burned to death in the holocaust. Knowing that it would be useless, they made no attempt to fight the flames but watched in horror as the city was destroyed around them.

MUSEUM FUR VOLKERKUNDE, HAMBURG

*Linked at the shoulder to a male partner, and distinguishable from him only by her longer hair, the pipe-playing silver dancer above once decorated a Chimú ear plug.*

STAATLICHES MUSEUM FUR VOLKERKUNDE, MUNICH

*Inca foot soldiers probably fought in mufti, but the Lord Inca's own men wore black and white checkered ponchos with red yokes (above). Typically, de Bry has failed to clothe the besieging Indians at Cuzco at all—let alone with checkered ponchos. The nakedness of the natives, like the carnage itself (right), was a product of the engraver's cheerful ignorance.*

NEW YORK PUBLIC LIBRARY

The fire raged for several days, until the splendid capital of the Incas was a burned-out shell. Smoldering beams and rubbish from the houses hindered the cavalrymen as they rode out of the square to drive back the attackers. When the infantry and Indian allies had cleared a path for the horses, the Inca warriors planted stakes and put up barricades. To clear these was a perilous task, for archers and slingers lay in wait among the ruined houses, and they killed many of the unprotected Spanish infantry. When at last the way was open, the horsemen burst out furiously and slaughtered dozens of the enemy. But the Indians, though beaten back for a moment, rushed in again. Attacking the flanks of the Spaniards as they withdrew, the natives brought down several cavaliers with stones and arrows, and lassoing others, they hauled them away in triumph.

The siege dragged on, with Indian attacks and retaliatory strikes by the Spaniards occurring almost daily. To strike terror into their enemy, or sometimes just to bring in food, the besieged conquerors made frequent forays outside the city. Constantly under fire from the fortress of Sacsahuamán, the men in the square had little rest day or night. When at last Hernando saw that his followers were losing heart under the strain, he determined to take action.

First, he led a surprise attack on the Inca camp, his horsemen and harquebusiers doing terrible damage to the Indians, who were caught off guard. The natives soon rallied, however, and fought with desperate ferocity. Some used steel weapons that they had captured from the Spaniards, and a few, led by Manco, even fought on horseback. But the unequal struggle did not last long. The Indians were unable to withstand the repeated charges of the Spanish cavalry and the murderous fire that was poured into their massed companies by the Spanish infantry. As the Inca warriors retreated, the horsemen followed, killing until they were too exhausted to go on with the slaughter. Hernando withdrew, hoping that this dreadful lesson would deter the Indians and give his men some rest.

An assault on the Indian fortress of Sacsahuamán was planned next, and Juan Pizarro was chosen to lead a contingent of troopers one evening, as though on a foraging expedition. At nightfall he made a detour, and being fortunate enough to find the mountain passes unguarded, he reached the first wall of the fortification unnoticed. In silence his men pried out the great stones that barred the entrance to the outer enclosure and rode through toward the second wall. By the time the Spaniards had reached this

ROYAL LIBRARY, COPENHAGEN

*Brandishing torches, three of the Lord Inca's warriors—one of whom is clad in a checkered poncho like the one opposite—make their way to a festival of Inti, the sun god.*

# ANDEAN MYSTERY

Abandoned for centuries, the Andean citadel of Machu Picchu (above), built on a mountain saddle between two Amazon tributaries, has puzzled archaeologists ever since its rediscovery in 1911. No one can say with certainty when it was built and by whom or why it was abandoned. Its location, 10,000 feet up in the Andes, and the presence of a strong perimeter wall—rare in Inca architecture—suggest that the fortress may have served as a final Indian refuge during the conquest. If so, this remarkable product of mortarless masonry (above right) probably was a spiritual, rather than a military, retreat—for the lack of arable land and the vulnerability of the city's aqueduct would have made Machu Picchu hard to garrison and easy to besiege. The enigmatic, many-leveled stone at right (set atop the terraced hillock at the far left in the view above) may have been an altar or possibly a sundial.

AMERICAN MUSEUM OF NATURAL HISTORY

JOSEPH L. GARDNER

PEDRO ROJAS PONCE

defense, however, the garrison was aroused, and a hail of missiles struck the tiny Spanish attack force.

Juan had been wounded in the jaw on a previous occasion. Finding that his helmet was paining him, he rashly took it off and relied on his shield for protection. As he led the assault, a great stone from the tower struck him on the head. Yet even when prostrate on the ground, he cheered his men on until they had won the second enclosure. Then he was carried back to Cuzco, where he died about two weeks later. As brave as his brothers, Juan Pizarro nevertheless had been of a gentler and more courteous nature and was popular with many of his fellow countrymen.

Hernando now took command of the attacking party and at last captured Sacsahuamán's central tower by escalade. The Inca noble in command there fought like a hero, killing several Spaniards with his heavy copper mace and overturning some of the scaling ladders to send the men on them hurtling to the ground below. But seeing that his followers had given up the fight, he, too, decided that the end had come, and flinging away his mace, he wrapped himself in his cloak and leaped to his death.

Although the energy and resourcefulness of Hernando had relieved the strain on his men, the Spaniards grew despondent as more weeks and months passed with no sign of help from outside Cuzco. Feeling that they might even be the last Spanish survivors in Peru, they began to talk of cutting their way to the coast and escaping by sea.

Scores of isolated settlers, indeed, had been killed, but the state of affairs was not quite so desperate as the garrison imagined. Another Inca army had invested Lima, but Francisco Pizarro had led his forces out and slaughtered so many Indians on the plain outside the city that the natives had kept their distance after that. Still, the Inca forces continued to hold all the passes of the Cordilleras. Several detachments of troops that had been sent to relieve Cuzco had been trapped in the mountains, and only a few soldiers had straggled back to Lima. Soon Pizarro was in such desperate straits that he sent to other Spanish governors in the New World for help, offering to share with those who came the rewards in land and gold from future conquests.

The Spaniards in Cuzco were on the verge of despair, but the hosts of Indians camped around the city were in danger of starvation. It was not usual for Inca armies to keep the field for months at a time, and by August, 1536, Manco was compelled to send the greater part of his forces home to tend to the sowing of the following year's crops.

*The triple-walled Inca fortress of Sacsahuamán, built on a hilltop above Cuzco, is treated more as a tourist attraction than as a military stronghold in the eighteenth-century water color below. Groups of Inca nobles tour the battlements (G, above and below) while children frolic along the approach to the fortress itself.*

With the several thousand troops that remained, he withdrew to a strong hill fortress some miles from the city, leaving only detachments of observers outside Cuzco.

The Spaniards in the city witnessed the melting away of the besieging forces with enormous relief. Since they could gather supplies more easily now, all danger of starvation for them was at an end. Not content with this, the indomitable Hernando determined to finish the war with a single, fierce blow—by capturing the Lord Inca. The expedition was planned with great care, but for once the Indian troops were on their guard. Hernando's party was repulsed and forced to retreat, with the exultant native warriors harrying the Spaniards as they returned to Cuzco. The Lord Inca, however, had won his last victory.

ARCHIVO DE INDIAS, SEVILLE: MAS

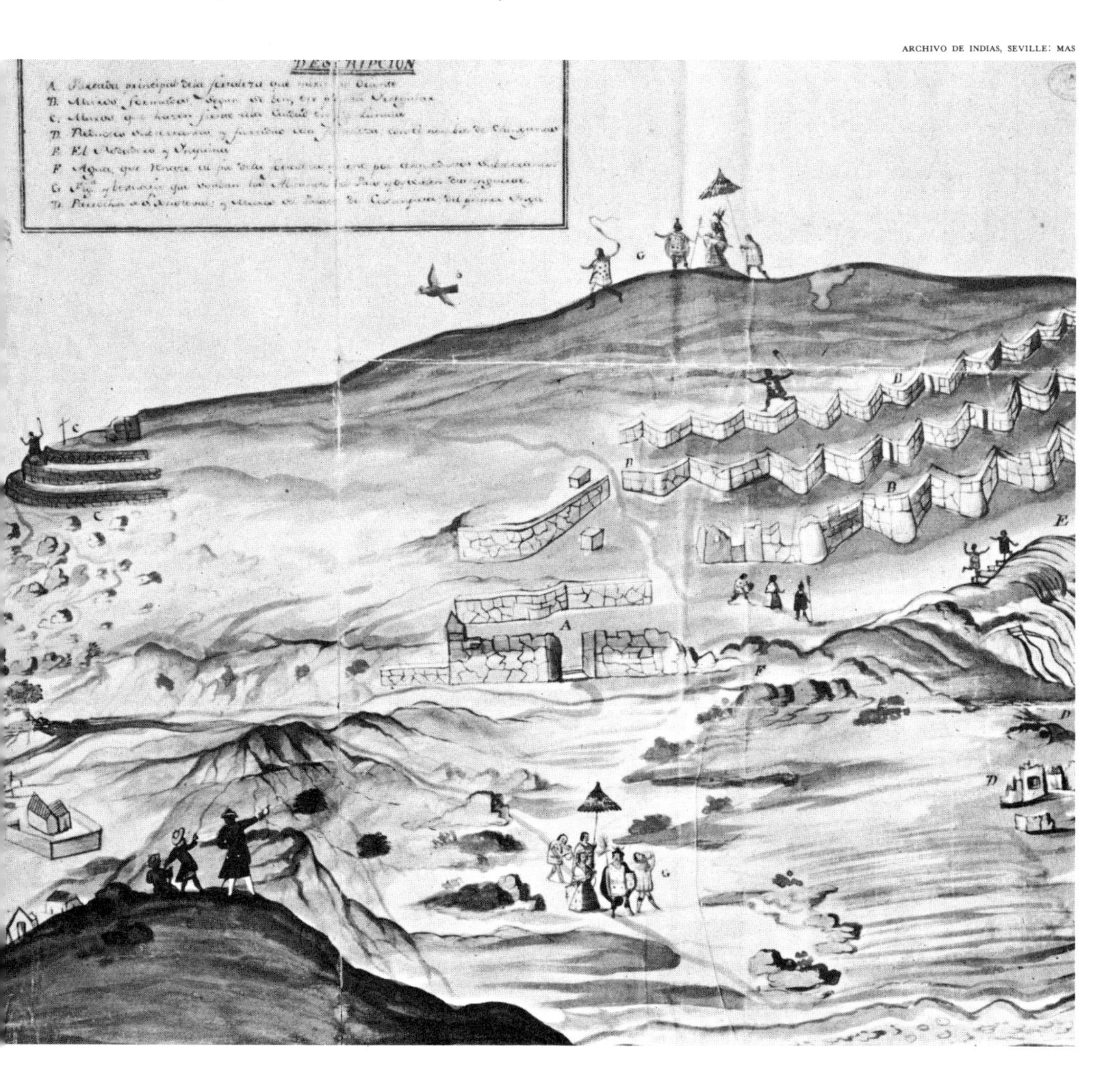

VI

# THE CONQUERORS FALL OUT

Hernando Pizarro's house in Cuzco was on fire. Its flames lit up the square as though it were day and reflected against the steel breastplates of the Spanish infantrymen who stood outside in the rain. Eight months had passed since Manco's siege of the capital had been raised, and now Hernando, his brother Gonzalo, and several of their bodyguards found themselves surrendering to a Spanish force.

ROYAL LIBRARY, COPENHAGEN

By a single treacherous stroke on the dark night of April 8, 1537, Almagro made himself master of the city that had defied Manco and his armies for six months. A thirst for power and a smoldering distrust of his former partner Francisco Pizarro had led Almagro to this disastrous decision to take Cuzco by force. Its end result would be civil war among the conquistadors.

After suffering great hardships in Chile and finding no gold to reward their pains, the old adventurer and his men had returned to Peru more determined than ever to enjoy its wealth and comforts. Part of their resolve stemmed from the fact that Almagro's official land grant had at last reached him during the expedition, and the explorer—with the help of his officers—had gradually persuaded himself that the grant included Cuzco. That done, the decision to snatch the city from the grasp of the Pizarros came easily. Still, Almagro's armies knew nothing of Manco's rebellion until they were within leagues of their goal. By then, they, too, were embroiled. Some fifteen thousand of the Inca's warriors had launched a side attack against them, and only after slaughtering many Indians did they manage to get free and carry out their plan.

*Almost certainly acquired after 1532—when his share of the spoils of the conquest enabled him to afford such luxuries—Pizarro's dress sword (left), with its velvet grip and delicate filigree, is now preserved in Spain. Ayala's drawing (above) gives the young Almagro, rather than his henchmen, credit for assassinating Francisco Pizarro with a similar sword.*

The first step in Almagro's scheme was to send a messenger to Cuzco, demanding that Hernando Pizarro turn over the city to him. Hernando, of course, refused. Hearing of this and preferring not to suffer another long and bloody siege, the local magistrates immediately interceded to establish a truce between the longtime adversaries. But the peace was uneasy at best. Much grumbling filled Almagro's camp from the start, as well as the inevitable, though periodic, rumors that Hernando had broken the terms of the agreement. When a scout brought word that a relief force, sent against Manco by Francisco Pizarro, was on its way from Lima, Almagro struck quickly. Under cover of night he and his forces invested the city, besieged the home of Hernando, and set it to the torch. Almost without bloodshed all was won, and the Pizarros were led away unharmed under the heavy guard of the Almagrists, or Men of Chile, as they often were called.

With Cuzco taken and two Pizarros imprisoned, civil war was unavoidable. Diego de Almagro and Francisco Pizarro were to be the chief opponents in the contest, but each had on his side a number of lieutenants who were equally anxious for power and gold and willing to use almost any ruthless tactic to satisfy these desires. Moreover, Pizarro and Almagro both were nearly seventy years old by 1537. Thus, it became more and more common in the years that followed for the two commanders to rely on younger men to lead their forces in the field.

BIBLIOTECA NACIONAL, MADRID: MAS

*Possibly influenced by the dress of Spain's onetime Moorish rulers, the illustrator of a sixteenth-century treatise on hand weapons has given his subject a scimitar and baggy pants. Flames burn at the tip—rather than along the shaft—of the arrow he carries.*

Almagro's good right hand was a loyal and fearless officer named Rodrigo de Orgóñez, who had learned his trade under the most distinguished commanders in Europe. It was Orgóñez who ordered the firing of Hernando's house; the battle won, it was Orgóñez who urged his commander to strike off the heads of his captives, arguing that "a Pizarro was never known to forget an injury" and quoting the proverb "Dead men never bite." When Almagro refused to take action against either Hernando or Gonzalo, it was Orgóñez again who spoke up, warning his commander that they would all live to regret such mercy to the Pizarros.

By July, the relief force sent by Francisco Pizarro was within one hundred thirty miles of Cuzco. Alonso de Alvarado was in command. Almagro, or the Marshal, as his men affectionately called him, marched out to meet him and came upon his enemy at the Abancay River. There Alvarado was in a strong position, having placed most of his troops in defense of a ford and a bridge on the river.

GOMARA, *Historia General . . .*, 1555: BIBLIOTECA NACIONAL, MADRID

*Illustrations for Francisco López de Gómara's 1555 history of the New World include the one above, which shows Almagro laying siege to Cuzco. His nattily attired troops gather just outside the city's walls (which Gómara has embellished with a European drawbridge and crenelated towers).*

The encounter that followed was typical of many that would come later, in which self-interest rather than loyalty seemed to motivate most of the participants. Unfortunately for Alvarado, one of his most distinguished officers, Pedro de Lerma, had a grudge against him and chose this critical moment to strike a deadly blow. At the first opportunity, De Lerma wrote to Almagro and advised him that if he wished to win, he should send a detachment of troops across the ford in the dead of night but keep his main army ready to attack the bridge. The Marshal agreed and dispatched Orgóñez with an assault party. Reaching the shallows of the swiftly flowing stream, Orgóñez and his men waded across, opening a fierce assault on the Spaniards stationed there and setting up a great clatter. His purpose was greatly aided by De Lerma and some other traitors.

Hearing the noise, Alvarado naturally believed that the Almagrists were attacking in full strength at the ford and

*Early in 1538, Almagro's forces marched unopposed through Santo Domingo (seen above in a detail of a seventeenth-century map) to the River Chincha, in hopes of establishing a sea link with Spain.*

hurried to oppose them, leaving only a few men to hold the bridge. As the traitor had promised, the way was open. After a short confrontation, Alvarado was forced to surrender his army, and Almagro returned to Cuzco in triumph.

Still in Lima, Francisco Pizarro knew nothing of these happenings. But an earlier, urgent plea to other colonial governors for help to resist Manco had not gone unanswered. Detachments of well-equipped troops and military stores had begun to pour into the new city from Panama, Santo Domingo, and even Mexico. When the first warning of Almagro's seizure of Cuzco reached him, the Captain-General was thus prepared to march out with a force of four hundred fifty men. News of the disaster at the Abancay came soon after, however, and Pizarro decided to follow a prudent path and strengthen his defenses in Lima before taking further action.

Never very scrupulous, the Captain-General always was prepared to negotiate—until he was strong enough to take what he wanted by force. This time he hoped to lure his old partner into a trap. As a first maneuver, he sent envoys to Almagro with a mild reproach for seizing Cuzco and suggestions for a friendly settlement. But Orgóñez was suspicious of the offer and persuaded Almagro to refuse it point-blank. Instead, preparations were made to carry the war into Pizarro's own territory.

To make Cuzco safe from Indian attack during his absence, Almagro first sent Orgóñez against Manco. With the Indians subdued, the Marshal joined forces with Orgóñez and began the march westward, forcing Hernando to accompany him and leaving Gonzalo and Alvarado under strong guard in Cuzco. Without incident, the army reached the coast at Chincha, about a hundred miles south of Lima, and Almagro proceeded to establish a Spanish settlement there to open communications with Spain. Not until then did he receive news that Gonzalo and his fellow prisoners had escaped. Orgóñez wasted no time in pointing out the Marshal's folly in not heeding his advice and once more pressed Almagro to get Hernando out of the way. Once more the Marshal refused.

Francisco, who knew that to open hostilities probably would mean the death of his brother, now made concessions. In return for Hernando's freedom, Almagro was to keep Cuzco until Charles V had decided the boundary question. Moreover, Hernando was to leave Peru within six weeks of his release. Foolishly, Almagro believed the Captain-General and agreed to the terms. Orgóñez, however, viewed the settlement with despair. Realizing that no argument would alter this fatal decision, he merely passed his fingers across his throat, exclaiming prophetically, "What has my fidelity to my commander cost me!"

Within a few hours of Hernando's release, Francisco repudiated the treaty. With his brother at his side, he now had a captain on whose loyalty, ability, and courage he could rely implicitly. Immediately he began rallying men to his cause and soon sent off a warning to Almagro that he must either surrender Cuzco or fight for it.

Almagro was astonished, for he had been confident that Pizarro would honor his agreement. Now far too ill to command, he had little choice left but to place Orgóñez at the head of his troops and return to Cuzco at once. However, Orgóñez mismanaged his command, failing to man properly the passes between the coast and the high sierra as he retreated. It was a mistake that signaled the Almagrists' doom even before a shot was fired; Francisco Pizarro's forces—swelled to a total of seven hundred—were able to make the long march to the Inca capital without fighting a single engagement. Now the Captain-General also abandoned the field. Declaring that he was too old for active service, he relinquished his command to his brothers Hernando and Gonzalo and withdrew to Lima.

News had spread among the Indians that the Spaniards were quarreling and that a battle was at hand. From miles around, they came to enjoy the sight of their conquerors killing one another. To the natives, massing by the thousands on the slopes of the Cordilleras outside Cuzco, the Pizarros' army, with banners waving and Spanish armor glistening in the sunlight, looked gay and splendid as it filed through the mountain passes. But it was all a mirage; there was no gaiety in the ranks. Every man knew that this clash with his countrymen would be bitter. It was in gloomy silence that the Pizarros' forces advanced across the plains toward Almagro's army, waiting at Las Salinas, on April 25, 1538.

At Cuzco a council of war had chosen the plains as the battle site, and Orgóñez had drawn up his troops in a defensive position, protected in front by a shallow river flowing through marshy ground. With cavalry and falconets on the wings and pikemen and harquebusiers in the center of his battle line, he coolly awaited his opponents.

By the time the Pizarros' army reached the river, the sun already had set, and there was little choice but to biv-

HERRERA, *Decades*, 1728: INSTITUTO DE CULTURA HISPANICA, MADRID: PHOTO, ORONOZ

*While this portrait of Almagro, which appears in Herrera's 1728 history of the Americas, is for the most part pure fabrication, the artist has discreetly shown the crusty soldier in profile—to conceal his missing left eye.*

ouac for the night. With heavy silence brooding over the plain, the two forces made camp, and no sound came from the expectant Indians.

Early on the morning of April 26, the trumpeters sounded the call to arms. Mass was said, and Hernando gave a short speech, stirring his men to fury by reminding them that Almagro and his followers had stolen what was rightfully theirs. Pointing to Cuzco, he declared that there lay the prize for the victors, and his men, determined to die fighting rather than see others enjoy the riches and property to which they had first laid claim, welcomed the signal to advance.

Gonzalo Pizarro commanded the infantry in the center of the battle line, and at the sound of the trumpets, he made straight for the enemy, wading through marsh and water to the river's far bank without serious resistance. Not until they pressed on through the bog on the opposite side did his files feel the murderous storm of shot from Orgóñez's heavy guns. Men were mowed down in swaths, and Gonzalo and his officers had to work like madmen to prevent a rout. Striking some with their swords to encourage them to hold and rallying others by example, they at last regrouped their troops and carried them to firm ground. Now, it was Gonzalo's harquebusiers who did the pelting, letting loose their

DUKE OF ALBA COLLECTION, MADRID: PHOTO, ORONOZ

*Purportedly a representation of the fall of Jericho, this scene from a fifteenth-century Spanish Bible combines elements of both the military encampments and open-air produce markets of that era's Spain. Mounts as well as their riders are bedded down in the brightly colored tents above.*

sprays of chain shot and sending the enemy reeling in disorder. Then the harquebusiers attacked the flanks of Orgóñez's cavalry, striking with deadly effect.

Under cover of heavy small arms fire, Hernando led his cavalry forth. To the war cry of "*El Rey y Pizarro*," "The King and Pizarro," they galloped out to meet the Almagrist horsemen, a challenge that was eagerly accepted. Like the cheers from spectators in a modern sports arena, wild yells from the watching Indians drifted down the mountain until the din of steel meeting steel drowned out even these frenzied sounds. Both horses and riders rolled in the dust, while flashing lances and glistening swords pierced and slashed brutally.

Orgóñez fought like a hero. Singling out a horseman whom he mistook for Hernando, he rode toward him at full gallop and struck him down with his lance, then he killed a second horseman and after that a third. At that instant a shot from a harquebus pierced his visor, the heavy bullet grazing his forehead and blinding him for a moment. Before he could recover, his horse went down under him, and he was forced to stand, sword in hand, to fight off the enemy quickly surrounding him. Realizing that his situation was hopeless, he asked only if there was a knight present to whom he might surrender. One of Hernando's men-at-arms instantly claimed the rank, and Orgóñez handed him his sword. But the man was no knight; he immediately drew a dagger and plunged it into his prisoner's heart. Then the ruffian struck off the brave commander's head and held it up on a pike for all to see.

Hernando, too, was wounded, but not fatally. He managed to continue until Almagro's forces broke and fled. Almagro himself watched the battle from a distance, and seeing his men defeated, he climbed onto a mule and rode back to Cuzco, where he hid in the fortress until discovered by Hernando's pursuing horsemen.

The contest had been hard fought and had lasted almost two hours. More than one hundred fifty men lay dead, with many more wounded, but for the most part the fight had been honorable. What followed was not. Even before the battle had begun, when it was reported to Hernando Pizarro that Almagro was ill and might die, Pizarro had commented, "Heaven forbid that this should come to pass before he falls into my hands." Now Almagro was in Pizarro's hands—a prisoner, broken in body and spirit. But Hernando could not bear to see his old enemy escape by dying a natural death. To cheer Almagro from his deep

HERRERA, *Decades*, 1601–15: INSTITUTO DE CULTURA HISPANICA, MADRID: PHOTO, ORONOZ

*A key factor in the Battle of Las Salinas was the inevitable need to ford the river, under fire, in order to engage the opposition at close quarters. That action can be seen clearly in this rendering from the title page of the 1601–15 edition of Herrera's history.*

depression, Hernando visited him in prison and led him to think that he soon would be set free. For weeks Hernando sent the prisoner choice foods from his own table to strengthen the old man and buoy his spirits, and slowly Almagro began to recover. Yet all the time Hernando was working to bring Almagro to a shameful end.

Everyone who had a complaint against the Marshal was induced to make a formal statement of his grievance, and two and a half months later, with formal charges drawn up, Hernando struck. Without being allowed to appear before his judges or even being informed of the crimes he was charged with committing, Almagro was found guilty of treason on July 8, 1538, and sentenced to be beheaded in the great square of Cuzco.

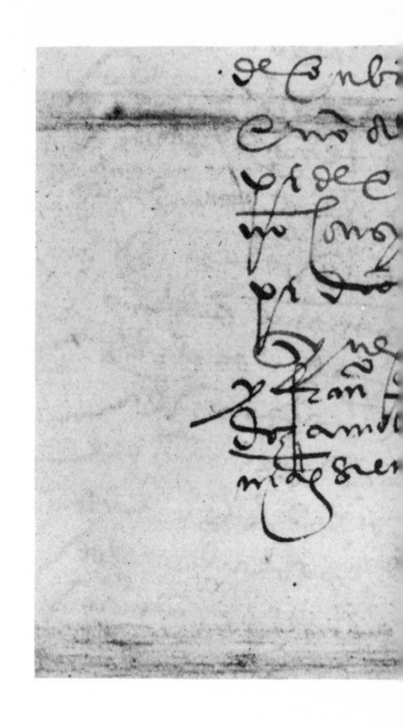

Almagro could scarcely believe that even a Pizarro would be so barbaric. He begged to see Hernando, and Hernando came—but only to gloat over his victim. Reminding Hernando of the loyal support he had given his brother Francisco through the years of discovery and of the fact that, only a short time before, he had saved Hernando from being led from the very same cell to the executioner's block, Almagro asked for mercy. But no hint of sympathy warmed Pizarro's bitter face. He savored each moment, as the Marshal broke down and begged to be spared. Then, sneering at the old man's fear of death, Hernando departed, advising Almagro to make his peace with God, for he was about to die.

NEW YORK PUBLIC LIBRARY

*Garroted at the stake, Almagro then was cut down—de Bry's view notwithstanding—and carried to the block before being beheaded.*

Word of Almagro's fate caused a great stir in Cuzco. Many had admired the Marshal's courage and skillful leadership and loved him for his free and easy nature. Even his enemies, fresh from the bloody Battle of Las Salinas, were sorry that such a man should die a traitor's death. Some called on Hernando to reprieve Almagro, but Hernando was adamant. His only concession was to allow Almagro to be garroted in prison, rather than be beheaded in public.

On the day prescribed by the Spanish commander, the executioner, with a friar by his side, secretly entered the building where Almagro was imprisoned. There the old man received the sacraments of confession and holy communion and accepted his sentence, brave soldier that he was, making no protest when the strangler's cord encircled his neck. But when Almagro was dead, his body was carried to the great square so that Hernando might enjoy his final revenge. In full sight of a large, silent crowd, the headsman's axe was seen to rise and fall, and the gray head was sent rolling in the dust.

HENRY E. HUNTINGTON LIBRARY, SAN MARINO, CALIFORNIA

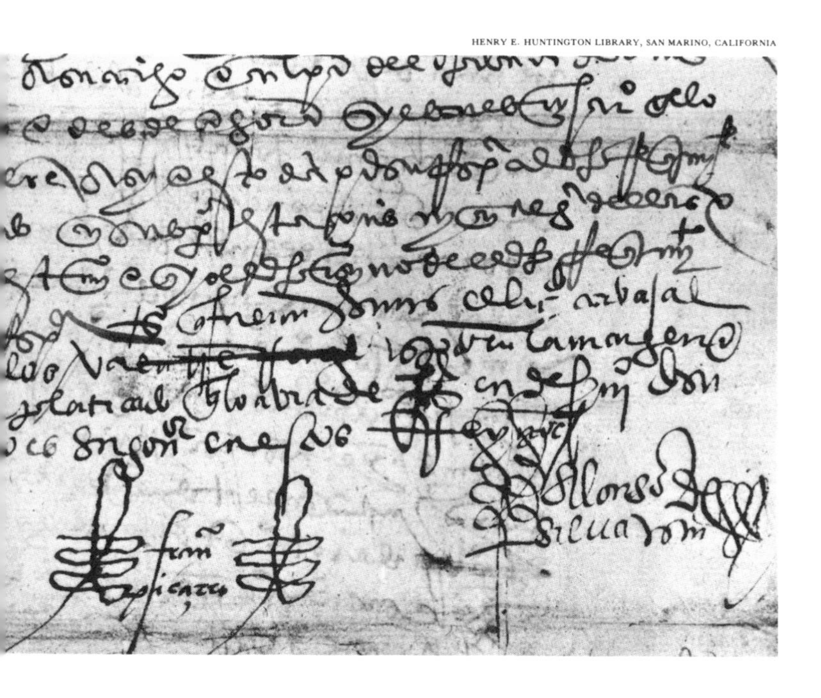

*Unable to read or write, Pizarro could only affix his mark to the bottom of this affidavit, which offers up his side of the dispute between himself and Almagro over the division of conquered lands.*

Only the Indians were ecstatic over the outcome. The sight of their oppressors slaughtering one another seemed to them a fitting end to an unhappy decade. Even after the Spanish armies had retired from the field at Las Salinas, hordes of Incas had swarmed down on the plain to pillage the dead and wounded. Naked and helpless, the Spanish figures lying there seemed to provide the appropriate, shameful conclusion to this prolonged episode of greed. Yet the final acts remained to be played.

In 1539 Francisco Pizarro returned to Cuzco. He came like a conqueror—to reward the men who had fought with him and with his brothers and to punish those who had stood with Almagro against them. If he was touched by the tragedy that preceded his victorious entrance, he failed to show it. Almagro's men lost all vestiges of their material well-being—land, homes, gold—while Pizarro's men enjoyed generous endowments of all of these. Some Almagrists, too proud to bear the shame of poverty and fearing Pizarro's next move, sneaked from the city into the hills, where they took refuge and brooded on revenge.

Almagro had bequeathed all his holdings to his son, Diego, then eighteen—a legacy that was well supported by Charles V, who had granted the Marshal the right to name his successor. The territories left to young Almagro were to be ruled by Diego de Alvarado, brother of the governor of Guatemala and the Marshal's most trusted captain, until

ROYAL LIBRARY, COPENHAGEN

*The Spaniards' roughshod treatment of Indians of both sexes is underscored in the Ayala drawing above, which shows Pizarro's men beating a helpless native woman.*

young Almagro came of age. Pizarro, of course, disregarded the will, declaring that "his own territory covered all on this side of Flanders" and making clear by this proud boast that he would tolerate no rival in Peru, even if Charles V himself ordered it.

As unfortunate as this greedy determination was to prove for Francisco, it was Hernando Pizarro who first and last suffered for his brother's hasty decision. In July, 1539, Hernando returned to Spain, hoping to clear his name of charges brought against him by his enemies. To regain the emperor's favor, the cavalier brought with him a vast sum of gold and silver, collected specifically for that purpose. But Diego de Alvarado had preceded Hernando Pizarro to court, and Alvarado was a highly connected nobleman. So inflamed was opinion against Hernando by the time he arrived that even the enormous bribes offered at court failed to induce Charles V to overlook the tyrannous acts in Peru. Instead of enjoying his fame and fortune, Hernando was imprisoned in one of the royal fortresses, where he was to remain for more than twenty years.

Francisco Pizarro's fate was even harsher. Before sailing for Spain, Hernando had cautioned his brother "to beware the Men of Chile." But the Captain-General paid little attention to the warning. He had allowed young Almagro to keep a big house in Lima, and the dwelling soon became the rallying point for Pizarro's bitterest enemies.

In addition, the Spaniards had lost their once-firm grip on the Indians. Seeing their masters cutting one another's throats in open defiance of both governor and emperor, they had become uncertain as to whom to follow and to whom to look for protection. They were used to the rule of one Lord Inca whom everyone obeyed implicitly and found it difficult to understand how the Spaniards could overlook so easily the rulings of their own leaders. It was not surprising, then, that Manco, poised in the high sierras between Cuzco and the coast, was never short of warriors.

Regularly the Inca leader swooped down on the unsuspecting Spaniards who lived on plantations outside the towns, killing them and driving off their cattle. He waylaid travelers when the opportunity arose and defeated the troops sent against him. He was a brilliant guerrilla leader, choosing well his places for battle—on mountain slopes and in narrow passes, where the Spanish cavalry could not operate easily. Eventually, Gonzalo Pizarro was dispatched to put an end to Manco's reign of terror and did manage in a number of desperate encounters to defeat the Inca, but always Manco eluded him. Neither torture nor the promise of reward could induce an Indian to betray his new leader. As soon as the Spaniards withdrew, Manco and his warriors reassembled and again began to raid and massacre their enemy.

BY GRACIOUS PERMISSION OF HIS GRACE THE DUKE OF WELLINGTON, K.G.

*Fray Murua's portrait of an Inca* coya *clutching a bouquet in one hand and a long-eared animal in the other suggests a demureness not seen in the maiden opposite.*

Hoping to solve the problem once and for all and regain Manco's confidence, the Captain-General sent one of his slaves to the Inca with a rich present. But the unfortunate envoy was ambushed and cruelly butchered on the way, and the precious gift was carried off. Without hesitation, Pizarro took a dreadful revenge. Among his prisoners was Manco's favorite wife, a young and beautiful girl. Pizarro had her stripped naked, bound to a tree, and beaten with rods in the presence of his army. Then he ordered her shot to death with arrows. So pitiful was the sight that even the most callous of Spanish soldiers were filled with shame for the act and with admiration for the victim, who suffered the tortures with scarcely a groan. Peaceful settlement was now impossible.

Besides the Indians, there were other problems with which the Captain-General was forced to deal. As news of Peru's conquest and its rivers of gold and silver reached Spain, hundreds of adventurers from all classes of society embarked from Seville to win fame and riches in the New World. To keep such newcomers from causing trouble, Pizarro wisely set most of them to work at building new towns and garrisons, while he enlisted others for a conquest of Chile and for an exploration of the lands in the Amazon basin, east of modern Ecuador. The latter expedition was led by Gonzalo Pizarro, who set out from Quito early in 1540 with some three hundred fifty Spaniards and four thousand Indians.

The Men of Chile, however, remained without land or honorable employment through it all, desperately poor and with no prospect of relief. One Spanish writer noted that twelve of them, living in the same lodging, shared one cloak. Too proud to be seen poorly dressed, each took his turn wearing the garment outdoors, while the other eleven remained at home. To make matters worse, these men, and many more like them, were affronted daily in the streets of Lima by Pizarro's arrogant supporters, who flaunted their own riches and splendid apparel.

For a while, the Almagrists' hands were stayed by news that an emissary of Charles V was on his way to Lima to inquire into the state of the country. They were prepared to lay their grievances before him and trust to his justice. But

CASAMIENTO
Ylo probenido deste Ma
rimonio q. prebaleze en
Europa e sus d Sendient.
D. Martin d Loyola Cavall.
lorden d Calat. S. de la Casa
Loyola, Gov. y Cap. Ge.
Reyno d Chile Sob. d l Glo
oso S. Ygnacio Patriarca de la
agrada Comp. de Jhes. Con la
eñora D. Beatris Clara Coya
Yndia del Peru;
D. Beatris Clara Coya Inga


*Intermarriage between the Spanish and Indian upper. classes was a sign of successful colonization and peaceful assimilation —proof that relations between the conquerors and the conquered were not always hostile. This carefully*

NUESTRA SENORA DE COPACABANA, LIMA: PHOTO, JOSE CASALS

*constructed eighteenth-century canvas shows the double wedding of two Spaniards and two Inca* coyas*—one (at left) in native costume, standing before Peruvian nobles, the other (at right) in European ceremonial dress.*

rumors that the emissary and his ship had been lost at sea changed all that, and Pizarro's antagonists suddenly felt forced to act on their own. Francisco Pizarro must die, they decided.

Diego de Almagro was no active conspirator, although he must have known something of what was afoot. An elderly officer named Juan de Rada led the desperate band of assassins, about twenty in all. De Rada had risen to high rank in the Marshal's army through ability and loyal service, and he, more than anyone, was determined that the son of his old friend should have justice. The assassination was planned for Sunday, June 26, 1541. The conspirators were to lie in wait at young Diego's house, next to Lima's cathedral, until Pizarro appeared in the street after Mass, then rush out and strike him down. When the deed was accomplished, a white flag, waved from Almagro's window, was to summon other Men of Chile to their aid.

An uneasy conscience led one of the conspirators to tell the whole plot to his confessor, however, and the priest immediately warned Pizarro's secretary. The business might have ended there, but with incredible folly, Pizarro brushed aside the report.

"It is a device of the priest," he said. "He wants to be made a bishop."

Nevertheless, Pizarro related the story to the chief judge, who was equally unconcerned, declaring that while the rod of justice was in his hand, no harm could come to the Captain-General. Neither tried to investigate the accusation, and the only precaution taken by Pizarro was to avoid the cathedral on Sunday.

When De Rada and his companions, lurking nearby, saw that Pizarro was not at Mass, they became alarmed and guessed that their plot had been betrayed. Expecting every moment to hear the tramp of Pizarro's harquebusiers coming to arrest them, they tried to decide what to do next. Some were for escaping, others urged that they attack the tyrant in his palace. While the debate raged, one of the cavaliers suddenly rushed to the door, declaring that he would betray the entire plot if the others did not follow him to the palace. The argument was ended. De Rada led the way, his fellow conspirators followed close behind, shouting, "Long live the King! Death to the tyrant!"

People came out of their houses to discover the cause of the uproar. Yet no one attempted to stop the assassins. They crossed the square unhindered, then marched through the massive gates into Pizarro's outer courtyard, striking down

COURTESY OF MR. AND MRS. GEORGE RENCHARD, WASHINGTON

*Collaboration between European cabinetmakers and Peruvian craftsmen was necessary to produce the magnificent chest above. The former contributed the scrollwork border; the latter, the intricately inlaid cityscape within.*

a servant who tried to bar their way. A guard rushed into the palace shouting, "Help, help! the Men of Chile are all coming to murder the Marquis!"

Pizarro was dining with about twenty friends when the cries were heard. Some of the guests left him to find out what was happening, but as soon as they realized the danger of the situation, they climbed out a window and escaped through the garden. Had any of them so much as fastened the strong doors, the conspirators probably could have been barred until help came. Why these men, some of whom had proved their mettle in a score of fights, should desert their leader now in such cowardly fashion is difficult to understand. But it does seem to suggest that Pizarro, though feared and respected, inspired little love or devotion among his followers.

*De Rada's plan—to assassinate Pizarro as he returned from the cathedral (2) to his palace (1)—fell through when Pizarro sensed the danger and remained at home.*

Only a few men stood ready to defend the Captain-General. One of these was his half-brother, Martín de Alcántara. Calmly Pizarro ordered that the door at the head of the stairs leading to his apartments be locked, while he and his brother armed themselves. But someone half opened the door in an attempt to parley with the assassins, and the Men of Chile burst in, striking down two or three who opposed them. Quickly they made for Pizarro's private room.

The Captain-General, aided by his brother, was trying to fasten the straps of his cuirass, while others fought desperately to hold the narrow passage leading to his room. Martín left Francisco to help stay the assassins in the corridor. Two of the conspirators were killed in the cut and thrust that followed, but it soon was clear that the odds were heavily against Pizarro and his companions. By then the Captain-General was rushing to help his brother. Pizarro had thrown his breastplate aside, wrapped one arm in a cloak, and seized his sword. Before he could reach the younger man's side, however, Martín fell, fatally wounded.

All the old vigor and skill that had made Francisco Pizarro so feared in battle was now displayed for the last time. Two of the assassins went down before his assault, and the rest drew back—but only momentarily. They pressed forward again, and the unequal fight was renewed. Although Pizarro soon was quite alone, his menacing blade still held the enemy at bay.

"Why are we so long about it?" De Rada shrieked. "Down with the tyrant!" As he spoke, he thrust one of his comrades forward against their victim, and Pizarro instantly ran the man through with his sword. It was a fatal exchange. Before Pizarro could disengage, he was wounded in the throat. He fell to the floor, where his body was hacked by a dozen swords. With his finger Pizarro reached out, drew a cross on the bloodied surface where he lay, and bent to kiss it as the last, fatal blow descended.

The Men of Chile ran into the street, shouting, "The tyrant is dead! The laws are restored! Long live our master the emperor and his governor Almagro!" Their cries were answered by scores of the Marshal's old followers, who flocked to join them. The palace was looted, and some of Pizarro's supporters were taken prisoner.

De Rada and his men showed unusual restraint. Some, it is true, wished to display Pizarro's body in public, but young Almagro forbade it. Instead, he allowed Pizarro's friends to arrange a secret burial in a corner of the cathedral, and with a few servants and a priest looking on, the body of the conqueror of Peru was lowered unceremoniously into its grave.

*Once again, de Bry has combined several separate actions into a single scene. At right, de Rada and the conspirators gather for the assault upon the palace—and (right center) force their way into Pizarro's private chambers. Their path is blocked by two of Pizarro's cohorts (left) while a third escapes through a window.*

VII

# THE KING TAKES ALL

In the months that followed Almagro's death, Charles V and the Council of the Indies became increasingly disturbed over the state of affairs in Peru. Reports brought home by cavaliers convinced them that Francisco Pizarro was not protecting the Indians or meting out justice to all Spanish settlers. Yet even Charles V realized that great care was needed in dealing with the fiery conquistador, who was thousands of miles from Spain and well entrenched with a powerful army. He therefore moved with extreme caution and appointed one of his judges, Cristóbal Vaca de Castro, to visit the country. Since Vaca de Castro was a man of great discretion, Charles V felt that he could be relied upon to look into the reports of cruelty and injustice without arousing Pizarro's suspicions. More important, in case of Pizarro's death, Vaca de Castro was to make public a royal warrant naming himself as Peru's new governor.

In the autumn of 1540 Vaca de Castro set sail for the Indies. It was he about whom the Men of Chile had heard so many rumors and for whose arrival they believed they had waited in vain. Vaca de Castro had come as far as Popayán, in modern Colombia, when news reached him of Pizarro's death. By then, the conquistadors were once again at one another's throats.

With no army and only the Spanish monarch's word to support his claims, Vaca de Castro sent messengers ahead to the chief cities, ordering all commanders to acknowledge him as the new governor. It was a bold policy, but it was successful. Pizarro's two chief captains, each with two hundred or more first-rate troops, declared for him. One seized



*Bowlegged and worn with age, the angel above supports a 1601 candlestick—one of the earliest pieces of dated Peruvian silver.*

*Gold, almost a common metal in Spain's wealthiest colony, was lavishly used to embellish Peruvian art of the Viceregal period. Both Saint James, patron saint of the conquistadors and prototype of the soldier-missionary, and his elaborately caparisoned steed (left) wear gold-encrusted trappings.*

BIBLIOTECA DEL PALACIO, PATRIMONIO NACIONAL, MADRID: PHOTO, ORONOZ

*Horses rapidly proved to be of both psychological and practical advantage to the conquistadors, giving them greater speed and mobility on the battlefields and highways, while striking terror into the horseless Inca troops. In a scene from Compañon's history, Spaniards round up their mounts.*

Cuzco, which, since Pizarro's assassination, had been in the hands of Almagro's party, and the other urged Vaca de Castro to hasten his march south to a place called Huaura, just north of Lima, where they could join forces.

It was at such an inauspicious moment, in June, 1542, that Gonzalo Pizarro returned to Quito from his Amazon expedition. He was astounded to find that his brothers had been murdered, that young Almagro was calling himself governor of Peru, and that an appointed minister of the Crown was marching in dispute of Almagro's claim.

Gonzalo, meanwhile, had had problems of his own. He had set out from Quito two years earlier at the head of a company of three hundred fifty Spaniards and four thousand Indians. Now, after suffering the bitter cold of the Cordilleras and the suffocating heat and torrential rain of the jungle to reach the Amazon River, about a thousand miles away, only eighty Spaniards remained. Their limbs were wasted, their hair was long and matted, and their weapons were rusted. Yet, Gonzalo prudently offered their support to Vaca de Castro as soon as they had rested.

Vaca de Castro, however, had his hands full controlling the officers he already had, and as badly as he needed experienced troops, he politely suggested that Gonzalo not take the field until he was fully recovered from his recent hardships. The proud Gonzalo Pizarro, mortally offended, was forced to remain inactive in Quito.

Young Almagro and his followers, on the other hand, had committed a crime for which there could be no pardon. Their situation left them little choice. Their only chance for success—indeed, for survival—would have been to strike swiftly, before Vaca de Castro could concentrate his scattered supporters. Still, they continued to hope that some agreement could be reached with the new governor. Instead of attacking without delay, the Almagrists led their forces from Lima toward Jauja, simply to prevent a junction between the army of Cuzco and one of their enemy's main forces in the north.

Even in this they failed. The untimely death of their commander, Juan de Rada, and violent quarrels between two senior officers proved their undoing. At a critical moment, the army of Cuzco had slipped away to join its ally.

Almagro, moving his forces to Cuzco, soon took command himself and showed that the time spent at his father's side had not been wasted. Only twenty-two years old, he proved himself a firm and able leader. By the Peruvian winter of 1542, he had amassed sufficient armaments and an army of five hundred men—described as the finest yet seen in Peru. In addition, he had won an important ally in the Inca Manco.

While Almagro was preparing his garrison in Cuzco, Vaca de Castro was, with some difficulty, establishing his authority over Francisco Pizarro's ambitious ex-commanders. Though trained in the law and not in arms, he took charge of the army himself and slowly moved toward Huaura and finally on to Lima, recruiting as he went. At Lima he was given a warm welcome, and within a few months he, too, had a fine fighting force, his army numbering some seven hundred. But the Almagrists had seized every weapon and horse they could lay hands on, and a shortage of equipment became Vaca de Castro's biggest problem.

As soon as he felt that his troops were ready, he set out after Almagro. It was a chase that taxed the endurance and wits of both forces. Finally, late on the afternoon of September 16, 1542, after weeks of marching and maneuvering, the two armies stood face to face on the plains of Chupas, about halfway between Lima and Cuzco. Both forces were drawn up in battle order, but Almagro had the more advantageous position. His forces were excellently placed on the crest of a gentle slope, which could be swept by shot from his sixteen powerful cannon. In addition, his Inca ally had kept his word: In the hills to Almagro's right stood several thousand of Manco's warriors ready to help.

As Vaca de Castro's army approached the rise, Almagro's battery opened fire with deadly effect, cutting down the new governor's forces like hay in a field. It was quite obvious that a frontal attack might well end in disaster. One of Vaca de Castro's veteran officers, Francisco de Carvajal, a peppery old soldier nearing eighty years of age but still in excellent condition, offered to lead the army by a safer route, detouring to the left of the hill. The governor agreed. But as the men advanced, Manco's warriors fell on their flank. It was a tense moment until Vaca de Castro's harquebusiers dispersed the Indians.

On up the slope the royal forces drove, but as they breasted the hill, they felt once again the sharp sting of battle, and their front ranks staggered under a storm of shot from Almagro's cannon. Unaccountably, however, the gunners' aim seemed suddenly to lose accuracy, and the deadly hail screamed over their heads, allowing the royal troops a momentary reprieve. Almagro's artillery officer had betrayed him. Galloping toward the man, Almagro ordered him to lower his guns, and when the officer refused,

BOTH: BIBLIOTECA NACIONAL, MADRID: PHOTO, ORONOZ

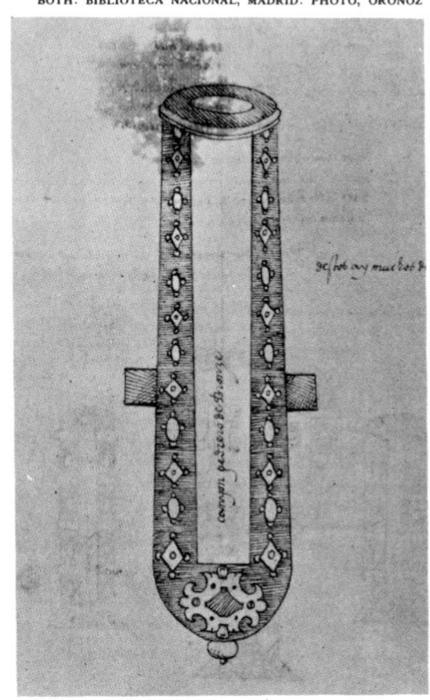

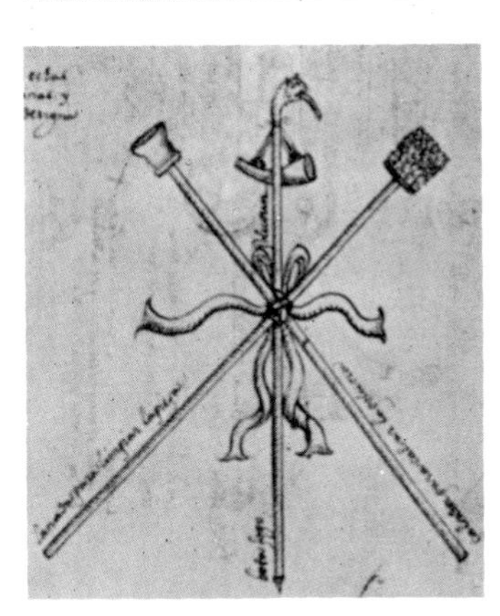

*Looking more delicate than deadly, the cannon shown in cross-section (top) is one of a series of ink drawings of military weapons from a sixteenth-century manuscript. Paired with it is a rendering of implements used to load (left), light (center), and clean (right) the weapon—neatly tied together.*

INTERNATIONAL FILM FOUNDATION, INC.: PHOTO, JULIEN BRYAN

CUZCO CATHEDRAL: PHOTO, SMITHSONIAN INSTITUTION

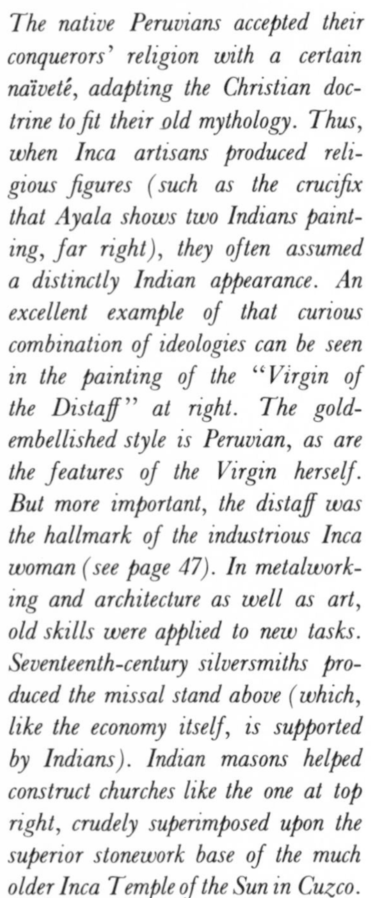

*The native Peruvians accepted their conquerors' religion with a certain naïveté, adapting the Christian doctrine to fit their old mythology. Thus, when Inca artisans produced religious figures (such as the crucifix that Ayala shows two Indians painting, far right), they often assumed a distinctly Indian appearance. An excellent example of that curious combination of ideologies can be seen in the painting of the "Virgin of the Distaff" at right. The gold-embellished style is Peruvian, as are the features of the Virgin herself. But more important, the distaff was the hallmark of the industrious Inca woman (see page 47). In metalworking and architecture as well as art, old skills were applied to new tasks. Seventeenth-century silversmiths produced the missal stand above (which, like the economy itself, is supported by Indians). Indian masons helped construct churches like the one at top right, crudely superimposed upon the superior stonework base of the much older Inca Temple of the Sun in Cuzco.*

MARIA ENGARCIA FREYER COLLECTION, COLUMBUS

Almagro killed him on the spot, flung himself on one of the big cannons, resighted it, and fired, sweeping away several of the approaching horsemen. After that his gunners continued to fire with accuracy, and Vaca de Castro's infantry, advancing on the left under Carvajal, suffered such terrible losses that they were thrown into disorder.

A cavalry charge was ordered in an attempt to silence the cannon. It was then that Almagro made a fatal mistake: Instead of maintaining his strong position and allowing the guns to do the work, he led his horsemen out to meet the charge. Although the superior weapons of his cavaliers almost carried them to victory in the first half hour of the vicious struggle, their opponent's superior numbers overwhelmed them in the end.

While the horsemen battled on, Carvajal managed to regroup his shattered forces. Then, throwing away both helmet and breastplate, he led his men straight for the cannon. The contest was desperate, but the royal forces held, and finally they captured the guns. A second well-timed cavalry charge by Vaca de Castro completed the rout and drove Almagro from the field.

As night fell, the trumpets of the royal troops sounded the recall, and the victors mustered on the battlefield, where they stood to arms all night. In the morning, while the priests were performing their last services for the dying and the surgeons were tending the hundreds who were injured, four large graves were dug for the bodies of the Spanish dead of both sides—over three hundred in all.

ROYAL LIBRARY, COPENHAGEN

Vaca de Castro's army had suffered the heaviest losses, owing to the destructive fire of Almagro's cannon. But at least half the survivors of the defeated army, including their leader, had been captured. Forty were condemned to death, after being tried by a commission, and thirty others were sentenced to mutilation and banishment. The new governor, at the head of a council of war, sat in judgment on Diego de Almagro, who, in spite of his youth, was sentenced to death as a traitor. With his execution the party known as the Men of Chile came to an end.

Perhaps the most tragic postscript to the battle, however, was the death of the Inca Manco, murdered by Spaniards to whom he had given shelter after the defeat of young Almagro. Manco had steadfastly refused to come to terms with the conquistadors, and by always fighting on the weaker side, he had helped to prolong the civil wars. The last of the Lord Incas to live up to their heroic traditions, he died still in arms against the hated foreigners.

After his victory at Chupas, the new governor worked hard to establish justice and order in Peru. He rid Lima of scores of restless firebrands by sending them off to explore and conquer new lands, and he encouraged the missionaries to build churches and schools for the Indians. He also had the *curacas* restock the old Inca storehouses so that Spanish travelers would no longer have an excuse for robbing the natives, and finally, he seized for Charles V large tracts of the estates that Francisco Pizarro had granted to his men. Although he was a harsh, cold man, Vaca de Castro's rule was so impartial and efficient that many of the most influential colonists petitioned the emperor to retain him as permanent governor.

But news traveled slowly between the New World and the Old, and plans had been made to supersede Vaca de Castro even before he had smashed Almagro's revolt. In 1541 Charles V returned to Spain after some months in Germany and with deep distress learned of the upheavals in Peru. It seemed to him that the shocking events that had brought death to thousands of natives in the West Indies might soon be repeated in South America. Animals raised for food and clothing were being destroyed to satisfy the European passion for meat; the whole system of agriculture, once carefully controlled by the Incas, was being disrupted by Spanish demands for Indian labor. Furthermore, the rough soldiers who had won Peru by the sword accepted no responsibility for the well-being of these people. The conquistadors had lived hard and dangerously themselves, and few felt pity for the Indians dying around them from ill-treatment, disease, and hard labor.

It was with all these injustices in mind that Charles V summoned a council to advise him on the drafting of a code of laws for his South American subjects, and by 1543, the *New Laws* for the Indies had been published in Madrid. They decreed that all Indian slaves be set free on the death of their masters; that Spaniards who had ill-treated any slaves lose them immediately, and in the case of gross ill-usage, lose their estates as well; and that slaves held by public officials, monasteries, town councils, or men who had taken part in the civil wars revert to the Crown.

To make certain that his orders were obeyed, Charles V appointed Blasco Núñez Vela, a cavalier of a well-placed family, as his viceroy, with a grant of extraordinary powers. In addition, a royal audience, or high court, consisting of four judges, was appointed to accompany Blasco Núñez to Lima, which was to take the place of Panama as the center of justice and chief seat of government for Spain's South American colonies. In November, 1543, Blasco Núñez and his magnificent retinue set sail.

When news of these changes reached Peru, angry crowds gathered in the towns and cities. Rugged old soldiers, who had spent years in the desperate struggle for gold, faced ruin. While riches poured into the king's treasury, they yelled, the men who had won them were to die in poverty. Realizing that nearly all of them would suffer from the new laws, their initial protests soon turned to action. What they had won by the sword they would hold by the sword.

Vaca de Castro's admirable mixture of firmness and sympathy held the Spaniards in check for a time. He suggested that a deputation be sent to the viceroy to ask him to suspend the laws until the settlers' grievances were laid before Charles V. But Blasco Núñez was an arrogant and unbending man, and no sooner had he arrived in the New World than he began to act. At Nombre de Dios he seized a shipload of Peruvian silver on the grounds that it had been mined with slave labor; he freed slaves at Panama and

*An early copy of the* New Laws *—governing treatment of his South American subjects—bears Charles V's coat of arms. His escutcheon (center), capped by the Hapsburg double eagle, is dominated by the lions of León and the castles of Castile, symbols of the kingdoms united by the marriage of Ferdinand and Isabella, grandparents of the Holy Roman Emperor.*

BIBLIOTECA NACIONAL, MADRID

BIBLIOTECA NACIONAL, MADRID: PHOTO, ORONOZ

Tumbes and actually paid his Indian porters on his march to Lima. When he was formally installed as viceroy, he let it be known that he thought the laws unfair to the colonists, although he refused to suspend them.

His actions played right into the hands of the man whom Vaca de Castro regarded as the most dangerous in all of Peru—Gonzalo Pizarro. Up to this point, Gonzalo had been biding his time. Now, supported by most of the Spanish settlers, he induced Cuzco's magistrates to name him as Captain-General and to authorize the raising of an army. His professed intention was to march to Lima to plead the settlers' case before the viceroy. The last of the Pizarros in South America, Gonzalo finally had his chance.

As the new Captain-General moved westward with his four hundred well-armed troops, men from the northern and coastal regions joined him, swelling his ranks. In Lima

*Yoked dray horses strain under the weight of cannon (right) and siege machinery (left) in this detail from a sixteenth-century view of a triumphal military procession. Pikes and halberds (at center), like the cannon and siege weapons, also were unknown in the lands the conquistadors invaded.*

Blasco Núñez, trusting no one, placed Vaca de Castro under arrest in a ship in the harbor. Too weak to take the offensive against Gonzalo Pizarro and skeptical of his pretended peaceful mission, the viceroy at first decided to defend Lima, but he quickly lost his nerve and ordered the evacuation of the capital instead. The four judges of the audience, who were quite disenchanted with the way in which Blasco Núñez was handling things, found the evacuation order intolerable. Hastily they raised their own armed force, entered the palace without opposition, and arrested the viceroy. They then declared themselves to be the provisional government and elected one of their number, named Cépeda, as president. Cépeda's first act was to suspend the hated ordinances; immediately after, he sent an envoy to Gonzalo, informing him of the suspension and ordering him to disband his forces.

*The concave sides of this ornate silver stirrup take their shape from the original leather stirrups worn by the conquistadors, on which the one above is modeled.*

CONSTANTE LARCO, HACIENDA CHICLIN: PHOTO, SMITHSONIAN INSTITUTION

Gonzalo's reply was curt: He said that the people had called on him to save the country and that he meant to do so. He threatened to turn his soldiers loose on Lima unless Cépeda immediately acknowledged him as the legal governor of Peru. To show that he was serious, he sent Francisco de Carvajal, the formidable old officer who had played such a decisive role in the defeat of Almagro at Chupas, into the city at night with a small band of soldiers. There Carvajal dragged from their beds several cavaliers who had deserted Gonzalo when he had begun his march, and he hanged three of the captives from a tree just outside the city.

Realizing that their lives were at stake, the audience yielded, and on October 28, 1544, the last Pizarro entered Lima in great splendor, with twelve hundred Spaniards and several thousand Indians at his side. Amid wild celebration, Gonzalo was proclaimed governor of Peru.

He was not the undisputed master of Peru for long, however. Within days of Gonzalo's inauguration, Blasco Núñez, set free by the judge who was to have escorted him to Spain, raised his standard at San Miguel. Gonzalo was soon on the march again. The clash between the two adversaries came on the plains of Añaquito, about a mile from Quito, on January 18, 1546. Gonzalo's chase had been a long one, but not without reward. Blasco Núñez's inexperienced army fell quickly before his staunch veterans, and the ex-viceroy himself was savagely slain on the field.

This time the dignitaries of both Church and State awaited the "Liberator of the People" at Lima, ready and willing to escort Gonzalo into the city. With two of his captains leading his horse and the Archbishop of Lima and three lesser bishops riding beside him, the new governor moved at a regal pace along the streets under gay, flowered arches raised in his honor. Delegates from every chief city in Peru had come to offer their allegiance. His ships controlled the Pacific, and his admiral, Pedro de Hinojosa, had taken Panama and Nombre de Dios in his name. In addition, Carvajal soon reported that the south was completely subjugated. Peru seemed to be more firmly in the grip of a Pizarro than it had ever been before.

But Carvajal knew better. He and Gonzalo had overthrown the forces of Charles V, and a shameful death awaited them if ever they should fall into the hands of the emperor's captains. It was the old soldier's plan that Gonzalo should entirely throw off his allegiance to Charles and establish a Peruvian royal house by marrying the *coya*, the female representative of the Lord Incas.

Gonzalo, unhappily, could not swerve from his inborn loyalty to the Crown. He persuaded himself that he had been fighting for his sovereign's best interests. He knew so little about the ways of princes that he hoped for pardon—and official appointment as governor of Peru.

Instead, the things most feared by Carvajal happened. News of the conflict that again had set Peru alight reached Spain, and Philip II, ruling in his father's absence, summoned another council. All indignantly agreed that Gonzalo was a rebel. Their only problem was to find a way to destroy him. Knowing that it was impossible to raise a force strong enough to challenge Gonzalo Pizarro, the councilors reluctantly advised Philip to try persuasion.

The man chosen for this delicate mission was a priest named Pedro de La Gasca. La Gasca already had shown great skill and nerve in the service of his monarch. He accepted the appointment, but he boldly asked for powers never before granted to a servant of the Crown, powers

HERRERA, *Decades*, 1726, NAVAL MUSEUM, MADRID: PHOTO, ORONOZ

*The broad-brimmed hats worn by the Spanish foot soldiers in this view from a 1726 edition of Herrera's* Decades *betray the artist's ignorance of the styles of the 1500's, when this civil war clash supposedly occurred.*

*The Spaniards' innate fondness for spectacle was quickly asserted, even in the remote corners of the empire—as is clear in this busy painting of the entry of the viceroy into the silver-mine center at Potosí, Bolivia, in 1716.*

almost equal to those of his sovereign: the powers to control every department in the colony, to make peace or war, to appoint or dismiss whom he chose, and to pardon any offense. So great was Charles V's confidence in La Gasca that when the request reached the monarch in Flanders, he granted it without hesitation. In May, 1546, the royal envoy set out for Peru, dressed as a simple cleric and accompanied by only a few attendants. His title was simply president of the royal audience.

After an uneventful voyage he landed at Santa Marta,

MUSEO DE AMERICA, MADRID: PHOTO, ORONOZ

where he was shocked by the news of Blasco Núñez's defeat and death. But undeterred by the knowledge that Gonzalo held all Peru, La Gasca headed for Nombre de Dios. So little was the governor of that port impressed by the priest and his small party that he disregarded Gonzalo's orders and gave permission for them to enter the town. Even when La Gasca's rank and mission were made known, the rough soldiers laughed. "If this is the sort of governor His Majesty sends over to us," sneered one, "Pizarro need not trouble his head much about it."

Still, within a few days, La Gasca's friendly manner and promises to repeal the new laws and to pardon all those who submitted to his legal authority had won the governor of Nombre de Dios, one of Pizarro's most trusted officials. La Gasca's greatest triumph came later, however, when after months of patient diplomacy he persuaded Hinojosa, governor of Panama and admiral of the fleet, to join him. The ports of Peru were thus in La Gasca's hands.

During his stay on the Isthmus, La Gasca sent secret agents to Lima and other important cities to distribute manifestos stating that the grievances of the colonists had been redressed and that the colonists themselves had nothing to fear if they returned their allegiance to Charles V. Then he wrote a conciliatory letter to Gonzalo, promising him pardon and asking his help in settling the country.

Gonzalo realized that he could no longer shun the fateful decision that he had been so unwilling to face for months: He must either accept the emperor's pardon, give up the land won by the conquistadors, and sink into obscurity or he must turn his sword against his sovereign. Realizing that their chance of holding Peru had gone, Carvajal now advised Gonzalo to accept the offer. Cépeda, on the other hand, urged him to resist. Recklessly, Gonzalo threw all to the wind, and trusting to the strength of his forces and the loyalty of his commanders, he bluntly rejected La Gasca's offer.

La Gasca, meanwhile, was preparing to use force. Funds were raised, men were recruited in Panama, and demands

HENRY E. HUNTINGTON LIBRARY

*This caricature was included in a letter written in 1549 by an embittered Spanish settler, who drew La Gasca (right) awarding slave grants to the Bishop of Lima. La Gasca's virtually limitless powers as the King's representative in Peru naturally made him enemies, and the accusation leveled against him may be considerably distorted.*

for troops were sent to other Spanish colonies in the New World. A squadron of ships left port to cruise off the coast of Peru to encourage and protect those who were ready to take up arms against Gonzalo Pizarro.

Disaster upon disaster soon crowded Gonzalo. First he learned of the desertion of Hinojosa and the loss of the fleet, then of the murder of his commander at Quito and the loss of the northern towns, and lastly of the reappearance of Diego Centeno, a young captain who had served with Blasco Núñez at Añaquito. Centeno had seized Cuzco, and before long he controlled all the southern highlands as far as the shores of Lake Titicaca.

Gonzalo held Lima, but his commanders were falling away on all sides. Decisive action was imperative to reestablish his authority, and with the aid of Carvajal he recruited an armed force of a thousand men. But Gonzalo's personality had changed. Gone were his usual high spirits and easy camaraderie; he had become violent and capricious. Scenting treachery around him, he turned fiercely on anyone against whom there was the faintest breath of suspicion. Lima became a place where men whispered in corners, fearful lest a careless word bring them ruin or death.

At the appearance of La Gasca's squadron off the coast, Gonzalo's fears reached their peak. He set guards along the shore to prevent his men from deserting to the enemy. When this stratagem failed, the despairing commander abandoned the city for Arequipa, some five hundred miles to the southeast. By then, his magnificent army had shrunk to a bare

*Vaca de Castro enclosed the crude drawing at right, of the Panamanian town of Nombre de Dios, in a letter urging the construction of a sturdy fortress (which he has drawn in, bottom center) to guard the port city. His 1541 letter suggests that the town be made the embarkation point for treasure ships bound for Spain.*

ARCHIVO DE INDIAS, SEVILLE

NEW YORK PUBLIC LIBRARY

*A wholly imaginary European city has been added to the shores of Lake Titicaca—highest lake in the world—by the creator of this sixteenth-century woodcut. The artist has depicted manned Inca balsas with similar inaccuracy.*

five hundred men. Moreover, the moment Gonzalo withdrew, Lima opened its gates to La Gasca's commander.

By late winter, 1547, La Gasca and his forces were at Jauja, and with Centeno holding the mountain passes in the southern highlands, Gonzalo was trapped at Arequipa. Still, danger always seemed to lift the spirits of the Pizarro brothers, and Gonzalo suddenly regained his confidence.

"It is misfortune that teaches us who are our friends," he exclaimed. "If but ten only remain true to me, fear not but I will again be master of Peru!" Clearly, however, he could not wait at Arequipa until the two royal armies converged on him. Thus, he determined to retire to Chile until the inevitable quarrels broke out between the settlers and the government and he could again strike a blow for empire.

Diego Centeno and a thousand men barred his way. Gonzalo first tried to negotiate. He sent a message to Centeno reminding him of their former friendship and asking for free passage across the mountains. Centeno replied courteously that he could not in honor grant the request. Yet, if Gonzalo surrendered to him, he promised to use all his influence with the government to win favorable terms for the Captain-General and his men. Gonzalo was infuriated. He had not marched hundreds of miles to hand his sword to one of his turncoat captains, he said.

Still, he did try to avoid a battle against such odds. Fate was against him. On October 26, 1547, the armies of Centeno and Gonzalo Pizarro stood face to face on the plains of Huarina, southeast of Lake Titicaca. Centeno had been ill for some time, and on this day, unable to keep his saddle, he retired from the field in a litter after he had drawn up his men for battle. Gonzalo, on the other hand, appeared in the flower of manhood, superbly mounted, with shining mail and a crimson velvet surcoat. He attracted all eyes. Carvajal, whose bulky figure was protected by a serviceable if dull suit of steel, rode a war horse, strong but almost as ungainly as its master, and was quite undistinguishable from the common soldier. At a position about six hundred paces apart, the two armies came to a halt.

The battle that raged on the sun-drenched plains that day was the bloodiest yet fought by the conquerors in Peru, and the victory for Gonzalo Pizarro's forces was complete. Viewing the heaps of dead around him, the brave Gonzalo was seen to make the sign of the cross and heard to exclaim, "Jesus! What a victory!" More than half of Centeno's army lay critically or mortally wounded on the field, and the victors had lost a hundred men.

PEDRO ROJAS PONCE

*Built high in the Andes, like Machu Picchu (pages 104–105), the fortress at Jauja was used both by the Incas and later, during the civil wars, by the Spaniards as a stronghold and surveillance point. The stonework, probably destroyed during the conquest and rebuilt sometime after by less skilled masons, is considerably cruder than that at Machu Picchu.*

Carvajal, whose leadership and well-trained troops had made Gonzalo's victory possible, tarnished the honor he had won in the struggle by his heartless cruelty to his captives, as well as to all deserters from Gonzalo's ranks. Seeing the fate of his troops, Centeno quickly fled. Despite his illness, he mounted a horse and evaded his pursuers. Undoubtedly, his effort was spurred on by the knowledge that he could expect no mercy from Carvajal if he were caught.

There was no further talk of withdrawing to Chile. Gonzalo had completely regained his confidence. Marching north to Cuzco, he declined the triumphal entry offered by its citizens and instead went on foot to the cathedral to offer thanksgiving for his victory.

Shocked by the news of Centeno's utter defeat, La Gasca nevertheless remained undaunted. In March, 1548, he put his army in motion. He now commanded the largest Spanish force yet seen in Peru, about two thousand men in all. Since he was no soldier, he placed Hinojosa in command, but other distinguished officers had joined him too. Centeno had recovered from his illness and was eager to retrieve his reputation. Sebastián de Belalcázar also had come with a force from north of Quito. La Gasca's was a powerful army, excellently led.

That summer Gonzalo lived a careless, pleasure-filled life in Cuzco. The victory at Huarina seemed to have convinced him that he could smash any army La Gasca might bring against him. Carvajal, on the other hand, never relaxed. Every day he could be seen supervising the manufacture of arms and military stores or drilling his infantry. More and more disgruntled with his indolent leader, the old soldier became convinced that the odds they now faced were too great. He advised Gonzalo to disband those of Diego Centeno's men who had joined them after Huarina, believing that such men were unreliable, and he suggested that they take the loyal core of their army into the mountains to outwait La Gasca. But Gonzalo was unconvinced.

At last, La Gasca advanced. Crossing the Abancay and Apurímac rivers, he was able to approach Cuzco without serious opposition. An incompetent commander of Gonzalo's was accountable for that success. Gonzalo had but one decision to make: where to give battle to the enemy. He chose the valley of Jaquijaguana, about fifteen miles from the city, and on April 8 the royal troops arrived.

It is pointless to describe the battle order and preliminary movements of the two armies, since no fighting took

*This 1555 woodcut from Gômara's history of the conquest clearly favors La Gasca's field commander, Diego Centeno (standing benignly at left). In reality, Centeno was just as ruthless in battle and cruel in the treatment of his prisoners as Gonzalo Pizarro's second-in-command, Carvajal (at right, splitting a man's skull).*

place. Carvajal had been right, as usual. As the enemy lines approached, first Cépeda, who held a high command on that day, then a group recruited from Centeno's harquebusiers rode over to the enemy. When a troop of horsemen were dispatched to intercept the fugitives, they, too, deserted. Convinced that the rebel force was disintegrating, La Gasca halted his army.

Even Gonzalo's own troops, who never had flinched before, melted away in the confusion caused by the treachery of the new recruits. Some surrendered to La Gasca, hoping for pardon, while others fled to Cuzco or into the hills. Only a handful of Gonzalo's most devoted followers stood firm.

Stunned by this rapid change in his fortunes, Gonzalo turned to one of them and asked, "What remains for us?"

"Fall on the enemy . . . and die as Romans," was the reply.

"Better to die as Christians," retorted Gonzalo, and he rode off in the direction of his enemy's camp.

NAVAL MUSEUM, MADRID: PHOTO, ORONOZ

*Accompanied by a large band of well-armed soldiers, and a small, unarmed puppy, Pedro de La Gasca is shown landing in Peru in 1546, in this central detail from an illustration for Herrera's* Decades.

Meeting a royalist officer, Gonzalo Pizarro surrendered his sword and was led before La Gasca, who received him coldly. Accused of killing Blasco Núñez, of seizing Peru's government, and of remaining under arms against the sovereign's forces when he had been offered a free pardon, Gonzalo defended his conduct: "It was my family who conquered the country—and as their representative here, I felt I had a right to the government."

The priest's tone was stern. "Your brother did, indeed, conquer the land, and for this the emperor was pleased to raise both him and you from the dust. He lived and died a true and loyal subject, and it only makes your ingratitude to your sovereign the more heinous." Refusing to hear another word, La Gasca then placed Gonzalo under arrest.

Carvajal, meanwhile, had remained at his post, cynically humming his favorite ballad as he saw his troops deserting: "The wind blows the hairs off my head, Mother; Two at a time, it blows them away!" When there was no further point in holding his ground, he put spurs to his horse and galloped away. But climbing the steep bank of the stream near which Gonzalo had pitched his camp, the horse slipped and threw its rider. Carvajal was seized before he could regain himself. Some of his own men led him to La Gasca, hoping to win a pardon with such a prize.

As the turncoat troopers marched the old soldier along, they were surrounded by loyalists who jeered the old man who had for so long been the terror of their army; Carvajal, preferring death to insult, infuriated them further with contemptuous replies. Only the arrival of Centeno prevented them from robbing the executioner of his victim.

Still, Carvajal could not resist a sardonic joke. Pretending not to know the officer who had protected him, he courteously asked his name.

"Do you not know me?—Diego Centeno!" the young commander answered in surprise.

"I crave your pardon," said Carvajal. Then, taunting Centeno about his wise but uncourageous retreat at Huarina, he continued: "It is so long since I have seen anything but your back that I had forgotten your face!"

ROYAL LIBRARY, COPENHAGEN

Since the rebels had been taken with arms, there was no need for a lengthy trial. A court-martial sentenced Gonzalo, Carvajal, and three or four other cavaliers to death, and the executions were carried out the next day at the battle site. After a night spent in prayer, Gonzalo met his death with dignified courage. Mounting the scaffold, he made a short speech to the soldiers, begging them in their charity to buy masses for his soul since he would die penniless. Many were deeply moved by his last words. After kneeling to pray, he bent forward to receive the executioner's blow. Soon thereafter, Gonzalo's estates were confiscated, and his house was razed.

Four of the brothers who had conquered Peru were now dead, and the fifth was imprisoned in Spain. The two Almagros lay in a common grave. The land of the Incas had brought little reward to the rough leaders who had faced so many perils to win it. For the first time, however, Charles V held undisputed sway over his new realm in South America. Riches poured into his treasury and he appointed governors to enforce his laws. But although the authority of the Spanish Crown was not seriously challenged for almost three hundred years, royal attempts to control the unruly New World colonists never were completely successful. And the natives of Peru themselves, a gifted, industrious, and law-abiding people, who under a more enlightened rule than that of the conquerors might have led the march of civilization in South America, instead sank into a poverty and despair from which they have scarcely emerged to this day.

*Used by the Spaniards as slave laborers (above) to strip their own country of its resources, the once highly civilized Incas soon sank into listless imitation of their former ways. In Bishop Compañon's water color of the 1780's at right, a group of long-exploited Indians observe an ancient festival—but without the gaiety that once marked such ritual celebrations.*

BIBLIOTECA DEL PALACIO, PATRIMONIO NACIONAL, MADRID: PHOTO, ORONOZ

ARCHIVO DE INDIAS, SEVILLE

*The coat of arms awarded Francisco Pizarro and his descendants by Charles V included discovery ships, natives fighting wild beasts, and an imaginary Spanish view of the coastal town of Tumbes.*



# ACKNOWLEDGMENTS

The Editors would like to thank the following individuals and organizations for their valuable assistance:

The Duke of Alba, Madrid
American Museum of Natural History, Department of Anthropology, New York—Junius Bird
Archivo de Indias, Seville—Doña Rosario Parra, Curator
Archivo MAS, Barcelona—Doña Monserrat Blanch de Alcolea
Biblioteca Nacional, Manuscript Division, Madrid—D. Ramon Paz, Curator
Jose Casals, Lima
International Film Foundation, Inc., New York—Julien Bryan
Museo de America, Madrid—Doña Pilar de Ferrandis, Curator
Naval Museum, Madrid
Oronoz, Madrid
His Excellency Celso Pastor, Peruvian Ambassador to the United States, Washington
Peruvian Embassy to the United States, Washington—Antonio Lulli, Minister Councillor
Pedro Rojas Ponce, Lima
Mr. and Mrs. George Renchard, Washington
Smithsonian Institution, Division of Cultural History, Washington—Richard E. Ahlborn

# FURTHER REFERENCE

Ever since its publication well over a century ago, William Hickling Prescott's *The Conquest of Peru* has been recognized as the authoritative work on its subject. Such durability is in itself remarkable—though no more so than the manner in which the two-volume work was assembled. A dining hall accident in Prescott's junior year at Harvard left the young Boston Brahmin blind in one eye—and turned him from social life to scholarship. Working in semidarkness to conserve the rapidly failing vision in his one good eye, Prescott gradually gathered information about the conquest from archives on three continents. Each copied manuscript was read aloud to Prescott while he made notes on carbon paper with an ivory stylus. Those notes, and the handwritten manuscript that grew out of them, eventually were transformed into printed pages—a Herculean exercise in patience and diligence that consumed three full years of Prescott's life and almost cost him what little vision remained in his right eye. Nearly blind but nonetheless indefatigable, Prescott was working on a biography of Charles V's son and heir, Philip II, when he died in 1859.

For further information on the conquest of Peru and the Indian cultures of the Andean region, the following books are recommended:


Baudin, Louis, *Daily Life in Peru Under the Last Incas*, translated by Winifred Bradford. Macmillan, 1962.

Bennett, Wendell C., and Bird, Junius, *Andean Culture History*. Doubleday, 1964 (paperback).

Bingham, Hiram, *Lost City of the Incas*. Meredith, 1948.*

Brundage, Burr C., *Empire of the Incas*. University of Oklahoma Press, 1963.

———, *Lords of Cuzco*. University of Oklahoma Press, 1967.

Cieza de León, Pedro de, *The Incas of Pedro de Cieza de León*, edited by Victor W. von Hagen, translated by Harriet de Onis. University of Oklahoma Press, 1959.

Disselhoff, Hans Dietrich, *Daily Life in Ancient Peru*, translated by Alisa Jaffa. McGraw-Hill, 1967.

Dockstader, Frederick J., *Indian Art in South America*. New York Graphic Society, 1967.

Garcilasco de la Vega, El Inca, *Royal Commentaries of the Incas and General History of Peru*, 2 vols., translated by H. V. Livermore. University of Texas Press, 1965.

Jones, Julie, *Art of Empire: The Inca of Peru*. Museum of Primitive Art, New York, 1964 (paperback).

Mason, J. A., *The Ancient Civilizations of Peru*. Penguin, 1957 (paperback).

Means, Philip A., *Ancient Civilizations of the Andes*. Gordian, 1964.

———, *Fall of the Inca Empire and the Spanish Rule in Peru, 1530–1780*. Gordian, 1964.

Prescott, William H., *Conquest of Mexico and Conquest of Peru*. Modern Library, 1936.*

Von Hagen, Victor W., *Realm of the Incas*. New American Library, 1961 (paperback).

*Also available in paperback.

*With llama fat burning in the center hole, this lifelike carved armadillo was used in Inca ceremonies.*

# INDEX

**Boldface** indicates pages on which maps or illustrations appear

# B



# C



# D



# E



# F



# G



# H



# I

# Q



# R



# S



# T



# U



# V



# W



# Y

EFIGIES DE LOS INGAS O REYES
del PERU con su Origen y Serie: Y de los
Catholicos Reyes de Castilla, y de Leon, que les
han sucedido hasta el presente que Dios guarde.
Algunos Historiadores ponen antes de estos Sr.es Ingas, quatro Edades, en que florecieron quatro famosos Capitanes: el 1.º Huari Viracocha Runa casado con Mama Huarmi: el 2.º Huari Runa casado con Mama Pucillo: el 3.º Purun Runa casado con Mama Sicac: el 4.º Auca Runa casado con Mama Panchiri Sisac: Otros cuentan por sus nombres, desde el Diluvio, hasta el primer Inga, ciento, y quatro Reyes, por noticia de sus Quipus ó Anales, escritos, ó formados con ñudos en hilos de varios colores.
REY DE REYES, Y AL REY INMORTAL
que os llamó de las
SINCHI ROCA INGA II.º
LLOQUIYUPANGUI INGA III
MAYTA CAPAC INGA
El Pacifico. Con el Culto y Prudencia Estendió su Imperio. Puso leyes cōtra la Pereza de los Indios. Vivio 155 años La Coya su Muger fue Mama Chimbo Urma. En su tiempo nacio Nro Salvador IESUS Sucediole su Hijo.
El Famoso: aunque es varia su fama. Con valor, y benignidad conquistó las grandes Provincias de los Ayavíris Collas y Chucuytos. Vivio 150 años La Coya su Muger fue Mama Cora Ocllo Sucediole su Hijo.
El Melancolico Conquistó á los
hasta el famoso Cerro de Potosí
el celebre Puente de Apurimac
Sucediole su Hijo.
PACHACUTIC INGA IX
INGA YUPANGUI INGA X
TOPA YUPANGUI INGA XI
El Hazañoso Constituyó un Sumo Sacerdote y otros menores. Solía arrojar con vna honda pedazos d oro al Sol. Conquistó á los Taramas, y otras Provin. Vivio 88 años. La Coya su Muger Mama Huaicu, Sucediole su Hijo.
El Bueno Penetró la Cordillera. Descubrió la Provincia de los Moxos. Dió princi.º á la Gr. Fortaleza del Cuzco. Erigio grandes Templos y Casas d Virgenes. La Coya fue Mama Chimbo Ocllo. Algunos lo confunden cō su Pe. No se sabe su edad. Suced.º su Hijo.
El Sabio Conquistador
FELIPE II REY XVII del PERU
FELIPE IV REY XVIII del PERU
CARLOS II REY
El Bueno Nacio a 10 de Mayo 1578. Reynó desde 1598 23 años. Expidio la Cedula sobre el servicio Personal de los Indios 1609. Con otras muchas a fav. de ellos. Murio a 31 de Marzo 1621 de 43 años. Sucediole su Hijo.
El Grande Nacio Viernes S.to 8 de Abril 1605. Reynó desde 1621, 44 años. Embio grandes Ministros y Prelados al Perú. Murio a 17 de Setiembre de 1665 de 60 años de edad. Sucediole su Hijo.

# WHEN MOM AND DAD DIVORCE

Also by Steven Nickman

THE ADOPTION EXPERIENCE

# WHEN MOM AND DAD DIVORCE

by Steven L. Nickman

ILLUSTRATED BY
Diane de Groat

JULIAN MESSNER  NEW YORK

Published by Julian Messner,
A Division of Simon & Schuster, Inc.
Simon & Schuster Building
Rockefeller Center
1230 Avenue of the Americas
New York, New York 10020

JULIAN MESSNER and colophon are trademarks
of Simon & Schuster, Inc.

Manufactured in the United States of America

10 9 8 7 6 5 4 3 2 1 (trade)
10 9 8 7 6 5 4 3 2 1 (pbk.)

Library of Congress Cataloging in Publication Data
Nickman, Steven., when mom and dad divorce.

Includes index
Summary: Focuses on the thoughts, feelings, and emotional problems of children whose parents are divorcing, accompanied by brief stories of young people in such situations.
1. Divorce—Juvenile literature. 2. Separation (Psychology)—Juvenile literature. [1. Divorce] I. Title.
HQ814.N53 1986 306.8'9 86-5295
ISBN 0-671-60153-9 (trade)
ISBN 0-671-62878-X (pbk.)

# Contents

# To the Reader

When parents divorce, a lot of kids have a hard time for a while. Sometimes they feel sad, sometimes lonely, sometimes both. They may wish they could get a little advice or comfort. Adult friends or relatives will help, but a book can help too.

If your parents are divorced or living apart and planning to divorce, this book can help you through a tough time. If you have a friend whose parents are divorcing, it can help you understand what your friend is going through.

You'll read about boys and girls who were shocked and frightened when they learned their family was breaking up. After each story I've put down what I've learned about these things that happen to families. I believe my ideas will help you understand why your parents act the way they do. I've also given you a lot of ideas about what you can do to help yourself.

When divorce happens to a family, most kids feel confused. These are some of the questions they wonder about:

Our family used to be like a team. Now it's like two teams playing against each other. Or it's even like two armies in a war. What side am I supposed to be on? Do I have to be on a side at all?

My parents always cared a lot about what I did. Sometimes they disagreed about me. Could this thing have gotten started because of something I did? Something I didn't do? Could it be my fault?

Mom and Dad seem different now. Sometimes she's really sad-looking, and he has a bad temper, and sometimes it's the other way around. I want to help them when they're sad, but I don't know how. And when they yell at me, I know I don't always deserve it. I feel so alone these days, and so awful, and I don't know if I'll ever feel better.

Mom tells me about bad things Dad does. I think she wants me to be as mad at him as she is, but mostly I don't feel mad at him at all. What should I say?

Mom and Dad want me to go on living with each of them half the time. I've been switching back and forth every few days for a year now and it's good in some ways, but I get mixed up about when I'm supposed to be where. I'd rather be in one place most of the time. But if I said that, somebody's feelings would be hurt. Who can I tell? Is there anybody who would listen?

These are just a few of the questions kids worry about. Some come from the beginning of a divorce, some from later on. There are no magic answers to any of them. But this book can help you figure out what's going on in your life and what you can do to make it run more smoothly.

Steven L. Nickman, M.D.

# WHEN MOM AND DAD DIVORCE

# 1

# The Problem

One afternoon when Sam came home from school, his mother looked different. Her eyes were red and her cheeks were wet. He headed toward the kitchen for a snack, as usual, but Mom steered him into the living room instead.

"Sam, I have to tell you something."

Whatever it is, it doesn't look good, thought ten-year-old Sam. Maybe Barney had been run over? The old mutt was getting confused lately, and Sam worried about him sometimes. Maybe Grandpa'd had another heart attack?

"Daddy won't be home tonight."

That's not so awful, thought Sam. He's gone away on business trips lots of times. Why the big deal? "When'll he be back?"

"I don't know, Sam. Maybe not ever, to live here with us. He'll visit with you, because he loves you a lot. But Daddy and I haven't been getting along for more than a year now, and it keeps getting worse. It looks as though the best thing we can do is get a divorce."

Sam didn't know what to say. He was overwhelmed. "Why didn't you tell me before?"

"We didn't want to upset you, Sam. And we weren't sure until this week."

Sure, Sam had heard his parents arguing, and sometimes it got pretty loud. But even though he'd been scared, he thought everybody's parents did that. A *divorce*? What did that mean? That he didn't have a family anymore? Would he have to go live in an orphans' home somewhere? A divorce didn't sound like the best thing to Sam. It sounded like the worst thing he could possibly imagine.

Why couldn't his parents keep things going the way they were supposed to? And why did Dad leave without saying good-bye to him?

Tears came to his eyes. Just then Barney started to lick his face. He pushed the dog away roughly. Barney whined, and Sam felt even worse. His mother tried to hug him, but he got up and ran into his room.

He banged the door closed, threw himself down on his bed, and said over and over: "I'm not here. This is not happening. It *can't* be happening!"

A lot of boys and girls never guess that something big is about to happen in their family. Even if they've heard loud arguments or seen one parent hit the other, they still don't suspect that Mom and Dad may not stay together. When these kids are told that their parents are going to separate, they're as surprised as if they'd suddenly come face to face with a dinosaur in the street.

It's bad enough if kids haven't been given any preparation. What can make it even worse is that many times parents are so upset themselves that they don't explain the situation well enough to their kids. The children still have questions and may be afraid to ask them, because Mom looks as though she doesn't want to talk anymore, or because Dad seems about to cry.

In the beginning, when you first learn that your parents are about to separate, you may wonder how you can get through the weeks and months ahead. It might be useful to remember tough times you've had before—times you got through successfully. (Maybe, like Sam, you've had a pet or a favorite relative that was very sick or even died.) You may have a friend whose parents split up, a friend you could talk to who would understand. And remember too your aunts, uncles, or grandparents. They'll be around for you to visit and have fun with, and to give you advice if you need it.

You'll need to talk to your parents a lot. After they've told you the bad news, that shouldn't be the end of talking about what's happening. Most likely you'll keep having new questions that need to be answered.

When you can put the questions into words inside your head, wait for a time when you can talk privately with each parent. Tell Mom that you need to talk to her for a while, and could she please put aside what she's doing, or find another time that day when she can really talk.

Then do the same with Dad. (Maybe they can talk to you together too—at least, it's worth asking.)

One thing this book will tell you again and again is that in a divorce, kids often have to stick up for their rights. You have a right to know what's going to happen. There's probably a lot that your parents don't know yet themselves, but they should at least share with you some of what they do know.

And asking questions and getting answers can do something else for you, too, besides giving you information. Finding out what's going on can make you feel that you're not just a lump, or a possession, or a victim of your parents' quarrel. You're *somebody,* and you can *do something.*

# 2

# What Leads to Divorce

Amy was awakened from a pleasant dream by the sound of shouts. She burrowed under her pillow and pulled the quilt around and over her, trying to get back to the place she'd been in. But she couldn't find it. She wondered if the shouts could have happened in the dream, which up to then had been so enjoyable.

Instead she realized that the voices she'd heard were Mom's and Dad's voices. She'd never heard them like this before, and she was frightened.

Mom sounded very different in the dark of this strange night. Amy couldn't hear her words clearly through the wall, but she seemed very cold, very angry at Dad. She couldn't hear his replies well either. Mostly he spoke in a low tone, as if he were asking her to calm down or forgive him for something. But once or twice his voice got louder, with a hard edge.

Amy wished this were a dream too, but she knew it wasn't. She could see all the furniture and stuffed animals in her room and remember how they looked in the daylight. If only it were a dream I could wake up from it, she thought. Instead, she took the extra pillow she kept at the foot of her bed and put it on top of the first, then stuck her head under both. Eventually she slept again.

In the morning her parents were both at the breakfast table in the kitchen. The toast was burnt. Dad was silent and looked down at the table most of the time, and Mom spilled coffee on the floor. They both gave her a kiss as she left for school, but neither really seemed to notice she was leaving. It was like last night all over again—just quieted down a bit in the daylight.

Amy wondered if sunlight turned loud, angry fights into silent ones.

She had a hard time thinking about her work in school that day, and she didn't have much appetite for lunch.

---

Kids who hear their parents fighting usually get worried. Especially if they hear nasty words, angry words, insults—and even more if they see one parent hit the other. (This doesn't happen often, but it can.) They don't know what to expect. And when

there've been many fights, lasting over months or years, the kid has had time to develop a lot of fears.

A common fear of children is that they'll be left alone with no one to care for them. That happens in fairy tales, but almost never in real life. If your parents are both really upset, they may pay less attention to you for a while—but usually they're both eager to have you with them and take care of you. Once in a while, relatives or friends might have to step in and help out, but that's almost always temporary.

Other kids are afraid mother and father may be badly hurt or may even die. Remember that they're both grown up, and can take care of themselves as they've always done. That is true even if you may have seen them act childish at times when they're upset.

Still others believe that somehow the trouble between their parents was their own fault, for not behaving well—even for not eating their broccoli or spinach. They're likely to think this way especially if they've heard their parents argue about them. But the way a kid behaves is almost never the reason for parents separating. Children are often too ready to take the blame.

Did you ever have a friend, a really good friend, someone you could talk to about everything? Or someone you went fishing with or jumped rope with and saw just about every day? And if he or she wasn't around to play with after school, it was like a big black hole in the middle of the afternoon?

If you've had a friend like that, maybe it happened that eventually a third person came along, and your friend started to play with that person sometimes, and you felt left out and angry. Or you and your friend had an argument over who won a game, or where you were going to have a sleep-over, and the argument didn't end, and you stayed mad at each other instead. You saw each other less often. And eventually you and your friend had lost each other: you were both still around, but the magic of close friendship was gone.

That's pretty much what happens in a divorce.

A man and a woman get married because they believe they have a kind of magic between them, an excitement and love that will last their whole lives. They want children—partly because they feel so wonderful that they want to share the feeling with a child of their own. And partly so their children will be special and carry the parents' special feeling about each other and their new family out into the world. Sometimes this special feeling lasts a whole lifetime.

Sometimes it gradually changes into a deep friendship between the man and the woman. They still love each other, but they learn that it takes some work to keep the marriage going. (That's what happens with a lot of marriages that last.)

But sometimes the woman and man (they may be parents now) are disappointed in each other and blame each other often. Maybe the man is spending his time with other people and doesn't hug his wife and praise her the way she hoped he would, and she feels he doesn't love her. Or maybe the woman is busy with a job and doesn't cook her husband's favorite dinners the way his mother did, and *he* feels unloved. Or maybe they can never agree about how to spend money or whether to go out or stay home on the weekends.

Little things like this can get it started. Gradually it may get worse, until two people who loved each other and wanted to stay together forever have turned into a couple who are always angry at each other. They are always annoyed or fighting, and both unhappy most of the time.

In a way it's like two close friends who drift apart. But it's also different, because the husband and wife are grown up with a house or apartment and children, and that's where you come in.

# 3

# How It Feels at the Beginning

Darryl was shocked. So many times his parents had told him, "Watch your language!" Now here was his father using words Darryl never would have thought he'd hear him say. Not only that—he was saying them to Darryl's mother!

Dad had taken Darryl and his younger sister Rhonda out for the afternoon. They saw a movie and then had pizza and sodas. The air was chilly as they emerged from the brightly lighted restaurant into the dark street. Five-year-old Rhonda was unusually quiet. Darryl guessed what was in her mind: this was the first time the three of them had gone to a restaurant without Mom, and it felt strange.

Dad had moved out two weeks before. He'd rented a small apartment, a few minutes' drive from the house. Darryl didn't like being there; it made him feel sad. None of the family furniture was there, and it seemed weird that this place without memories was where his daddy slept now.

When Dad drove up to the house, Mom was standing on the porch. Her hands were on her hips and her eyes were blazing. She hardly seemed to notice the children. Instead she said loudly to their father: "You're an hour late. You're supposed to have them back by six." Her voice was so chilly it sent a shiver up Darryl's spine.

"I just took them for a snack after the movie. I guess I lost track of the time. . . . And I wanted to be with them, Jean."

"You had lots of chances for that, but you blew it. You weren't supposed to have them for dinner. I'll have to talk to my lawyer about this."

At this point Darryl's father swore and called Mom a name. She turned away toward the house without answering. The boy looked up at her face. Her expression was hard, but her eyes were wet with tears. She put one hand on Darryl's back and one on Rhonda's and walked them into the house. As he crossed the threshold Darryl remembered that he hadn't said goodnight to his father.

"'Bye, Dad! Thanks a lot! The movie was great!" He was suddenly aware of the pizza he'd eaten. Now his stomach felt as though the pizza were made of lead. It seemed to drag him down so he could hardly stand. He wanted to get rid of it. He smelled the roast chicken Mom had prepared. He resolved to eat some even though he was so full.

Rhonda had tears in her eyes now. And through the still-open doorway he heard his father's voice:

“ ’Bye, kids. See you on Wednesday.” He sounded tired and far away. Not angry anymore, and not like the Dad that Darryl knew.

Mom didn't seem like herself either—at least not right now.

Darryl didn't understand.

I must have done something wrong, he thought. I'm only nine years old, but kids my age can do bad things. I remember how mad they were when I broke the china lamp last year. Maybe all of this is my fault.

He tried to eat the chicken, but he felt like throwing up after the first few bites.

“I always love your chicken, Mom. I'm sorry.”

Mom smiled. “It's O.K., Darryl,” she said. “I understand.” Darryl found himself sobbing, tears running down his cheeks. He couldn't stop. Mom carried him into his bedroom. She stroked his head. After he changed into his pajamas and brushed his teeth, she read to him.

He slept.

---

What actually happens when parents decide to separate?

Almost always one of them moves out. The house is emptier, quieter. Mom (if she's the one who goes on living in the house) may look sad a lot of the time, or she may be irritable and yell more than usual. If she didn't have a job before, chances are she has one now or is looking for one. She may come home tired from her job and not have much time to spend with you.

You may feel confused about what happened, and you may be angry that one of your parents isn't living with you anymore. You may blame the parent who left, or you may feel that the one who stayed made the other one go away. You've probably never felt so awful or so lonely in your life before, and your eyes may fill with tears when you least expect it.

Maybe you wish you could talk to your best friend about the family trouble, but you're not sure it's safe, or you can't find the words. You may find that certain adults outside the home give you support and help you talk about what's going on: grandparents, a teacher, a family friend. Or you may not have that kind of help at all. That kind of thing is mainly a matter of luck.

If you're living with your mother, Dad usually comes to pick you up for weekend visits, or to take you out to dinner during the week. Your parents may still be very angry at each other and not want to see each other, but they pretty much have to if Dad's going to pick you up at the house where Mom lives. So when it's time for Dad to pick you up for a visit, Mom may act tense and angry, or Dad may be like that when he comes. You may think it's you that Mom or Dad is annoyed at, but it almost never is.

Even if you know it's not your fault, though, it's hard to begin the visit with Dad if anybody is upset—Mom, Dad, or you.

Usually visits with the parent who isn't living with you seem strange at first. But most kids find that after a few weeks they're more used to it, and they can enjoy seeing Dad or Mom for just a day or two, even though it's so different from the way things used to be. (Some kids even enjoy the time they spend with the visiting parent more than they used to when that parent lived with them. The time spent together is short, but the parent may actually pay more careful attention to the kid during visits than he used to when they were in the same house.)

When parents first separate, things usually get worse before they get better. Everyone feels pretty miserable. The grown-ups may not show it much; they've been around longer, and they've learned ways to protect themselves from painful feelings.

Kids can protect themselves too: by losing themselves in a game of football or a video game, by listening to a favorite record or tape, by playing an instrument or reading a book, by spending time with a good friend. It's terribly important for kids to be able to put out of their minds, just for a while, whatever troubles they and their families are going through.

# 4

# When Mom and Dad Both Want You

Laura felt as though she were being torn into two pieces. "No," she said, "I don't want to talk about that! It's private. I can solve it myself." She turned away from the woman at the desk.

Laura was nine, in the fourth grade. She was pretty and popular, and she usually got A's and B's in school.

Sometimes my friends say I'm lucky, she thought as she sat there avoiding the court officer's eyes. But here I am in this huge building with a woman asking me questions: What do I do with Daddy? Where do I like to go with Mom? I think she wants me to say I like one better than the other. It's *stupid*! I love them both. I want them to get back together. Mom and Daddy could work it out if they really tried. I'm not going to help with the divorce!

"I don't want to pry into your private business, Laura," said Miss Rossi. "I work here in the family service office of the Court, and my job is to get to know you so I can help the judge figure out what's best for you. Your parents couldn't agree, so it's really up to the judge now. You don't have to talk to me if you don't want to. But the only way the judge can know what *you* want is if you tell me how you feel about the divorce and the different things that are going on."

Laura wondered what to do. This woman didn't seem too bad really. Her skirt and blouse were bright and colorful, her voice was gentle, and Laura liked her jet-black hair.

Laura was silent for a minute or so. "O.K. You can ask me questions. But there's one thing I want you to know."

"What's that, Laura?"

She said it quickly, in one breath: "I-love-both-my-parents-and-I-*hate*-the-divorce!"

"Fair enough. I'll remember that."

They talked for half an hour. It wasn't as bad as Laura had expected. At the end, Miss Rossi said,

"It sounds to me like a problem that can be solved. Tell me if you think I understand. You have some great times with your father, but there aren't enough of them, so you're a little annoyed at him. You and your Mom get along well most of the time, even though you have arguments once a week or so.

"Your Dad's worried about these fights you have with Mom. He thinks there are too many of them and they're bad for you. And your Mom says your father makes things worse by talking so much about what goes on between you and Mom. But apart from all these fireworks that are going on now, they both sound like pretty good parents."

Laura sighed with relief. Miss Rossi had understood the way things were.

"You know, I think there might be a way out of this. A way that won't look like you chose one parent over the other. I know that since your parents separated, your time with your dad has been mostly on weekends. But before that, when did you mainly see him?"

"Well, I saw him just about every night, and most mornings—but he'd get home late a lot from work, and sometimes he'd leave before I got downstairs for breakfast. On the weekends he spent pretty much all his time with the family. Either we'd go someplace together, or he'd take Aaron and me for a walk in the woods near our house."

"So even before the split it was mainly weekends that you saw him! What would you think about a setup where you'd be with him mainly on the weekends? Say, from Friday after school 'til just before dinnertime on Sunday. And there could be one time during the week when you'd be with him for dinner —say on Wednesday, and then you'd go back to your Mom's house."

"But . . . I wouldn't want to be at his house every weekend for the whole time, 'cause I couldn't see any of my friends in my old neighborhood very

much. A lot of us are pretty busy after school with activities and stuff, so we usually can only play on Saturday or Sunday."

"I thought of that, Laura. There are a couple of ways around it. One would be that you'd only go with him every other weekend—"

"I don't think he'd be too happy with that. I might not either, really."

"You didn't let me finish. I was going to say on the in-between weekends—"

"I know! I'd just go for a kind of day visit. Maybe from lunch on Sunday till after dinner. That way I could still have some Saturday sleep-overs at my Mom's."

"You read my mind, Laura. Maybe we can work it out like that. But remember, it might not be easy for your Dad to agree to this. I think his feelings are really hurt when he thinks you might not be with him most of the time. We'll both have to talk to him. If he sees it's what you really want, maybe he'll agree."

The girl shook the woman's hand enthusiastically as they made plans to meet again in a week. "I like your hair," Laura said shyly.

"Thanks, Laura. I've always thought I was lucky to have it. But you have a lot of good things going for you too: you're pretty, and you seem like a smart kid, and you're not afraid to say just what you think."

Laura smiled as she walked out the door. Inside she thought things might not work out too badly after all.

When parents divorce, someone has to decide which one of them the children will live with. Most of the time the parents manage to agree about that, but sometimes they can't.

Then the judge who decides how the parents should divide their money and furniture has to decide a harder question. Is it better for the kids to spend most of their time with their father or with their mother?

This question is called *custody*. You may hear people say, "John's father wants custody of him," or "Alicia's parents share legal custody, but her mother has physical custody." This sounds pretty confusing, but it's not too hard to understand.

The word custody comes from Latin, the language of the old Roman Empire: *custos* means a guard or watchman. (Your school probably has a *custodian*, which comes from the same Latin word.)

In a divorce, the parent with *legal* custody has the right to make important decisions about a child: where the child will go to school, what church the child will attend, what medical or dental treatment is given. Many states now have laws that say both parents should share this kind of custody whenever possible. That way both parents make the important decisions.

*Physical* custody means which parent the boy or girl actually lives with. It used to be that this was almost always the mother, but things have changed a lot in the last ten or fifteen years. Now many children—especially boys—wind up living with their fathers. The other parent has what are called "visitation rights." The child usually spends every weekend, or every other weekend, and an evening during the week, and part of the school holidays, with the second parent.

Many parents work out an arrangement where the child spends roughly the same amount of time in each house. Sometimes this works out well, but sometimes it confuses the kid too much. Then the family has to go back to the more common kind of setup, with most of the time being spent with one parent.

Being in the middle of a divorce is probably the hardest time you've had in your life so far. And it's even harder when parents fight over custody. You may feel as though the two parents who loved you and paid attention to you have suddenly turned into a pair of fighters in the middle of a boxing ring. What *you've* turned into is the cash prize at the end of the match.

What do you do when your parents both want you?

*The first thing to remember is that you're a person with feelings*. You're not a piece of furniture or a valuable old clock or a bunch of papers in the safe-deposit box in the bank. Parents can argue over *possessions* like that until the cows come home, and the sofa or clock still won't have any idea what's going on. But as a human being, you're not a possession. *You don't belong to anybody* (except yourself, and maybe God). So the decisions about who has legal custody and which parent you stay with have to depend on what's best for *you*. Your own opinions should be listened to.

You may wonder, Who's going to ask me what I want? Even if I'm pretty sure I know, what person will I tell it to? Usually a probation officer (sometimes called a family service officer), like Miss Rossi in the story, is the first person outside the family who talks to you. Later on the judge may ask your whole family to see a psychiatrist or psychologist or social worker, someone who's an expert on families. That person will talk to each member of your family privately and see you with each of your parents. These are your chances to talk about how you feel and what you think should happen.

The older you are, the more likely it is that what you ask for is what you'll get. A child of four or five may state a choice, but young kids often don't know what's best for them, and they change their minds a lot. A teenager of thirteen or fourteen often has a lot of influence on the decision. If you're nine or ten, you're somewhere in the middle. You may have things pretty well sorted out in your mind about which parent would make a better home for you. But you might also be so angry at one of your parents—maybe because the other parent is angry and has drawn you into that anger—that you can't see things very clearly. Chances are, though, that if you have a preference, the judge will consider it carefully.

*The second thing to remember is that no one can make you choose.* If you love both parents a lot and just think the whole divorce is a terrible idea, you're probably not about to say "I'd rather live with Dad (or Mom)." Probably you don't even know, and you just wish (like Laura) that they'd get together again.

That's one reason kids have for not wanting to choose.

Another one is that even if they're pretty sure which parent they want to live with, they may not want to say it for fear of hurting the other one.

So sometimes it's hard to be honest when people talk to you about it. Here's an idea that may help: Think about the things you do with each parent, and when you do them. Is one parent pretty much a "weekend parent" already? If so, that parent might not be around a lot during the week if he or she had custody. You might wind up spending a fair amount of time with relatives or a housekeeper.

And think about which parent you have fun with, and when you have it. Maybe Mom is more fun on trips and weekend outings, but tends to be a nag around the house, while Dad is more relaxed and easygoing at home. Or maybe it's the other way around. You'd probably want to spend more of your time with the parent who's more relaxed and fun at home and spends more time with you when you're there.

You don't have to tell anybody which parent you'd rather live with, even if you know. But you can talk about the kind of stuff that's in the last few paragraphs, and it may help people decide what custody setup you'll be happiest in.

*The third thing to remember is that if your parents are fighting with each other, you don't have to help them fight.* That may sound silly at first, but there are different ways parents get kids involved in their disputes—usually without meaning to.

When you come home from a visit with Mom, your Dad may ask a lot of questions: What did you eat? Where did you go? Was anyone else there? It's fine for a parent to be interested in what happens on visits. But one parent may try to find out whether a man or woman friend is sleeping over in the house. That kind of thing should be strictly between the adults. If you feel that you're being used as a source of information, here's an answer you can try:

Just say, "Mom, I'd really rather you asked Dad about that yourself." Or, "Dad, I'm glad to tell you about what *I* do when I'm with Mom, but it makes me uncomfortable when you ask me about the things *she* does and who *her* friends are."

Another thing parents sometimes do is put each other down. Dad may criticize the clothes Mom buys for you. Mom may talk about how Dad doesn't stick to the visiting schedule or is late with the support checks. It may seem that they're trying to get you to agree with them that the other parent has done something bad.

You can do the same kind of thing in this situation that you did in the last one. What you need to say is something like this:

"Mom (or Dad), I love both of you. It gets me really upset when you say bad things about him (or her)."

This may not work the first few times, but if you stick to your guns, it will probably sink in pretty soon.

*The fourth thing to remember is that things usually work out eventually.* Even though everybody may feel miserable for the first six months or a year, most divorced families have settled down into a peaceful life within two or three years. Sometimes parents remarry, and that can help a lot, though it takes some getting used to. We'll talk about that in the next chapter. And the sad mood that so many kids feel is usually a lot better. In a few families, problems go on; we'll talk about that in Chapter 6.

# 5

# Stepfamilies

Sean stood in the bedroom holding his trumpet at ten o'clock on Sunday morning. He put it in its case. Then he took it out again. He blew a few bars of "The British Grenadiers." The music sounded cheerful, determined. It seemed to say, "Whatever happens, I'll be O.K." Sean's expression changed from worry to a little bit of a grin.

I don't care *what* they think of my trumpet, he said to himself. I'm taking it with me.

"They" was the people his father had married two weeks ago. Because he hadn't just married Cindy. She was all right, but along with her went Andrew and Steven and Alicia. Andrew was eleven, a year younger than Sean. Steven was thirteen and acted like he knew everything. Alicia was only eight. Sean's own younger brother Robbie was ten. It was going to be a regular zoo.

Sean and Robbie spent half their time with each

parent; once a month they changed houses. Dad getting married again had been hard to get used to. Before that there were other things, like the sick feeling he got every so often. Sometimes it came when he woke up in the morning. He expected to wake up in the Oak Lane house, but sometimes he'd look out the window and see cars going by on Huron Avenue. Then he had a nasty feeling in his stomach. It only lasted a few seconds and then he was all right. Or he'd be in school, and he'd make plans to play with friends. Then he'd realize he wasn't going to be in their neighborhood after school for another two weeks unless he called Mom at work to see if she could pick him up before dinner. Then he'd get the feeling again. It was no huge thing really. But at times it seemed to him that he had two ropes tied around his waist, leading in opposite directions.

That was old stuff. But today he was going to Dad's whole new family for the first time. Cindy and her kids had moved in right after the wedding—into the house where Sean used to live. Of course Sean and Robbie knew them all, but they'd never all spent more than a weekend together.

He had butterflies in his stomach and a headache, and he was sure he'd sprained his right knee. He had a bright thought: couldn't they just put it off a day, or a week, or maybe a year? He was worried about what the others would think of him. Cindy was nice, but maybe she was just polite because he was Dad's son. Her kids were another matter. . . . As he

STATE
30

thought about it, Sean realized that Andrew and Alicia weren't really a problem. It was Steven he was worried about.

All Steven ever did, as far as Sean could see, was lift weights. His muscles were huge. Steven was already five feet six, five inches taller than Sean. He was into what he called "heavy metal"—musical groups that sounded to Sean like gangs of rioting prisoners banging on trash can lids and yelling swear words. What would happen when Sean pulled out his trumpet and played classical pieces by Purcell and Gabrieli? He could picture the disgusted look on Steven's face.

Dad came to pick the boys up at noon. He seemed cheery, but Sean thought he looked a little worried too. As they pulled up to the old house they saw Cindy and the three kids on the lawn waiting for them. Steven looked incredibly large. He was smiling, but Sean was thinking that if it ever came to a fight, he would last about ten seconds with this hulk. He managed to greet everybody, but his "Hi, Steven!" sounded more like a growl.

Cindy had Sunday dinner ready. There was a big pot of delicious-smelling chili on the dining room table, with corn chips and hot sauce and chopped onions on the side. And a big bowl of rice, and soda to drink. Everybody made small talk, and Steven asked Sean and Robbie a few questions about school. Robbie had to answer most of them, since Sean couldn't do much more than mumble.

After dinner Sean went to his old room, the room he was to share with Steven, to practice his trumpet. It's now or never, he thought. Halfway through a piece, Steven came in and flopped down on his bed, hands under his head, looking up. He seemed to be listening, but Sean couldn't see his face. Sean kept playing. It took all his effort to keep concentrating on the music. When he'd finished, Steven asked, "Do you have any Louis Armstrong records?"

"Only eleven or so. Want to see them?" Steven looked through them carefully and finally put on "Memphis Blues." He looked really happy. Sean couldn't believe it.

"I used to play trumpet too, you know."

"Really? Why'd you give it up?"

"I get asthma pretty bad sometimes. I didn't have enough wind. That's when I began to lift—I wanted something to be proud of. Weight lifting doesn't bother my lungs much. . . . You sounded pretty good. But you didn't quite make that high F sharp on the Gabrieli."

Sean was speechless for a minute. "I thought you were different. I didn't know about the trumpet, or the asthma. I guess I was worried what you'd think about my music." After a pause he added quietly, "Maybe what you'd think about me, too."

"Hey, listen, Sean. Did you think I don't know this used to be your room? I mean your *exclusive* room? You looked nervous when you came this morning, and I figured it was mostly because of me. But did you stop to think that me and Andrew and Alicia were worried too?"

"Not really. . . . Maybe everybody was, huh?"

"Even the grown-ups, I bet. Hey, Sean, I have an idea. I still have my trumpet. It's a really good one, a silver Benge. You can play it sometime if you want. What if we surprise everybody now and give them a little concert? I can play if it's not more than twenty minutes or so."

Sean's face broke into a grin. "Great idea, Steven. Do you know 'The British Grenadiers'"?

"One of my favorites. Let's do it."

"In C."

"Right. They're out on the porch."

So the new roommates marched out together. No one could have criticized the sound; it was as though they'd played together for years. Everyone looked up amazed, and they applauded after the boys played the song through twice.

Sean's father and Steven's mother caught each other's eye.

"It might turn out easier than we thought," said Cindy. They both broke out laughing, and got up to give the boys a big hug in the middle of the porch.

---

The divorce story has to include stepfamilies, or it's not complete. Sooner or later many Dads and Moms find another man, another woman to share their lives with. Often this person lives with his or her own children, or has them come to visit. How do people who've never lived together suddenly start acting like a family?

They don't. First of all the parent has to get to know the other adult. The two of them have to figure out whether they want to get married or not. Often kids get to know this other person (and the other children, if there are any) pretty well before there's any decision to marry. We could call this the "Boyfriend or Girlfriend Stage." Let's talk about this stage first, before we talk about real stepfamilies.

Kids often have strong feelings when their mother or father begins to "date." They may resent the time their parent spends with this new person. They may not like it if the person tells them what to do. And sometimes a parent has several "boyfriends" or "girlfriends" one after the other—or even at the same time! This can confuse kids too, and make them wonder if Dad or Mom loves them the same as before.

What you can do at this stage, if you're upset, is talk to your Mom and ask if she's serious about the person. (She may not know yet herself.) Or talk to Dad. Tell him you wish he wouldn't have his new friend along on all your visits. Explain that you don't always like to share him with someone you don't even know. He may not agree completely, but he may change what he does because you told him how you felt. Maybe he didn't know before.

When the parent and the other adult get to know

each other really well, they may decide they want to get married. If they're sensible, they'll listen to what their kids say about the other person and the other person's kids. They won't push their kids too fast to like the new people. It's best if they don't plan to get married before everybody has had a chance to get to know each other. It takes time to find something we like in other people.

Stepfamilies where each parent brings one or more children are sometimes called "blended families." They mix together people who started out separately. You probably know that some things mix together better than others. In the kitchen you can easily mix milk and chocolate syrup, for example. But oil and water are a different story. No matter how much you stir them, they'll always separate again. How can a man and a woman and their kids mix together comfortably and enjoy the time they spend together?

*What the grown-ups can do.*

1. Not rush you into loving the new adult (the stepparent) or accepting him or her as someone who can tell you what to do. You need time to get used to the new adult. You need to find out over a period of months, or even a year or two, that this is someone who likes you and whom you can trust.

2. Not insist that the new adult be called any name like "mom" or "mother" or "dad." You may want to say "Ellen" or "Sam," and Ellen or Sam may be comfortable with that. If there's a "parent-word"

like Mom or Ma that doesn't sound like what you call your other parent, and everybody is comfortable with it, maybe you'll want to use it after you get to know and like the new person.

3. One thing a stepparent can do that's sometimes very useful is to just be a new adult friend for you. A lot of kids don't know many adults really well, especially if they don't have close relatives nearby. And a stepparent can sometimes help out when a kid and his or her mother or father are having a problem getting along. The stepparent—not having always lived with this family—may notice things the kid and the parent do that rub each other the wrong way.

*What kids can do.*

1. Remember that your father or mother deserves to have another adult to feel close to. This may be hard for you if you still miss the old family setup a great deal.

2. If you feel rushed into acting like a son or daughter to your stepparent, talk to your parent. It doesn't matter whether it's the parent you live with or the one you visit. Either way you could have stepparent problems. Explain how you feel. It will help if you don't start out by saying "I *hate* Sally. She *always* cooks this horrible food, and she *never* lets me watch T.V. after dinner, and she's *always* nagging." Certain words like hate, and always, and never, can get people angry fast. It's better if you say something like, "Dad, I wish you could talk to Sally.

I like her O.K., but there are some things that get me upset. . . ." Then go on and say what the things are.

3. When you have a new stepbrother or stepsister, it's really important to face the new situation and work it out. You may be jealous of the stepchild, and the stepchild is probably jealous of you. In the story, Sean and Steven didn't know how things were going to go, but they didn't avoid dealing with each other. They found they had something to share together.

If you can't manage to get along with a kid that you have to share a home with, this is another situation you should talk to your parent about. Then your Mom or Dad can talk to your step-Dad or step-Mom and together they can try to help the kids get along better.

# 6

# Kids Have a Right to Be Kids

It was Kristin's fifth grade teacher who noticed. Mr. Phillips was concerned, he told her parents. Kristin was a nice girl, a smart girl. Her classmates seemed to like her. But she wasn't really *there* most of the time.

Mrs. Sherman asked impatiently, "What do you mean, not there?" She and her husband didn't exchange glances, the teacher noticed. They hadn't said hello or smiled at each other when they arrived.

"She spends a lot of time looking out the window," he said. "When I call on her, nine times out of ten she doesn't know what we've been doing. She tries to cover up. Usually she gives an answer that's in the right ballpark. But she's bright enough to be learning much more."

"We don't need to be told our daughter is smart," said Mrs. Sherman. "Or that she's not doing well in school. That's obvious from her last report card. In my opinion she's just lazy."

Mr. Phillips tried again. "Mr. and Mrs. Sherman—"

"We're not Mr. and Mrs. anymore," interrupted the girl's mother. "He's Mr.," she said, jerking a thumb toward her ex-husband, "I'm Mrs."

"Mr. Sherman. Mrs. Sherman. I know you're divorced. That's none of my business unless it's affecting Kristin's life in school. I don't know if it's the divorce that's bothering her, but something is. I have more to tell you. I didn't ask you to come just to tell you she looks out the window."

Both parents looked up, alert.

"Last month Mrs. Stokes, the other fifth grade teacher, was walking down the basement corridor. She saw your daughter coming the other way. It was during class time, but Kristin had been excused to go to the bathroom. Mrs. Stokes knows her and said hello, but Kristin didn't seem to hear or notice at all. She just kept walking.

"And she's often confused. Three days last week she didn't know what the social studies assignment was. She's gotten lost in the school building twice this month that I know of. She was looking for this classroom she's been coming to all year.

"But what made me call you both yesterday was more serious. I was walking around the room while the kids did a math test. Everyone was working away at the problems except Kristin.

"They were adding fractions. But Kristin was busy making a sketch. It took some convincing before she let me have it, but I thought you should see."

He pulled out a piece of lined paper and placed it on the table. It was done in pencil: a peaceful hillside with grass, trees, and flowers. In the center was a gravestone with the simple inscription:

K. S.
1975 - 1985

Mr. Sherman looked at his ex-wife. There were tears in his eyes. She darted a glance at him, then looked away and blew her nose noisily.

"I'm only Kristin's teacher. I'm not a psychiatrist. But it's clear that your daughter is depressed. She needs to see a professional person as soon as possible."

Mrs. Sherman's unfriendly shell began to crack. "Perhaps you're right," she said. "I've noticed a difference myself lately. She doesn't seem to enjoy anything."

"I'll do whatever I can to help," Mr. Phillips said. "Why don't you come back in a few days?"

As they left Mr. Phillips thought they both looked scared to death.

The guidance counselor, Mrs. Sheehan, came to the next meeting. Kristin's parents told Mr. Phillips and the counselor that they'd talked it over and agreed to take Kristin to a therapist. "Can you recommend someone who's well qualified?" asked the father.

Mrs. Sheehan was prepared. She gave them the

name of an experienced social worker, Mrs. Cowan, who had an office nearby.

The social worker asked to meet with Kristin's parents. Mrs. Sherman thought as she sat down, She looks so young. What could she know about problems like ours? And Mr. Sherman thought, I hope we don't have to go into a lot of detail about the troubles we had. It's just too hard to go back over it again.

But the woman put them at ease right away. Smiling, she said, "Having a ten-year-old daughter isn't easy, I'll bet. Divorce or no divorce."

They could tell she wasn't going to blame them for all Kristin's troubles. Then they felt more relaxed and answered her questions.

The social worker learned a lot about the family. She found the parents hard to like at first, but she could tell they'd both had a hard time. They were both depressed, and both were hanging on to Kristin, their only child. She was their life raft in the open ocean, the one person that would help each of them stay afloat. The worst part was that Kristin herself was beginning to go under.

The first meeting with Kristin wasn't easy. The pretty, black-haired girl walked into the office mechanically, taking no notice of Mrs. Cowan's cheerful pictures and colorful ornaments. She sat obediently in the chair the therapist pointed out. And then she just sat, looking down. For a full twenty minutes she said nothing. Mrs. Cowan gently encouraged her to talk about her interests, her worries, anything at all that could open a conversation, but Kristin said nothing.

Finally, still looking at the floor, Kristin asked in a dull voice, "They're both mad at me, aren't they?"

"No, Kristin—though I suspect your mother can have a sharp tongue at times. They're worried about you. Why would they be mad?"

"Because I mess everything up. I get bad marks in school, and that bothers Mom. I don't have enough friends, and that bothers Dad. I don't like tennis or Brownies, and that bothers Mom because she was always a star at everything. When I'm with Dad I watch TV a lot, and that bothers him, because he thinks I should be reading good books. I never do anything either of them likes."

Mrs. Cowan said softly, "It must be really hard to feel like that."

Kristin began to cry. First it was just a tear or two down each cheek. After a minute she was sobbing, her shoulders shaking.

Mrs. Cowan wanted to hold her tight, but she held herself back. She didn't want Kristin to feel embarrassed. She hadn't asked for affection, and Mrs. Cowan had to respect that. So all she did was reach over for the Kleenex box and put it near the weeping girl's chair.

As she sat silently, she thought Kristin was letting herself feel her awful sadness for the first time. Before she had been trying to hide it. Mrs. Cowan tried to figure out what might have happened to make her so miserable.

From what Kristin's parents had told her, she knew the girl had always been shy. She was bright and athletic, and she'd always had some good friends, but she'd also doubted herself a lot. She used to ask her parents for comfort: Am I pretty? Do you think they'll be nice to me at Sarah's party? Her parents had always reassured her, but never seemed to wonder why she needed support so often.

They'd been able to, that is, until the divorce.

When that happened, both parents hit rock bottom and stayed there. Neither had many friends. Kristin was their only child. So the girl, at the age of eight, became the friend her mother needed, and the friend her father needed at the same time. She learned to cheer each of them up. She became the most considerate, sensitive child anyone had ever met.

But inside she gradually stopped feeling like a kid. Her worries were the worries of her mother and father, who talked to her so much about their lives. So when she was with other kids she felt out of place. They'd be talking about a rock group, or which boy in the class was cutest. Sometimes Kristin joined in, but often her thoughts were elsewhere. She thought about what she could do to make dinner with her mother more fun, or what she could do with her father that weekend that might make him smile and laugh a little.

And one of the girls might notice and say, "Kris, you're looking really spaced out. Or maybe you're too smart to talk about this stuff, huh?" But Kristin couldn't tell them what was on her mind. She was sure they wouldn't understand, not even the few girls she was closest to. So she gradually saw them less. Eventually she felt cut off from them. She couldn't see a way to get back in. No wonder she was so lonely and sad.

That was Mrs. Cowan's theory. She finished putting it together around the time Kristin finished crying. The social worker said, "I'm glad you could cry here, Kristin. I have a feeling it's been a while."

Kristin smiled shyly. When Mrs. Cowan stood up and offered her hand to shake, the girl's grip was surprisingly firm. Mrs. Cowan said, "I'll see you day after tomorrow. Maybe you'll be able to talk more then. I have some ideas about how all this got started, but I'll need your help to find out if I'm on the right track. And I need you to promise that you'll call me anytime, day or night, if you ever decide life isn't worth living. I don't want you to hurt yourself."

She gave the girl a card with her name, address, and phone number. Kristin put it in her pocket.

They were at the door by now. Kristin said, "I don't think you need to worry about that. I guess they showed you that drawing. But there must be better ways to solve things. At least I'll give it a good try."

We all have a generous streak. In all of us there's a part that wants to give to other people, to help. In family life it's like that: parents giving to kids, and kids giving to parents. (I don't just mean things like money or food or Father's Day presents, but the words and thoughts that show a person cares. "Is your headache better?" "How did your day go at work?")

If parents and kids didn't do this kind of thing for each other, families probably wouldn't exist as we know them. As children get older and able to do more things, the bond that parents feel for them begins to change. Parents depend on their children being able to help them when it's needed.

Adults usually live together and help each other in many different ways. What kids do to help their parents is only part of the picture. A kid can't keep a parent feeling good if that parent isn't getting support from the other parent, or from other adults.

With Kristin's parents, it looked as though neither of them was getting much help or support from anyone except their daughter. That's too much to expect from a child. Especially one who, like Kristin, didn't have a lot of confidence in herself to begin with.

How happy we feel, how well our lives are going, how much we're able to get done, all depend on our "emotional energy balance." Imagine a set of kitchen scales. On one side is what we're getting from other people (hugs, compliments, smiles) and from feeling good about things we do. Doing well on a test or winning a swimming race makes us feel good.

The other side—the side that makes us feel drained—comes from giving to others who don't appreciate what we give. Or from doing badly at something when we want to do really well, or from times when people are mean or cruel to us.

If the first and second set balance each other out, most people feel O.K.—not super, not miserable. If the first set outweighs the second, we're usually in a good mood and full of energy—for an hour, a day, a week or more, depending on how important the good stuff is.

But if the second set (whatever makes us feel bad or disappointed or not loved enough) is heavier than what's coming in, we may be headed for trouble. Usually when this happens we feel empty, sad. In a really bad case we may feel hopeless.

Nobody likes to feel that way. Adults can usually stand it better than kids, especially if there's an end in sight. But kids aren't as strong as grown-ups, and they haven't developed the patience grown-ups have. They want to feel better *now*. If they can't they may cry, or do something bad to call attention to themselves. Or after a while they may just give up and fall into a silent depression.

That was what Kristin did—except for one thing. She sent a signal that she needed help, by making the drawing. And that was lucky, because it got her the help she needed.

When kids of divorced parents are upset, the trouble sometimes doesn't get noticed soon enough. A parent may see that something isn't right, but he or she may not realize that the kid has a real problem. Instead it may be blamed on something the other parent is doing. Then the parent gets mad at the other parent instead of helping the kid.

Or, as with Kristin, the child's problem might not be picked up because the parents have so many problems of their own that keep them from noticing. And they may need so much for the child to be strong that they can't recognize the truth.

*If you feel even a quarter as bad as Kristin did, you should tell your parents you need help.* And if they don't listen, find another adult who will: a relative, a family friend, a teacher; a doctor or priest or minister or rabbi. Ask the adult to convince your parent that you're hurting and need to see an expert in that kind of pain.

Finally, a word about treatment (also called therapy or counseling). Sometimes people think getting help from a psychiatrist or psychologist or social worker means a person is seriously disturbed or crazy. Nothing could be more wrong.

Most people who go to therapists aren't badly disturbed at all. They're normal kids and adults who are having trouble getting through a tough situation. They need the expert knowledge of someone outside the family. Usually the help they need is not advice (though most good therapists will give advice once in a while). What they need instead is a person who can help them figure out how they got so confused or depressed in the first place. They need to know how they can get over it, and how they can avoid feeling so terrible in the future.

You might want to know what happened to Kristin. Mrs. Cowan met with her often, and also saw her father and her mother from time to time. Mrs. Cowan asked Kristin about the things she liked to do, what she hoped to be when she was older, what music and TV shows and movies she enjoyed. Sometimes she told the girl about her own favorite things: places she'd visited on vacation when she was a child, good times she'd had with friends. And she shared with Kristin what it felt like when she'd been unhappy, too. Gradually they got to know each other, talking, and sometimes drawing or playing a card game when they couldn't find anything to talk about.

After a few sessions Kristin felt comfortable enough to talk about more private feelings—feelings about her parents' divorce. She talked about how often they were grouchy and down in the dumps and about how hard it was for Kristin to relax with her friends. It turned out that Mrs. Cowan's guesses had all been right.

It was hard for Kristin to talk about these things. She still felt like a bad, unworthy person. She thought a lot of things were her fault that weren't (including her parents' divorce). So naturally she was afraid the therapist would stop liking her if she really knew what she was like. But Mrs. Cowan knew what was going on. She kidded Kristin: "Who are you, Wonder Woman? Are you supposed to be perfect?"

Eventually Kristin was able to relax more. She spent time with her friends again and began to enjoy her schoolwork once more. Meanwhile her parents had found therapists of their own, to help them work out their own problems. Eventually they both began seeing their adult friends more often. That helped Kristin, because she didn't have to listen to their worries much now; they were talking to their friends, and they weren't as worried anyway.

It's been two years since Kristin drew the picture of the gravestone. She's getting ready to say goodbye to her therapist, but she knows her good friend will still be there if she needs her again. She doesn't think she will, though, because she feels like a different girl from the one who walked mechanically, with slumped shoulders, into the social worker's brightly colored office and sat crying for half an hour.

She has a gift planned for their final session: half a dozen white roses. It will be her way of thanking Mrs. Cowan for making her life easier. And it's a safe bet that Mrs. Cowan will have a present for Kristin too.

# 7

# What You Can Do

Martin was nine when his parents divorced. His brother Adam was fourteen and wanted to live with their father, but Martin wasn't so sure. He felt close to both parents, and besides, he and his brother weren't getting along.

Finally the court decided Martin should stay with his mother, and he and Adam would see each other mainly on weekends and vacations.

A year later, one gloomy November day, Martin's mother broke the news that Dad and Adam were moving away from New Jersey. Four states away, to be exact. Instead of being half an hour away by car, they would now be four hundred miles distant, in Charlotte, North Carolina—a seven-hour car trip, or a couple of hours by plane.

It would mean he couldn't see them anywhere near as often. Mom was upset. She even asked him if he would want to go live with Dad now.

Martin hesitated. Dad did a lot of things he liked. Dad knew about baseball and soccer and was Martin's biggest fan when he played on the town teams. But Mom was fun to be around. She laughed and smiled more than his father did, and she had a lot of friends. And there was Dave, her boyfriend, who helped Martin with his pitching.

He finally decided to stay with Mom and see how it went.

By the time he was eleven, Mom and Dave had stopped seeing each other. Martin noticed another change: his mother worried about him a lot. With Dave not around anymore she tried to play ball with him, which didn't work out too well because he was much better than she was. He didn't want to offend her, so he didn't say anything. She often asked him about his friends and seemed upset if he'd had a fight with Jack or Larry, his two buddies. Mom wanted to know about everything in his life. She seemed to be afraid some terrible thing was going to happen to him.

Martin always managed to calm his mother down when she worried too much. But he realized that he missed his father. As time went on he began to wonder whether he'd be happier living with his Dad and Adam in North Carolina.

That summer Martin joined his father and brother at Dad's vacation home on Lake Norman. They always fished and swam and practiced archery for the whole month of July. But both boys' favorite was sailing. Martin owned a little fiberglass Sunfish, just for one person, and Adam had a catboat rigged with a mainsail and a little jibsail in front.

One lazy morning they were out in Adam's boat. Although Martin had sailed quite a lot, Adam liked to teach his younger brother. They were out in the middle of the lake—it was several miles across at that point—when the wind died down. They waited there a while and worked on their suntans but eventually got hungry for lunch.

"How will we get back, Adam? There's no wind."

"A bit, Skipper, a bit." (Skipper was Martin's nickname.) "Coming from Dad's dock, seems like. We'll have to sail into the breeze, what there is of it."

Sailing into the wind means going on a series of "tacks"—first diagonally toward the left, then toward the right, zigzagging until you finally reach your destination. Each time you change direction you have to turn the tiller so that the boat is aimed at a point some distance away from where you have been heading. Then the sail flaps until it catches the wind coming from the other side. It's a slow and tiresome way to travel, but you have to do it when there's hardly any wind.

Adam said, "You'll hardly ever have a day when there's no wind at all. The secret is to find where it's coming from and catch it in your sails. I've seen people stuck out here for an hour or more—people Dad knows, who used to talk about what hotshot sailors they were. I've gone out and helped them come in a few times."

Martin felt a vague touch of air on his face. "Here it is," said Adam tensely and adjusted the tiller's position. The mainsail was pointed almost straight back, but the light breeze bellied it out on the right side, and they began moving through the water. Adam smiled.

They went about sixty yards until they were aimed at a point far distant from their dock. "Ready about," said Adam in sailor's jargon, to announce a new tack; then "Hard alee." Sure enough, after the sails luffed helplessly for a moment, they caught the wind again from the other side.

The boys were home in plenty of time for lunch. Martin was full of admiration for what his brother had done with almost no wind at all.

Through the afternoon he kept thinking about tacking into the wind. It seemed such a neat way to get somewhere if you couldn't go directly. But it was also frustrating to keep aiming at places you didn't want to go to. And there was something else about it that Martin couldn't put his finger on, something that connected up with his own life.

That night he had a dream. He was alone in his Sunfish on a huge lake. The waves were a foot high, and he was far from land. He was terrified that the boat would be swamped. Two houses stood on the shore, a mile or so apart. The one on the left was his mother's, and the other his father's. He licked a finger and held it up: the side where it dried first showed him that the wind was coming from a point midway between the two houses.

"I'll have to sail into the wind," thought Martin in the dream. Then he paused. "But I won't wind up at either house." He looked straight ahead: the shoreline between the houses was a rocky coast that would chew up a little boat like his. "What do I do now?" he cried out, terrified.

Suddenly Adam was there at his elbow, crouching down so as not to get hit by the boom when the sail moved, and they were in Adam's catboat, and his brother was giving instructions. "You've done well," he said. "You've caught the wind. You don't have to worry about a thing."

"But if I sail straight into the wind, look where we'll end up!"

"Don't be silly. Nobody wants you to get killed. You don't have to keep tacking forever and wind up on the rocks. Just aim for the nearest safe port. You can sail just as well with the wind coming from the side. I taught you about that: it's called reaching. There's plenty of wind for reaching. Just look at those waves!"

"But Adam, then I have to choose between Mom's house and Dad's house!"

"No you don't, Skip. Not while you're in danger. You just have to get to one of them temporarily, and step out on the dock safe and sound. There's a hot shower and towels and food in both places. After that you go wherever you want—wherever your wind blows. Don't panic. You'll do fine!" And Adam turned into a big gray-and-white seagull and flew away.

The following night Martin told Adam his dream. He and his older brother had gotten a lot closer since the divorce. Adam was lying in bed listening to the Rolling Stones. He turned the music down when Martin came in.

When the younger boy had finished, Adam said, "Wow. That's exactly what it felt like!"

"What do you mean? What *what* felt like?"

"When I decided to go with Dad. I thought I must be making a terrible mistake, and all kinds of terrible things would happen—to me, to you, to Mom. And I was fourteen—a lot older than you are now. I bet you've had a real hard time of it too."

Martin, who thought boys should never cry, felt tears on his cheeks. His brother's understanding had touched him.

"I wish we could be together more, Adam."

The teenager gave his brother a rough hug. "We can, dumbbell. Don't you see nothing is written in stone? You just have to know what your goal is and be smart and fast enough to catch the wind, even if it's just a little breeze. Like, if you want to spend the whole summer down here instead of just July, wait till Mom starts talking about how she wishes she could go on a long vacation in Europe. She always brings that up: how she's never done it, and all. That would be the perfect opening! Or convincing her that we get along great now, and she should have me up there more on school vacations. You could probably even live with Dad and me if you really want to."

He punched Martin in the ribs playfully, and Martin punched him back. Soon they were rolling around on the floor practicing wrestling holds until Dad opened the door and asked, "What's going on here, guys?"

"Nothing, Dad," Adam said. "We were just going over some of the finer points of navigation."

Their father stood there a moment, scratched his head, and walked out.

---

What can you do if you're a kid and your parents are divorcing, or already divorced? You can't keep them from splitting up, or get them back together again. But there are a lot of things you can do to look out for yourself and make sure you get a fair deal.

1. *Keep your confidence in yourself.* Don't think you caused the grown-ups' troubles—even if they argue about you. And don't think there's anything wrong or different about you just because you have parents who are divorced.

2. *Don't bite off more than you can chew.* That means don't take on a job bigger than you can handle. For example, if your parents are divorcing, and someone asks you straight out which parent you want to live with, you may be old enough and sure enough to answer that question comfortably. But if you're not, say so. It's the court's job to get somebody

who can find out what's best for you without putting you in the uncomfortable position of having to choose one parent over the other when you don't want to do that. And after the divorce, if parents lean on you too much, like Kristin's in the last chapter, don't take on the responsibility for keeping adults happy. It's enough for kids to carry out their own responsibilities.

3. *Say what you feel.* Don't be afraid to talk to your parents about what you think and what you want. You may not get things exactly the way you want them, but your chances are better if you speak up than if you never let people know how you feel. Talk to brothers and sisters, too. Find out how they feel about what's going on, and tell them your thoughts. It may help all of you.

4. *Don't let anybody interfere with your childhood.* To grow up happy and emotionally strong, a kid needs:

—A parent of the same sex as a model, to learn how a grown man or woman lives and solves problems and has fun, and to teach the kid the things a man or woman is supposed to know.

—A parent of the opposite sex to help the kid in various ways. It's nice when a boy can have his manliness appreciated by his mother, or when a father tells his daughter that he thinks she's really pretty and really smart. These things can mean a lot.

—A lot of time to just be a kid, alone or with

friends. Boys and girls need time that isn't spent doing what adults want them to do.

If visitation arrangements are so tight that you don't have time to be a kid, YELL!

If your parents keep putting you in the middle, YELL!

If any of the other problem situations that we've mentioned comes up, and talking about it doesn't help, YELL!

If you don't have enough time with your father, YELL!

If you don't have enough time with your mother, YELL!

Divorced parents can be "hard of hearing," so you may have to yell a bit. I don't mean actually screaming in their ears, of course; I mean saying the same thing over and over until they finally catch on.

There's one thing you should be warned about. You may say something you really feel, and the parent you say it to may answer, "I think your father (or mother) put those ideas into your head." Now there's possibly nothing that makes us angrier than when someone else doesn't believe our ideas are really ours. Maybe the other parent did influence you a little. But if what you're saying is mainly your own genuine feeling, you should set the record straight with the parent you're talking to. You could say, "I know you and Mom don't agree about tennis lessons for me. But what I just said doesn't come from her. A lot of my friends get them, and I'm really interested in tennis. It makes me mad when you treat what I say as if it weren't really me talking!"

Most of all, try to remember that in the vast majority of divorcing families, both parents go on caring deeply for their children, and the kids usually come through the experience without serious trouble. Sometimes they even wind up with an extra little bit of "smarts" about how other people feel. It helps them to manage in a situation where people don't agree with each other.

This can be an advantage, later on in your life—if you're quick enough to catch the wind in your sail.

# Index

# ABOUT THE AUTHOR

Steven L. Nickman is a pediatrician and a psychiatrist who specializes in children and their parents. He has a private practice and is on the staff of the Massachusetts General Hospital in Boston. He has worked with over a hundred divorcing families.

Dr. Nickman lives in Massachusetts with his wife and two sons, twelve and eighteen years old. He enjoys travel, learning foreign languages, bird watching, music, and cooking. He is the author of *The Adoption Experience*.